Lecture Notes in Earth Sciences   66

Editors:
S. Bhattacharji, Brooklyn
G. M. Friedman, Brooklyn and Troy
H. J. Neugebauer, Bonn
A. Seilacher, Tuebingen and Yale

Springer-Verlag Berlin Heidelberg GmbH

Helmut Wilhelm   Walter Zürn
Hans-Georg Wenzel (Eds.)

Tidal Phenomena

With 111 Figures and 30 Tables

 Springer

Editors

Prof. Dr. Helmut Wilhelm
Universität Karlsruhe, Geophysikalisches Institut
Hertzstraße 16, D-76187 Karlsruhe, Germany

Dr. Walter Zürn
Observatorium Schiltach
Heubach 206, D-77709 Wolfach, Germany

Prof. Dr. Hans-Georg Wenzel
Universität Karlsruhe, Geodätisches Institut
Englerstraße 7, D-76128 Karlsruhe, Germany

"For all Lecture Notes in Earth Sciences published till now please see final pages of the book"

Cataloging-in-Publication data applied for

Die Deutsche Bibliothek - CIP-Einheitsaufnahme

Tidal phenomena / ed.: Helmut Wilhelm ...

(Lecture notes in earth sciences ; Vol. 66)
ISBN 978-3-540-62833-0   ISBN 978-3-540-68700-9 (eBook)
DOI 10.1007/978-3-540-68700-9

ISSN 0930-0317
ISBN 978-3-540-62833-0

© Springer-Verlag Berlin Heidelberg 1997
Originally published by Springer-Verlag Berlin Heidelberg New York in 1997

Typesetting: Camera ready by editors
SPIN: 10552261   32/3142-543210 - Printed on acid-free paper

Preface

The idea for this book came up during a seminar on 'Tidal Phenomena' held at Oberwolfach (Black Forest) from October 17 - 21, 1994, sponsored by the German Geophysical Society (Deutsche Geophysikalische Gesellschaft, DGG). The contributions here represent essentials of the lectures given at this seminar in the form of either upgraded manuscripts or extended abstracts. The aim of this seminar was to give tidal reseachers, who are working as experienced experts on special problems, an opportunity to extend their views and to get a wider impression of the broad spectrum of tidal effects occurring in different regions of the earth, and moreover in the planetary system and the universe. It was also intended to introduce the results of modern tidal research to open-minded scientists who like to cast a glance beyond their personal fields of work.

The programme of the seminar has been published in the DGG-Mitteilungen, Special Volume II/1995. Some authors, who have not been able to contribute the full text of their lecture, were persuaded by the editors, to deliver at least an abstract and a list of references in order to give the interested reader a chance to obtain original information about the subject of the corresponding lecture. The scientific content of each article lies in the responsability of the authors alone.

After an expansive phase during the seventies, tidal research has an internationally appreciated position in Germany. A great part of this work is performed within the frame of a geodetic-geophysical working group which meets once per year. Interest in tidal phenomena on a larger scale, as decribed in this volume, however surpasses the subjects of this working group at least partially.

The fascination of tides arises from the regularity of their appearence and their conspicuous influence on very precise measurements of many natural physical phenomena. We hope that this book will mediate at least a faint image of the enthusiasm which tides can arise in people who have gotten in close contact with them.

Karlsruhe, April 1997

Helmut Wilhelm, Walter Zürn, Hans-Georg Wenzel

TABLE OF CONTENTS

Introduction

Introduction

The goal of this book is to provide insight into the very different aspects of modern tidal research which are normally treated in separated scientific disciplines. Earth tides, ocean tides, atmospheric tides are pertaining to geophysics and geodesy, oceanography and meteorology, respectively. Tidally induced phenomena are investigated in geomagnetism, hydrology, climatology, and geology. Tides on planets are discussed in planetology, and tides of stars and galaxies are part of astronomy and astrophysics. Finally, a fundamental understanding of tides is provided by general relativity, i. e. by the theoretical foundations of physics.

The first section concentrates on the tidal potential and the tides of the solid earth. In the first contribution the earth-bound tide-generating potential and its spectrum is derived. This spectrum contains periods up to 20942 years, i. e. the period of the mean ecliptic longitude of the sun's perigee. The tidal spectrum also contains the gravitational influences of the planets Mercury, Venus, Mars, Jupiter, and Saturn. The tidal potential for the earth is, of course, relevant for all but the last three articles and the one about the Chandler wobble.

In the strict sense, tides on earth are caused by the gravitational forces related to this potential. In a wider sense, all measured temporal variations caused by the celestial motion of the earth and other heavenly bodies can be called tidal variations. To be more distinctive, this term is however restricted to the effects which are related to the gravitational variations derived from the tide-generating potential, whereas the other variations which are not related to *gravitational* effects of the earth's celestial motion are sometimes called tide-related or tidal-like variations. In this book both kinds of variations are subsummed in the term 'tidal phenomena' which also comprises tides in larger spatial expanses like the solar or the galactic system.

The second contribution gives a state-of-the-art introduction to the theory of the deformation of a rotating, elliptical, elastic, gravitating earth by tidal forces. Special attention is paid to different descriptions (Langrangian, Eulerian and a mixed combination of both) of the incrementals. Perturbations to the results of this theory which are caused by lateral heterogeneities and by anelasticity in the earth's mantle are also discussed.

The analysis of tidal observations has long been necessary for the prediction of tides in harbours. A substantial increase in accuracy of the tidal models became necessary when the precision of earth tide measurements was enhanced by orders of magnitude, especially due to the advent of superconducting gravimeters. The analysis of earth tides must be able to extract tiny effects from the data, and the third article describes a widely used modern analysis method based on least squares.

The fourth contribution discusses earth tide observations and the difficulties of their geophysical interpretation. Only a few characteristic examples could be selected from a multitude of experimental results and with those an attempt is made to give insight into the controversial views of different research groups towards the meaning of (apparent) spatial variations of the earth's response to tidal forcing. Systematic disturbances arise from ocean attraction and loading in all measurements and from local heterogeneities in tilt and strain records.

Earth tides are the response of the earth to gravitational forcing by the moon, the sun and the planets, and the tidal deformation of the earth is physically a forced oscillation. If the period of the forcing lies in the vicinity of the period of a free mode of the earth, a resonance effect is to be expected in the response. The nearly-diurnal free-wobble (NDFW), associated with the free core nutation (FCN), has its eigenperiod in the diurnal frequency band. This resonance arises due to the ellipticity of the core-mantle boundary. The effect from this mode in the spectrum of measured earth tides and the conclusions, which can be drawn, are the subject of the fifth article.

To the public tides are generally associated with ocean tides, because the regularity of the appearance of tidal currents and sea level changes is inevitably noticed by everyone who ever lived for some time at the sea-side. The practical aspects of navigation of vessels near the coasts implied early continuous observation of the tides in ports and necessitated the prediction of tidal heights and current velocities. Ocean tides are even used to generate electrical power. In order to understand why the marine tides show a specific appearance at different locations they need to be modelled. The first article in the second section gives an overview about specific ocean tide phenomena and current methods of ocean tide modelling. It is interesting to note that ocean tides and earth tides are intimately interacting.

The influence of ocean tides on earth tide measurements is described in the next article. While the effect on tilt and strain is declining rapidly with increasing distance from the coast, gravity is considerably affected to very large distances. This systematic, so-called indirect effect, has to be precisely modelled in order to be reliably separable from the data. On the other hand this effect can be used to improve local ocean tide or earth tide models, if it can be determined with sufficient accuracy.

The moon's orbit around the earth has changed during the 4.5 billion years of the earth's existence. But the angular momentum of the two-body system must approximately be conserved. The earth's spin is decelerated by tidal friction, so that the number of days during a month or a year decreases with time. The orbital angular momentum of the moon correspondingly increases, so that the moon is moving away from the earth and, consequently, the number of days during a month should increase. Hence, there are two counteracting effects regarding the development of the number of days during a lunar month. Actually, the second effect is smaller than the first one and the increase in the length of the day is accompanied by a decrease in the number of days per month since the Proterozoic. There is no discussion about the fact that the main part of tidal dissipation occurs in the oceans but it is still not completely resolved how large the effect of viscous dissipation in the earth's mantle is. The friction of ocean tides at the sea bottom is the main cause for the increase in the length of the day and in the moon's distance with time. The third contribution in this section explains these effects in detail and describes the results of modern observation techniques like VLBI.

The last article of this section is devoted to the so-called pole tide associated with the Chandler wobble. The major questions here concern the excitation mechanism of this free mode and the problem whether the pole tide is an equilibrium tide or not; both are addressed in the article. Note that the differential forces for the pole tide are centrifugal, not gravitational.

The third part of the book describes effects which mainly arise from non-gravitational variations related to the celestial motions of the earth which are also responsible for the tide-generating gravitational forces. Atmospheric tides, i. e. tidal variations in pressure, temperature or wind direction and velocity, are predominantly excited by thermal input from the sun into the atmosphere, i. e. by the solar radiational flux. However, the gravitational signal, although very small, is also detectable by refined time series analysis. The first article gives a review of the excitation mechanisms, the observations and the theoretical interpretation of the measured atmospheric signals.

The second contribution touches palaeoclimatic varitations like the ice ages which may have been caused by variations of the solar radiation incident on the earth's atmosphere, resulting from the precession of the earth's celestial orbit and from variations of its obliquity and eccentricity. The last two parameters are held constant in tidal theory although they show long period changes which are taken into account in climate research. Actually there is observational evidence for corresponding periods in climatic indicators, but certain discrepancies imply that other mechanisms related to earth-sea-atmosphere interactions or astrophysical phenomena must be investigated also as possible candidates for a source of climate variations.

The third article in this section describes geomagnetic variations which, to a great part, are also caused by varations of solar irradiance, but also by regularly varying particle fluxes in the ionosphere and magnetosphere. In addition, they contain a small fraction of gravitationally induced lunar tides, whereas the solar gravitational tidal influence is overwhelmed by the thermal driving mechanisms of the ionospheric dynamo.

The fourth section of the book is devoted to special tidal phenomena which largely depend on local conditions. Boreholes in sediments or crystalline rocks often exhibit tidal variations in water level which are investigated with respect to hydrogeological parameters. The first contribution in this part introduces the relations between hydrology and tides and exemplifies the corresponding effects for some aquifer models. An increasing number of corresponding investigations of tidal water level variations shows a rising interest of hydrologists in this research branch.

In the second article, the controversial discussion about tidal triggering of earthquakes and volcanic events is critically reviewed. It is generally well known that a positive outcome of a statistical analysis does not necessarily prove a physical relation between the corresponding events, but as it is shown in this paper, this rule is often neglected in superficial investigations.

The third contribution is a report about the influence of the so-called geological effect on earth tides, especially on tidal tilt measurements. A temporal variation of the tidal tilt response in the vicinity of an earthquake fault is considered to be indicative for regional stress build-up before an earthquake happens, and corresponding investigations using a tilt station array at the North-Anatolian Fault Zone are described.

In the last section of the book tidal effects in physical systems located at increasing distance from the earth are considered. The first contribution gives a short summary and literature of tidal perturbations of satellite orbits. The second article contains an account of tidal phenomena on Jupiter's satellite Io, which displays strong volcanism and obviously extracts the necessary energy for this internal activity from tidal friction. The third contribution is a summary introducing the reader to the important tidal interaction between binary stars, and the last article extends the view to tidal interactions between galaxies.

The main purpose of this book is to give a glimpse on the fascinating variety of tidal phenomena happening in nature. The human experience is naturally concentrated on the earth, and so earth-bound phenomena are the predominating effects compiled in this lecture script. However, the last chapter which looks far out beyond the earth's neighbourhood, demonstrates that tides are existing everywhere in the universe and are of outmost importance for ist past and future.

Helmut Wilhelm
Walter Zürn
Hans-Georg Wenzel

Earth Tides

Tide-Generating Potential for the Earth

Hans-Georg Wenzel

Geodätisches Institut, Universität Karlsruhe, Englerstr. 7, D-76128 Karlsruhe,
Germany

Abstract. We will discuss in the following the definition and computation of tidal accelerations and of the tidal potential, the expansion of the tidal potential in spherical harmonics, the spectral analysis of the spherical harmonic expansion yielding a tidal potential catalogue, the accuracy of the currently available tidal potential catalogues, and the computation of the tidal forcing function (tidal potential and tidal accelerations) for a rigid Earth using a tidal potential catalogue.

1 Introduction

For all bodies of the Universe moving in a stationary orbit (which is in first approximation a Kepler ellipse), the **gravitational accelerations** produced by other bodies (planets, satellites) are completely compensated in their centers of mass by **centrifugal accelerations** due to the orbital motion of the body. Because of the spatial extension of the body (e.g. the Earth), the gravitational accelerations due to other celestial bodies (e.g. Moon, Sun) are slightly **position dependent**, whereas the centrifugal accelerations are constant within the body and on the surface of the body. The difference between the gravitational accelerations and the centrifugal accelerations are small **tidal accelerations**; on the Earth, the tidal accelerations are less than $\pm 1 \mu m/s^2 = 10^{-7}$ of the Earth's gravity g.

The computation of functionals of the tidal potential (e.g. tidal accelerations, tidal tilt, tidal strain) at a specific station and epoch can be carried out using one of the following two methods:

- Using ephemerides (coordinates) **of the celestial bodies** (Moon, Sun and planets), functionals of the tidal potential can be computed for a rigid oceanless Earth model with very high precision. This method is used to compute tidal potential catalogues and so-called benchmark series to check tidal potential catalogues (e.g. Wenzel, 1996a). But its practical application is restricted to less precise demands because it is impossible to compute accurately tidal effects for an elastic Earth covered with oceans using the ephemeris method.
- The tidal potential can be expanded in solid spherical harmonics; the spectral analysis of the tidal potential's spherical harmonic expansion yields a **tidal potential catalogue** (a table of amplitudes, phases and frequencies for some tidal waves). There are currently available several tidal potential catalogues with different total number of tidal waves and different accuracy (see Tab.

1). Using such a tidal potential catalogue and additional information like models of the elasticity inside the Earth from seismology and models of the ocean tides, tidal phenomena can be computed accurately for an elastic Earth covered with oceans. Nowadays, this method is mainly used.

Table 1. Available tidal potential catalogues.

Author(s)	no. of waves	accuracy[*] $[nm/s^2]$
Doodson (1921)	377	1.0408
Cartwright et al. (1971, 1973)	505	0.3844
Büllesfeld (1985)	656	0.2402
Tamura (1987)	1 200	0.0834
Xi (1989)	3 070	0.0642
Tamura (1993)	2 060	0.0308
Roosbeek (1996)	6 499	0.0200
Hartmann and Wenzel (1995a,b)	12 935	0.0014

[*] rms difference of gravity tides computed from the catalogue to gravity tide benchmark series BFDE403A, see below.

2 Definition of the Tidal Accelerations

The tidal acceleration $\vec{b}$ in an observation point P on the Earth's surface (see Fig. 1) results from the sum of the gravitational acceleration $\vec{a}_p$ generated by a celestial body at point P and of the orbital acceleration $-\vec{a}_0$ due to the motion of the Earth around the barycenter of the two-body system (the Earth and the celestial body). The orbital acceleration equals in good approximation the negative gravitational acceleration of the celestial body in the geocenter (we will consider later on effects of the extended mass distribution of the Earth, the so-called Earth's flattening effects). Using Newton's gravitational law, the tidal acceleration vector $(\vec{b})$ is given by:

$$\vec{b} = \vec{a}_p - \vec{a}_0 = \frac{GM_b}{d^2} \cdot \frac{\vec{d}}{d} - \frac{GM_b}{s^2} \cdot \frac{\vec{s}}{s} \tag{1}$$

with $G = 6.6672 \cdot 10^{11}$ m^3 = Newtonian gravitational constant, M_b = mass of the celestial body, $\vec{d}$ = topocentric distance vector, $\vec{s}$ = geocentric distance vector. In the geocenter, the distances d and s to the celestial body are identical and therefore the tidal acceleration vanishes: $\vec{b} = 0$!

One can easily understand that (1) results in tidal accelerations on the surface of the Earth by the rotation of the Earth around the Sun. But the Moon and the nearby planets of our solar system also generate tidal accelerations, resulting from the gravitational acceleration due to the celestial body and from the motion of the Earth around the **barycenter of the two-body system**. For the Earth-Moon system the barycenter is located inside the Earth's body. The Earth as well as the Moon are in motion approximately on an ellipsoidal orbit around the barycenter of the Earth-Moon system, and all particles of the Earth move on parallel orbits. The orbital motion of the Earth around the Earth-Moon barycenter generates orbital accelerations, which are completely compensated in the geocenter by the gravitational accelerations due to the Moon. The difference between the gravitational acelerations and the orbital accelerations are the tidal accelerations, just as given by eq. (1). Similar considerations are valid for the other celestial bodies, e.g. the planets of the solar system.

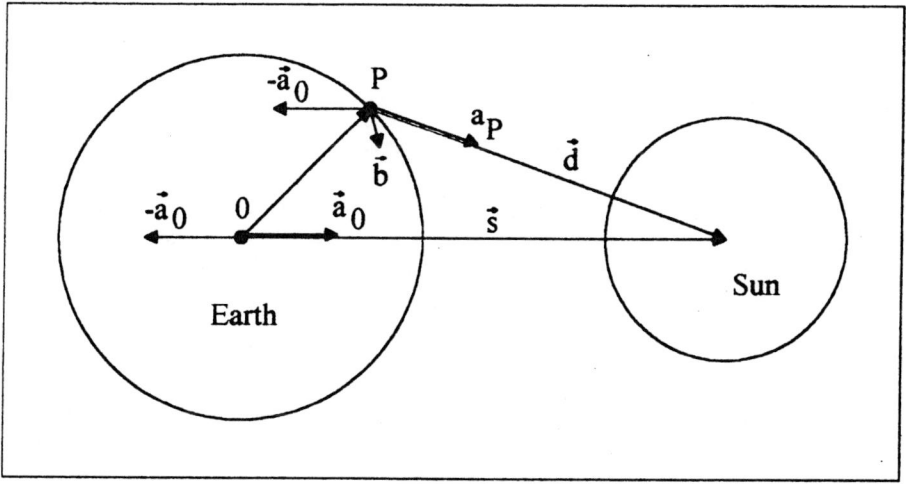

Fig. 1. Gravitational acceleration, orbital acceleration and tidal acceleration.

The maximum tidal accelerations on the surface of the Earth are

due to the Moon:	$1.37 \cdot 10^{-06}$	m/s^2
due to the Sun:	$0.50 \cdot 10^{-06}$	m/s^2
due to the Mercury:	$3.64 \cdot 10^{-13}$	m/s^2
due to the Venus:	$5.88 \cdot 10^{-11}$	m/s^2
due to the Mars:	$1.18 \cdot 10^{-12}$	m/s^2
due to the Jupiter:	$6.54 \cdot 10^{-12}$	m/s^2
due to the Saturn:	$2.36 \cdot 10^{-13}$	m/s^2
due to the Uranus:	$3.67 \cdot 10^{-15}$	m/s^2
due to the Neptune:	$1.06 \cdot 10^{-15}$	m/s^2
due to the Pluto:	$7.61 \cdot 10^{-20}$	m/s^2

The resolution of high quality recording gravimeters is below 10^{-11} m/s² $= 10^{-12}$ g; therefore, the tidal accelerations due to the nearby planets have to be considered for the analysis of such records.

The tidal accelerations are symmetrical to three orthogonal axes (A - A' and B - B' in Fig. 2, the third axis goes through the Earth's center of gravity and is orthogonal to the plane of the paper); therefore, the tidal accelerations do not generate any net acceleration of the Earth. Because the Earth rotates around axis C - C', we have mainly diurnal and semidiurnal variations of the tidal accelerations (see below).

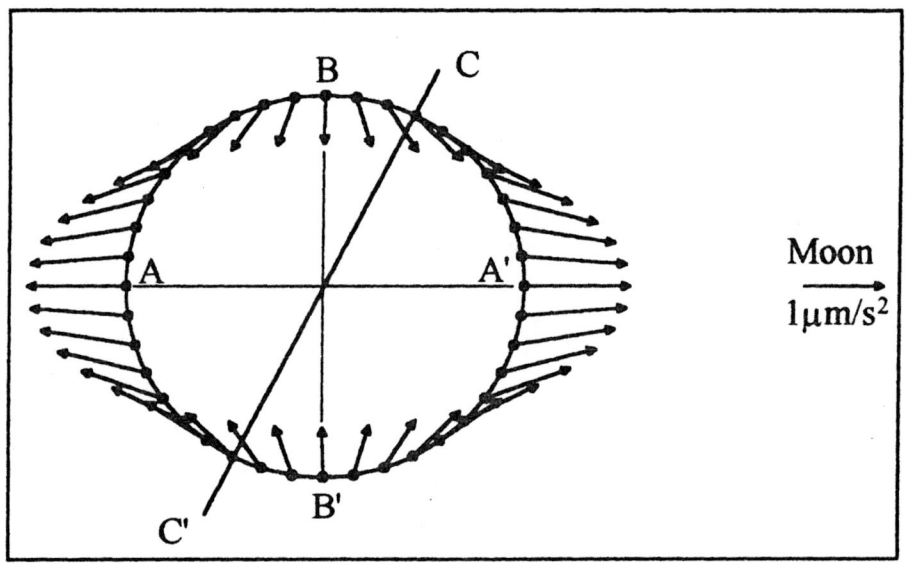

Fig. 2. Tidal accelerations due to the Moon at the surface of the Earth

3   Tidal Potential

We will now introduce the scalar **tidal potential** V instead of the vectorial tidal accelerations $\vec{b}$ in order to enable an expansion of the tidal potential into scalar spherical harmonics. This will allow us the separation of the tidal potential into latitude dependent terms and time/longitude dependent terms, and the **spectral representation** of the tidal potential by a **tidal potential catalogue**.

The tidal acceleration vector $\vec{b}$ is by definition the gradient of the tidal potential V:

$$\vec{b} = \vec{grad}\, V = \frac{\partial V}{\partial \vec{r}} \tag{2}$$

with the additional constraint that the tidal potential vanishes in the geocenter: $V = 0$ for $\vec{r} = \vec{0}$. The solution of eq. (2) is

$$V = GM_b\left(\frac{1}{d} - \frac{1}{s} - \frac{r \cdot \cos\psi}{s^2}\right) \tag{3}$$

with ψ = geocentric zenith angle. Eq. (3) can be expanded into a series of Legendre polynomials $P_\ell(\cos\psi)$:

$$V = \frac{GM_b}{s} \cdot \sum_{\ell=2}^{\infty} \left(\frac{r}{s}\right)^\ell P_\ell(\cos\psi) \tag{4}$$

Because the relation r/s is about $1.6 \cdot 10^{-2}$ for the Moon and about $4 \cdot 10^{-5}$ for the Sun, the series expansion (4) converges rapidly. For the most accurate tidal potential catalogues, we use $\ell_{max} = 6$ for the Moon, $\ell_{max} = 3$ for the Sun and $\ell_{max} = 2$ for the planets. The largest contribution to the tidal potential results from degree 2 with about 98 % of V. Fig. 3 shows the tidal potential due to the Moon at the surface of the Earth.

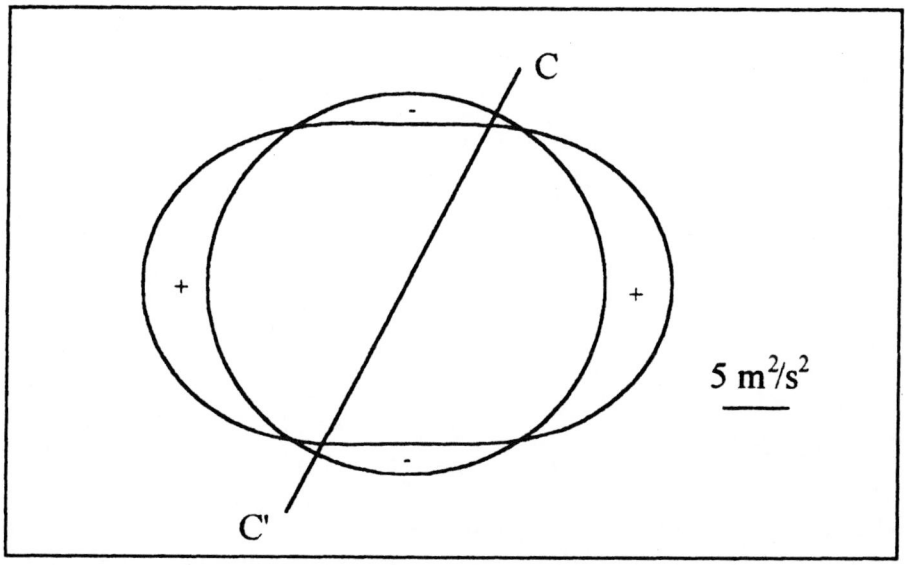

Fig. 3. Tidal potential at the Earth's surface due to the Moon

The geocentric zenith angle ψ can be expressed by geocentric spherical co-ordinates of the station and of the celestial body (e.g. Hartmann and Wenzel, 1995a,b)

$$\cos\psi = \cos\theta \cdot \cos\Theta_b + \sin\theta \cdot \sin\Theta_b \cdot \cos(\lambda - \Lambda_b) \tag{5}$$

with ψ = geocentric zenith angle of the celestial body, θ = geocentric spherical polar distance of the station, λ = geocentric spherical longitude of the station, with ψ = geocentric zenith angle of the celestial body, θ = geocentric spherical polar distance of the station, λ = geocentric spherical longitude of the station, Θ_b = geocentric spherical polar distance of the celestial body, Λ_b = geocentric spherical longitude of the celestial body.

Eq. (5) enables the expansion of the Legendre polynomials into **fully normalized spherical harmonics** $\overline{P}_{\ell,m}$ (exactly, the $\overline{P}$ are the fully normalized associated Legendre functions of 1st kind, see e.g. Heiskanen and Moritz, 1967):

$$P_\ell(\cos\psi) = \frac{1}{(2\ell+1)} \sum_{m=0}^{\ell} \overline{P}_{\ell,m}(\cos\theta) \cdot \overline{P}_{\ell,m}(\cos\Theta_b) \cdot \cos(m\lambda - m\Lambda_b) \qquad (6)$$

with ℓ = degree, m = order. Inserting (6) into (4) yields the **spherical harmonic expansion** of the tidal potential on the Earth by a specific celestial body:

$$V = \frac{GM_b}{s} \sum_{\ell=2}^{\infty} (\frac{r}{s})^\ell \frac{1}{(2\ell+1)} \sum_{m=0}^{\ell} \overline{P}_{\ell,m}(\cos\theta) \cdot \overline{P}_{n,m}(\cos\Theta_b) \cdot \cos(m\lambda - m\Lambda_b) \quad (7)$$

where s, Θ_b and Λ_b are time dependent.

The fully normalized spherical harmonics are defined by (e.g. Heiskanen and Moritz, 1967)

$$\overline{P}_{\ell m}(x) = (-1)^m (1-x^2)^{m/2} \frac{d^m}{dx^m} \left[\frac{1}{2^\ell \ell!} \frac{d^\ell}{dx^\ell}(x^2-1)^\ell \right] \sqrt{(2\ell+1)\frac{(\ell-m)!}{(\ell+m)!}(2-\delta_{m,0})}$$
$$(8)$$

The fully normalized spherical harmonics and their derivations with respect to θ can be computed to high degree and order from recursion formulas (e.g. Wenzel, 1985). Explicit formulas for the fully normalized spherical harmonics and their derivatives to θ up to degree and order 6 are given in Tab. 2. The fully normalized spherical harmonics of degree 2 are shown in Fig. 4 as function of latitude. Because of the latitude dependence of the spherical harmonics, the **amplitudes** of the tidal waves are **latitude dependent**. For the tidal potential and for the radial tidal acceleration (approximately the negative gravity tide, see below)

the longperiodic tidal waves have a maximum at the poles,

the diurnal tides have a maximum at $\pm 45^0$ latitude,

the semidiurnal tides have a maximum at the equator.

Table 2. Closed formulas for the fully normalized Legendre functions and their derivatives.

ℓ	m	$\overline{P}_{\ell m}(\cos\theta)$	$\partial\overline{P}_{\ell m}(\cos\theta)/\partial\theta$
1	0	$\sqrt{3}\cos\theta$	$-\sqrt{3}\sin\theta$
1	1	$\sqrt{3}\sin\theta$	$\sqrt{3}\cos\theta$
2	0	$\sqrt{\frac{5}{4}}(3\cos^2\theta - 1)$	$-\sqrt{45}\sin\theta\cos\theta$
2	1	$\sqrt{15}\sin\theta\cos\theta$	$\sqrt{15}(1 - 2\sin^2\theta)$
2	2	$\sqrt{\frac{15}{4}}\sin^2\theta$	$\sqrt{15}\sin\theta\cos\theta$
3	0	$\sqrt{\frac{7}{4}}\cos\theta(5\cos^2\theta - 3)$	$\sqrt{\frac{63}{4}}\sin\theta(1 - 5\cos^2\theta)$
3	1	$\sqrt{\frac{21}{8}}\sin\theta(5\cos^2\theta - 1)$	$\sqrt{\frac{21}{8}}\cos\theta(4 - 15\sin^2\theta)$
3	2	$\sqrt{\frac{105}{4}}\sin^2\theta\cos\theta$	$\sqrt{\frac{105}{4}}\sin\theta(3\cos^2\theta - 1)$
3	3	$\sqrt{\frac{35}{8}}\sin^3\theta$	$\sqrt{\frac{315}{8}}\sin^2\theta\cos\theta$
4	0	$\sqrt{\frac{9}{64}}(3 - 30\cos^2\theta + 35\cos^4\theta)$	$\sqrt{\frac{225}{4}}(3 - 7\cos^2\theta)\sin\theta\cos\theta$
4	1	$\sqrt{\frac{45}{8}}\sin\theta\cos\theta(7\cos^2\theta - 3)$	$\sqrt{\frac{45}{8}}(3 - 27\cos^2\theta + 28\cos^4\theta)$
4	2	$\sqrt{\frac{45}{16}}\sin^2\theta(7\cos^2\theta - 1)$	$\sqrt{45}(7\cos^2\theta - 4)\sin\theta\cos\theta$
4	3	$\sqrt{\frac{315}{8}}\sin^3\theta\cos\theta$	$\sqrt{\frac{315}{8}}\sin^2\theta(4\cos^2\theta - 1)$
4	4	$\sqrt{\frac{315}{64}}\sin^4\theta$	$\sqrt{\frac{315}{4}}\sin^3\theta\cos\theta$
5	0	$\sqrt{\frac{11}{64}}\cos\theta(15 - 70\cos^2\theta + 63\cos^4\theta)$	$\sqrt{\frac{2475}{64}}\sin\theta(-1 + 14\cos^2\theta - 21\cos^4\theta)$
5	1	$\sqrt{\frac{165}{64}}\sin\theta(1 - 14\cos^2\theta + 21\cos^4\theta)$	$\sqrt{\frac{165}{64}}\cos\theta(29 - 126\cos^2\theta + 105\cos^4\theta)$
5	2	$\sqrt{\frac{1155}{16}}\sin^2\theta\cos\theta(3\cos^2\theta - 1)$	$\sqrt{\frac{1155}{16}}\sin\theta(1 - 12\cos^2\theta + 15\cos^4\theta)$
5	3	$\sqrt{\frac{385}{128}}\sin^3\theta(9\cos^2\theta - 1)$	$\sqrt{\frac{3465}{128}}\sin^2\theta\cos\theta(15\cos^2\theta - 7)$
5	4	$\sqrt{\frac{3465}{64}}\sin^4\theta\cos\theta$	$\sqrt{\frac{3465}{64}}\sin\theta(-1 + 6\cos^2\theta - 5\cos^4\theta)$
5	5	$\sqrt{\frac{693}{128}}\sin^5\theta$	$\sqrt{\frac{17325}{128}}\cos\theta\sin^4\theta$
6	0	$\sqrt{\frac{13}{256}}(-5 + 105\cos^2\theta - 315\cos^4\theta + 231\cos^6\theta)$	$\sqrt{\frac{5733}{64}}\sin\theta\cos\theta(-5 + 30\cos^2\theta - 33\cos^4\theta)$
6	1	$\sqrt{\frac{273}{64}}\sin\theta\cos\theta(5 - 30\cos^2\theta + 33\cos^4\theta)$	$\sqrt{\frac{273}{64}}(-5 + 100\cos^2\theta - 285\cos^4\theta + 198\cos^6\theta)$
6	2	$\sqrt{\frac{1365}{512}}\sin^2\theta(1 - 18\cos^2\theta + 33\cos^4\theta)$	$\sqrt{\frac{1365}{128}}\sin\theta\cos\theta(19 - 102\cos^2\theta + 99\cos^4\theta)$
6	3	$\sqrt{\frac{1365}{128}}\sin^3\theta\cos\theta(-3 + 11\cos^2\theta)$	$\sqrt{\frac{12285}{128}}\sin^2\theta(1 - 15\cos^2\theta + 22\cos^4\theta)$
6	4	$\sqrt{\frac{819}{256}}\sin^4\theta(11\cos^2\theta - 1)$	$\sqrt{\frac{819}{64}}\cos\theta\sin\theta(-13 + 46\cos^2\theta - 33\cos^4\theta)$
6	5	$\sqrt{\frac{9009}{128}}\sin^5\theta\cos\theta$	$\sqrt{\frac{9009}{128}}\sin^4\theta(6\cos^2\theta - 1)$
6	6	$\sqrt{\frac{3003}{512}}\sin^6\theta$	$\sqrt{\frac{27027}{128}}\sin^5\theta\cos\theta$

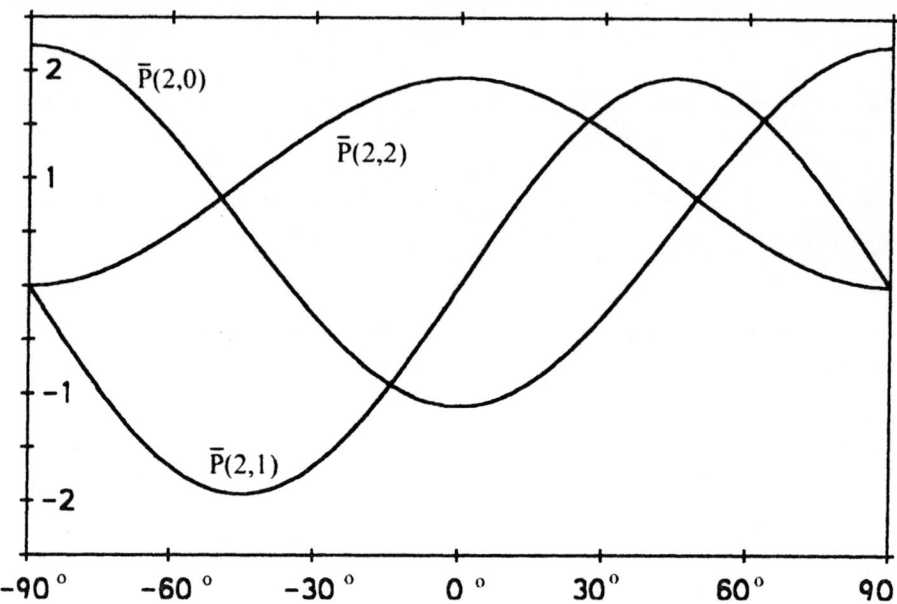

Fig. 4. Fully normalized spherical harmonics of degree 2.

In eq. (7), the geocentric longitude Λ_b of the celestial body is multiplied by the order m of the spherical harmonic expansion. Because the geocentric longitude Λ_b of the celestial bodies varies by about 2π per 24^h due to the rotation of the Earth, the tidal potential of a certain order varies in a certain station (see also Fig. 5) for

$$
\begin{aligned}
&m = 0: \text{with } 14^d \dots 18.6^a \text{ period} &&\rightarrow \text{long periodic waves} \\
&m = 1: \text{with} \sim 24^h \text{ period} &&\rightarrow \text{diurnal waves} \\
&m = 2: \text{with} \sim 12^h \text{ period} &&\rightarrow \text{semidiurnal waves} \\
&m = 3: \text{with} \sim 8^h \text{ period} &&\rightarrow \text{terdiurnal waves} \\
&m = 4: \text{with} \sim 6^h \text{ period} \\
&m = 5: \text{with} \sim 4.8^h \text{ period} \\
&m = 6: \text{with} \sim 4^h \text{ period}
\end{aligned}
$$

For the derivation of the tidal potential, we have until now assumed that the centrifugal accelerations due to the orbital motion of the Earth equals the gravitational acceleration due to the celestial body in the center of gravity of the Earth. This is an approximation only and neglects the orbital motion due to the so-called figure forces (forces due to the extended mass distribution of the Earth, e.g. Ilk (1983). These figure forces generate the so-called Earth's flattening effects (e.g. Wilhelm, 1982; Dahlen, 1993). Regarding the **Earth's flattening effects**, the tidal potential V due to a specific celestial body at the surface of the Earth is given by (e.g. Hartmann and Wenzel, 1995a,b)

$$V_{(t)} = GM_b \sum_{\ell=2}^{\ell=\ell_{\max}} \sum_{m=0}^{m=\ell} \frac{r^\ell}{r_b^{\ell+1}(t)} \frac{1}{2\ell+1} \overline{P}_{\ell m}(\cos\theta) \overline{P}_{\ell m}(\cos\theta_b(t)) \cos[m(\lambda - \Lambda_b(t))]$$

$$+ GM_\oplus \frac{r_\oplus^2 r}{r_b^4(t)} J_{2\oplus} \sqrt{\frac{3}{7}} \overline{P}_{10}(\cos\theta) \overline{P}_{30}(\cos\theta_b(t)) \qquad (9)$$

$$+ GM_\oplus \frac{r_\oplus^2 r}{r_b^4(t)} J_{2\oplus} \sqrt{\frac{2}{7}} \overline{P}_{11}(\cos\theta) \overline{P}_{31}(\cos\theta_b(t)) \cos(\lambda - \Lambda_b(t))$$

with $GM_\oplus$ = geocentric gravitational constant of the Earth, $J_{2\oplus}$ = second degree zonal coefficient of the gravity field of the Earth. The two last lines in (9) account for the Earth's flattening effect. The tidal potential catalogues of Hartmann and Wenzel (1995a,b) and Roosbeek (1996) have included coefficients accounting for the Earth's flattening effects.

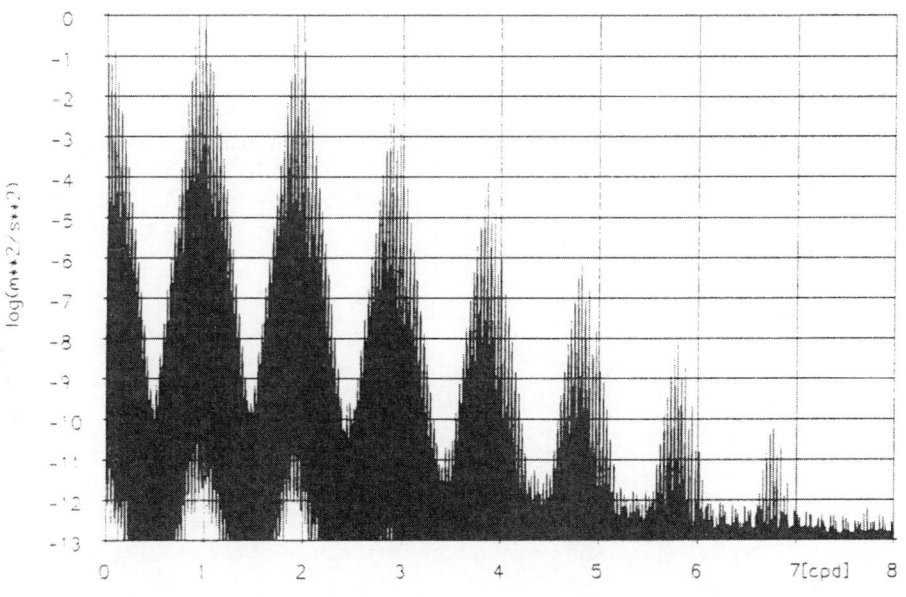

Fig. 5. Amplitude spectrum of the tidal potential for a rigid Earth computed from DE404/LE404 ephemerides between years 1563 and 2042 at station Schiltach (ϕ = 48.3306° N, λ = 8.3300° E, H = 589 m).

4   Tidal Potential Catalogues

Darwin (1883) has probably for the first time computed a catalogue of tidal waves, and he has given names to the main tidal waves contained in his catalogue, which are still in use. The currently available tidal potential catalogues (see Tab.

1) have either been computed by analytical spectral analysis (Doodson, 1921; Xi, 1987; Xi, 1989; Roosbeek, 1996) or by numerical spectral analysis (Cartwright and Tayler, 1971; Cartwright and Edden, 1973; Büllesfeld, 1985; Tamura, 1987; Tamura, 1993; Hartmann and Wenzel, 1995a,b) of the tidal potential generated by the celestial bodies. The analytical spectral analysis method requires analytical ephemerides of the celestial bodies (e.g. Bretagnon, 1981; Chapront-Touzé and Chapront, 1988), whereas the numerical spectral analysis method needs numerical ephemerides only. The numerical ephemerides computed by the Jet Propulsion Laboratory at Pasadena (e.g. Standish and Williams, 1981; Standish et al., 1995) are known as the most accurate ephemerides currently available, and they are accurate enough for the computation of tidal potential catalogues (e.g. Hartmann and Wenzel, 1995a,b).

In Fig. 5 is shown an amplitude spectrum of the tidal potential of the rigid Earth computed between years 1563 and 2042 (4.2 million samples) at station Schiltach (latitude 48.3306^0 N, longitude 8.3300^0 E, height 589 m). One can see the different tidal groups with $m = 0...7$ at 0...7 cpd frequency as well as the numerical noise of about $1 \cdot 10^{-12}$ m^2/s^2.

All tidal potential catalogues use a representation of the tidal potential on a rigid Earth similar to

$$V_{(t)} = D \sum_{\ell=1}^{\ell=\ell_{\max}} \sum_{m=0}^{m=\ell} \left(\frac{r}{a}\right)^{\ell} \Gamma(\theta) \cdot \overline{P}_{\ell m}(\cos\theta)$$
$$\cdot \sum_{i} [C_i^{\ell m}(t) \cos(\alpha_i(t)) + S_i^{\ell m}(t) \sin(\alpha_i(t))] \tag{10}$$

with $D, \Gamma(\theta) =$ normalization constants, $a =$ semi-major axis of the reference ellipsoid, $C_i^{\ell m}(t), S_i^{\ell m}(t) =$ time dependent coefficients of the catalogue.

For the Hartmann and Wenzel (1995a,b) catalogue, the normalization constants D and $\Gamma(\theta)$ have been set to unity. The arguments $\alpha_i(t)$ are given by

$$\alpha_i(t) = m \cdot \lambda + \sum_{j-1}^{j=jmax} k_{ij} \cdot arg_j(t) \quad \text{with } k_{i1} = m. \tag{11}$$

The integer coefficients k_{ij} are given in the specific catalogue, while the astronomical arguments $arg_j(t)$ can be computed from polynomials in time (see Tab. 3 and 4).

Table 3. Fundamental frequencies of astronomical arguments in °/hour after Simon et al. (1994) at J2000

j argument	symbol	frequency [°/hour]
1. mean local lunar time	τ	14.492 052 120 18
2. mean lunar longitude	s	0.549 016 519 73
3. mean solar longitude	h	0.041 068 639 91
4. mean longitude of lunar perigee	p	0.004 641 813 41
5. negative mean longitude of lunar ascending node	N'	0.002 206 406 87
6. mean longitude of solar perigee	p_s	0.000 001 961 51
7. mean longitude of Mercury	L_{Mer}	0.170 515 710 90
8. mean longitude of Venus	L_{Ven}	0.066 757 030 52
9. mean longitude of Mars	L_{Mar}	0.021 836 295 20
10. mean longitude of Jupiter	L_{Jup}	0.003 463 726 64
11. mean longitude of Saturn	L_{Sat}	0.001 395 746 14

Table 4. Polynomial coefficients for astronomical arguments after Simon et al. (1994), units are: ° and °/1000 yrs

j	Constant	t	t^2	t^3	t^4
1.	242.14980452999	127037328.88553056	0.17696111	−0.00183140	0.00008824
2.	218.31664562999	4812678.81195750	−0.14663889	0.00185140	−0.00015355
3.	280.46645016002	360007.69748806	0.03032222	0.00002000	−0.00006532
4.	83.35324311998	40690.13635250	−1.03217222	−0.01249168	0.00052655
5.	234.95544499000	19341.36261972	−0.20756111	−0.00213942	0.00016501
6.	282.93734098001	17.19457667	0.04568889	−0.00001776	−0.00003323
7.	252.25090551999	1494740.72172233	0.03034984	0.00001811	−0.00006532
8.	181.97980085000	585192.12953330	0.03101395	0.00001490	−0.00006532
9.	355.43299958002	191416.96370297	0.03105187	0.00001564	−0.00006532
10.	34.35151874003	30363.02774848	0.02232972	0.00003701	−0.00005214
11.	50.07744430000	12235.11068622	0.05190783	−0.00002985	−0.00009740

The time dependent coefficients $C_i^{lm}(t)$, $S_i^{lm}(t)$ are given by

$$C_i^{lm}(t) = C0_i^{lm} + t \cdot C1_i^{lm} \tag{12}$$
$$S_i^{lm}(t) = S0_i^{lm} + t \cdot S1_i^{lm} \tag{13}$$

The errors of the Cartwright et al. (1971, 1973) tidal potential catalogue have been estimated already by Wenzel (1976) to about 0.350 nm/s^2 in time domain, mainly because of neglection of the lunar tidal potential of degree 4. This was the reason to compute more accurate tidal potential catalogues (see Tab. 3) including the lunar tidal potential of degree 4 by Büllesfeld (1985), Xi (1987), Tamura (1987), Xi (1989), and Tamura (1993). The catalogue of Tamura (1993) includes coefficients due to the indirect tidal potential of the planets Venus and Jupiter. The catalogues of Roosbeek (1996) and Hartmann and Wenzel (1995a,b) include the lunar tidal potential of degree 5 and 6 respectively, and they also include coefficients due to the direct tidal potential of the nearby planets and due to the flattening of the Earth. The truncation level of the tidal potential catalogues has continuously been decreased, and the number of waves and the number of coefficients has continuously been increased with time (see Tab. 5).

5   Computation of Tidal Accelerations using a Tidal Potential Catalogue

By differentiation of the spectral representation of the tidal potential from (10) with respect to the spherical coordinates of the station, the tidal accelerations in a spherical coordinate system can be computed from a tidal potential catalogue. Using the Hartmann and Wenzel (1995a,b) normalization with $D = \Gamma(\theta) = 1$ we have for the radial direction (approximately the negative gravity tide)

$$
b_r = \frac{\partial V}{\partial r} = \frac{1}{a} \cdot \sum_{\ell=1} \ell \cdot (\frac{r}{a})^{\ell-1} \sum_{m=0}^{\ell} \overline{P}_{\ell,m}(\theta)
$$
$$
\cdot \sum_i [C_i^{\ell m}(t) \cos(\alpha_i(t)) + S_i^{\ell m}(t) \sin(\alpha_i(t))] \tag{14}
$$

for the south direction:

$$
b_\theta = \frac{\partial v}{r \cdot \partial \theta} = \frac{1}{a} \cdot \sum_{\ell=1}(\frac{r}{a})^{\ell-1} \sum_{m=0}^{\ell} \frac{\partial \overline{P}_{\ell,m}}{\partial \theta}
$$
$$
\cdot \sum_i [C_i^{\ell m}(t) \cos(\alpha_i(t)) + S_i^{\ell m}(t) \sin(\alpha_i(t))] \tag{15}
$$

and for the east direction:

$$
b_\lambda = \frac{\partial V}{r \cdot \sin \theta \cdot \partial \lambda} \tag{16}
$$

and because of (11) $\alpha_i(t) = m \cdot \lambda + c$

$$
\frac{\partial \cos(m\lambda + c)}{\partial \lambda} = -m \cdot \sin(m\lambda + c) \tag{17}
$$

$$
\frac{\partial \sin(m\lambda + c)}{\partial \lambda} = m \cdot \cos(m\lambda + c) \tag{18}
$$

$$b_\lambda = \frac{1}{a} \cdot \sum_{\ell=1} (\frac{r}{a})^{\ell-1} \sum_{m=1}^{\ell} m \cdot \frac{\overline{P}_{\ell,m}(\theta)}{\sin \theta}$$

$$\sum_i \left[-C_i^{\ell m}(t) \sin(\alpha_i(t)) + S_i^{\ell m}(t) \cos(\alpha_i(t)) \right] \qquad (19)$$

Because of the multiplication by m in (19) the east component b_λ of the tidal acceleration does not contain longperiodic tides. For precise tidal computations, the tidal accelerations have to be rotated from the spherical coordinate system into a coordinate system oriented into the local plumb line (e.g. Wenzel, 1974). This would require the knowledge of the directions of the plumb line in the station (i.e. the vertical deflections), which generally have to be observed by extensive astronomical methods. Instead of the coordinate system oriented to the local plumb line, a coordinate system oriented to the ellipsoidal normal is usually sufficient.

Table 5. Short description of tidal potential catalogues

catalogue	no. of waves	no. of coeff.	max. degree	truncation [m²/s²]	ephemerides
Doodson (1921)	378	378	3	$1.0 \cdot 10^{-4}$	BN
Cartwright et al. (1971, 1973)	505	1 010	3	$0.4 \cdot 10^{-4}$	BN
Büllesfeld (1985)	656	656	4	$0.2 \cdot 10^{-4}$	BN
Tamura (1987)	1 200	1 326	4	$0.4 \cdot 10^{-5}$	DE118/LE62
Xi (1989)	2 934	2 934	4	$0.9 \cdot 10^{-6}$	BN
Tamura (1993)	2 060	3 046	4	$0.4 \cdot 10^{-5}$	DE200/LE200
Roosbeek (1996)	6 499	7 202	5	$0.8 \cdot 10^{-7}$	CT/BR
Hartmann and Wenzel (1995a,b)	12 935	19 271	6	$0.1 \cdot 10^{-9}$	DE200/LE200

BN = Brown (1905), Newcomb (1897),
CT = Chapront-Touzé and Chapront (1988), BR = Bretagnon (1981)
truncation = smallest coefficient greater zero

In Tab. 6 are given the amplitudes and frequencies of the 17 major tidal waves at 48.3306^0 latitude. Gravity tides and tidal accelerations for north and east direction computed for a rigid Earth model using the Hartmann and Wenzel (1995a,b) tidal potential catalogue are shown in Fig. 6...8. We can see mainly diurnal and semidiurnal variations with monthly modulation.

Using a tidal potential catalogue and an elastic Earth model, gravity tides, radial and horizontal tidal deformations as well as all kinds of tidal strains can be computed (e.g. Wilhelm und Zürn 1985).

Table 6. The 17 largest tidal waves for latitude $\phi = 48.3^0\,\text{N}$.

name	ℓ	m	period [d]	origin M=Moon S=Sun	amplitudes for a rigid Earth potential $\bar{u} = \frac{1}{g} \cdot V$ [m²/s²]		gravity
						[mm]	[nm/s²]
M_m	2	0	27.554	M	0.072	7	23
M_f	2	0	13.660	M	0.136	14	43
Q_1	2	1	1.1195	M	0.188	19	59
O_1	2	1	1.0758	M	0.983	100	309
M_1	2	1	1.0347	M	0.077	8	24
P_1	2	1	1.0027	S	0.457	47	144
S_1	2	1	1.0000	S	0.011	1	3
K_1	2	1	0.9973	M+S	1.382	141	434
J_1	2	1	0.9624	M	0.077	8	24
OO_1	2	1	0.9294	M	0.042	4	13
$2N_2$	2	2	0.5377	M	0.027	3	8
N_2	2	2	0.5274	M	0.203	21	64
M_2	2	2	0.5175	M	1.061	108	332
L_2	2	2	0.5080	M	0.030	3	9
S_2	2	2	0.5000	S	0.494	50	154
K_2	2	2	0.4986	M+S	0.134	14	42
M_3	3	3	0.3450	M	0.009	1	4

$\bar{u}$ = radial deformation of the geoid

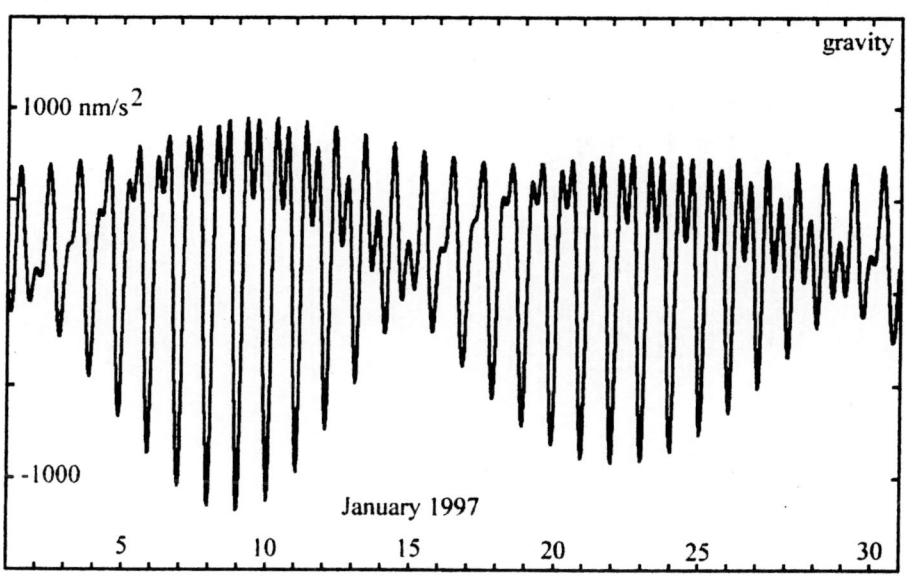

Fig. 6. Tidal gravity for a rigid Earth model computed at station Schiltach ($\phi = 48.3306^0$ N, $\lambda = 8.3300^0$ E, h = 589 m).

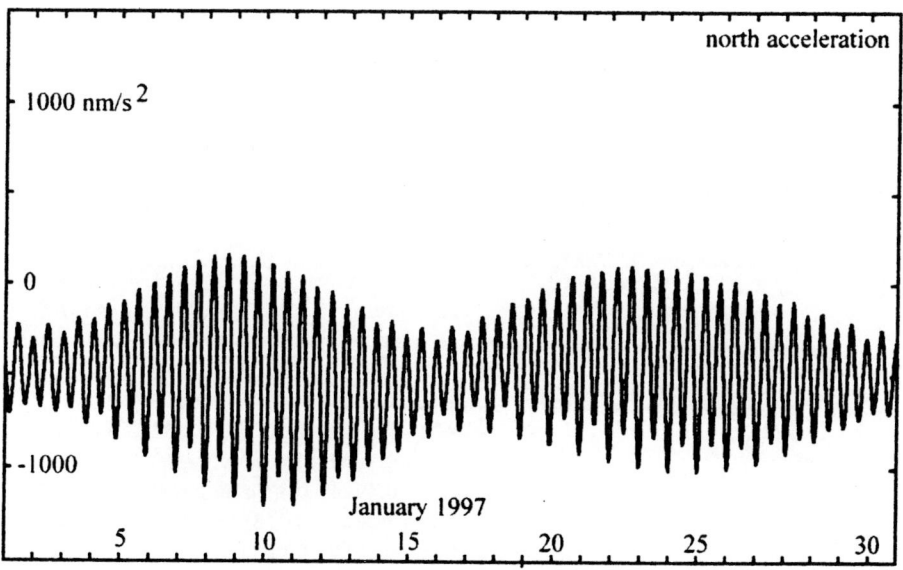

Fig. 7. Tidal acceleration for a rigid Earth model, north component, computed at station Schiltach ($\phi = 48.3306^0$ N, $\lambda = 8.3300^0$ E, h = 589 m).

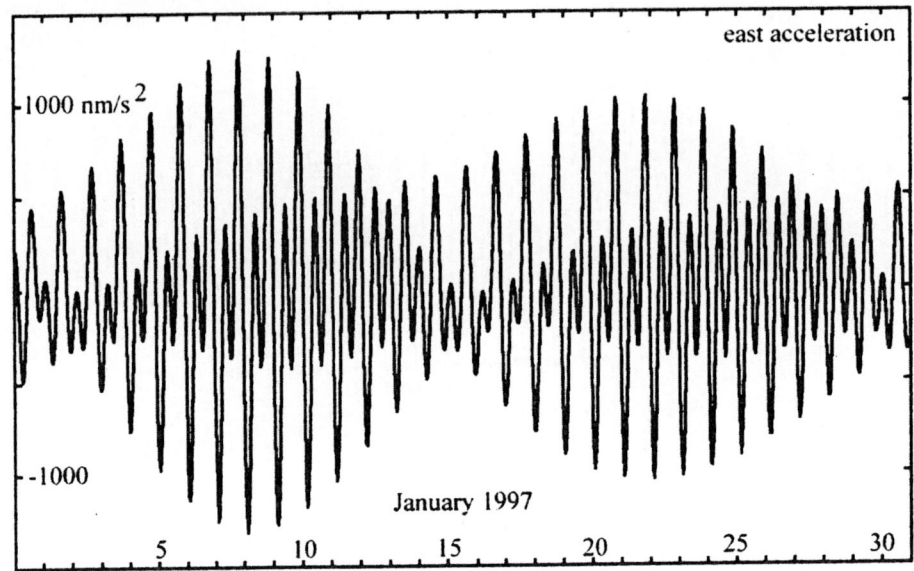

Fig. 8. Tidal acceleration for a rigid Earth model, east component, computed at station Schiltach ($\phi = 48.3306^0$ N, $\lambda = 8.3300^0$ E, h $=$ 589 m).

Table 7. Errors of gravity tides computed from different tidal potential catalogues derived from comparison with benchmark series BFDE403A.

catalogue	time domain rms [nm/s²]	min. [nm/s²]	max. [nm/s²]	frequency domain rms [nm/s²]	max. [nm/s²]
Doodson (1921)	1.0408	-4.0068	4.4896	0.01415	1.01204
Cartwright et al. (1971, 1973)	0.3844	-2.1401	2.4100	0.00565	0.13226
Büllesfeld (1985)	0.2402	-1.2234	1.3043	0.00334	0.06765
Tamura (1987)	0.0834	-0.4692	0.5839	0.00118	0.03459
Xi (1989)	0.0642	-0.4294	0.4768	0.00090	0.03525
Tamura (1993)	0.0308	-0.2237	0.2489	0.00046	0.01448
Roosbeek (1996)	0.0200	-0.0935	0.0933	0.00026	0.01598
Hartmann and Wenzel (1995a,b)	0.0014	-0.0094	0.0104	0.00002	0.00034

6   Accuracy of Tidal Potential Catalogues

Using different independent gravity tide benchmark series from ephemeris of different type (computed with program GTIDE of Merriam 1993, and using the high precision numerical ephmerides DE200 and DE403 from Jet Propulsion Laboratory) at two different time spans (1987···1993 and 2017···2023), the accuracy of gravity tides computed from different tidal potential catalogues has been estimated by Wenzel, 1996a. Using the most accurate benchmark gravity tide series BFDE403A (hourly gravity tides computed from DE403 ephemerides between 1987 and 1993), the errors of gravity tides computed from tidal potential catalogues are given in Tab. 7. The most accurate tidal potential catalogue is the catalogue of Hartmann and Wenzel (1995a,b). The errors of this catalogue are about ten times less than the errors of the two other recent catalogues by Tamura (1993) and Roosbeek (1996).

References

Bretagnon, P. 1982. Théorie du mouvement de l'ensemble des planètes, solution VSOP82. *Astron. Astrophys.* **144**: 278 - 288.

Brown, W.E. 1905. Theory of the motion of the Moon. *Mem. Royal Astron. Soc.* **57**: 136–141.

Büllesfeld, F.-J. 1985. Ein Beitrag zur harmonischen Darstellung des gezeitenerzeugenden Potentials. *Deutsche Geod. Komm.* **C 314**: 1 - 103.

Cartwright, D.E. and R.J. Tayler. 1971. New Computations of the Tide Generating Potential. *Geophys. J. R. astr. Soc.* **23**: 45 - 74.

Cartwright, D.E. and C.A. Edden. 1973. Corrected Tables of Tidal Harmonics. *Geophys. J. R. astr. Soc.* **33**: 253 - 264.

Chapront-Touzé, M. and J. Chapront. 1988. ELP-2000-85: A semi-analytical lunar ephemeris adequate for historical times. *Astron. Astrophys.* **190**: 342 - 352.

Dahlen, F.A. 1993. Effect of the Earth's ellipticity on the lunar potential. *Geophys. J. Int.* **113**: 250 - 251.

Darwin, G.H. 1883. Report of a committee for the harmonic analysis of tidal observations. *Brit. Ass. Rep.*: 48 - 118.

Doodson, A.T. 1921. The harmonic development of the tide generating potential. *Proc. Royal Soc. London* **A 100**: 306 - 328. Reprint in *Int. Hydrogr. Rev.* **31**

Hartmann, T. and H.-G. Wenzel. 1994a. Catalogue of the earth tide generating potential due to the planets. *Bull. Inf. Marées Terrestres* **119**: 8847 - 8880.

Hartmann, T. and H.-G. Wenzel. 1994b. The harmonic development of the earth tide generating potential due to the direct effects of the planets. *Geophys. Res. Lett.* **21**: 1991 - 1993.

Hartmann, T. and H.-G. Wenzel. 1995a. The HW95 tidal potential catalogue. *Geophys. Res. Lett.* **22**: 3553 - 3556.

Hartmann, T. and H.-G. Wenzel. 1995b. Catalogue HW95 of the tide generating potential. *Bull. Inf. Marées Terrestres* **123**: 9278 - 9301.

Heiskanen, W.A. and H. Moritz. 1967. Physical geodesy. W.H. Freeman and Co., San Francisco.

Ilk, K.H. 1983. Ein Beitrag zur Dynamik ausgedehnter Körper - Gravitationswechselwirkung. *Deutsche Geod. Komm.*, **C 288**.

Merriam, J.B. 1993. A comparison of recent tidal catalogues. *Bull. Inf. Marées Terrestres* **115**: 8515 - 8535.

Newcomb, S. 1897. A new determination of precessional constant with the resulting precessional motions. *Astron. Papers Am. Ephemeris VIII*, Part 1, Washington.

Roosbeek, F. 1996. RATGP95: A harmonic development of the tide generating potential using an analytical method. *Geophys. J. Int.* **126**: 197 - 204.

Simon, J.L., P. Bretagnon, J. Chapront, M. Chapront-Touzé, G. Francou and J. Laskar. 1994. Numerical expression for precession formulae and mean elements for the Moon and the planets. *Astron. Astrophys.* **282**: 663 - 683.

Standish, E.M. and J.G. Williams. 1981. Planetary and lunar ephemerides DE200 / LE200 (magnetic tape).

Standish, E.M., X.X. Newhall, J.G. Williams and W.F. Folkner. 1995. JPL planetary and lunra ephemerides DE403/LE403. Jet Propulsion Laboratory, Inter Office Memorandum 314.10-127, Pasadena.

Tamura, Y. 1987. A harmonic development of the tide-generating potential. *Bull. Inf. Marées Terrestres* **99**: 6813 - 6855.

Tamura, Y. 1993. Additional terms to the tidal harmonic tables. *Proc. 12th Int. Symp. Earth Tides* pp. 345 - 350. H.-T. Hsu (ed.). Science Press, Beijing.

Wenzel, H.-G. 1974. The correction of the tidal force development to the ellipsoidal normal. *Bull. Inf. Marées Terrestres* **68**: 3748 - 3790.

Wenzel, H.-G. 1976. Zur Genauigkeit von gravimetrischen Erdgezeitenbeobachtungen. *Wiss. Arbeiten d. Lehrstühle Geodäsie, Photogrammetrie u. Kartographie Techn. Univ. Hannover* **67**: 1 - 177.

Wenzel, H.-G. 1985. Hochauflösende Kugelfunktionsmodelle für das Gravitationspotential der Erde. *Wiss. Arbeiten Fachr. Vermessungswesen Univ. Hannover* **137**: 1 - 154. Hannover.

Wenzel, H.-G. 1996a. Accuracy assesment for tidal potential catalogues. *Bull. Inf. Marées Terrestres* **124**: 9394 - 9416.

Wilhelm, H. 1982. Earth's flattening effects on the tidal forcing field. *J. Geophysics*, **1983**: 131 - 135.

Wilhelm, H. and W. Zürn. 1984. Tides of the Earth. In: *Landolt-Börnstein, Neue Serie V/2a* **2.5.2**: 259 - 279. H. Soffel, K. Fuchs (eds.), Springer, Berlin.

Xi, Q. 1987. A new complete development of the tide-generating potential for the epoch J2000. *Bull. Inf. Marées Terrestres* **99**: 6766 - 6812.

Xi, Q. 1989. The precision of the development of the tidal generating and some explanatory notes. *Bull. Inf. Marées Terrestres* **105**: 7396 - 7404.

Xi, Q. 1993. On the comparison of the new developments of the tidal generating potential. *Bull. Inf. Marées Terrestres* **115**: 8439 - 8445.

Tidal Response of the Solid Earth

Rongjiang Wang

GeoForschungsZentrum Potsdam (GFZ),
Telegrafenberg A17,
D-14473 Potsdam, Germany

Abstract. In this note, the boundary-value problem for Earth tides is investigated. The tidal motion of the solid Earth is treated as an infinitesimal perturbation superimposed on the hydrostatic equilibrium of a rotating and self-gravitating Earth. Both Lagrangian (material-fixed) and Eulerian (space-fixed) incrementals are defined for describing the tidal perturbations. The linearized differential equations of motion and boundary conditions are derived and given in three different forms, which differ from each other in whether the pure Lagrangian incrementals, or the pure Eulerian incrementals, or a mixed combination of the two are chosen for describing variations in the potential and stress field. Analytical solutions for simple Earth models are discussed. In case of a rotating, elliptical, incompressible and homogeneous Earth, we have found inconsistency in Love's equations of motion and several calculation errors in his analytical expressions. Semi-analytical methods which are mostly used nowadays to determine the Earth tide parameters are presented and the results of different authors are discussed.

1 Introduction

The lunar-solar attraction is responsible not only for the orbital motion, but also for the nearly periodic tidal motion of the Earth. This is because the Earth can be regarded as a large sphere of radius about 6400 km, the lunar (solar) attraction, that changes in inverse proportion to square of the distance to the Moon (Sun), is not balanced throughout in the Earth by the inertia of its orbital motion. The difference between the attraction and the inertial force results in the tidal force that always tends to deform the Earth to a prolate ellipsoid aligned with the Earth-Moon (Sun) axis. The most obvious phenomenon which represents the Earth's response to the lunar-solar tidal force is the ocean tide, but also there are deformations of the solid Earth. The latter are called body tides or simply Earth tides.

Earth tides were indirectly observed as early as more than one century ago. Scientists first found that the equilibrium ocean tide is lower by about one third of what would be expected by assuming the solid Earth to be rigid. A reasonable explanation is the non-rigid behavior of the solid Earth. Today studies on Earth tides have become an important subject of geophysics (see e.g., Melchior, 1978; Harrison, 1985). The Earth tides, on one hand, are a source of noise to be corrected for many other geophysical measurements and, on the other hand, may contain useful information about the internal structure of the Earth. A decisive advantage of the Earth tide research over many other subjects in geophysics is that we know exactly the source function. This advantage allows us to be able to detect very small Earth tide signals from the observed data by spectrum analysis (see Wenzel "Tidal Analysis", this volume). By comparison with the theoretical model, Earth tide observations may supply important constraints to understanding of the Earth's interior.

The theory applied to a non-rotating, spherically stratified, elastic and self-gravitating Earth has been very successfully developed. Theoretical studies can be traced back to the beginning of this century. Since the fifties, tidal deformations have been computed for realistic Earth models (e.g. Longman, 1962, 1963 and Farrell, 1972). The theoretical tidal parameters (Love numbers) agree very well with the observation. The current difference is on the level of a few per cent.

Also the effect of rotation and ellipticity was already studied by Love (1911). Based on an incompressible and homogeneous model, the influence on the Love numbers was analytically estimated to be at the level of the ellipticity. Extended studies since the fifties (Jeffreys and Vicente 1957a, b; Molodensky, 1961; Shen and Mansinha, 1976; Sasao et al., 1980) showed that the elliptically stratified fluid core responds resonantly at diurnal frequencies. These results were greatly extended by Wahr (1979, 1981a, b, c, 1982) based on previous work by Smith (1974, 1976, 1977). In Smith-Wahr's theory, the effects of rotation and ellipticity are considered throughout the Earth including the mantle and, especially, they are able to compute simultaneously the body tides and the forced nutations and changes in rotation rate. While the predicted free core nutation has been confirmed observationally (see Zürn "Nearly-Diurnal Resonance", this volume), it is difficult to verify the latitude dependence of the Love numbers caused by the effect of rotation and ellipticity. For example, the latitude-dependence of the gravimeter factor calculated by Wahr is about 3% (peak to peak). This value has been halved later in the Wahr-Dehant model by redefinition of the normalization factor (Dehant, 1987; Dehant and Ducarme, 1987). Wang (1991, 1994) recomputed the effect of rotation and ellipticity with an independent method. His results showed a negligible latitude dependence of the gravimeter factor. The discrepancy to the Wahr-Dehant model seems to have been explained. There are indications of computation errors in older publications (Dehant, 1995). These errors could have been detected earlier if the numerical procedures were checked against Love's analytical solution for a rotating, elliptical and homogeneous model. Unfortunately, also Love's results are incorrect. Wang (1994) found not only some calculation errors in Love's formulation, but also inconsistency in his equations of motion. Only the corrected Love's solution helped to bring to light the numerical errors in the Wahr-Dehant model.

Other structural effects on Earth tides may come from mantle anelasticity and lateral heterogeneities. Viscoelasticity causes the frequency dependence of both amplitude and phase shift of tidal parameters. In the main tidal band, it produces an amplitude increase of about 1% and a phase delay of a few tenths of degree for Love numbers, but the effect on the gravimeter factor is much less significant (Zschau, 1979; Wang, 1986; Zschau and Wang, 1986; Dehant and Zschau, 1989; Wahr and Bergen, 1989; Wang, 1991). Compared with the effect of rotation and ellipticity, the predicted effects of anelasticity are strongly dependent on the model of the quality factor Q in the mantle.

Molodensky and Kramer (1980) first modeled the effect of large-scale lateral heterogeneities in the mantle on Earth tides by a perturbation method based on the variational principle. This method was extended by Wang (1991) to include viscoelasticity and any other small deviations from the spherical symmetry including rotation and ellipticity. Independently, Li et al. (e.g. Li and Hsu, 1989) developed the modeling theory of Earth tides for a laterally heterogeneous, viscoelastic Earth model based on the small parameter method. The application of their theory to a rotating, elliptical and elastic Earth model has given results which are in good agreement with those of Wang (Li, pers. comm. 1995). Based on the modern 3-dimensional reference model from seismic tomography (Dziewonski, 1984; Woodhouse and Dziewonski, 1984), Wang (1991) concluded that Earth tides cannot be significantly affected by the large-scale lateral heterogeneities in the mantle. However, it has been often reported that tidal measurements, in particular the tilt and strain measurements, can be strongly influenced by local inhomogeneities, for example, the known strain-tilt coupling effect in the vicinity of a cavity or a fault zone (see Westerhaus "Tidal Tilt modification" and Zürn "Earth Tide Observations", this volume). The local inhomogeneities are usually modeled using the finite element method (Beaumont and Berger, 1976; Gerstenecker et al., 1986; Meertens, 1987).

The purpose of this lecture is to give an introduction to the fundamental structure of the theory of Earth tides. In the first part we will derive different forms of the governing differential equations of motion and boundary conditions. In the second part, we will discuss some semi-analytical methods to estimate the global tidal effects.

2   Differential Equations of Motion

2.1   The Unperturbed Initial Equilibrium

The theory of Earth tides is based on classical continuum mechanics. We will consider a rotating, self-gravitating Earth and suppose that initially it is in a hydrostatic equilibrium state described by the following equations:

$$\left.\begin{array}{r} \rho_o(\mathbf{x})\,\nabla V_o(\mathbf{x}) - \nabla P_o(\mathbf{x}) = \rho_o(\mathbf{x})\,\boldsymbol{\Omega} \times (\boldsymbol{\Omega} \times \mathbf{x}), \\ \nabla^2 V_o(\mathbf{x}) = -4\pi G \rho_o(\mathbf{x}), \end{array}\right\} \tag{1}$$

where ρ_o is the density, V_o the gravitational potential, P_o the hydrostatic pressure, $\boldsymbol{\Omega}$ the constant angular velocity of Earth's rotation, $\mathbf{x}$ the position vector with its coordinate origin in the Earth's centre of mass, and G the gravitational constant. The subscript o indicates the unperturbed initial state.

The first equation of (1) is Newton's law of acceleration ($\mathbf{f} = m\mathbf{a}$). The term $\rho_o \nabla V_o$ expresses the attraction due to self-gravitation and $-\nabla P_o$ the pressure gradient force on the mass per unit volume in the position $\mathbf{x}$. The vector product $\boldsymbol{\Omega} \times (\boldsymbol{\Omega} \times \mathbf{x})$ on the right-hand side of the first equation is the centrifugal acceleration in a reference fram that rotates with constant angular velocity $\boldsymbol{\Omega}$ around the Earth's polar axis.

Since we are interested in the tidal deformations in the Earth's interior and on the Earth's surface, it is convenient for our purpose to define a rotating coordinate system. Its x_3 or z axis coincides with the Earth's north polar axis, and its angular velocity of rotation remains constant, $\boldsymbol{\Omega}$. From this rotating reference system, we would not observe any motion in the initial state of equilibrium. The term $\rho_o \, \boldsymbol{\Omega} \times (\boldsymbol{\Omega} \times \mathbf{x})$ will be called centrifugal force which is an apparent inertial force in the rotating frame of reference. So we can also say, the first equation of (1) expresses equilibrium among the gravitational attraction, the hydrostatic pressure gradient and the centrifugal force.

The second equation of (1) is the well-known Poisson's equation governing the gravitational potential of density ρ_o. Thus, this equation is the mathematical expression of Newton's law of gravitation.

2.2   The Tidal Perturbations

Supposing the initial equilibrium to be perturbed by a tidal force, which can be described by the gradient of the tidal potential W varying with time, ∇W, a particle will be forced to move away from its equilibrium position $\mathbf{x}$. We assume this particle occupies a new position $\mathbf{r}$ at time t. The distance that it has moved is denoted by displacement $\mathbf{u}$,

$$\mathbf{r} = \mathbf{x} + \mathbf{u}(\mathbf{x}, t). \tag{2}$$

Because the displacement of the particle varies with time, its total acceleration in position $\mathbf{r}$ and at time t consists of three parts:

$$\frac{d^2 \mathbf{r}}{dt^2} = \frac{\partial^2}{\partial t^2} \mathbf{u}(\mathbf{x}, t) + 2\boldsymbol{\Omega} \times \frac{\partial}{\partial t} \mathbf{u}(\mathbf{x}, t) + \boldsymbol{\Omega} \times (\boldsymbol{\Omega} \times \mathbf{r})$$

$$= \ddot{\mathbf{u}} + 2\boldsymbol{\Omega} \times \dot{\mathbf{u}} + \boldsymbol{\Omega} \times (\boldsymbol{\Omega} \times \mathbf{r}). \tag{3}$$

The last two terms are the Coriolis and centrifugal acceleration.

Also the density will be changed. The new density distribution ρ differs from the unperturbed density in the initial equilibrium (called the initial or background density) by a perturbation part varying with time, ρ',

$$\rho = \rho_o + \rho'. \tag{4}$$

As stated above, the Earth is self-gravitating, that means, each redistribution of the mass due to interior deformation will change the gravitational field which influences the deformation in turn. So the total gravitational potential has background and variation parts. In the literature, the tide-generating potential is usually added to the gravitational potential, though it originates from the Earth's exterior. Denoting the tide-generating part and the deformation-induced part by ψ^p and ψ^s, respectively, we can express the perturbed gravitational potential by

$$V = V_o + \psi^s + \psi^p. \tag{5}$$

ψ^p and ψ^s are usually called the primary and secondary part of tidal potential, respectively. For simplicity, we sum ψ^p and ψ^s up to ψ and denote

$$\psi = \psi^s + \psi^p. \tag{6}$$

Instead of the pressure as a single scalar parameter we need in general a tensor of second order for describing the stress field in the perturbed state. We denote the stress tensor by $\mathbf{T}$, consisting of the pre-stress part $\mathbf{T}_o = -P\mathbf{I}$ and the tidal stress part $\mathbf{T}'$,

$$\mathbf{T} = -P\mathbf{I} + \mathbf{T}', \tag{7}$$

where $\mathbf{I}$ is the unit tensor of second order.

For attentive readers it may be remarkable that no arguments $(\mathbf{x}, t)$ or $(\mathbf{r}, t)$ have been written to variables in (4)-(7). In fact, there are two possibilities to describe a varying state of material: the material-fixed or space-fixed description. In the material-fixed or Lagrangian description, we are interested in the motion of a fixed material particle and in the state of its surroundings. That means, we observe what is happening by following individual particles. In the space-fixed description or Eulerian description, we observe what is happening by looking at individual positions fixed in space.

In (2), for example, we already used the Lagrangian description for expression of the displacement. To avoid any confusion, we will distinguish the Eulerian and Lagrangian descriptions by subscript E and L, respectively. The equation

$$\mathbf{u}_L(\mathbf{x}, t) = \mathbf{r}(\mathbf{x}, t) - \mathbf{x} \tag{8}$$

then defines displacement of a material particle that occupied position $\mathbf{x}$ in the initial state. In the Lagrangian description, the coordinate argument, $\mathbf{x}$, is only a label for particle. It only answers for "which particle" and does not give explicit information about its current position. Therefore, we can say "particle $\mathbf{x}$" for such particle whose initial position was $\mathbf{x}$.

The displacement in the Eulerian description is defined by

$$\mathbf{u}_E(\mathbf{r}, t) = \mathbf{r} - \mathbf{x}(\mathbf{r}, t). \tag{9}$$

Equation (9) states for displacement of certain particle whose current position at time t is $\mathbf{r}$. The Eulerian description here answers only for "where" this displacement is given, but does not say for "which particle". In order to understand the difference between the two descriptions, we consider the following example: Let's observe an individual particle $\mathbf{x} = \mathbf{a}$ and an individual position $\mathbf{r} = \mathbf{b}$. In case that $\mathbf{b} = \mathbf{r}(\mathbf{a}, t_o)$ or $\mathbf{a} = \mathbf{x}(\mathbf{b}, t_o)$, that means, at time $t = t_o$ particle $\mathbf{a}$ just occupies position $\mathbf{b}$, we find that

$$\mathbf{u}_L(\mathbf{a}, t_o) = \mathbf{u}_E(\mathbf{b}, t_o),$$

since both give displacement of the same particle at time $t = t_o$. In case that $\mathbf{b} = \mathbf{a}$, however, we know in general

$$u_L(a,t) \neq u_B(b,t)$$

because it concerns displacements of two different particles.

In general, the relationship between the Lagrangian and Eulerian displacement is expressed by

$$u_B(r,t) = u_L(x(r,t),t) = u_L(r - u_B(r,t),t).$$

If the displacement is small, we can write approximately

$$u_B(r,t) = u_L(r,t) - u_B(r,t) \cdot \nabla u_L(r,t). \tag{10}$$

This means, only an error of second order results if we ignore the difference between the initial and current position in the coordinate arguments of the Eulerian or Lagrangian displacement.

In practice, the tidal motions are treated as an infinitesimal perturbation. All quantities which are of higher order in the displacement can be ignored. In the first order approximation we then have

$$\left.\begin{aligned} u_B(r,t) &= u_L(r,t), \\ u_L(x,t) &= u_B(x,t), \end{aligned}\right\} \tag{11}$$

i.e., there is no need to differentiate the Eulerian displacement from Lagrangian displacement. In the following, we will omit subscript E and L and consider an infinitesimal tidal displacement and denote it simply by u.

Now we discuss the linearization of the density, potential and stress field. The density in the Lagrangian description, simply called Lagrangian density, is defined by

$$\rho_L(x,t) = \rho_o(x) + \rho_L'(x,t). \tag{12}$$

It is the density of the surrounding material moving with particle x and consists of the time-independent initial part, ρ_o, and a time-dependent perturbation part, ρ_L'. The latter is called the incremental Lagrangian density. The incremental Lagrangian density is caused by the volume change in the neighbourhood of particle x and given, to first order, by

$$\rho_L'(x,t) = -\rho_o(x) \nabla \cdot u. \tag{13}$$

The Eulerian density is the density of the material in the neighbourhood of a fixed space position, r. It is defined by

$$\rho_B(r,t) = \rho_o(r) + \rho_B'(r,t). \tag{14}$$

where ρ_o is the time-independent background density and ρ_B' the incremental Eulerian density describing the temporal density variation in position r. Because

$$\begin{aligned} \rho_B(r,t) &= \rho_L(r - u, t) \\ &= \rho_o(r - u) + \rho_L'(r - u, t) \\ &= \rho_o(r) - u(r,t) \cdot \nabla \rho_o(r) - \rho_o(r) \nabla \cdot u(r,t) \\ &= \rho_o(r) - \nabla \cdot (\rho_o u), \end{aligned} \tag{15}$$

we find that the incremental Eulerian density can be expressed by

$$\rho'_B(\mathbf{r}, t) = -\nabla \cdot (\rho_o \mathbf{u}). \tag{16}$$

Here we see a difference of $\mathbf{u} \cdot \nabla \rho_o$ between the two incremental densities ρ'_L and ρ'_B. This difference is understandable when we keep in mind that in the definition of the Lagrangian and Eulerian incrementals in (12) and (14), respectively, two different reference positions for the initial density have been considered. Since the initial density, ρ_o, is a quantity of zeroth order, its value increases just by $\mathbf{u} \cdot \nabla \rho_o$ if the two reference positions differ by $\mathbf{u}$.

Analogically, we define the Lagrangian potential by

$$V_L(\mathbf{x}, t) = V_o(\mathbf{x}) + \psi_L(\mathbf{x}, t), \tag{17}$$

and the Eulerian potential by

$$V_B(\mathbf{r}, t) = V_o(\mathbf{r}) + \psi_B(\mathbf{r}, t). \tag{18}$$

Here ψ_L is the incremental Lagrangian potential and ψ_B the incremental Eulerian potential. Both incremental potentials are related by

$$\begin{aligned}
\psi_B(\mathbf{r}, t) &= V_B(\mathbf{r}, t) - V_o(\mathbf{r}) \\
&\quad - V_L(\mathbf{r} - \mathbf{u}, t) - V_o(\mathbf{r}) \\
&= \psi_L(\mathbf{r} - \mathbf{u}, t) + V_o(\mathbf{r} - \mathbf{u}) - V_o(\mathbf{r}).
\end{aligned}$$

Since in the first order $V_o(\mathbf{r} - \mathbf{u}) = V_o(\mathbf{r}) - \mathbf{u} \cdot \nabla V_o(\mathbf{r})$ and $\psi_L(\mathbf{r} - \mathbf{u}, t) = \psi_L(\mathbf{r}, t)$, we obtain

$$\psi_B(\mathbf{r}, t) = \psi_L(\mathbf{r}, t) - \mathbf{u} \cdot \nabla V_o(\mathbf{r}). \tag{19}$$

The Lagrangian and Eulerian stress tensors are defined by

$$\mathbf{T}_L(\mathbf{x}, t) = -P_o(\mathbf{x})\mathbf{I} + \mathbf{T}'_L(\mathbf{x}, t), \tag{20}$$

and

$$\mathbf{T}_B(\mathbf{r}, t) = -P_o(\mathbf{r})\mathbf{I} + \mathbf{T}'_B(\mathbf{r}, t), \tag{21}$$

respectively. Also we can write the relation between Lagrangian and Eulerian incrementals for stress tensor by

$$\mathbf{T}'_B(\mathbf{r}, t) = \mathbf{T}'_L(\mathbf{r}, t) + [\mathbf{u} \cdot \nabla P_o(\mathbf{r})]\mathbf{I}. \tag{22}$$

In the literature, one comes across the Piola-Kirchhoff pseudo stress tensor, whose components are given by the force per unit directional area in the initial undeformed configuration. In order to understand the physical meaning of the different stress tensors introduced above, let us consider the net force on an arbitrary infinitesimal surface element whose initial area is ds_o. Assume that the initial position of this surface element is marked by $\mathbf{x}$ and its unit outward normal by $\mathbf{n}_o$. At time t, the mass point will have moved from $\mathbf{x}$ to a new location $\mathbf{r} = \mathbf{x} + \mathbf{u}$, the surface element will have changed from ds_o to ds and its unit outward normal from $\mathbf{n}_o$ to $\mathbf{n}$. The net force df on the deformed surface element ds is given by Cauchy's fundamental theorem

$$df = \mathbf{n}(\mathbf{r}, t) \cdot \mathbf{T_B}(\mathbf{r}, t)\, ds. \tag{23}$$

To first order in $\mathbf{u}$, it may be shown that $\mathbf{n}(\mathbf{r}, t)ds$ of the deformed surface element is related to $\mathbf{n}_o(\mathbf{x})ds_o$ of the undeformed surface element by equation

$$\mathbf{n}(\mathbf{r}, t)ds = \mathbf{n}_o(\mathbf{x}) \cdot [(1 + \nabla \cdot \mathbf{u})\mathbf{I} - (\nabla \mathbf{u})^\dagger]ds_o \tag{24}$$

(e.g. Dahlen, 1972), where $(\nabla \mathbf{u})^\dagger$ denotes the tensor transpose of $\nabla \mathbf{u}$. By means of this relation we may rewrite (23) as

$$
\begin{aligned}
df &= \mathbf{n}(\mathbf{r}, t) \cdot \mathbf{T_B}(\mathbf{r}, t)ds \\
&= \mathbf{n}(\mathbf{r}, t) \cdot \mathbf{T_L}(\mathbf{x}, t)ds \\
&= \mathbf{n}_o(\mathbf{x}) \cdot [\mathbf{T_L}(\mathbf{x}, t) + (\nabla \cdot \mathbf{u})\mathbf{T}_o(\mathbf{x}) - (\nabla \mathbf{u})^\dagger \cdot \mathbf{T}_o(\mathbf{x})]ds_o \\
&= \mathbf{n}_o(\mathbf{x}) \cdot \mathbf{T_P}(\mathbf{x}, t)ds_o,
\end{aligned}
\tag{25}
$$

where

$$\mathbf{T_P}(\mathbf{x}, t) = \mathbf{T_L}(\mathbf{x}, t) + (\nabla \cdot \mathbf{u})\mathbf{T}_o(\mathbf{x}) - (\nabla \mathbf{u})^\dagger \cdot \mathbf{T}_o(\mathbf{x}) \tag{26}$$

is the Piola-Kirchhoff pseudo stress tensor. Equation (25) expresses that the net force on a deformed surface may be evaluated by integrating either the Eulerian or Lagrangian stress over the deformed surface or the Piola-Kirchhoff pseudo stress over the corresponding undeformed surface.

2.3   Different Forms of Equations of Motion

With the above definitions we begin to derive the differential equations governing the tidal motion. First we consider the gravitational force acting on an infinitesimal material element $\mathbf{x}$. The mass of the element is

$$
\begin{aligned}
dm &= \rho_o(\mathbf{x})\, dv_o \\
&= \rho_B(\mathbf{r}, t)\, dv \\
&= [\rho_o(\mathbf{r}) - \nabla \cdot (\rho_o \mathbf{u})]\, dv,
\end{aligned}
\tag{27}
$$

with $\mathbf{r} = \mathbf{x} + \mathbf{u}$,

where dv_o and dv are the initial and current volume of the element, respectively. Here the coordinate arguments of the quantities of first order have been omitted, because in the infinitesimal approximation there is no change in these quantities, no matter whether the initial position, $\mathbf{x}$, or the current position, $\mathbf{r}$, of the element is considered.

The gravitational acceleration is given by the space gradient of the Eulerian potential:

$$\nabla V_B(\mathbf{r}, t) = \nabla V_o(\mathbf{r}) + \nabla \psi_B, \tag{28}$$

and the force acting on the mass of the element

$$\nabla V_{\mathtt{B}}(\mathbf{r}, t)\, dm \;=\; [\,\rho_o(\mathbf{r})\nabla V_o(\mathbf{r}) + \rho_o(\mathbf{r})\nabla\psi_{\mathtt{B}} - \nabla\cdot(\rho_o\,\mathbf{u})\nabla V_o(\mathbf{r})\,]\, dv. \tag{29}$$

Additionally, there is interaction of the element through the interface to its surrounding material. This kind of force is also called the elastic feedback and can be given by space divergence of the Eulerian stress tensor,

$$\nabla\cdot\mathbf{T}_{\mathtt{B}}(\mathbf{r}, t) dv \;=\; [\,-\nabla P_o(\mathbf{r}) + \nabla\cdot\mathbf{T}'_{\mathtt{B}}\,]\, dv. \tag{30}$$

The acceleration of the element is calculated by

$$\ddot{\mathbf{r}} + 2\boldsymbol{\Omega}\times\dot{\mathbf{r}} + \boldsymbol{\Omega}\times(\boldsymbol{\Omega}\times\mathbf{r}) \;=\; \ddot{\mathbf{u}} + 2\boldsymbol{\Omega}\times\dot{\mathbf{u}} + \boldsymbol{\Omega}\times(\boldsymbol{\Omega}\times\mathbf{r}), \tag{31}$$

where $\dot{\mathbf{r}}$ $(\dot{\mathbf{u}})$ is the substantial or Lagrangian time derivative of $\mathbf{r}$ $(\mathbf{u})$.

Therefore, Newton's law $\mathbf{f} = m\mathbf{a}$ may be written in the following form

$$\nabla V_{\mathtt{B}}(\mathbf{r}, t)\, dm + \nabla\cdot\mathbf{T}_{\mathtt{B}}(\mathbf{r}, t)\, dv \;=\; [\,\ddot{\mathbf{u}} + 2\boldsymbol{\Omega}\times\dot{\mathbf{u}} + \boldsymbol{\Omega}\times(\boldsymbol{\Omega}\times\mathbf{r})\,]\, dm. \tag{32}$$

Substituting (27), (28) and (30) in (32) and considering that dv is arbitrary, we obtain

$$\rho_o(\mathbf{r})\nabla V_o(\mathbf{r}) - \nabla P_o(\mathbf{r}) + \rho_o(\mathbf{r})\nabla\psi_{\mathtt{B}} - \nabla\cdot(\rho_o\,\mathbf{u})\,\nabla V_o(\mathbf{r}) + \nabla\cdot\mathbf{T}'_{\mathtt{B}}$$
$$= \rho_o(\mathbf{r})\,[\,\ddot{\mathbf{u}} + 2\boldsymbol{\Omega}\times\dot{\mathbf{u}} + \boldsymbol{\Omega}\times(\boldsymbol{\Omega}\times\mathbf{r})\,] - \nabla\cdot(\rho_o\,\mathbf{u})\,\boldsymbol{\Omega}\times(\boldsymbol{\Omega}\times\mathbf{r}). \tag{33}$$

Assume that the tidal perturbation is slowly accumulated from a starting time, say $t = 0$. As time $t \to 0$, all time-dependent terms in (33) are going to vanish and (33) turns into

$$\rho_o(\mathbf{r})\,\nabla V_o(\mathbf{r}) - \nabla P_o(\mathbf{r}) \;=\; \rho_o(\mathbf{r})\,\boldsymbol{\Omega}\times(\boldsymbol{\Omega}\times\mathbf{r}). \tag{34}$$

This seems to be similar to the first equation of (1) for the initial equilibrium. The only difference is that argument $\mathbf{x}$ in (1) is now replaced by $\mathbf{r}$. Note that because both $\mathbf{x}$ in (1) and $\mathbf{r}$ in (33) stand for the space position which the Laplace operators in the equation directly operate upon, we are ensured that (33) is the same as the first equation of (1). Subtracting (33) from (32), we obtain the linearized equations for the Eulerian incrementals:

$$\rho_o\nabla\psi_{\mathtt{B}} - \nabla\cdot(\rho_o\,\mathbf{u})\,\mathbf{g}_o(\mathbf{r}) + \nabla\cdot\mathbf{T}'_{\mathtt{B}} \;=\; \rho_o\,[\,\ddot{\mathbf{u}} + 2\boldsymbol{\Omega}\times\dot{\mathbf{u}}\,]. \tag{35}$$

where

$$\mathbf{g}_o(\mathbf{r}) \;=\; \nabla V_o(\mathbf{r}) - \boldsymbol{\Omega}\times(\boldsymbol{\Omega}\times\mathbf{r}) \tag{36}$$

is the Earth's gravity.

Poisson's equation in the perturbed state is

$$\nabla^2 V_{\mathtt{B}}(\mathbf{r}, t) \;=\; -4\pi G\rho_{\mathtt{B}}(\mathbf{r}, t), \tag{37}$$

or in the linearized form:

$$\nabla^2 V_o(\mathbf{r}) + \nabla^2\psi_{\mathtt{B}} \;=\; -4\pi G\,[\,\rho_o(\mathbf{r}) - \nabla\cdot(\rho_o\,\mathbf{u})\,]. \tag{38}$$

Subtracting Poisson's equation for the initial equilibrium from (38), we obtain the linearized Poisson's equation for the incremental Eulerian potential in the following form

$$\nabla^2 \psi_{\mathbf{B}} = 4\pi G \nabla \cdot (\rho_o \mathbf{u}). \tag{39}$$

The complete equation system for the Eulerian incrementals, or equations of motion in the **Eulerian form**, is

$$\left.\begin{array}{c} \rho_o \nabla \psi_{\mathbf{B}} - \nabla \cdot (\rho_o \mathbf{u}) \, \mathbf{g}_o(\mathbf{r}) + \nabla \cdot \mathbf{T}'_{\mathbf{B}} = \rho_o \left[\ddot{\mathbf{u}} + 2\boldsymbol{\Omega} \times \dot{\mathbf{u}}\right], \\[2mm] \nabla^2 \psi_{\mathbf{B}} = 4\pi G \nabla \cdot (\rho_o \mathbf{u}). \end{array}\right\} \tag{40}$$

Now we derive the equations of motion for the Lagrangian incrementals, called equations of motion in the **Lagrangian form**. Note that operations, such as the gradient of a Lagrangian scalar or the divergence of a Lagrangian vector, are more complicated here than in case of Eulerian variables. This is because in the Lagrangian description the coordinate arguments stand for the initial position of a particle rather than the current position where the variable value is valid. The current position is related to the initial position by the displacement field. As an example, we consider the Lagrangian potential and calculate its space gradient in an individual position b:

$$\begin{aligned} \left[grad\, V_{\mathbf{L}}\right]_{\mathbf{b}} &= \left[\nabla_{\mathbf{r}} V_{\mathbf{L}} \left(\mathbf{x}(\mathbf{r}, t), t\right)\right]_{\mathbf{r}=\mathbf{b}} \\[2mm] &= \left[\nabla_{\mathbf{x}} V_{\mathbf{L}} \left(\mathbf{x}, t\right)\right]_{\mathbf{x}=\mathbf{x}(\mathbf{b}, t)} \cdot \left[\nabla_{\mathbf{r}} \mathbf{x}(\mathbf{r}, t)\right]^{\dagger}_{\mathbf{r}=\mathbf{b}} \\[2mm] &= \left[\nabla_{\mathbf{x}} V_{\mathbf{L}} \left(\mathbf{x}, t\right)\right]_{\mathbf{x}=\mathbf{x}(\mathbf{b}, t)} \cdot \left[\mathbf{I} - \nabla \mathbf{u}\right]^{\dagger} \\[2mm] &= \left[\nabla_{\mathbf{x}} V_{\mathbf{L}} \left(\mathbf{x}, t\right)\right]_{\mathbf{x}=\mathbf{x}(\mathbf{b}, t)} - \nabla \mathbf{u} \cdot \nabla V_{\mathbf{L}} \\[2mm] &= \left[\nabla_{\mathbf{x}} V_{\mathbf{L}} \left(\mathbf{x}, t\right)\right]_{\mathbf{x}=\mathbf{x}(\mathbf{b}, t)} - \nabla \mathbf{u} \cdot \nabla V_o. \end{aligned} \tag{41}$$

Here $\nabla_{\mathbf{r}}$ and $\nabla_{\mathbf{x}}$ are Laplace operators with reference to space coordinates $\mathbf{r}$ and $\mathbf{x}$, respectively,

$$\left.\begin{array}{l} \nabla_{\mathbf{r}} = \displaystyle\sum_{i=1}^{3} \mathbf{e}_i \frac{\partial}{\partial r_i}, \\[4mm] \nabla_{\mathbf{x}} = \displaystyle\sum_{i=1}^{3} \mathbf{e}_i \frac{\partial}{\partial x_i}, \end{array}\right\} \tag{42}$$

† denotes the tensor transpose, and $\mathbf{e}_i$ ($i = 1, 2, 3$) are the three unit basis vectors of the Cartesian coordinate system.

We may call $\nabla_{\mathbf{r}} V_{\mathbf{L}} \left[\mathbf{x}(\mathbf{r}, t), t\right]$ the "true gradient" and $\nabla_{\mathbf{x}} V_{\mathbf{L}} \left(\mathbf{x}, t\right)$ the "apparent gradient" of $V_{\mathbf{L}}$. Equation (41) shows that the difference between the true and apparent gradients is proportional to the tidal strain, $\nabla \mathbf{u}$. Thus we arrive at a general rule: For a zeroth order Lagrangian variable, the difference between its true and apparent gradient is of first order; for a first order Lagrangian incremental, the difference is of second order and can therefore be ignored. With this knowledge we can simply transform the equations of motion from the Eulerian form into the Lagrangian form. What we need to do is to substitute all Eulerian incrementals in (40) by the Lagrangian ones. Using (19) and (22), we obtain immediately the linearized equations of motion in the Lagrangian form:

$$\left.\begin{array}{r} \rho_o\nabla\psi_L - \rho_o\nabla(\mathbf{u}\cdot\nabla V_o) - \nabla\cdot(\rho_o\,\mathbf{u})\,\mathbf{g}_o(\mathbf{r}) + \nabla(\mathbf{u}\cdot\nabla P_o) + \nabla\cdot\mathbf{T}'_L \\ = \rho_o\,[\ddot{\mathbf{u}} + 2\mathit{\Omega}\times\dot{\mathbf{u}}], \\ \nabla^2\psi_L = 4\pi G\,\nabla\cdot(\rho_o\,\mathbf{u}) + \nabla^2(\mathbf{u}\cdot\nabla V_o). \end{array}\right\} \qquad (43)$$

In practice, however, the Lagrangian potential is not convenient to deal with. The pure Lagrangian form of the equations of motion has therefore been never used. Most modern geophysicists prefer the so-called mixed form of the equations of motion in which the incremental Eulerian potential and the incremental Lagrangian stress tensor are used. The reason for introducing the mixed form is that this form will have simple boundary conditions as demonstrated later.

The **mixed form** of equations of motion is given by

$$\left.\begin{array}{r} \rho_o\nabla\psi_B - \nabla\cdot(\rho_o\,\mathbf{u})\,\mathbf{g}_o(\mathbf{r}) + \nabla(\mathbf{u}\cdot\nabla P_o) + \nabla\cdot\mathbf{T}'_L \\ = \rho_o\,[\ddot{\mathbf{u}} + 2\mathit{\Omega}\times\dot{\mathbf{u}}], \\ \nabla^2\psi_B = 4\pi G\,\nabla\cdot(\rho_o\,\mathbf{u}). \end{array}\right\} \qquad (44)$$

Dahlen (1972) derived the equations of motion under a more general assumption on the initial static equilibrium. In case of the hydrostatic pre-stress, the equations of motion given by Dahlen are identical with (44).

In order to determine the tidal deformations, however, the equation system (40), (43) or (44) is not enough. Here we have 4 independent equations, but the total unknowns are 10, viz., 3 displacement components, 1 potential and 6 independent components of the stress tensor. We need therefore an additional relationship between the incremental stress and strain, for example, Hooke's law for an isotropic material:

$$\mathbf{T}'_L = \lambda\,(\nabla\cdot\mathbf{u})\mathbf{I} + \mu\,[\nabla\mathbf{u} + (\nabla\mathbf{u})^\dagger], \qquad (45)$$

where λ and μ are Lamé constants. Substituting (45) in (40), (43) or (44), we finally set up a system of 4 differential equations of second order with 4 unknowns, $\mathbf{u}$ and ψ. In the next section we will discuss the boundary conditions.

2.4 Boundary Conditions

We know that both the initial and boundary conditions are necessary for a physically meaningful solution of the differential equations. The initial conditions give the tidal displacement and velocity field of the Earth at an arbitary starting time point. Since the tidal motion is a periodic movement, the initial conditions here are unimportant. In practice, the equations are solved in the frequency domain. The Fourier transformation of the motion equations may be simply realized by assuming that all variables have the same time-dependent factor, $e^{i\omega t}$, where ω is the circular frequency. The time derivative, $\partial/\partial t$, may be replaced by multiplication factor $i\omega$, and the relative and Coriolis acceleration are expressed by

$$\ddot{\mathbf{u}} + 2\boldsymbol{\Omega} \times \dot{\mathbf{u}} = -\omega^2 \mathbf{u} + 2i\omega\,\boldsymbol{\Omega} \times \mathbf{u} \tag{46}$$

The boundary conditions describe the external influence on the Earth's surface and the coupling through any interior interface, e.g. core-mantle boundary. For the 4 differential equations of second order, there are 8 independent conditions on each boundary needed.

Consider an interface which is smooth and whose surface normal is continuous and differentiable. In the initial equilibrium, the boundary equation may have the form

$$\mathbf{x} = \mathbf{a}, \tag{47}$$

and both inner and outer side of the boundary will be denoted by $\mathbf{x} = \mathbf{a}^+$ and $\mathbf{x} = \mathbf{a}^-$, respectively. In the perturbed state, the boundary equation becomes

$$\mathbf{r} = \mathbf{a} + \mathbf{u}(\mathbf{a}). \tag{48}$$

If the two sides of the boundary are welded together, all displacement components should be continuous:

$$\mathbf{u}(\mathbf{a}^+) - \mathbf{u}(\mathbf{a}^-) = \left[\mathbf{u}(\mathbf{a})\right]_-^+ = \mathbf{0}. \tag{49}$$

Otherwise, if a relative (frictionless) slip between the two sides is allowed, then there is only continuity of the component which is normal to the boundary:

$$u_n(\mathbf{a}^+) - u_n(\mathbf{a}^-) = \left[\mathbf{n} \cdot \mathbf{u}(\mathbf{a})\right]_-^+ = 0, \tag{50}$$

where $\mathbf{n}$ is the normal of the boundary (outwards positive). Here, it should be noted that the continuities, (49) and (50), hold strictly only for the Lagrangian displacement. In the Eulerian description, the deformed boundary can not be expressed explicitly. In this case, both boundary values of the displacement in (49) and (50), $\mathbf{u}(\mathbf{a}^+)$ and $\mathbf{u}(\mathbf{a}^-)$, should be understood as values linearly extrapolated from the outer and inner region, respectively. This fact shows that the Eulerian description in solid mechanics, where we are interested in the displacement and deformation, is only applicable for the infinitesimal deformations. For finite deformation it is rather inconvenient. However, in fluid dynamics where we are interested in the current velocity, the situation is exactly opposite. Because the displacement here can reach any arbitrary large value and becomes unimportant, the Eulerian description, instead of the Lagrangian description, should be employed. In the present case since we are considering infinitesimal deformations and both Eulerian and Lagrangian descriptions are valid.

If the boundary is a density discontinuity, its deformation causes a mass redistribution. Under the assumption of infinitesimal deformation, the density perturbation can be determined by the method of the so-called surface condensation. To first order, the density change induced by the boundary deformation can be described by a surface density σ defined by

$$\sigma(\mathbf{a}) = -\left[\rho_0(\mathbf{a})\right]_-^+ u_n(\mathbf{a}). \tag{51}$$

From potential theory we know the continuity of the Eulerian potential:

$$[V_{\text{B}}(\mathbf{a})]_-^+ = [V_o(\mathbf{a}) + \psi_{\text{B}}(\mathbf{a})]_-^+ = 0 \tag{52}$$

and the discontinuity of its normal gradient:

$$\left[\frac{\partial}{\partial n}V_{\text{B}}(\mathbf{a})\right]_-^+ = \left[\frac{\partial}{\partial n}V_o(\mathbf{a}) + \frac{\partial}{\partial n}\psi_{\text{B}}(\mathbf{a})\right]_-^+$$

$$= -4\pi G\sigma(\mathbf{a}) = 4\pi G\left[\rho_o(\mathbf{a})\right]_-^+ u_n(\mathbf{a}). \tag{53}$$

Because in the initial state

$$\left.\begin{aligned}[V_o(\mathbf{a})]_-^+ &= 0,\\ \left[\frac{\partial}{\partial n}V_o(\mathbf{a})\right]_-^+ &= 0,\end{aligned}\right\} \tag{54}$$

(52) and (53) are identical with conditions

$$\left.\begin{aligned}[\psi_{\text{B}}(\mathbf{a})]_-^+ &= 0,\\ \left[\frac{\partial}{\partial n}\psi_{\text{B}}(\mathbf{a}) - 4\pi G\rho_o(\mathbf{a})u_n(\mathbf{a})\right]_-^+ &= 0.\end{aligned}\right\} \tag{55}$$

Because the Eulerian potential is related to the Lagrangian potential by

$$\psi_{\text{B}} = \psi_{\text{L}} - \mathbf{u} \cdot \nabla V_o,$$

we can write the corresponding boundary conditions concerning the Lagrangian potential

$$\left.\begin{aligned}[\psi_{\text{L}}(\mathbf{a}) - \mathbf{u}(\mathbf{a}) \cdot \nabla V_o(\mathbf{a})]_-^+ &= 0,\\ \left[\frac{\partial}{\partial n}[\psi_{\text{L}}(\mathbf{a}) - \mathbf{u}(\mathbf{a}) \cdot \nabla V_o(\mathbf{a})] - 4\pi G\rho_o(\mathbf{a})u_n(\mathbf{a})\right]_-^+ &= 0.\end{aligned}\right\} \tag{56}$$

Only when the boundary is welded, or when its normal is parallel to the gravitation ∇V_o (not the gravity!), there is continuity of the incremental Lagrangian potential:

$$[\psi_{\text{L}}(\mathbf{a})]_-^+ = 0, \tag{57}$$

because the term $\mathbf{u}(\mathbf{a}) \cdot \nabla V_o(\mathbf{a})$ is continuous in both cases.

The remaining conditions are about the coupling through the surface force. In the Eulerian description, the surface density in (51) makes an additional loading on the boundary and causes discontinuity of the stress field. Consider the equilibrium of an arbitrary disc with one surface in the upper and the other in the lower material region. The disc is thin enough so that force on the side surface can be ignored. Assume that the surface area ΔS, then

$$\int_{\Delta S} \left\{[\mathbf{n} \cdot \mathbf{T}_{\text{B}}(\mathbf{a})]_-^+ + \sigma(\mathbf{a})g_o(\mathbf{a})\right\} ds$$

$$= \int_{\Delta S} \left\{[-P(\mathbf{a})\mathbf{n} + \mathbf{n} \cdot \mathbf{T}'_{\text{B}}(\mathbf{a}) - \rho_o(\mathbf{a})\, u_n(\mathbf{a})g_o(\mathbf{a})]_-^+\right\} ds$$

$$= \int_{\Delta S} \left\{[\mathbf{n} \cdot \mathbf{T}'_{\text{B}}(\mathbf{a}) - \rho_o(\mathbf{a})\, u_n(\mathbf{a})g_o(\mathbf{a})]_-^+\right\} ds = 0.$$

Because ΔS is arbitrary, it holds that

$$\left[\, \mathbf{n} \cdot \mathbf{T}'_{\mathbf{B}}(\mathbf{a}) - \rho_o(\mathbf{a})\, u_n(\mathbf{a}) \mathbf{g}_o(\mathbf{a}) \,\right]^+_- = \mathbf{0}. \tag{58}$$

Here we see a discontinuity of the normal components of the incremental Eulerian stress tensor. Substituting the incremental Lagrangian stress tensor, (22), in (58), we obtain

$$\left[\, \mathbf{n} \cdot \mathbf{T}'_{\mathbf{L}}(\mathbf{a}) \,\right]^+_- = \mathbf{0}. \tag{59}$$

The normal components of the incremental Lagrangian stress tensor, in contrast to the Eulerian components, are continuous through a boundary.

Furthermore, we know that shear stress vanishes on a frictionless boundary. For such a contact boundary, we write the following additional conditions:

$$\left.\begin{aligned}
(\mathbf{I} - \mathbf{nn}) \cdot &\left[\, \mathbf{n} \cdot \mathbf{T}'_{\mathbf{B}}(\mathbf{a}^+) - \rho_o(\mathbf{a}^+)\, u_n(\mathbf{a}^+) \mathbf{g}_o(\mathbf{a}^+) \,\right] \\
= (\mathbf{I} - \mathbf{nn}) \cdot &\left[\, \mathbf{n} \cdot \mathbf{T}'_{\mathbf{B}}(\mathbf{a}^-) - \rho_o(\mathbf{a}^-)\, u_n(\mathbf{a}^-) \mathbf{g}_o(\mathbf{a}^-) \,\right] = \mathbf{0}; \\
(\mathbf{I} - \mathbf{nn}) \cdot &\left[\, \mathbf{n} \cdot \mathbf{T}'_{\mathbf{L}}(\mathbf{a}^+) \,\right] \\
= (\mathbf{I} - \mathbf{nn}) \cdot &\left[\, \mathbf{n} \cdot \mathbf{T}'_{\mathbf{L}}(\mathbf{a}^-) \,\right] = \mathbf{0}.
\end{aligned}\right\} \tag{60}$$

There are totally 8 independent boundary conditions. They are summarized in the following sets of equations:

$$\left.\begin{array}{ll}
\text{N} & \text{Eulerian incrementals} \\[4pt]
3 & \left[\, \mathbf{u} \,\right]^+_- = 0, \\[4pt]
1 & \left[\, \psi_{\mathbf{B}} \,\right]^+_- = 0, \\[4pt]
1 & \left[\, \dfrac{\partial}{\partial n} \psi_{\mathbf{B}} - 4\pi G \rho_o u_n \,\right]^+_- = 0, \\[4pt]
3 & \left[\, \mathbf{n} \cdot \mathbf{T}'_{\mathbf{B}} - \rho_o\, u_n \mathbf{g}_o \,\right]^+_- = 0,
\end{array}\right\} \tag{61}$$

and

$$\left.\begin{array}{ll}
\text{N} & \text{Lagrangian incrementals} \\[4pt]
3 & \left[\, \mathbf{u} \,\right]^+_- = 0, \\[4pt]
1 & \left[\, \psi_{\mathbf{L}} \,\right]^+_- = 0, \\[4pt]
1 & \left[\, \dfrac{\partial}{\partial n}(\, \psi_{\mathbf{L}} - \mathbf{u} \cdot \nabla V_o \,) - 4\pi G \rho_o u_n \,\right]^+_- = 0, \\[4pt]
3 & \left[\, \mathbf{n} \cdot \mathbf{T}'_{\mathbf{L}} \,\right]^+_- = 0,
\end{array}\right\} \tag{62}$$

or on a frictionless contact boundary:

N   Eulerian incrementals

$$1 \qquad [u_n]_-^+ = 0,$$

$$1 \qquad [\psi_{\text{B}}]_-^+ = 0,$$

$$1 \qquad \left[\frac{\partial}{\partial n}\psi_{\text{B}} - 4\pi G\rho_o u_n\right]_-^+ = 0,$$

$$3 \qquad [\mathbf{n}\cdot\mathbf{T}'_{\text{B}} - \rho_o u_n g_o]_-^+ = 0,$$

$$2 \qquad (\mathbf{I}-\mathbf{nn})\cdot(\mathbf{n}\cdot\mathbf{T}'_{\text{B}} - \rho_o u_n g_o) = 0, \tag{63}$$

and

N   Lagrangian incrementals

$$1 \qquad [u_n]_-^+ = 0,$$

$$1 \qquad [\psi_{\text{L}} - \mathbf{u}\cdot\nabla V_o]_-^+ = 0,$$

$$1 \qquad \left[\frac{\partial}{\partial n}(\psi_{\text{L}} - \mathbf{u}\cdot\nabla V_o) - 4\pi G\rho_o u_n\right]_-^+ = 0,$$

$$3 \qquad [\mathbf{n}\cdot\mathbf{T}'_{\text{L}}]_-^+ = 0,$$

$$2 \qquad (\mathbf{I}-\mathbf{nn})\cdot(\mathbf{n}\cdot\mathbf{T}'_{\text{L}}) = 0, \tag{64}$$

where N gives the number of the scalar equations.

Here we see that the Eulerian potential and the Lagrangian stress tensor are the most convenient combination for a simple form of the continuity conditions. This is why the mixed form of the equations of motion is mostly used in practice.

On the free surface of the Earth, there is no restriction of displacement and potential. However, the normal gradient of the deformation-induced potential will become discontinuous because of the surface deformation.

N   Eulerian incrementals

$$1 \qquad \left(\frac{\partial}{\partial n}\psi_{\text{B}}^s - 4\pi G\rho_o u_n\right)(\mathbf{a}^-)$$
$$\qquad\qquad = \frac{\partial}{\partial n}\psi_{\text{B}}^s(\mathbf{a}^+)$$

$$3 \qquad (\mathbf{n}\cdot\mathbf{T}'_{\text{B}} - \rho_o u_n g_o)(\mathbf{a}^-) = 0, \tag{65}$$

and

N   Lagrangian incrementals

$$1 \qquad \left[\frac{\partial}{\partial n}(\psi_{\text{L}}^s - \mathbf{u}\cdot\nabla V_o) - 4\pi G\rho_o u_n\right](\mathbf{a}^-)$$
$$\qquad\qquad = \frac{\partial}{\partial n}\psi_{\text{L}}^s(\mathbf{a}^+), \tag{66}$$

$$3 \qquad (\mathbf{n}\cdot\mathbf{T}'_{\text{L}})(\mathbf{a}^-) = 0.$$

Note that in free space the deformation-induced tidal potential, $\psi_{\text{B}}^s(|\mathbf{r}| > |\mathbf{a}|)$ in (65) or $\psi_{\text{L}}^s(|\mathbf{r}| > |\mathbf{a}|)$ in (66), is a harmonic function and can be uniquely determined from its surface value $\psi_{\text{B}}^s(\mathbf{a})$ or $\psi_{\text{L}}^s(\mathbf{a})$. The first equation in (65) or

(66) stands for a homogeneous relationship of the inside value of the deformation-induced tidal force to the surface deformation.

There are totally 4 scalar conditions in (65) or (66), but our boundary-value problem requires 8 independent surface conditions. The other 4 conditions should be found at "the other end" of the surface, i.e., at the Earth's centre of mass. Because the Earth's centre is the equilibrium point of the tidal motion, the desired solution should additionally satisfy the following conditions

$$
\left.
\begin{array}{ll}
N & \text{Eulerian incrementals} \\[2mm]
3 & \mathbf{u}(|\mathbf{r}| = 0) = \mathbf{0}, \\[2mm]
1 & \psi_{\mathbf{E}}(|\mathbf{r}| = 0) = \text{const.}
\end{array}
\right\} \tag{67}
$$

and

$$
\left.
\begin{array}{ll}
N & \text{Lagrangian incrementals} \\[2mm]
3 & \mathbf{u}(|\mathbf{x}| = 0) = \mathbf{0}, \\[2mm]
1 & \psi_{\mathbf{L}}(|\mathbf{x}| = 0) = \text{const.}
\end{array}
\right\} \tag{68}
$$

The first equation in (67) or (68) excludes a global translation of the Earth. The constant in the second equation can be chosen freely. For simplicity, it will be set to zero.

2.5   Inconsistency in Love's Equations of Motion

In his book "*Some Problems of Geodynamics*" Love (1911) applied a first-order perturbation theory to estimate the effect of rotation and ellipticity on Earth tides. The Earth model used is a rotating, incompressible, homogeneous and elastic ellipsoid. His incremental stress tensor is given by

$$
\mathbf{T}' = -P'\mathbf{I} + \mu \left[\nabla \mathbf{u} + (\nabla \mathbf{u})^{\dagger} \right],
$$

where P' defines the infinitesimal pressure perturbation. In Love's text, we could not find out explicitly whether he used the Eulerian or Lagrangian description. Adopting our present notations, one may rewrite the boundary-value problem given by Love in the following form:

---

the equations of motion:

$$
\rho_o \nabla \psi - \nabla P' + \mu \nabla^2 \mathbf{u} = \rho_o \left[\ddot{\mathbf{u}} + 2\boldsymbol{\Omega} \times \dot{\mathbf{u}} + \boldsymbol{\Omega} \times (\boldsymbol{\Omega} \times \mathbf{u}) \right],
$$

$$
\psi^s = G \oint_S \frac{\rho_o u_n}{|\mathbf{r} - \mathbf{a}|} ds,
$$

the surface conditions:

$$
-P'\mathbf{n} + \mu \mathbf{n} \cdot \left[\nabla \mathbf{u} + (\nabla \mathbf{u})^{\dagger} \right] = \rho_o u_n g_o.
$$

Love (1911) states explicitly that his deformation-induced potential is harmonic in the Earth's interior. From (43) and (44), we know only the incremental Eulerian potential is a harmonic function in case of the homogeneity and incompressibility and the Lagrangian is not. So ψ^s here is an Eulerian incremental, (i.e., ψ^s_{B} in our notation), and his equations of motion are either of the Eulerian form or of the mixed form. The pure Lagrangian form can be excluded. From the knowledge of the last two subsections, Love's boundary-value problem should be formulated either in the Eulerian form,

---

the equations of motion:

$$\rho_o \nabla \psi_{\text{B}} - \nabla P'_{\text{B}} + \mu \nabla^2 \mathbf{u} = \rho_o \left[\ddot{\mathbf{u}} + 2\boldsymbol{\Omega} \times \dot{\mathbf{u}} \right],$$

$$\psi^s_{\text{B}} = G \oint_S \frac{\rho_o u_n}{|\mathbf{r} - \mathbf{a}|} ds,$$

the surface conditions:

$$- P'_{\text{B}} \mathbf{n} + \mu \mathbf{n} \cdot \left[\nabla \mathbf{u} + (\nabla \mathbf{u})^\dagger \right] = \rho_o u_n \mathbf{g}_o,$$

---

or in the mixed form,

---

the equations of motion:

$$\rho_o \nabla \psi_{\text{B}} + \rho_o (\mathbf{u} \cdot \nabla) \nabla V_o + (\nabla \mathbf{u}) \cdot \mathbf{g}_o - \nabla P'_{\text{L}} + \mu \nabla^2 \mathbf{u}$$
$$= \rho_o \left[\ddot{\mathbf{u}} + 2\boldsymbol{\Omega} \times \dot{\mathbf{u}} + \boldsymbol{\Omega} \times (\boldsymbol{\Omega} \times \mathbf{u}) \right],$$

$$\psi^s_{\text{B}} = G \oint_S \frac{\rho_o u_n}{|\mathbf{r} - \mathbf{a}|} ds,$$

the surface conditions:

$$- P'_{\text{L}} \mathbf{n} + \mu \mathbf{n} \cdot \left[\nabla \mathbf{u} + (\nabla \mathbf{u})^\dagger \right] = 0.$$

Here we see, Love's boundary condition has the Eulerian form, but his motion equations agree neither with the Eulerian form nor with the mixed form. He made probably a mistake in his formulation. Having read his work carefully, one could understand he did mean in fact the Eulerian description. So his P' must be the incremental Eulerian pressure, i.e., P'_{B}, and his equations of motion should be corrected by deleting term $\rho_o \boldsymbol{\Omega} \times (\boldsymbol{\Omega} \times \mathbf{u})$ on the right-hand side. We have found some other calculation errors which can be easily verified and corrected (Wang, 1994).

3   Methods to Determine Tidal Parameters

3.1   The General Solution Expression

We know that any partial constituent of the tide-generating potential, ψ^p, may be expressed in the form

$$\psi^p(r,\theta,\phi) = \psi_o \left(\frac{r}{a}\right)^{n_o} Y_{n_o m_o}(\theta,\phi), \tag{69}$$

where (r,θ,ϕ) are the spherical coordinates, ψ_o a coefficient determining the tide amplitude, a the mean Earth's radius and $Y_{n_o m_o}$ the spherical harmonics of degree n_o and order m_o.

The response of the Earth to this tidal forcing, i.e., the displacement u and the deformation-induced tidal potential ψ^s, can in general be expressed in terms of spherical harmonics:

$$\left.\begin{aligned}
\mathbf{u}(r,\theta,\phi) &= \sum_{n,m} \left[H_{nm}(r)\mathbf{e}_r + T_{nm}(r)\nabla_1 \right. \\
&\quad \left. + W_{nm}(r)\mathbf{e}_r \times \nabla_1 \right] Y_{nm}(\theta,\phi), \\
\psi^s(r,\theta,\phi) &= \sum_{n,m} R_{nm}(r)Y_{nm}(\theta,\phi),
\end{aligned}\right\} \tag{70}$$

where ∇_1 is differential operator

$$\nabla_1 = \mathbf{e}_\theta \frac{\partial}{\partial\theta} + \mathbf{e}_\phi \frac{1}{\sin\theta}\frac{\partial}{\partial\phi}, \tag{71}$$

and $H_{nm}, T_{nm}, W_{nm}, R_{nm}$ are unknown coefficients as functions of radius r. Symbolically, we may introduce a 4-dimensional solution vector:

$$\mathbf{y} = (\mathbf{u}, \psi^s). \tag{72}$$

The solution vector can be divided into two parts:

$$\mathbf{y} = \sum_{n,m} \sigma_n^m + \sum_{n,m} \tau_n^m, \tag{73}$$

where σ_n^m is called spheroidal mode and τ_n^m the toroidal mode of degree n and order m,

$$\left.\begin{aligned}
\sigma_n^m &= (H_{nm}Y_{nm}\mathbf{e}_r + T_{nm}\nabla_1 Y_{nm}, \ R_{nm}Y_{nm}), \\
\tau_n^m &= (W_{nm}\mathbf{e}_r \times \nabla_1 Y_{nm}, \ 0).
\end{aligned}\right\} \tag{74}$$

Making use of the orthogonality of the spherical harmonics, we are able to transform the partial differential equations of motion to a system of ordinary differential equations. This is, in general, an infinite system in which all equations are coupled with each other. The unknowns to be solved for are all the coefficients of both spheroidal and toroidal modes, $H_{nm}, T_{nm}, W_{nm}, R_{nm}$, for $n = 0, 1, 2, ...$, and $-n \leq m \leq n$.

3.2   Response Functions of a Non-Rotating and Spherical Earth

In case of a non-rotating and spherical Earth there are three important features of the equation system because of the spherical symmetry: (i) all coefficients are independent of their harmonic order, m; (ii) modes of different harmonic degrees are decoupled and (iii) the spheroidal and toroidal modes are decoupled from each other. One can find out from these features that the Earth's response can be described by a single spheroidal mode of the same degree and order as the tide-generating potential:

$$\mathbf{y} = \sigma_{n_o}^{m_o}. \tag{75}$$

All other modes do not exist in the solution vector. The coefficients $H_{n_o m_o}$, $T_{n_o m_o}$, $R_{n_o m_o}$ are independent of the harmonic order m_o and can therefore be denoted simply by H_{n_o}, T_{n_o}, R_{n_o}. The system has totally 3 ordinary differential equations of second order or, equivalently, 6 equations of first order. An analytical solution for H_{n_o}, T_{n_o}, R_{n_o} is only possible for a simple model, such as an incompressible homogeneous sphere. For a realistic reference model, in which stratification in the Earth's interior has to be considered, only a numerical solution can be achieved.

On the Earth's surface, the tidal response can be represented by three dimensionless parameters. They are Love numbers h and k, and Shida number l:

$$\left.\begin{aligned}
\mathbf{u}(a,\theta,\phi) &= h_{n_o} \frac{\psi^p(a,\theta,\phi)}{g_o} \mathbf{e}_r + l_{n_o} \frac{1}{g_o} \nabla_1 \psi^p(a,\theta,\phi), \\
\psi^s(a,\theta,\phi) &= k_{n_o} \psi^p(a,\theta,\phi),
\end{aligned}\right\} \tag{76}$$

or in our present notation,

$$\left.\begin{aligned}
h_{n_o} &= \frac{g}{\psi_o} H_{n_o}, \\
l_{n_o} &= \frac{g}{\psi_o} T_{n_o}, \\
k_{n_o} &= \frac{1}{\psi_o} R_{n_o},
\end{aligned}\right\} \tag{77}$$

where g is the surface gravity and a the Earth's radius.

In case of an incompressible homogeneous Earth, Love and Shida numbers have a simple analytical form:

$$\left.\begin{aligned}
h_{n_o} &= \frac{n_o(2n_o+1)\rho g a}{2(n_o-1)[(2n_o^2+4n_o+3)\mu + n_o\rho g a]}, \\
l_{n_o} &= \frac{3}{n_o(2n_o+1)} h_{n_o}, \\
k_{n_o} &= \frac{3}{2n_o+1} h_{n_o},
\end{aligned}\right\} \tag{78}$$

where ρ is the constant density and μ the shear modulus.

The body tides were observed already in the last century. For example, it was discovered that ocean tides amount only to about 70% of what would be expected for a rigid Earth. Because the relative sea level change ΔH is given by

$$\Delta H = \frac{\psi}{g} - u_r, \tag{79}$$

and a dominant part of more than 95% of the tides is given by the harmonic term of degree 2, we can conclude from the ocean tide observation that

$$1 + k_2 - h_2 = \frac{19\mu}{19\mu + 2\rho g a} \sim 0.7, \tag{80}$$

or

$$\mu \sim 0.25\rho ga \sim 8.5 \times 10^{11}\,cgs, \tag{81}$$

a value which corresponds to stiffness of steel.

For a stratified Earth model like the seismic reference models 1066A (Gilbert and Dziewonski, 1975) or PREM (Dziewonski and Anderson, 1981) consisting of a solid mantle, a liquid outer and a solid inner core, Love and Shida numbers have been calculated numerically. Table 1 shows our results calculated for the Earth model PREM (modified). In the calculation we have made two small modifications of the original PREM (see the caption of Tab. 1). The reason for the first modification is because we are considering body tides. The second modification is necessary for the existence of the static Love numbers.

Table 1. Static Love numbers calculated for the modified PREM (1 sec). In the calculation the ocean and upper crust of the original PREM have been replaced by a single solid layer (15 km), whose seismic velocities are the same as the original upper crust, but the density is given by the average value so that the surface gravity remains unchanged. Additionally, the slightly non-neutral fluid core of PREM has been modified to be neutrally stratified by way of re-determination of the bulk modulus by means of the Adams-Williamson condition.

n_o	h_{n_o}	l_{n_o}	k_{n_o}	δ_{n_o}	γ_{n_o}
2	0.603264	0.083892	0.298009	1.156251	0.694745
3	0.287947	0.014741	0.092025	1.069265	0.804078
4	0.174999	0.010180	0.041425	1.035718	0.866426

Earth tides are usually observed by gravimeters or tiltmeters. A gravimeter measures the vertical component of gravity changes, δg (downwards positive):

$$\delta g = -\mathbf{n} \cdot \delta \mathbf{g} = -\mathbf{n} \cdot \left\{ \nabla \psi + (\mathbf{u} \cdot \nabla)\mathbf{g} + \omega^2 \mathbf{u} - 2i\omega \, \boldsymbol{\Omega} \times \mathbf{u} \right\}. \tag{82}$$

The first term in the bracket describes the direct tidal force acting on the mass sensor of the gravimeter. The second term is the gravity change due to movement of the instrument in the Earth's gravity field. The last two terms are the inertial and the Coriolis acceleration, respectively. The effect of both acceleration terms is small. Ignoring the effects of rotation and ellipticity, the gravity tide can also be represented by a dimensionless parameter:

$$\delta g = -\delta_{n_o} \frac{\partial}{\partial r} \psi^p(a, \theta, \phi), \tag{83}$$

where

$$\delta_{n_o} = 1 + h_{n_o} - \frac{n_o + 1}{n_o} k_{n_o} \tag{84}$$

is called gravimetric factor.

A tiltmeter, e.g. a vertical pendulum tiltmeter which is usually installed in a borehole, measures the plumb-line deviation relative to the borehole wall. A tilt signal has two components, e.g., the north-south and the east-west component:

$$\delta t_{s,e} = \frac{\mathbf{e}_{s,e} \cdot \delta \mathbf{g}}{g} + u_{s,e}\,(\mathbf{e}_{s,e} \cdot \nabla)\mathbf{n} + (\mathbf{e}_{s,e} \cdot \nabla)u_n, \tag{85}$$

where $\mathbf{e}_s$ and $\mathbf{e}_e$ are the southward and the eastward unit vector, respectively. The first term in (84) describes the plumb-line variation caused by the tidal acceleration. The second term is the change of the local surface normal because of the tangential movement of the Earth's surface. The last term is the normal change due to surface deformation. In case of a spherical Earth surface,

$$\mathbf{e}_s = \mathbf{e}_\theta, \qquad \mathbf{e}_e = \mathbf{e}_\phi,$$

both tilt components can be represented by a single dimensionless parameter, the tilt (or diminishing) factor γ,

$$\left.\begin{aligned}
\delta t_\theta &= \gamma_{n_o} \frac{1}{ga} \frac{\partial}{\partial \theta} \psi^P(a,\theta,\phi), \\
\delta t_\phi &= \gamma_{n_o} \frac{1}{ga \sin\theta} \frac{\partial}{\partial \phi} \psi^P(a,\theta,\phi),
\end{aligned}\right\} \tag{86}$$

where

$$\gamma_{n_o} = 1 + k_{n_o} - h_{n_o}. \tag{87}$$

Both Love numbers, h and k, and the gravimetric factor, δ, which have been calculated for a seismic reference Earth model agree very well with the global observations. However, the other Earth tide parameters, e.g. the tilt factors, γ, have been found to be strongly influenced by the local geological structure, e.g. topography, cavity effect or lateral inhomogeneities in the underground. The relation of the tilt factor to the Love numbers, (87), only holds for a laterally homogeneous Earth model.

3.3 Effect of Rotation and Ellipticity

The calculation of Earth tide parameters will become much more complicated if the effect of rotation and ellipticity of the Earth is considered. In this case, besides the main spheroidal mode of degree n_o and order m_o there exist many other small terms of both spheroidal and toroidal modes. They are coupled with different degrees subject to a certain rule, but decoupled for different orders because of the rotation symmetry. In fact, we have to determine all coefficients of the series

$$\mathbf{y} = \sum_{n=m_o}^{\infty} (\sigma_n^{m_o} + \tau_n^{m_o}). \tag{88}$$

This is in practice impossible without introducing any approximation. In fact, the real Earth is only slightly elliptical. The hydrostatic surface ellipticity calculated for modern Earth models is very close to the observed ellipticity of geoid, $\sim \frac{1}{298.3}$. The hydrostatic ellipticity of Earth model PREM, for example, is approximately

$$\epsilon = \frac{1}{299.9}. \tag{89}$$

To first order, all interior boundaries of discontinuity including the surface deviate from their mean sphere by an undulation of

$$\delta h(r, \theta, \phi) = -\frac{2}{3} r \epsilon(r) P_2(\cos \theta), \tag{90}$$

where $\epsilon(r)$ is the radius-dependent ellipticity and P_2 the Legendre polynomial of second degree. Suppose the density and the potential as well as both Lamé constants are elliptically stratified throughout the Earth's interior and their contours coincide with each other. In the first order approximation we can express them by a small latitude dependence superimposed on their mean spherical distribution:

$$\left.\begin{aligned}
\rho(r, \theta, \phi) &= \rho_o(r) + \frac{2}{3} r \epsilon(r) \frac{d\rho_o}{dr} P_2(\cos \theta), \\
\lambda(r, \theta, \phi) &= \lambda_o(r) + \frac{2}{3} r \epsilon(r) \frac{d\lambda_o}{dr} P_2(\cos \theta), \\
\mu(r, \theta, \phi) &= \mu_o(r) + \frac{2}{3} r \epsilon(r) \frac{d\mu_o}{dr} P_2(\cos \theta), \\
V(r, \theta, \phi) + W(r, \theta, \phi) &= V_o(r) + \frac{1}{3} \Omega^2 r^2 + \frac{2}{3} r \epsilon(r) \frac{dV_o}{dr} P_2(\cos \theta),
\end{aligned}\right\} \tag{91}$$

where $(\rho_o, \lambda_o, \mu_o, V_o)$ are taken from a spherical reference model and W is the centrifugal potential.

Substituting the general solution expression (88) and the parameters given by (91) in the equations of motion discussed in the last section, we find that there exists the following infinite series of modes:

$$\cdots \quad \sigma_{n_o-4}^{m_o}, \; \tau_{n_o-3}^{m_o}, \; \sigma_{n_o-2}^{m_o}, \; \tau_{n_o-1}^{m_o}, \; \sigma_{n_o}^{m_o}, \; \tau_{n_o+1}^{m_o}, \; \sigma_{n_o+2}^{m_o}, \; \tau_{n_o+3}^{m_o}, \; \sigma_{n_o+4}^{m_o} \quad \cdots$$

The coefficients of these modes are not of the same order. From the perturbation theory, we can classify them by their order in ellipticity:

$$\epsilon^0 \quad \sigma_{n_o}^{m_o};$$

$$\epsilon^1 \quad \sigma_{n_o-2}^{m_o}, \; \tau_{n_o-1}^{m_o}, \tau_{n_o+1}^{m_o}, \; \sigma_{n_o+2}^{m_o};$$

$$\epsilon^2 \quad \sigma_{n_o-4}^{m_o}, \; \tau_{n_o-3}^{m_o}, \tau_{n_o+3}^{m_o}, \; \sigma_{n_o+4}^{m_o};$$

$$\vdots \quad \cdots$$

Thus, we know how to truncate the infinite series of modes. In fact, it is enough to consider terms of up to the first order in the ellipticity. The truncated form of the desired solution is then

$$y = \sigma_{n_o-2}^{m_o} + \tau_{n_o-1}^{m_o} + \sigma_{n_o}^{m_o} + \tau_{n_o+1}^{m_o} + \sigma_{n_o+2}^{m_o}. \tag{92}$$

There are totally 11 independent coefficients satisfying a system with 11 ordinary differential equations of second order or, equivalently, 22 equations of first order.

To solve this equation system, many semi-analytical methods have been developed. All these methods have a common principle in the treatment of the boundary conditions. To each elliptical boundary, an equivalent spherical surface is taken as reference. All functions on a boundary ellipsoid are linearly extrapolated to the corresponding sphere according to the rule: the inner side to the inner side and the outer side to the outer side, and all conditions which hold for the elliptical boundary are transformed to the spherical reference boundary. The equation system is first integrated over the equivalent spherical configuration. The final results are then transformed back to the elliptical configuration.

In the following we give a brief introduction to the methods developed by different authors:

(i) **Smith-Wahr's method.** In this method all 11 coefficients are solved simultaneously by directly integrating the coupled system of the 22 equations. The advantage of this method is that both the deformation and the forced nutation can be determined simultaneously. It is known that there is a resonant response of the outer liquid core to tidal forcing in the diurnal band, the so-called nearly-diurnal free wobble (also called the free core nutation). The diurnal tide-generating potential is described by the spherical harmonics of degree $n_o = 2$ and order $m_o = 1$. In this case, the resonance function consists mostly of the toroidal mode T_1^1. In the coupled system, the interaction between the deformation and the nutation is automatically included. The disadvantage of the method, however, is also obvious. The formulae are very complicated and the numerical procedure requires a large effort. Furthermore, numerical instability may arise because quantities of different orders have to be determined together in the same equation system.

(ii)**Love-Li's method.** This is a method of small parameter perturbation. Suppose the solution vector can be developed into a Taylor series in ellipticity ϵ:

$$\mathbf{y} = \mathbf{y}_0 + \epsilon \mathbf{y}_1 + \epsilon^2 \mathbf{y}_2 + ..., \tag{93}$$

where all $\mathbf{y}_{0,1,2,...}$ have the same form given by (88). Substituting (93) into the equations of motion and comparing coefficients of the same power of ϵ, one obtains a hierarchically coupled equation system. $\mathbf{y}_0$ satisfies the unperturbed equation system in the spherical case and is, therefore, the zeroth order solution without effect of rotation and ellipticity. The modes for the first order correction, $\mathbf{y}_1$, depend on $\mathbf{y}_0$; $\mathbf{y}_2$ for the second order correction depend on $\mathbf{y}_0$ and $\mathbf{y}_1$; and so on. In practice, only the first order correction is needed. The advantage of this method is the partial decoupling of the modes of different order. In the beginning of this century, Love applied this method to a rotating, elliptical, incompressible and homogeneous Earth and achieved an analytical estimation of the effect of rotation and ellipticity on the Love numbers. Although some mistakes in Love's formulation should be corrected, his idea of the first order perturbation is clear and elegant. Based on this idea, Li et al. (see e.g., Li and Hsu, 1989) established the Earth tide theory for a laterally heterogeneous Earth model.

(iii)**Molodensky-Wang's method.** Molodensky and Kramer (1980) developed a method based on the variation principle to estimate influences of large-scale mantle inhomogeneities on Earth tides. This method was extended by Wang (1991) to include the effect of rotation and ellipticity. In the following we give an outline of their idea. Denote the equation system of the rotating and elliptical Earth symbolically by

$$\mathbf{L}(\mathbf{y}) = \mathbf{0} \tag{94}$$

and that without the effect of rotation and ellipticity by

$$\mathbf{L}^o(\mathbf{y}^o) = \mathbf{0}, \tag{95}$$

where $\mathbf{L}$ is the differential operator depending on the model parameters ($\mathit{\Omega}$, ω, ρ, λ, μ, δh) and $\mathbf{L}^o$ depending on (ρ_o, λ_o, μ_o).

There are 4 fundamental solutions of the equation system (95),

$$\mathbf{L}^o(\mathbf{y}^{(\alpha)}) = \mathbf{0}, \qquad \alpha = 1, 2, 3, 4. \tag{96}$$

Three of them are spheroidal and one is toroidal. These fundamental solutions describe the response of a spherically symmetric Earth to body force, surface loading and surface shearing in both tangential directions, respectively.

Any solution of (95), $\mathbf{y}^o$, can be expressed by a linear composition of the fundamental solutions. Let us assume $\mathbf{y}^o$ is the zeroth order approximation of the desired solution $\mathbf{y}$ satisfying (94) and define

$$\left.\begin{array}{l} \delta\mathbf{L} = \mathbf{L} - \mathbf{L}^o, \\ \delta\mathbf{y} = \mathbf{y} - \mathbf{y}^o. \end{array}\right\} \tag{97}$$

The first-order variation of (94) has then the following form:

$$\delta\mathbf{L}(\mathbf{y}^o) + \mathbf{L}^o(\delta\mathbf{y}) = \mathbf{0}. \tag{98}$$

We try to express $\delta\mathbf{y}$ by an integration of the fundamental modes weighted by the parameter perturbations. For this purpose, we integrate $\delta\mathbf{y}\cdot(96) - \mathbf{y}^{(\alpha)}\cdot(98)$ over the whole volume of the spherical reference configuration,

$$\int_V \left[\delta\mathbf{y} \cdot \mathbf{L}^o(\mathbf{y}^{(\alpha)}) - \mathbf{y}^{(\alpha)} \cdot \mathbf{L}^o(\delta\mathbf{y})\right] dv = \int_V \mathbf{y}^{(\alpha)} \cdot \delta\mathbf{L}(\mathbf{y}^o) \, dv. \tag{99}$$

The right-hand side of (99) is a volume integral whose integrand is dependent on only the fundamental modes and the parameter perturbations, and can be treated as a known quantity. Because the differential operator $\mathbf{L}^o$ is self-conjugate, we can apply the Green's formula to the left-hand side of (99) and convert it to a surface integral:

$$\int_V \left[\delta\mathbf{y} \cdot \mathbf{L}^o(\mathbf{y}^{(\alpha)}) - \mathbf{y}^{(\alpha)} \cdot \mathbf{L}^o(\delta\mathbf{y})\right] dv = \oint_S \delta\mathbf{y} \cdot \mathbf{T}^{(\alpha)} \, ds, \tag{100}$$

where $\mathbf{T}^{(\alpha)}$, $\alpha = 1, 2, 3, 4$ are components of the generalized 4-dimensional surface force, which determine the 4 fundamental solutions, $\mathbf{y}^{(\alpha)}$, respectively. Thus,

$$\oint_S \delta\mathbf{y} \cdot \mathbf{T}^{(\alpha)}\, ds = \int_V \mathbf{y}^{(\alpha)} \cdot \delta\mathbf{L}(\mathbf{y}^o)\, dv. \tag{101}$$

Making use of orthogonality of the spherical harmonics, one can analytically solve the surface integral and determine all coefficients of $\delta\mathbf{y}$ by the volume integral on the right-hand side of (101) independently from each other(!).

The advantage of Molodensky-Wang's method is that (a) all coefficients of the solution perturbation are totally decoupled and (b) the computational effort is minimized because only one-dimensional integrals in the radial direction need to be numerically carried out. Once all required fundamental modes have been determined for the spherical reference model, the effects due to rotation, ellipticity as well as lateral inhomogeneities can be calculated together or separately.

The normalization of the surface solution in the elliptical case is much more complicated than in the spherical case. We need to introduce more dimensionless parameters, for example, for $n_o = 2$:

$$u_r(\theta, \phi) = \frac{\psi_o}{g}\left[\Delta h_- Y_{0m_o} + (h_2 + \Delta h_0)Y_{2m_o} + \Delta h_+ Y_{4m_o}\right], \tag{102}$$

$$\psi^s(r, \theta, \phi) = \psi_o\left[(k_2 + \Delta k_0)\left(\frac{a}{r}\right)^3 Y_{2m_o} + \Delta k_+ \left(\frac{a}{r}\right)^5 Y_{4m_o}\right], \tag{103}$$

$$\delta g(\theta, \phi) = -\frac{2\psi_o}{a}\left[\Delta\delta_- Y_{0m_o} + (\delta_2 + \Delta\delta_0)Y_{2m_o} + \Delta\delta_+ Y_{4m_o}\right]. \tag{104}$$

Here, h_2 and k_2 are the (static) Love numbers and δ_2 the (static) gravimetric factor without the effect of rotation and ellipticity. As in the spherical case, the $\Delta\delta$-parameters in (104) are not independent, but related to the Love numbers in (102) and (103). The relations can be given in an analytical form (Wang, 1994):

$$\Delta\delta_- = \Delta h_- - \frac{\sqrt{5}}{15}\left[-4\epsilon + \alpha h_2 + 6\epsilon k_2 + 3\beta l_2\right], \tag{105}$$

$$\Delta\delta_0 = \Delta h_0 - \frac{3}{2}\Delta k_0 + \frac{1}{6}(2 + 3\frac{\omega^2}{\Omega^2})m^* h_2 - m_o m^* \frac{\omega}{\Omega} l_2$$
$$+ \frac{(m_o^2 - 2)}{42}\left[-2\epsilon + 2\alpha h_2 + 18\epsilon k_2 + 3\beta l_2\right], \tag{106}$$

$$\Delta\delta_+ = \Delta h_+ - \frac{5}{2}\Delta k_+$$
$$- \frac{\sqrt{5(m_o^2 - 16)(m_o^2 - 9)}}{210}\left[6\epsilon + \alpha h_2 + 16\epsilon k_2 - 2\beta l_2\right], \tag{107}$$

where ϵ is the surface ellipticity, Ω the Earth's rotation rate, ω frequency of the tidal forcing, J_2 the coefficient of the Earth's zonal gravitational potential of second degree, $m^* = \frac{\Omega^2 a}{g}$, $\alpha = m^* - 6\epsilon + 18J_2$, and $\beta = m^* + 2\epsilon - 12J_2$. Note some differences to the corresponding formulae given in Wang (1994). This is because the Δh-parameters here are used for describing the radial displacement, rather than the vertical displacement as defined in Wang (1994).

Analytical expressions for the Δ-parameters in (102) - (104) are only possible for some simple Earth models. Here we write the corrected Love's formulae for the semi-diurnal tides ($n_o = 2$, $m_o = 2$) of an incompressible and homogeneous Earth model:

$$\Delta h_0 = \frac{2\left(4170 + 9225\hat{\mu} - 3166\hat{\omega} + 1659\hat{\omega}^2\right)}{1575(2 + 19\hat{\mu})}\epsilon h_2, \tag{108}$$

$$\Delta k_0 = \frac{3}{5}\Delta h_0 + \frac{62}{175}\epsilon h_2, \tag{109}$$

$$\Delta h_+ = -\frac{2\sqrt{3}\,(36 + 221\hat{\mu})}{105(4 + 51\hat{\mu})}\epsilon h_2, \tag{110}$$

$$\Delta k_+ = \frac{1}{3}\Delta h_+ - \frac{24\sqrt{3}}{105}\epsilon h_2, \tag{111}$$

where $\hat{\mu} = \frac{\mu}{\rho a g}$ and $\hat{\omega} = \frac{\omega}{\Omega}$ are the normalized shear modulus and normalized tidal frequency, respectively. In (108) - (111) the first order relation between the ellipticity and the rotation rate, $\epsilon = \frac{5\Omega^2 a}{4g}$, has been used. Also (108)-(111) are different from those given in Wang (1994) for the same reason as stated above.

Table 2. Effects of rotation and ellipticity. The basic Earth model is the modified PREM as stated in Tab. 1. In parameters Δh_0, Δk_0 and $\Delta \delta_0$, also the non-static effect due to the acceleration term (ω^2-term) in the equations of motion has been included.

(n_o, m_o)	ω	Δh_-	Δh_0	Δh_+	Δk_0	Δk_+	$\Delta \delta_-$	$\Delta \delta_0$	$\Delta \delta_+$
$(2, 0)$	M_f	$-.00018$	$-.00013$	$-.00017$	$-.00033$	$-.00087$	$+.00077$	$+.00092$	$-.00287$
$(2, 1)$	O_1	——	$+.00040$	$-.00016$	$-.00011$	$-.00080$	——	$+.00152$	$-.00262$
$(2, 2)$	M_2	——	$+.00215$	$-.00011$	$+.00165$	$-.00057$	——	$+.00379$	$-.00185$

Table 2 shows our numerical results calculated for the reference Earth model PREM (modified). While these results differ significantly from the former Wahr-Dehant model (Wahr, 1981b; Dehant, 1987), they have been verified by Li (pers. comm., 1995). His results agree remarkably up to the last digit with ours, though our methods are quite different from each other. In the meanwhile, certain computation errors in the old Wahr-Dehant model have been found, and the International Working Group on "Theoretical Tidal Model" (chaired by Dehant) is working on the new Earth tide model. The independent computations for cross check by term-to-term are being done by Dehant (Belgium), Li (China), Mathews (USA) and Wang (Germany).

It should be noted that in Table 2 the effect of the forced nutation is not included. In the middle of the diurnal band, this effect is significant due to the resonant response of the outer core. The variation of the Earth's rotation such as polar motion, change of the length of day, etc. is a special topic of this seminar (see Plag "Chandler Wobble", this volume).

3.4   Influences of Mantle Viscoelasticity and Lateral Heterogeneities

As stated above, we can estimate influences of lateral heterogeneities in the mantle on Earth tides by both Love-Li's and Molodensky-Wang's perturbation methods. The procedure is similar to modeling ellipticity, but many more harmonic degrees and orders should be considered. Also viscoelasticity can be treated as a special "inhomogeneity" of zeroth degree. According to the correspondence principle, the viscoelasticity may be described by a complex variation in the Lamé constants.

Mantle viscoelasticity causes an amplitude increase and a phase delay of the Earth's response functions. The effect in the Love numbers is on the level of 1% in the amplitude increase and of some tenths of a degree in the phase delay, depending on the viscoelastic models used. However, most model calculations have shown a negligible effect in the gravity tides. This is because the viscoelasticity affects both Love numbers in the same direction such that the total contribution to the gravimetric factor $\delta_2 = 1 + h_2 - 1.5k_2$ is reduced almost to zero.

The first quantitative estimation of the influences of lateral mantle inhomogeneities on Earth tides was done by Molodensky and Kramer (1980). Based on a simple continent-ocean model, in which a common contrast in the seismic velocities between the continental mantle and the oceanic mantle was supposed, they concluded that the effect in the gravimeter factor is in the order of half per cent if the contrast in the seismic velocities is not larger than 5%.

Since about last ten years, more realistic 3D mantle models have been constructed by seismic tomography (Dziewonski, 1984; Woodhouse and Dziewonski, 1984). In the long wavelength range, lateral variations of the seismic velocities are in the order of $1-2\%$, except in the low velocity zone (asthenosphere) where the variation can reach up to 8%. Our model calculations based on realistic mantle models have shown that the predicted global anomalies in Earth tides due to large-scale mantle heterogeneities (up to harmonic degree and order 8) are small. The M_2 residuals, for example, are in the order of one millimeter for the surface displacement and of one tenth of a microgal for gravity. These effects are too small to be verified by observation with the present accuracy of measurements.

While the theory of the global Earth tides has been successfully developed, the modeling of the local influences is still not satisfactory. For instance, it has been reported that the strain and tilt measurements are strongly influenced by the local inhomogeneities such as fault zones, cavities (tunnels) where the instruments have been installed, etc. The cause has been known as the local strain-strain and strain-tilt coupling effect (see Westerhaus "Tidal Tilt Modification", this volume). Modeling local inhomogeneities is more difficult because of large irregularities of the model parameters. In practice, finite element methods are widely employed (Beaumont and Berger, 1976; Gerstenecker et al., 1986; Meertens, 1987). The main difficulty here arises from the lack of knowledge of the underground geology as well as the complicated deformation process which might be quite different from what has been supposed so far.

4   Discussion

We have derived the linearized differential equations and boundary conditions governing the tidal motion that is supposed to be an infinitesimal elastic-gravitational disturbance from the initial hydrostatic equilibrium of a rotating, self-gravitating Earth. On choosing different kinds of incrementals of potential and stress, we get different forms of equations of motion. The Eulerian description, beyond all doubt, is convenient to depict the tidal effect on the Earth's gravity field, since we are interested particularly in the tidal potential in the free space. On the other hand, however, either the Eulerian or Lagrangian incremental can serve well in the determination of the tidal variation of stress field in the Earth interior. While Love chose the pure Eulerian description for both potential and stress field, here we prefer the so-called mixed form in which the Eulerian incremental of potential and the Lagrangian incremental of stress are used. It is important to keep in mind that the Lagrangian and Eulerian incrementals of the potential and stress are different even though the tidal motion is supposed to be infinitesimal. This is due to the presence of their a-priori existing fields: the gravitational potential and the hydrostatic pressure in the static equilibrium. The difference between the Eulerian and Lagrangian displacement, however, vanishes in the case of infinitesimal motion, because no initial displacement has been supposed.

The three independent numerical methods introduced in Section 3.3 are all based on the first order approximation. An important advantage of the Smith-Wahr method is that the variations of the Earth's rotation (nutation and change of the rotation rate) induced by tides can be computed simultaneously with the tidal deformation. The other two perturbation methods, which were developed originally for modeling the effects of lateral heterogeneities on Earth tides, do not include influences on the Earth's rotation. To include these effects, these methods should be extended by adding an additional component allowing differential rigid motions (rotations) between the mantle, outer core and inner core to the attempted first order solution perturbation. Because of the resonance effect, this rigid motion can become very large and should be treated in general as a quantity of zeroth order. There is still a lot of work to be done in order to determine the coefficients of the rigid motions, but it may contribute to an independent and complete check of the previous results.

Acknowledgments: The author is grateful to Prof. H. Wilhelm of the Institute of Geophysics, University of Karlsruhe, Germany, and his colleague Dr. D. Wolf for their interest in the present work. Their comments especially on the correction of the equations of motion in Love's book were very helpful. I am also thankful to Prof. C. Denis of the Institute of Astrophysics, University of Liège, Belgium. The motivation for this work was partly due to his critical comments on my previous paper on the effects of rotation and ellipticity on Earth tides. Dr. G. Li and Dr. D. Han, both of the Institute of Geodesy and Geophysics, the Chinese Academy of Sciences, Wuhan, China, read the original manuscript and gave helpful suggestions for corrections.

References

Beaumont, C. and J. Berger. 1976. Earthquake prediction: modification of earth tide tilts and strains by dilatancy. *Geophys. J. R. Astr. Soc.* **39**: 111-118.

Dahlen, F.A. 1972. Elastic dislocation theory of a self gravitating elastic configuration with an initial stress field. *Geophys. J. R. Astr. Soc.* **28**: 357-383.

Dehant, V. 1987. Integration of the gravitational motion equations for an elliptical uniformly rotating Earth with an inelastic mantle. *Phys. Earth Planet. Inter.* **49**: 242-258.

Dehant, V. 1995. Theoretical tidal parameters: state of the art. *Bull. Inf. Marées Terrestres* **121**: 9027 - 9029.

Dehant, V. and B. Ducarme. 1987. Comparison between the theoretical and observed tidal gravimetric factors. *Phys. Earth Plant. Inter.* **49**: 192-212.

Dehant, V. and J. Zschau. 1989. The effect of mantle anelasticity on tidal gravity: a comparison between the spherical and the elliptical Earth model. *Geophys. J.* **97**: 549-555.

Dziewonski, A.M. and D.L. Anderson. 1981. Preliminary Reference Earth Model. *Phys. Earth Planet. Inter.* **25**: 297-356.

Dziewonski, A.M. 1984. Mapping the lower mantle: determination of lateral Heterogeneity in P velocity up to degree and order 6. *J. Geophys. Res.* **89**: 5929-5952.

Farrell, W.E. 1972. Deformation of the Earth by surface loads. *Rev. Geophys. Space Phys.* **10**: 767-797.

Gerstenecker, C., J. Zschau and M. Bonatz. 1986. Finite element modeling of the Hunsrück tilt anomalies - a model comparison. Proc. 10. Int. Symp. Earth Tides, R. Vieira (ed.), Cons. Sup. Invest. Cient., Madrid: 797-804.

Gilbert, F. and A.M. Dziewonski. 1975. An application of normal mode theory to the retrieval of structural parameters and source mechanisms from seismic spectra. *Phil. Trans. R. Soc.* **278A**: 187-269.

Harrison, J.C. (ed.). 1985. Earth Tides. Benchmark Papers in Geology Series, Van Nostrand Reinhold, New York. 419 pp.

Jeffreys, H. and R.O. Vicente. 1957a. The theory of nutation and the variation of latitude. *Mon. Not. R. astr. Soc.* **117**: 142-161.

Jeffreys, H. and R.O. Vicente. 1957b. The theory of nutation and the variation of latitude: the Roche model core. *Mon. Not. R. astr. Soc.* **117**: 162-173.

Li, G. and H.T. Hsu. 1989. The tidal modeling theory with a lateral inhomogeneous, inelastic mantle. *Proc. 11th Int. Symp. on Earth tides*. Helsinki, Finland. E. Schweizerbart'sche Verlagsbuchhandlung, Stuttgart.

Longman, I.M. 1962. A Green's function for determining the deformation of the Earth under surface mass loads: 1. theory. *J. Geophys. Res.* **67**: 845-850.

Longman, I.M. 1963. A Green's function for determining the deformation of the Earth under surface mass loads: 2. computations and numerical results. *J. Geophys. Res.* **68**: 485-496.

Love, A.E.H. 1911. Some problems of Geodynamics, Dover, New York. 180 pp.

Meertens, C. 1987. Tilt Tides and Tectonics at Yellowstone National Park. PhD Thesis. University of Colorado, Boulder, USA. 236 pp.

Melchior, P. 1978. The Tides of the Planet Earth. Pergamon, Oxford. 609 pp.

Molodensky, M.S. 1961. The theory of nutation and diurnal Earth tides. *Communs Obs. r. Belg.* **288**: 25-56.

Molodensky, S.M. and M.V. Kramer. 1980. The influence of large-scale horizontal inhomogeneities in the mantle on earth tides. *Bull. Acad. Sci. USSR, Earth Physics.* **16**: 1-11.

Sasao, T., S. Okubo and M. Sasao. 1980. A simple theory on dynamical effects of stratified fluid core upon nutational motion of the Earth. *Proc. IAU Symp. No. 78.* Nutation and the Earth's Rotation. Kiev. May 1977. Edi. E.P. Fedorov, M.L. Smith and P.L. Bender.

Shen, P.-Y. and L. Mansinha. 1976. Oscillation, nutation, and wobble of an elliptical rotating Earth with fluid outer core. *Geophys. J. R. astr. Soc.* **46**: 467-496.

Smith, M.L. 1974. The scalar equations of infinitesimal elastic-gravitational motion for a rotating, slightly elliptical Earth. *Geophys. J. R. astr. Soc.* **37**: 491-526.

Smith, M.L. 1976. Translational inner core oscillations of a rotating, slightly elliptical Earth. *J. Geophys. Res.* **81**: 3055-3065.

Smith, M.L. 1977. Wobble and nutation of the Earth. *Geophys. J. R. astr. Soc.* **50**: 103-140.

Wahr, J.M. 1979. The Tidal Motion of a Rotating, Elliptical, Elastic and Oceanless Earth. PhD Thesis. University of Colorado, Boulder, USA. 216 pp.

Wahr, J.M. 1981a. A normal mode expansion for the forced response of a rotating Earth. *Geophys. J. R. Astr. Soc.* **64**: 651-675.

Wahr, J.M. 1981b. Body tides on an elliptical, rotating, elastic and oceanless Earth. *Geophys. J. R. Astr. Soc.* **64**: 677-703.

Wahr, J.M. 1981c. The forced nutations of an elliptical, rotating, elastic and oceanless Earth. *Geophys. J. R. Astr. Soc.* **64**: 705-727.

Wahr, J.M. 1982. Computing tides, nutations and tidally-induced variations in the Earth's rotation rate for a rotating elliptical Earth. Lecture at the Third Int. Summer School in the Mountains on "Geodesy and Global Geodynamics". Admont, Austria, 1982. Edi. H. Moritz and H. Sünkel. 327-379.

Wahr, J.M. and Z. Bergen. 1989. The effects of mantle anelasticity on nutations, Earth tides, and tidal variations of rotation rate. *Geophys. J. R. astr. Soc.* **87**: 633-668

Wang, R. 1986. Das viskoelastische Verhalten der Erde auf langfristige Gezeitenterme. Diplomarbeit. Math.-Nat. Fak. Universität Kiel. Germany. 94 pp.

Wang, R. 1991. Tidal Deformations on a Rotating, Spherically Asymmetric, Viscoelastic and Laterally Heterogeneous Earth, European University Studies. Series XVII. Earth Sciences. Vol. 5. Peter Lang, Frankfurt am Main. 139 pp.

Wang, R. 1994. Effect of rotation and ellipticity on earth tides. *Geophys. J. Int.* **117**: 562-565.

Woodhouse, J.H. and A.M. Dziewonski. 1984. Mapping the upper mantle: three-dimensional modeling of earth structure by inversion of seismic waveforms. *J. Geophys. Res.* **89**: 5953-5986.

Zschau, J. 1979. Auflastgezeiten. Habilitationsschift. Math.-Nat. Fak. Universität Kiel, Germany. 243 pp.

Zschau, J. and R. Wang. 1986. Imperfect elasticity in the Earth's mantle: implications for the Earth tides and long period deformations. *Proc. 10th Int. Symp. Earth Tides.* Madrid, Spain. Edi. R. Vieira. 379-384.

Analysis of Earth Tide Observations

Hans-Georg Wenzel

Geodätisches Institut, Universität Karlsruhe, Englerstr. 7, D-76128 Karlsruhe, Germany.

Abstract. We will give in the following a short description of the analysis of earth tide observations. Earth tide analysis is usually carried out in order to estimate the frequency transfer function of the system Earth - station - sensor, where the parameters of the transfer function (i.e. the tidal parameters) are determined from a least squares adjustment of the earth tide observations. We will describe the mathematical model of the earth tide analysis program ANALYZE (Wenzel, 1996a,b) and we will give two examples.

1 Introduction

The objective of earth tide analysis is to obtain information on the reaction of the system

Earth - station - sensor

to the accurately known tidal forcing function (see Fig. 1), or on deviations of the reaction of this system from a model. In general, the reaction of the Earth is of interest only, but in special cases the reaction of the station is also of interest. The properties of the sensor (i.e. its frequency transfer function) are usually to be determined by a calibration procedure.

The general model which is applied in earth tide analysis is a multiple input single output (MISO) linear time invariant system (LTIS) (see below). The usual realization is by formulation of the MISO-LTIS using an observation equation and the estimation of the unknown parameters by a least squares adjustment procedure.

The analysis of ocean tide observations requires an extended analysis model (because of non-linear shallow water effects) compared to the earth tide analysis and is not discussed here (e.g. Munk and Cartwright, 1966).

There are currently available different earth tide analysis methods and their implementations in software packages, respectively:

- Venedikov (1966a,b): Concept out of date, violates the sampling theorem (e.g. Schüller, 1978), single input channel concept, used tidal potential catalogue out of date.
- Chojnicki (1973): Single input channel concept, used tidal potential catalogue out of date.

- HYCON (Schüller, 1976): Single input channel concept, used tidal potential catalogue out of date.
- HYCON-MC (Schüller, 1986): Multiple input channel concept, used tidal potential catalogue out of date.
- De Toro et al. (1993): Similar to Venedikov (1966a,b), out of date, but multiple input channel concept.
- BAYTAP-G (Tamura et al., 1991): Multiple input channel concept, used mainly in Japan.
- ANALYZE of the ETERNA 3.30 software package (Wenzel, 1996a): Multiple input channel concept, all recent tidal potential catalogues available.

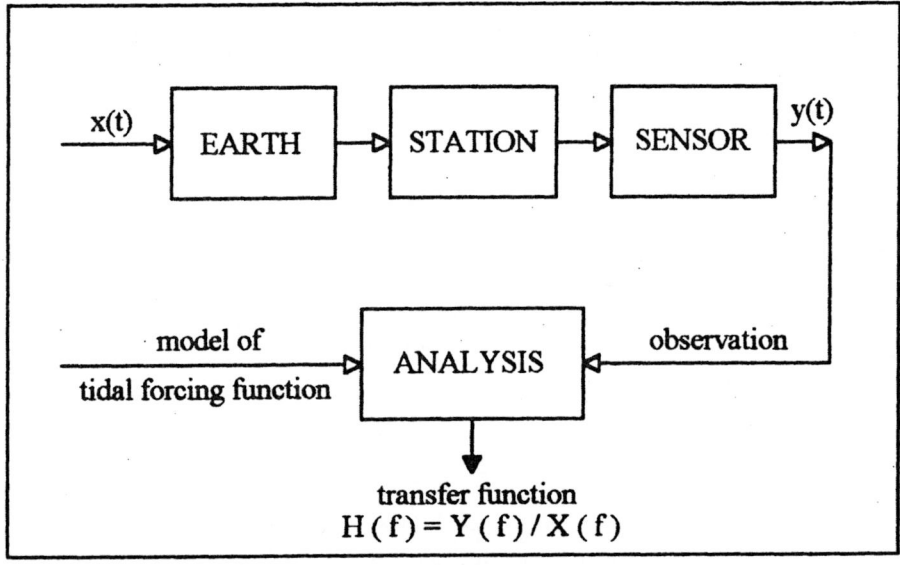

Fig. 1. Objective of the analysis of earth tide observations.

2   Least Squares Adjustment Model

The following mathematical description of the earth tide analysis by least squares adjustment is mainly referring to the concept which has been introduced by Chojnicki (1973), and is also used in the programs HYCON, HYCON-MC, and ANALYZE.

The reaction of the system Earth (solid body of the Earth plus oceans) + station + instrument to the earth tide forcing function (tide generating potential, tidal accelerations) or to other (e.g. atmospheric) forcing functions can be described for a linear time invariant single channel system (SISO = Single Input, Single Output) in the frequency domain by

$$\underline{Y}_{(f)} = \underline{H}_{(f)} \cdot \underline{X}_{(f)} \tag{1}$$

with $\underline{X}_{(f)}$ = complex spectrum of the earth tide forcing function, $\underline{H}_{(f)}$ = complex frequency transfer function, $\underline{Y}_{(f)}$ = complex spectrum of the observed earth tide signal. The frequency transfer function $H_{(f)}$ is defined by

$$\underline{H}_{(f)} = H_{(f)} \cdot e^{-j\Delta\Phi_{(f)}} \tag{2}$$

with $H_{(f)}$ = transfer function magnitude, $\Delta\Phi_{(f)}$ = transfer function phase (positive transfer function phase means that $y_{(t)}$ leads $x_{(t)}$).

Equation (1) can be formulated in time domain by

$$x_{(t)} = \sum_{i=1}^{n} A_i \cdot \cos(2\pi f_i t + \Phi_i) \tag{3}$$

$$y_{(t)} = \sum_{i=1}^{n} H_i \cdot A_i \cdot \cos(2\pi f_i t + \Phi_i + \Delta\Phi_i) \tag{4}$$

where the input function $x_{(t)}$ is the earth tide forcing function, and the output function $y_{(t)}$ is the observed earth tide signal. The amplitudes A_i, frequencies f_i and phases Φ_i of the earth tide forcing function can be computed using a tidal potential catalogue (see Wenzel "Tide-Generating Potential", this volume).

A linear time invariant system with multiple input channels (one earth tide forcing function $x_{(t)}$, several additional input channels $z_{m(t)}$) (e.g. air pressure, ground water, gravity pole tide) and one single output channel $y_{(t)}$ (MISO = Multiple Input, Single Output) can be described in frequency domain by

$$\underline{Y}_{(f)} = \underline{H}_{(f)} \cdot \underline{X}_{(f)} + \sum_{m} \underline{R}_{m(f)} \cdot \underline{Z}_{m(f)} \tag{5}$$

with $\underline{Z}_{m(f)}$ = complex spectrum of channel number m, $\underline{R}_{m(f)}$ = complex frequency transfer function for channel number m. The MISO-LTIS concept can be formulated in the time domain by the model equation

$$y_{(t)} = \sum_{i=1}^{n} H_i \cdot A_i \cdot \cos(2\pi f_i t + \Phi_i + \Delta\Phi_i) + \sum_{m} \int_{-\infty}^{\infty} r_{m(\tau)} \cdot z_{m(t-\tau)} d\tau \tag{6}$$

with $r_{m(\tau)}$ = weighting function for additional parameter no. m, τ = time lag.

For a constant real frequency transfer function $\underline{R}_{m(f)} = R_m$, the model equation (6) is simply the **multiple regression model**

$$y_{(t)} = \sum_{i=1}^{n} H_i \cdot A_i \cdot \cos(2\pi f_i t + \Phi_i + \Delta\Phi_i) + \sum_{m} R_m \cdot z_{m(t)} \tag{7}$$

The unknown parameters H_i, $\Delta\Phi_i$ und $R_{m(\tau)}$ can be determined by least squares adjustment with observation equations

$$\underline{\ell}+\underline{v} = \underline{A}\cdot\underline{x} \rightarrow \underline{v}^T\cdot\underline{P}\cdot\underline{v} = Minimum \rightarrow \underline{x} = (\underline{A}^T\cdot\underline{P}\cdot\underline{A})^{-1}\cdot\underline{A}^T\cdot\underline{P}\cdot\underline{\ell}(8)$$

with $\underline{\ell}$ = observation vector, $\underline{v}$ = vector of residuals, $\underline{x}$ = vector of unknown parameters $(H_i, \Delta\Phi_i, R_m)$, $\underline{A}$ = matrix of observation equation coefficients, $\underline{P}$ = weight matrix. The matrix $\underline{A}$ contains the derivatives of the observations ℓ_i with respect to the unknown parameters x_j

$$a_{ij} = \frac{d\ell_i}{dx_j} \tag{9}$$

computed from the applied observation equations.

The analysis of earth tide observations is usually carried out by least squares adjustment, because

- least squares adjustment can be applied if the observation series contains gaps; other methods of time series analysis require data series without gaps;
- least squares adjustment provides **residuals** (discrepancies between observations and model) in time domain. The residuals can be used for the **detection and elimination of blunders**, as well as for the investigation of systematic errors (e.g. of instrumental origin), and last not least additional unmodeled signals (like core modes etc.);
- least squares adjustment provides **standard deviations of the observations and of the adjusted parameters**. Many other methods of time series analysis do not provide error estimations.

Earth tide observations often contain long-periodic signals usually called drift e.g due to instrumental or meteorological influences (see Fig. 2), which partly are hard to model in the obervation equation by time varying functions as e.g. polynomials. Such longperiodic signals are therefore either

- eliminated by **digital highpass filtering** of the observations before entering the least squares adjustment, or
- approximated by a **drift polynomial**, where the coefficients of the drift polynomial are determined within the adjustment by adding corresponding terms on the right hand side of eq. (7).

The drift polynomial has to be chosen in such a way that it cannot absorb significant energy in those frequency bands in which tidal parameters are estimated. Drift approximation is usually applied for the analysis of observations performed with superconducting gravimeters, for which a small number of drift coefficients (2...10) is sufficient to approximate the drift. The program ANALYZE uses drift approximation by Tschebyscheff-polynomials $T_k(t)$ because of their nice numerical properties

$$d_{(t)} = \sum_k D_k \cdot T_k(t_n) \tag{10}$$

with $d_{(t)}$ = adjusted drift model, D_k = adjusted polynomial drift coefficient, t_n = normalized time (normalized to $t_n = \pm 1$). The Tschebyscheff-polynomials $T_k(t_n)$ can be computed by a recursion algorithm

$$T_0(t_n) = 1, \quad T_1(t_n) = t_n, \quad T_{k+1}(t_n) = 2t_n \cdot T_k(t_n) - T_{k-1}(t_n) \tag{11}$$

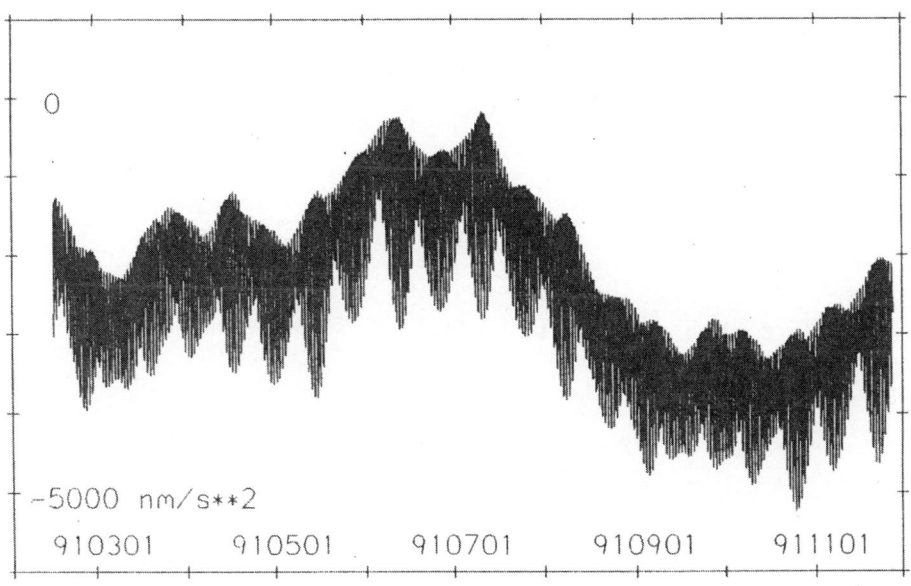

Fig. 2. Gravity tides observed using gravimeter LCR-ET19 at station BFO Schiltach.

Digital highpass filtering can be described by

$$\ell^*_{(t)} = \sum_{\tau=-\Delta t}^{\Delta t} g_{(\tau)} \cdot \ell_{(t+\tau)} \tag{12}$$

with $\ell_{(t)}$ = observed earth tide signal, $g_{(\tau)}$ = digital highpass filter, $\ell^*_{(t)}$ = highpass filtered earth tide signal.

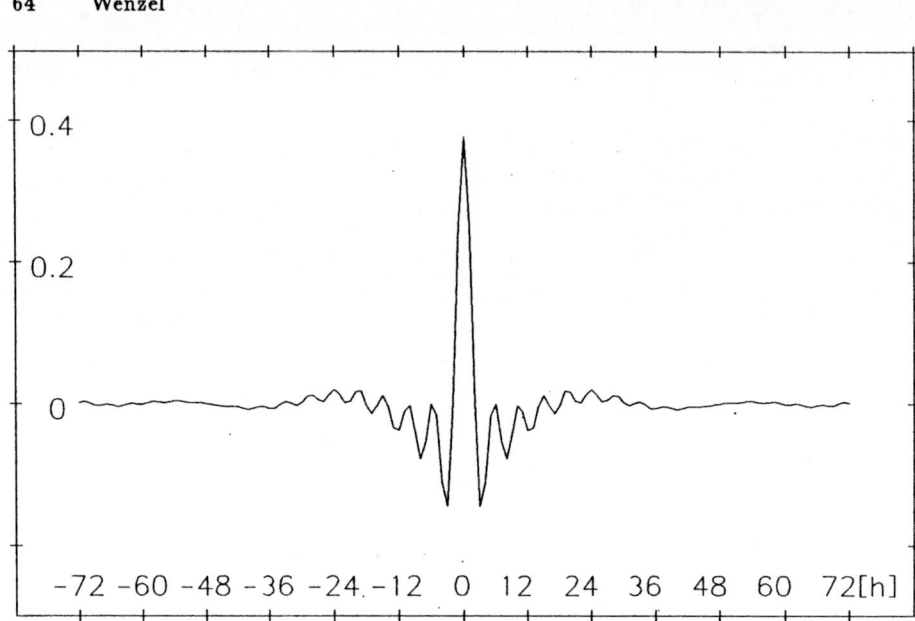

Fig. 3. Digital highpassfilter (Nr. 7) of the ETERNA 3.30 package, 145^h length.

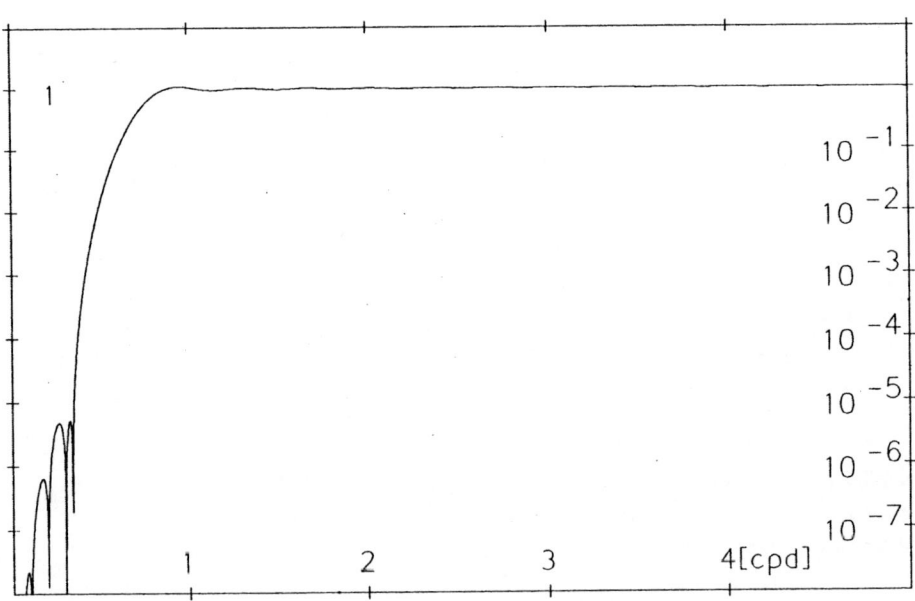

Fig. 4. Frequency transfer function of highpass filter (Nr. 7) of the ETERNA 3.30 package, 145^h length.

The **frequency transfer function** $\underline{G}_{(f)}$ of the digital highpass filter has to be considered for the observation equation of the least squares adjustment. Usually, zero phase symmetrical filters with $g_{(-\tau)} = g_{(\tau)}$ are used, for which the frequency transfer function has no imaginary component, i.e. the filter does not produce any phase shift:

$$G_{(f)} = \sum_{\tau=-\Delta t}^{\Delta t} g_{(\tau)} \cdot \cos(2\pi f \tau) \tag{13}$$

The cut-off frequency of the highpass filters used for the analysis of earth tide observations is usually slightly below 1 cpd, which means that the spectrum of the earth tide observations below 1 cpd is suppressed. Therefore, the earth tide analysis is restricted to short-periodic tidal parameters with frequencies ≥ 0.5 cpd. Fig. 3 shows the numerical filter Nr. 7 of the ETERNA 3.30 package with a length of 145 hours; in Fig. 4 is shown the frequency transfer function magnitude of the same filter. The quality of a numerical highpass filter (rejection within the stop band, slope of the filter flank at the cut-off frequency, rolloff) generally increases with increasing filter length. But unfortunately, at each gap in the observation series the gap in the highpass filtered series increases by the filter length (see Fig. 5). Therefore, short filters $(50^h \cdots 200^h)$ with limited filter quality are generally used in earth tide analysis.

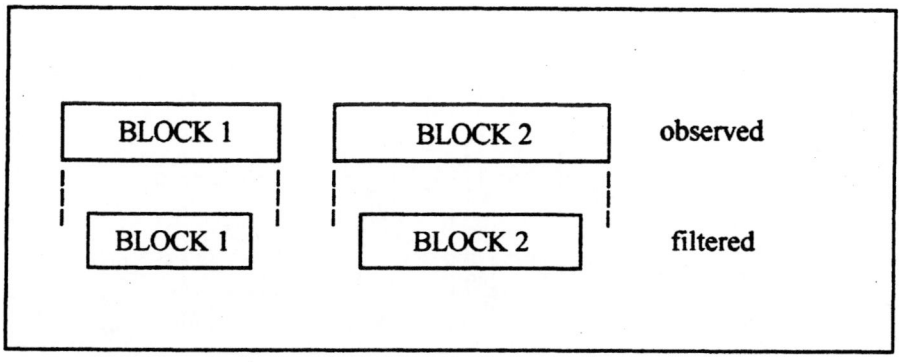

Fig. 5. Increasing data gaps by numerical highpass filtering.

Another problem which is not specially related to the least squares adjustment is produced by the **restricted recording length** T. The finite recording length T with the fundamental frequency $f_0 = 1/T$ generally restricts the separation of waves to $\Delta f \geq f_0$ (Rayleigh criterion, frequency resolution of the discrete Fourier transform). But the Rayleigh criterion should be used as a rule of thumb only; for the least squares adjustment method, where the frequencies are known a-priori, the separation depends on the recording length T and on the signal-to-noise ratio (e.g. Munk and Hasselmann, 1964). For high signal to noise ratios, waves with frequency differences of $\Delta f \leq f_0$ can be separated by least squares adjustment.

Because the smallest frequency difference in tidal potential catalogues is about $1/20942^a$ it is impossible to determine individual tidal parameters for all tidal waves contained in the tidal potential catalogue. Instead, average tidal parameters for **wave groups** can be determined, where the definition of the wavegroups (see Fig. 6) depend on the recording length T with $\Delta f > f_0$. The frequency difference Δf is counted between the maximum wave of neighbouring wavegroups. The tidal parameters H_i and $\Delta \Phi_i$ are assumed to be constant for all tidal waves within a wavegroup j; for each wavegroup, two tidal parameters H_j and $\Delta \Phi_j$ are adjusted. This wavegroup model has been invented by Venedikov (1961).

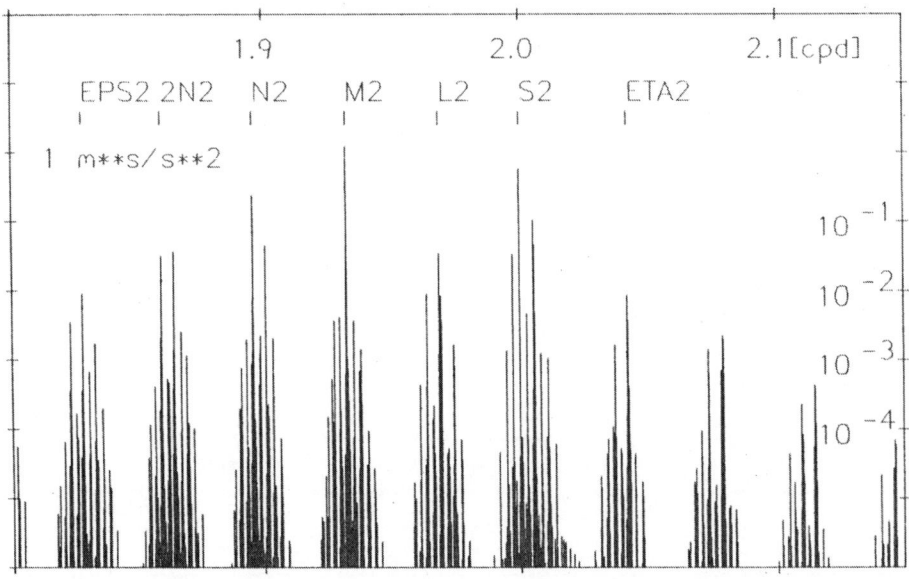

Fig. 6. Definition of wavegroups.

The observation equation for highpass filtered earth tide observations is

$$\ell_{(t)}^{*} + v_{(t)}^{*} = \sum_{j=1}^{q} \overline{H}_j \sum_{i=a_j}^{e_j} G_i \cdot A_i \cdot \cos\left(2\pi f_i t + \phi_i + \Delta\overline{\phi}_j\right)$$

$$+ \sum_m \overline{R}_m \cdot z_{m(t)} \tag{14}$$

and the observation equation for drift approximation is

$$\ell_{(t)} + v_{(t)} = \sum_{j=1}^{q} \overline{H}_j \sum_{i=a_j}^{e_j} A_i \cdot \cos\left(2\pi f_i t + \phi_i + \Delta\overline{\phi}_j\right)$$

$$+ \sum_k \overline{D}_k \cdot T_k(t_n) + \sum_m \overline{R}_m \cdot z_{m(t)} \tag{15}$$

with q = number of wave groups, $\overline{H}_j, \Delta\overline{\Phi}_j$ = adjusted amplitude factors and phase leads, $\overline{D}_k$ = adjusted drift coefficients, $\overline{R}_m$ = adjusted regression coefficients.

The above observation equations are non-linear with respect to the unknown tidal parameters $\overline{H}_j, \Delta\overline{\Phi}_j$. By introducing the auxiliary unknowns $\overline{X}_j$ und $\overline{Y}_j$

$$\overline{X}_j = \overline{H}_j \cdot \cos\Delta\overline{\Phi}_j, \overline{Y}_j = -\overline{H}_j \cdot \sin\Delta\overline{\Phi}_j \tag{16}$$

we can obtain the linear observation equation

$$v_{(t)}^{*} = \sum_{j=1}^{q} \left\{ \overline{X}_j \cdot CO_j + \overline{Y}_j \cdot SI_j \right\} - \ell_{(t)}^{*} + \sum_m R_m \cdot z_{m(t)} \tag{17}$$

or

$$v_{(t)} = \sum_{j=1}^{q} \left\{ \overline{X}_j \cdot CO_j + \overline{Y}_j \cdot SI_j \right\} - \ell_{(t)} + \sum_k \overline{D}_k \cdot T_k(t_n) + \sum_m R_m \cdot z_{m(t)} \tag{18}$$

with

$$CO_j = \sum_{i=a_j}^{e_j} G_i \cdot A_i \cdot \cos\left(2\pi f_i t + \Phi_i\right), \tag{19}$$

$$SI_j = \sum_{i=a_j}^{e_j} G_i \cdot A_i \cdot \sin\left(2\pi f_i t + \Phi_i\right) \tag{20}$$

The adjustment of the unknown parameters can be carried out using the standard method of least squares adjustment. The weight matrix of the observations $\underline{P}$ has to be chosen as a diagonal matrix in order to achieve reasonable computation times, although the errors usually show weak correlations (e.g. Wenzel, 1976).

The adjusted tidal parameters can be obtained from:

amplitude factors : $\overline{H}_j = (\overline{X}_j^2 + \overline{Y}_j^2)^{\frac{1}{2}}$ $\qquad$ (21)

phase leads : $\Delta\overline{\Phi}_j = -\arctan\dfrac{\overline{Y}_j}{\overline{X}_j}$ $\qquad$ (22)

The standard deviations of the adjusted tidal parameters estimated from error propagation in the adjustment procedure are usually too optimistic because of the neglected correlations of the observations in the adjustment. A more realistic error estimation can be carried out using the amplitude spectrum of the residuals (Wenzel 1976, 1977). The standard deviations of the tidal parameters as computed from the least squares adjustment are based on the assumption, that the residuals have a white noise spectrum (i.e. no frequency dependence of the noise). From the amplitude spectrum of the residuals, we can estimate

- the broadband noise amplitude β
- and the frequency dependent noise amplitude α_f at the specific tidal band.

For long-periodic tides, the noise amplitude α_f has to be modeled by a red noise function:

$$\alpha_f = \frac{a}{f} \qquad (23)$$

Within program ANALYZE (e.g. Wenzel 1996a), the standard deviations of the tidal parameters as obtained from the least squares adjustment procedure are scaled by a scaling factor s obtained from the relation of the frequency dependent noise amplitude to the broadband noise amplitude:

$$s = \frac{\alpha_f}{\beta}. \qquad (24)$$

3 Examples

In Tab. 1 we show the results of an earth tide analysis using program ANA-LYZE of the ETERNA 3.30 package for observations carried out with LaCoste-Romberg earth tide gravity meter ET19 at station BFO Schiltach (286^d in 1991). The analogue signal of the gravimeter and of a barometer have been digitized at 5 s interval using a 21 bit data acquisition system. The digitized data have been despiked and decimated to hourly samples using the PRE-TERNA 3.0 data preprocessing package (Wenzel, 1994). We have used the tidal potential catalogue of Hartmann and Wenzel (1995a,b) for the earth tide analysis, and we have applied a digital highpass filtering in order to eliminate the longperiodic drift of the instrument (see Fig. 2). The residuals of the adjustment are given in Fig. 7, and the amplitude spectrum of the residuals is shown in Fig. 8. The standard deviations of the adjusted tidal parameters have been estimated using the amplitude spectrum of the residuals. The signal to noise ratios given in Tab. 1 of up to 24 000 clearly show the high precision of the observations. The standard deviations of the amplitude factors and phase leads are approximately **inversely proportional** to the amplitudes of the waves. The average noise level is

$$\sim 0.050 \text{ nm/s}^2 \; \hat{=} \; 5 \cdot 10^{-12} \text{ g for diurnal waves,}$$
$$\sim 0.027 \text{ nm/s}^2 \; \hat{=} \; 3 \cdot 10^{-12} \text{ g for semidiurnal waves, and}$$
$$\sim 0.013 \text{ nm/s}^2 \; \hat{=} \; 1 \cdot 10^{-12} \text{ g for terdiurnal waves.}$$

The adjusted tidal parameters show the resonance of the liquid outer core of the Earth (see Zürn "Nearly-Diurnal Resonance", this volume) in the diurnal band, and moderate ocean tide effects (gravitation plus loading, see Jentzsch, "Ocean Loading", this volume) in the semidiurnal band of $\sim 3\%$ and 2^0 respectively.

Table 1. Example 1: LaCoste-Romberg earth tide gravimeter ET19, station BFO Schiltach, 286^d, 1991. Program ANALYZE of the ETERNA 3.30 package, digital highpass filtering, tidal potential catalogue of Hartmann and Wenzel (1995a,b.)

Welle	Amplitude $[nm \cdot s^{-2}]$	S/N	amplitude factor δ	phase lead $\Delta\Phi$
Q1	67.800	2798.	1.14710 ± 0.00041	-0.175 ^{0}± 0.020
O1	354.291	14346.	1.14767 ± 0.00008	0.065 ^{0}± 0.004
M1	27.680	696.	1.14073 ± 0.00164	0.092 ^{0}± 0.082
P1	164.973	4994.	1.14872 ± 0.00023	0.211 ^{0}± 0.012
S1	4.306	81.	1.26815 ± 0.01570	6.504 ^{0}± 0.712
K1	492.944	14199.	1.13589 ± 0.00008	0.246 ^{0}± 0.004
ψ1	4.493	125.	1.32264 ± 0.01059	1.294 ^{0}± 0.454
ϕ1	7.214	204.	1.16724 ± 0.00572	0.333 ^{0}± 0.281
J1	28.046	1216.	1.15536 ± 0.00095	0.258 ^{0}± 0.047
001	15.378	681.	1.15807 ± 0.00170	0.095 ^{0}± 0.084
2N2	11.702	968.	1.15165 ± 0.00119	2.544 ^{0}± 0.059
N2	74.419	5086.	1.16971 ± 0.00023	2.439 ^{0}± 0.011
M2	393.949	23711.	1.18556 ± 0.00005	1.994 ^{0}± 0.002
L2	11.383	493.	1.21185 ± 0.00246	3.298 ^{0}± 0.116
S2	183.525	11872.	1.18721 ± 0.00010	0.533 ^{0}± 0.006
K2	49.966	3604.	1.18947 ± 0.00033	0.823 ^{0}± 0.016
M3..M6	4.639	651.	1.06775 ± 0.00164	0.228 ^{0}± 0.088

S/N: signal-to-noise ratio
standard deviation: 0.558 nm/s^2
air pressure regression coefficient: -3.267 ± 0.016 nm/s^2 per hPa

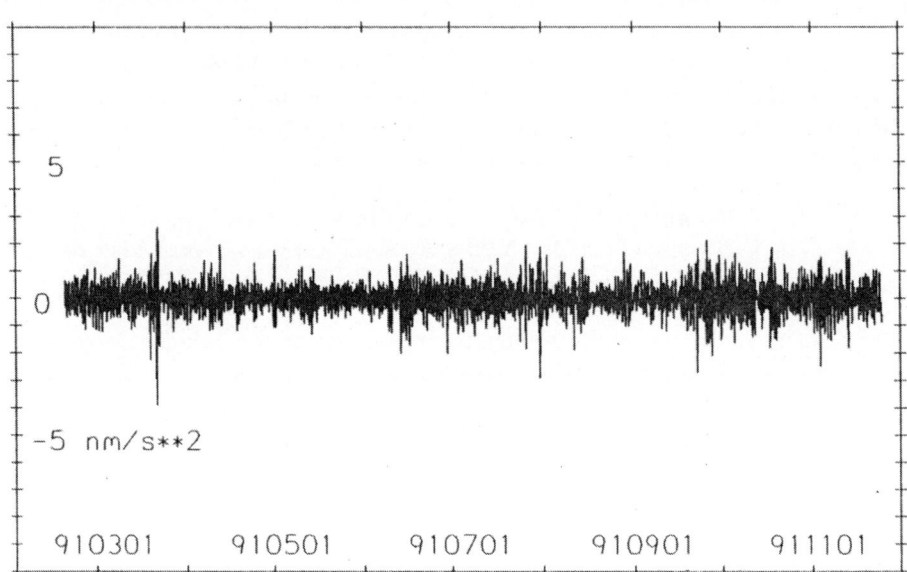

Fig. 7. Residuals of gravity tides, observed with gravimeter LCR-ET19 at station BFO Schiltach in 1991.

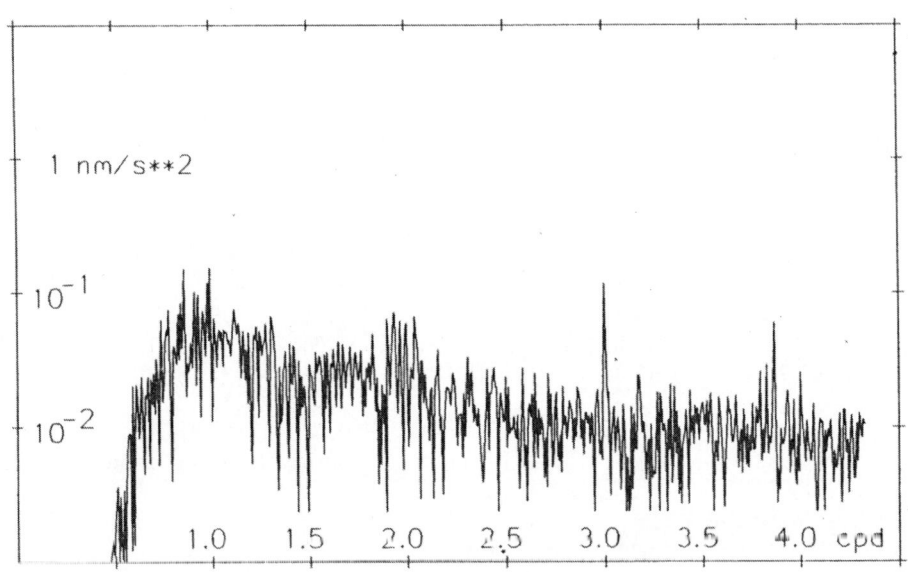

Fig. 8. Fourier amplitude spectrum of residuals of gravity tides, observed with gravimeter LCR-ET19 at station BFO Schiltach in 1991.

Table 2. Example 2: Superconducting gravimeter TT40, station Bad Homburg, 1004.5^d, 1981-1984, (Richter 1987), program **ANALYZE** of the **ETERNA** 3.30 package, drift approximation, tidal potential catalogue of Hartmann and Wenzel, 1995.

wave	Amplitude $[nm \cdot s^{-2}]$	S/N	amplitude factor δ	phase lead $\Delta\Phi$
SA	27.160	1.1	7.43298 ± 6.60730	-7.080^0± 56.984
SSA	25.981	3.7	1.12945 ± 0.30250	-1.129^0± 14.786
MM	29.504	30.5	1.12993 ± 0.03707	0.000^0± 1.881
MF	56.745	115.0	1.14768 ± 0.00998	0.428^0± 0.500
MTM	10.714	33.9	1.13176 ± 0.03335	-1.276^0± 1.684
Q1	67.156	1471.1	1.14748 ± 0.00078	-0.172^0± 0.039
O1	350.930	7175.5	1.14808 ± 0.00016	0.099^0± 0.008
M1	27.586	466.7	1.14814 ± 0.00246	0.351^0± 0.123
P1	163.330	3281.7	1.14858 ± 0.00035	0.206^0± 0.017
S1	3.936	56.0	1.17027 ± 0.02090	2.891^0± 1.023
K1	488.168	9467.3	1.13608 ± 0.00012	0.251^0± 0.006
ψ1	4.208	85.2	1.25068 ± 0.01467	0.592^0± 0.672
ϕ1	7.168	144.6	1.17117 ± 0.00810	-0.299^0± 0.396
J1	27.754	634.5	1.15471 ± 0.00182	0.129^0± 0.090
OO1	15.136	320.6	1.15102 ± 0.00359	0.300^0± 0.179
2N2	10.904	610.0	1.15894 ± 0.00190	2.712^0± 0.094
N2	69.101	2932.5	1.17300 ± 0.00040	2.421^0± 0.020
M2	365.141	14834.6	1.18677 ± 0.00008	1.966^0± 0.004
L2	10.470	279.9	1.20376 ± 0.00430	2.118^0± 0.205
S2	170.113	6602.7	1.18849 ± 0.00018	0.532^0± 0.009
K2	46.327	1777.6	1.19096 ± 0.00067	0.757^0± 0.032
M3..M6	4.124	443.9	1.06539 ± 0.00240	0.531^0± 0.129

S/N: signal-to-noise ratio
standard deviation: 6.884 nm/s^2
air pressure regression: -2.914 ± 0.005 nm/s^2 per hPa
pole tide amplitude factor: 1.40 ± 0.68
pole tide time lead: 40. ± 51. days

adjusted Tschebyscheff drift polynomial coefficients:

degree	drift coefficient $[nm/s^2]$	standard deviation $[nm/s^2]$
0	1810.82	0.06
1	-312.99	0.09
2	70.40	0.08

For the 2nd example of earth tide analysis (Tab. 2), data of the superconducting gravimeter GWR-TT40 recorded at station Bad Homburg (1004.5^d, 1981-1984, Richter, 1987) have been used. One should know that this data set has for the first time demonstrated that gravity pole tides can be observed using a superconducting gravimeter. We have used the program ANALYZE of the ETERNA 3.30 package (Wenzel, 1996a) and the Hartmann and Wenzel (1995a,b) tidal potential catalogue as in the first example. But in contrast to the first example, we have solved for gravity pole-tide regression parameters (amplitude factor and time lead), for long-periodic tidal parameters (up to wave SA with 1 year period) and short-periodic tidal parameters. Therefore we could not apply digital highpass filtering but had to adjust a drift model (Tschebyscheff polynomial of degree 2). The standard deviation of the observations therefore is considerably larger than in the 1st example. But because the standard deviations of the adjusted earth tide and pole tide parameters have been estimated from the amplitude spectrum of residuals, the standard deviations of the short periodic tidal parameters are similar to those obtained in the 1st example. The average noise level is

$$\sim 11.00 \text{ nm/s}^2 \; \hat{=} \; 1 \cdot 10^{-10} \text{ g at 1 year period,}$$
$$\sim 0.042 \text{ nm/s}^2 \; \hat{=} \; 4 \cdot 10^{-12} \text{ g for diurnal waves,}$$
$$\sim 0.022 \text{ nm/s}^2 \; \hat{=} \; 2 \cdot 10^{-12} \text{ g for semidiurnal waves,}$$
$$\sim 0.008 \text{ nm/s}^2 \; \hat{=} \; 8 \cdot 10^{-13} \text{ g for terdiurnal waves.}$$

which is for the short periodic tidal bands very similar to example 1.

One can again see the resonance effect caused by the liquid outer core (Nearly Diurnal Free Wobble) in the 1 cpd frequency band and moderate ocean tide effects of about $\sim 3\%$ and 2^0, respectively in the semidiurnal band.

The residuals after analysis (Fig. 9) show partly strong unmodeled long-periodic signals of probably instrumental origin; the amplitude spectrum of the residuals (Fig. 10) shows a clear red noise structure (1/f noise amplitude function). Although the residuals are much larger for this example in the time domain compared the example 1, the amplitude spectrum of the residuals (Fig. 8) above 0.5 cpd is similar to example 1. This is the reason for having obtained similar standard deviations for the short periodic tidal parameters as with example 1.

Before using the adjusted tidal parameters for further interpretations, they have to be corrected for ocean tide gravitation and loading (see Jentzsch, "Ocean Loading", this volume).

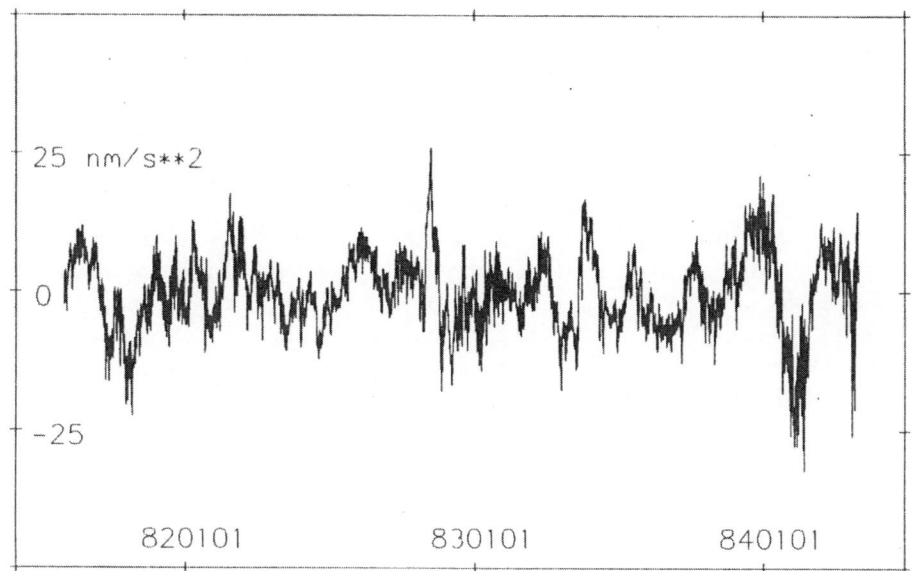

Fig. 9. Residuals of gravity tides, observed with superconducting gravimeter GWR-TT40 at station Bad Homburg in 1981-1984.

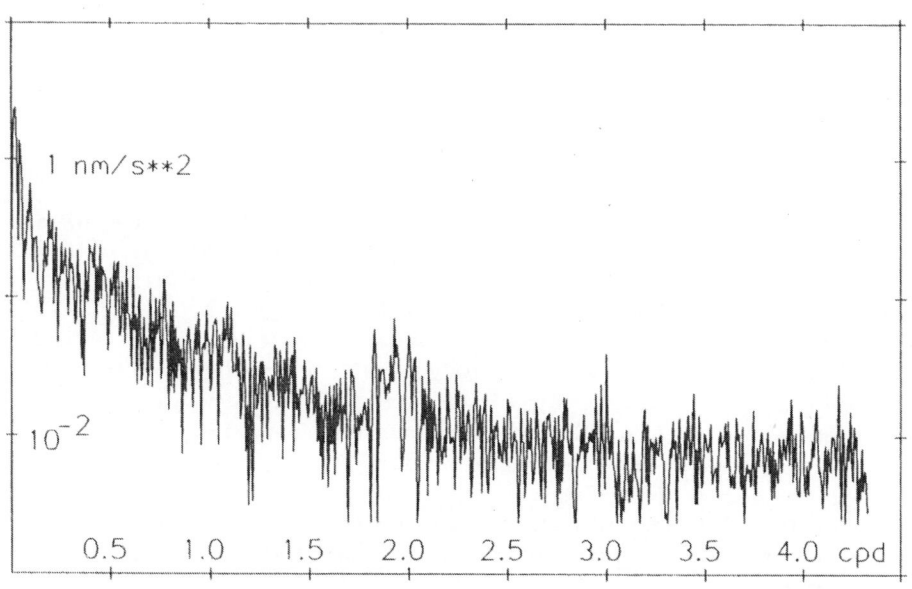

Fig. 10. Fourier amplitude spectrum of residuals of gravity tides, observed with superconducting gravimeter GWR-TT40 at station Bad Homburg in 1981-1984.

4   Conclusions

The analysis of earth tide observations has nowadays reached a high standard, partly due to the development of electronic data processing. The complete data preprocessing and data analysis can be carried out on a PC. The earth tide analysis is usually formulated as a MISO-LTIS problem (Multiple Input - Single Output Linear Time Invariant System) and solved using the least squares adjustment technique. Realistic error estimations for the adjusted parameters are obtained using the amplitude spectrum of the residuals. The adjustment of regression parameters to other phenomena like e.g. gravity pole tides makes a modern earth tide analysis package (e.g. Wenzel, 1996a) a rather universal tool for the analysis of gravity variations or deformations of the Earth's crust observed by a stationary instrument.

References:

Chojnicki, T. 1973. Ein Verfahren zur Erdgezeitenanalyse in Anlehnung an das Prinzip der kleinsten Quadrate. *Mitt. Inst. Theor. Geodäsie, Univ. Bonn* 15.

De Toro, C., A.P. Venedikov and R. Vieira 1993. A new method for earth tide data analysis. *Bull. Inf. Marées Terrestres* 116: 8557 - 8586.

Hartmann, T. and H.-G. Wenzel 1995a. The HW95 tidal potential catalogue. *Geophys. Res. Lett.* 22: 3553 - 3556.

Hartmann, T. and H.-G. Wenzel 1995b. Catalogue HW95 of the tide generating potential. *Bull. Inf. Marées Terrestres* 123: 9278 - 9301.

Munk, W.H. and D.E. Cartwright 1966. Tidal spectroscopy and prediction. *Phil. Trans. Royal Soc. London* A 259: 533 - 581.

Munk, W. and K. Hasselmann 1964. Super-resolution of tides. *Stud. Oceanogr.* 339 - 344.

Richter, B. 1987. Das supraleitende Gravimeter. *Deutsche Geod. Komm.* C 329: 1 - 126.

Schüller, K. 1976. Ein Beitrag zur Auswertung von Erdgezeitenregistrierungen. *Deutsche Geod. Komm.* C 227: 1 - 86.

Schüller, K. 1978. About the sensitivity of the Venedikov tidal parameter estimates to leakage effects. *Bull. Inf. Marées Terrestres* 78: 4635 - 4648.

Schüller, K. 1986. Simultaneous Tidal and Multi-Channel Input Analysis as Implemented in the HYCON-Method. *Proc. 10th Int. Symp. on Earth Tides.* pp. 515 - 520. R. Vieira (ed.). Cons. Sup. Invest. Cient., Madrid.

Tamura, Y. , T. Sato, T. Ooe, and M. Ishigiro. 1991. A procedure for tidal analysis with a Bayesian information criterion. *Geophys. J. Int.* 104: 507 - 516.

Venedikov, A.P. 1961. Application a l'analyse harmonique des observations des marées terrestres de la methode des moindres carrés. *Comptes rendus Acad. Bulgare Sciences* 14, Sofia.

Venedikov, A.P. 1966a. Une méthode pour l'analyse des marées terrestres a partir d'enregistrements de longueur arbitraire. *Comm. Observ. Royal Belgique* 71, Bruxelles.

Venedikov, A.P. 1966b. Sur la construction de filtres numeriques pour le traitement des enregistrements des marées terrestres. *Comm. Observ. Royal Belgique* 76, Bruxelles.

Wenzel, H.-G. 1976a. Zur Genauigkeit von gravimetrischen Erdgezeitenbeobachtungen. *Wiss. Arb. Lehrstühle Geodäsie, Photogramm. Kartogr. Techn. Univ. Hannover* **67**: 1 - 177.

Wenzel, H.-G. 1976b. Some remarks to the analysis method of Chojnicki. *Bull. Inf. Marées Terrestres* **73**: 4187 - 4191.

Wenzel, H.-G. 1977. Estimation of accuracy for the earth tide analysis results. *Bull. Inf. Marées Terrestres* **76**: 4427 - 4445.

Wenzel, H.-G. 1994. PRETERNA - a preprocessor for digitally recorded tidal data. *Bull. Inf. Marées Terrestres* **118**: 8722 - 8734.

Wenzel, H.-G. 1996a: The nanogal software: Earth tide data processing package ETERNA 3.30. *Bull. Inf. Marées Terrestres* **124**: 9425 - 9439.

Wenzel, H.-G. 1996b: Zum Stand der Erdgezeitenanalyse. *Zeitschr. f. Vermessungswesen* **121**: 242 - 255.

Earth Tide Observations and Interpretation

Walter Zürn

Black Forest Observatory, Heubach 206, D-77709 Wolfach, Germany

Abstract. Earth tide observations with gravimeters, tilt- and strainmeters were obtained at many stations on the globe. The original aim of the research was to determine the global response of the earth to the tidal forcing in the form of Love and Shida numbers. This goal could not be reached due to strong perturbations of the body tide signals by ocean tide loading and attraction and by local elastic effects for the two latter types of instruments. Higher accuracy is needed for the instruments and the corrections for these perturbations, before the true body tide signal can be gleaned from the observations and information about the elastic and anelastic threedimensional structure of the earth can be retrieved.

1 Introduction

By far the most excellent short introduction into the field of earth tide observations was given by Baker (1984). The classical textbook on earth tides by Melchior (1981) reports many observations of earth tides and contains abundant information about this field of research. For an extensive list of publications on earth tides and related research the reader is referred to that book and the bibliographies, which are regularly updated in the Bulletin d'Informations Marées Terrestres, published by the International Center for Earth Tides (ICET) at Brussels, Belgium. However, the book by Harrison (1985), a collection of benchmark papers up to that time with very enlightening discussions by the editor of every aspect treated is highly recommended. Most conclusions drawn then are still valid today. Tables and formulae can also be found in Wilhelm and Zürn (1984) and Zürn and Wilhelm (1984). At this point an apology is mandatory: it is impossible to give a perfectly balanced and complete account of all the work performed in this field and hopefully the colleagues will forgive the author for omissions.

The discussion here is restricted to tidal signals in gravity, tilt and strain as measured locally by corresponding instruments fixed to the earth's crust either in observatories or boreholes, other earth tide phenomena are treated elsewhere in this volume (see Kümpel "Well Tides" and Schwintzer "Satellite Orbit Perturbations"). Analyses of the resonant response of the earth to the tesseral tides in the diurnal frequency band due to the nearly-diurnal free-wobble are also discussed elsewhere (see Zürn "Nearly-Diurnal Resonance", this volume). The tidal displacements on the surface of the earth have decimeter amplitudes and must be accounted for in the analysis and interpretation of measurements by modern space geodetic methods, like VLBI (Very Long Baseline Interferometry), LLR and SLR (Lunar and Satellite Laser Ranging) and GPS(Global Positioning System): these techniques will not be discussed here. One example from physics, where tidal corrections are needed, are large particle accelerators (e. g. Melchior, 1995b).

Instrumentation for earth tide measurements will not be described in this article since excellent references are available. Gravity sensors consist of test masses suspended either on metal or quartz springs or in a magnetic field produced by highly stable currents through superconducting coils. Gravimeters are treated in textbooks on geodesy (e. g. Torge, 1989) and the most modern instrument of this kind, the cryogenic gravimeter, is extensively described by Prothero and Goodkind (1968), Richter (1987), Goodkind (1991) and Warburton and Brinton (1995).

Strainmeters employ either invar wires, quartz rods or laser beams as length standards against which the displacement of two points attached to the crust is measured. There are different kinds of tiltmeters: vertical pendulums, horizontal pendulums (the garden-gate suspension), level bubble instruments and long baseline fluid tiltmeters. In the textbook by Melchior (1981) essentially all the instrumentation used up to that time is discussed, however, with most of the space devoted to quartz horizontal pendulums and accessories. An excellent and extensive account of strain and tilt instrumentation and associated problems is provided by Agnew (1986), including all such instruments mentioned in the following examples, with ample references for the reader interested in detail.

Earth tides were recorded at many stations around the world (Melchior, 1981, 1994), with a heavy concentration in Europe. Extremely remote places like the Geographical South Pole (Rydelek and Knopoff, 1982) and other sites in Antarctica (Schneider, 1971; Shibuya and Ogawa, 1993), Greenland (Jentzsch et al., 1995), Spitzbergen (Bonatz et al., 1971), Kerguelen Island (Bonatz and Schüller, 1976) among others, were also selected for measurements of earth tides. However, most of these stations were occupied with gravimeters, while in comparison tidal tilt measurements are much less numerous and strain tide stations even less. Fig. 1 shows simultaneous records of different earth tide signals obtained with a gravimeter, two different tiltmeters and three strainmeters at the Black Forest Observatory Schiltach ($48.33^{o}N$, $8.33^{o}E$, 589 m a.m.s.l.). From top to bottom the corresponding instruments are: LaCoste-Romberg Earth Tide gravimeter ET 19 with electrostatic feedback, Askania borehole tiltmeter BLP 10 with two nearly orthogonal components in EW- and NS-direction, a 120 m version of a Horsfall differential fluid pressure tiltmeter (Emter et al., 1989) in the azimuth of $N151.2^{o}E$ and three 10 m Cambridge invar wire strainmeters (King and Bilham, 1976) with azimuths of $N2^{o}E$, $N300^{o}E$ and $N60^{o}E$, respectively. The units are 10^{-9} of local gravity, 10^{-9} rads and 10^{-9} for gravity, tilt and strain signals, respectively. The mean values (offsets) of the time series are meaningless and the signals visible in some traces with periods much longer than the daily tides are of instrumental origin and of no interest either.

Such time series, with usually much greater lengths and well calibrated, can be subjected to tidal analysis (see Wenzel "Tidal Analysis", this volume), which essentially compares the observed signal in very narrow frequency bands (groups of tidal constituents) with a certain theoretically predicted signal, either for a rigid or an elastic earth model. The results are tables of amplitude factors and phase shifts (or complex admittances) with respect to the predicted tidal variations for these narrow frequency bands, which are labelled with the Darwin symbol of the major tide(s) within that group (for example M_2). For gravity and tilt these amplitude factors are called gravimetric and diminishing factor, respectively, when the reference model corresponds to a rigid earth. For a spherical elastic earth model these factors and the corresponding amplitude factor for strain are simple linear functions of the Love numbers h and k and the Shida number l (see Wang "Solid Earth Response", this volume). Of course, the measured data contain noise, while the predicted signal does not and the quality of the results will necessarily depend on the signal-to-noise ratio in the data. Formal uncertainties are estimated in the tidal analysis from the noise in the residuals, these uncertainties do not incorporate systematic errors like calibration inaccuracies, instrumental phase shifts, timing inaccuracies, among others. Of course, the predictions are not perfect, however, the uncertainties arising from this side can be kept much smaller than the noise in the observations (see Wenzel "Tide-Generating Potential", this volume).

For a given group of tidal constituents in a narrow frequency band, the amplitude and phase derived from the data through tidal analysis can be compared in the complex plane with the predicted signal. Fig. 2 shows such a phasor diagram (d'Argand diagram) schematically (see also Fig. 1.1 of Jentzsch "Ocean Loading", this volume). The tidal driving forces not only deform the solid earth periodically (this response is called the body tide), they also produce the better known tides in the oceans (see Zahel "Ocean Tides", this volume) as well as tides in the atmosphere (see Volland "Atmospheric Tides", this volume). The solar atmospheric tides are not only caused by gravitation, but mainly by solar radiation (see Wilhelm "Solar Irradiance", this volume). Both oceanic and atmospheric tides lead to disturbances of the wanted body tide signal with identical periodicities (Jentzsch "Ocean Loading", this volume; Berger, 1975; Müller, 1977; Friederich and Wilhelm, 1986). Furthermore, the locally measured tilts and strains are strongly influenced by local conditions (cavities, topography, geology; see Westerhaus "Tidal Tilt Modification", this volume), which results in strong modifications of the observed signal with respect to the expected one for a seismologically constrained earth model. These disturbances also vary with tidal periodicity. There is no way to separate these different contributions to the total tidal signal by methods of time series analysis, the only way to extract a certain contribution from the data is to model all the others.

In Fig. 2 schematically the signals from the solid earth (including a small phase lag due to anelasticity in the earth's mantle), the attraction and loading effects of the ocean tides, a meteorological effect (important for solar constituents only) and the local elastic effect (important for tilt and strain, not for gravity) are vectorially added up and this sum can be compared to an observed tidal phasor. Of course, all the different phasors have uncertainties associated with them,

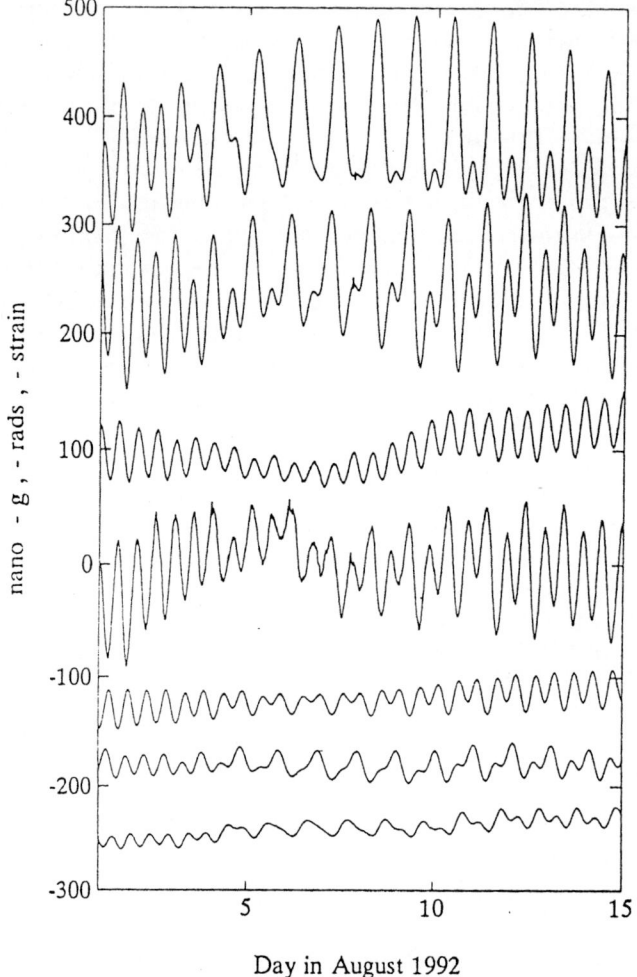

Fig. 1. 14 days of earth tide records simultaneously obtained digitally at the BFO starting August 1, 1992 at 0:00 h UTC. Original sampling interval of 5 seconds was reduced to 1 minute by simple averaging. See text for an identification of the corresponding sensors.

which in some cases are poorly known. If the observations deviate significantly from the predicted tide under consideration of all the uncertainties, the reason is either a poorly modeled or an unmodeled physical effect or the uncertainties were underestimated.

For example in the controversial interpretation of M_2 gravity-tide residuals by Melchior (1995a) on one hand and Rydelek et al. (1991) on the other hand, the first author claims an unmodeled physical effect (related to heat flow), while the latter authors suspect that erroneous ocean load modeling and underestimated

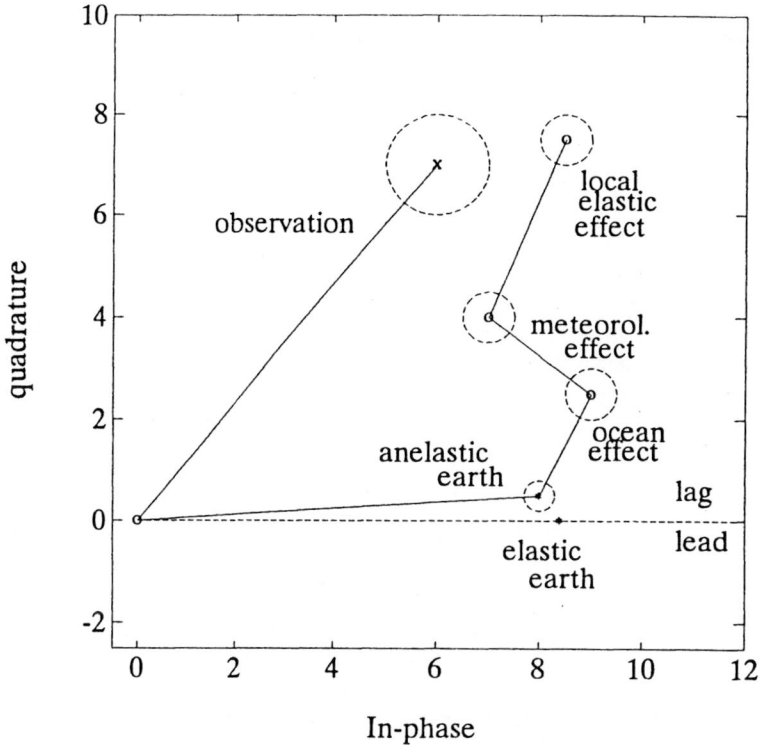

Fig. 2. Schematic phasor-diagram (d'Argand diagram) showing a standard type of interpretation of the observed amplitude and phase (or complex admittance) for a single constituent. For clarity some features are exaggerated. Broken circles indicate uncertainties of observations, theoretical predictions and corrections.

errors are the source of rather large residual phasors in the data bank accumulated by ICET. Different research projects in earth tides aim to determine different contributions to the model phasor. Some scientists try to determine the body tide (Baker et al., 1996), some others the ocean loading contribution in gravity, assuming the body tide to be known with higher accuracy (e. g. Baker 1980a, 1980b; Asch et al., 1987; Jahr et al. 1991). Still others (e. g. Melchior, 1995a) attempt to find the regional variations in tidal gravity due to lateral structure with the assumption, that the corrections for ocean loading are good enough and others (Gerstenecker et al., 1986) try to find lateral structure using tidal tilts assuming ocean loading to be known well enough and local elastic effects to be much smaller than the effects caused by a pronounced geological feature. Some of these attempts are discussed below in more detail.

The advantages of the local observation of earth tides in contrast to space techniques are the higher temporal resolution and precision which can be achieved with continuously recording gravimeters, tilt- and strainmeters. Major problems, however, are caused by the instrumental drift, local sources of random and systematic noise and the coupling to the earth. Because the most serious sources of systematic errors are different for tidal gravity versus tidal tilt and strain these observables are discussed separately in the following two sections.

2   Gravity Tides

The tidal variations in gravity can be measured with very high signal-to-noise ratio by superconducting gravimeters and LaCoste-Romberg gravimeters with electrostatic feedback (Zürn et al., 1991; Richter et al., 1995)). The response of the solid earth to the tidal driving forces surely contains information about the internal structure of the planet, both radially and laterally. This response is usually expressed as the gravimetric factor and phase difference (or as a complex admittance) for specific tidal frequencies (see Wang "Solid Earth Response", this volume). However, detailed knowledge about this structure exists from other fields, especially from seismology. Considering the large difference between tidal and seismic frequencies, the tidal response could shed much light especially on the problem of the rheology of the earth's mantle, if it could be determined with precision from the observations, the frequency dependence of the body tide amplitudes and the phase lags would be of great interest for earth structure determination. The phase lag of the body tide contributes to the braking of the earth's rotation. It is known for a long time that differences in the spherically symmetric structure of the earth's mantle within the constraints posed by seismological results cause only very small changes in the tidal response of the earth, i. e. the gravimetric factor (e. g. Varga, 1974; Wilhelm, 1978; Wahr, 1981; Varga and Denis, 1988). The largest deviations from spherical symmetry of the earth are its rotation and flattening. At present the effects of this symmetry breaking on the tidal response are being reinvestigated theoretically (e. g. Wang, 1994). While a latitude effect in the gravimetric factor of about 3 % (peak-to-peak) due to rotation and ellipticity seemed to be well established by previous theories (Wahr, 1981; Dehant, 1991), the new results by Wang (1991) indicate a negligible effect and an error was detected in the older computations (Dehant, 1995). Wang's (1991) and additional theoretical and numerical work on other lateral heterogeneities in the earth shows that for reasonable models of such asymmetries, the effects on tidal gravity are appreciably below 1 % (e.g. Zürn et al., 1976; Molodenskii and Kramer, 1980; Wang, 1991; Wang "Solid Earth Response", this volume).

Two serious problems make it at present impossible to extract these signatures from the observed gravity tides: a lack of absolute calibration accuracy better than a few tenths of a percent of even the best gravimeters and the large size and uncertainties of the indirect effects of the tides in the oceans on the observed earth tides. These problems are documented by the lengthy and complicated argumentation of Melchior (1994) with respect to the necessity of a change of the adopted tidal parameters for Brussels. Several successful efforts have been made to improve the calibration accuracy of recording gravimeters using basic physical principles (Van Ruymbeke, 1989; Richter, 1991; Varga et al., 1995;

Achilli et al., 1995; Richter et al., 1995; see also Goodkind et al., 1993) and the models of the global ocean tides will certainly be improved by the results from the TOPEX/POSEIDON mission (Agnew, 1995, Baker et al., 1996; Llubbes and Mazzega, 1996, Francis and Melchior, 1996). However, these new models of the global ocean tides still do not cover all the local seas and must therefore carefully be augmented with models of these seas, before the systematic signals from the oceans can be confidently removed from the observations.

Fig. 3 shows a plot of observed gravimetric factors for M_2 (the tide determined best) for some high quality tidal data recorded in central Europe against geographical latitude. These observations were obtained by Baker et al. (1989), Wenzel et al. (1991), Rydelek et al. (1991), Dittfeld and Wenzel (1993), Timmen and Wenzel (1994) and were corrected with models for both the oceanic tidal maps (Schwiderski, 1980) and the loading response of the earth.

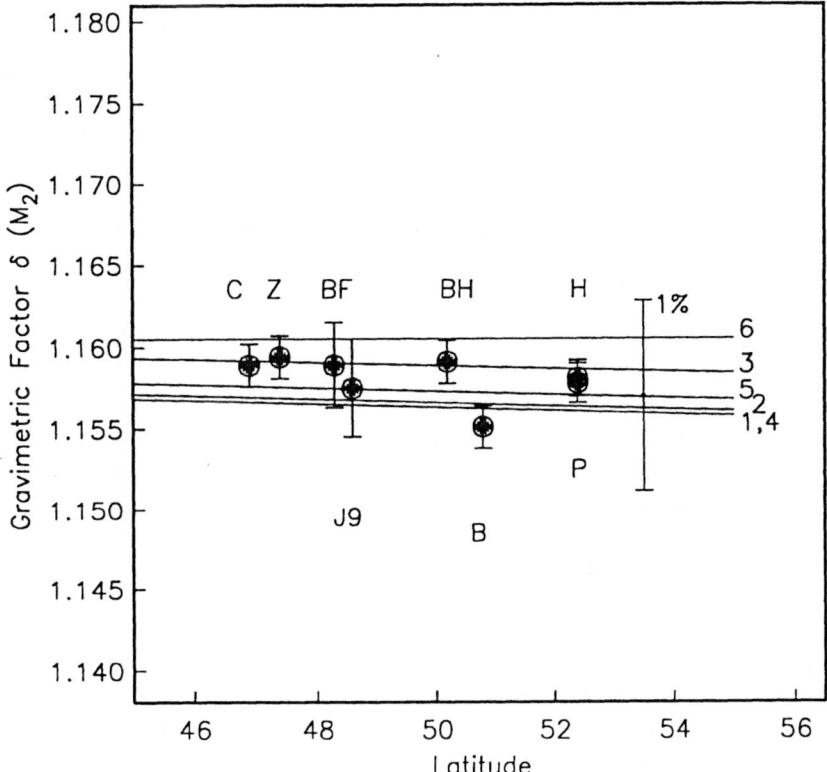

Fig. 3. Gravimetric factors for M_2, corrected for oceanic effects, vs. latitude in central Europe (see Timmen and Wenzel (1994) for similar plots). Stations: C - Chur, Z - Zürich, BF - BFO-Schiltach, J9 - Strasbourg, BH - Bad Homburg, B - Brussels, H - Hannover, P - Potsdam. All stations except J9 (superconducting gravimeter) were occupied with LaCoste-Romberg gravimeters with electrostatic feedback. Model calculations: 1/2 - Dehant and Zschau (1989), 3 - Dehant and Ducarme (1987), 4/5 - Dehant (1987), 6 - Wang (pers. comm. 1993). Uncertainty estimates contain calibration and noise-level, oceanic correction errors not included.

The error bars contain uncertainties from the noise level and errors of calibration, but no estimate of the uncertainty of the ocean correction is included. The data are compared with 6 different models for the gravimetric factor including latitude dependence. Note that these models are being reinvestigated carefully by several groups, while this article is written. Considering that appreciable additional uncertainties due to the oceanic corrections must exist, then the observations agree within their errors with all the model calculations and with each other. Similar statements can be made for other tides and for the observed phase differences as well. Note that the differences between the observations are well within 1 %. Given the errors a latitude effect cannot be discerned.

Actually no other effect can be discerned either, such as an effect of heatflow or lateral heterogeneities. More than 300 tidal measurements have been accumulated in the gravity tide data bank by the ICET (Melchior 1994). Yanshin et al. (1986) suggested from a correlation between M_2- residuals from this data bank and heat flow estimates at locations close to the tidal gravity stations that a physical effect exists connecting these two observables. While the statistical correlation exists in these data, Rydelek et al. (1991) questioned this conclusion. Their strongest argument is that the high quality data shown in Fig. 3, when plottet against heat flow show nothing of this sort for Central Europe, although the heat flow values obtained in the area of these sites range between 60 and 120 mW/m^2. In addition, the crustal structures and the surface geology for these stations differ dramatically as well, so there appears to be no discernible effect of local crustal structure on gravity tide measurements at the present level of accuracy. The scatter of the observed, corrected gravimetric factors in the ICET data bank is much larger than in Fig. 3 (Baker et al., 1989). Rydelek et al. (1991) also point out, that as long as ocean loading corrections are computed with a global model only and do not account for all the local seas, appreciable errors must be expected and with prudence no other effects can be expected to show up clearly. Actually, in his response to Rydelek et al. (1991), Melchior (1995a) himself pointed out, that the tides under the Ross and Filchner ice shelves must be taken into account when modeling the gravity tide observations at the South Pole. Agnew (1995) shows that the new ocean models including the tides under the ice shelves explain the anomalous South Pole observations (Knopoff et al., 1989) almost completely (the daily tides at the South Pole are theoretically absent). Another criticism by Rydelek et al. (1991) concerns the lack of appreciation of error estimates in all the work about correlation of tides with heat flow, notoriously error bars are not presented although estimates for the uncertainties due to noise are readily available from every program for tidal analysis. Of course, these would be minimal estimates because they do not include uncertainties in calibration and oceanic corrections, but they would give the reader an idea about the quality (or lack thereof) of the data. Also notoriously neglected is the fact, that other tides (O_1 for example) should be affected as well in case there is a physical cause for the anomalies. The controversy about this effect is continuing, since Melchior (1995a) responds to the criticism by Rydelek et al. (1991). Inspection of Fig. 3 in Rydelek et al. (1991) and Fig. 3 here tell the whole story. Obviously noise levels, calibrations and oceanic corrections must be improved, before effects of lateral heterogeneities become clearly visible even with the best data.

Baker et al. (1996) took a new look at the high quality tidal observations of M_2 and O_1 in central Europe using ocean tide corrections based on various ocean tide models, most of them referring to the TOPEX/POSEIDON results. Seven out of their eight tidal observations are identical to results shown in Fig. 3. They conclude, that these results are consistent with purely elastic, seismologically highly constrained models such as PREM (Preliminary Reference Earth Model) by Dziewonski and Anderson (1981). They further conclude that in a class of models, where due to anelasticity the quality factor for shear deformation Q_μ in the mantle changes with frequency f proportional to f^α (Wahr and Bergen, 1986), only models with $\alpha \leq 0.09$ are consistent with these observations. In addition, these authors do not find a station dependence of the gravimetric factors for M_2 and O_1 given their uncertainties, be it for reasons of geology, crustal structure, heat flow or latitude.

Warburton and Goodkind (1976) made an attempt to find evidence for deviations from general relativity using the gravity tides recorded by their superconducting gravimeter at the Pinion Flat observatory in California. Some tides (K_1, K_2 and R_2) should show an anomalous response if certain Post-Newtonian parameters deviate from the values they adopt for Einstein's general relativity (0 or 1), because additional signals at these frequencies arise from the corresponding effects (i.e. additional phasors in Fig. 2 for these tides). These authors concluded, that the uncertainties in the oceanic and atmospheric loading at the time when this analysis was undertaken, were too large to identify relativistic effects. This is still the case now.

3   Tidal Tilts and Strains

An excellent introduction into this field and several benchmark papers can again be found in Harrison (1985). The rapidly developing field of tides in aquifers will be discussed elsewhere in this volume (Kümpel "Well Tides"). It was stated above that the distortions of the tidal gravity field by local, regional and even large scale deviations of the earth from spherical symmetry are very small and hardly observable at present. In contrast, the tidal deformation field, as measured by tilt- and strainmeters, is heavily distorted by local heterogeneities to the extent, that the global or even regional response cannot be measured reliably with such instruments (however, see Weise et al., 1995 and Kohl and Levine, 1995 for a different view). These distortions are present in addition to tilts and strains due to ocean attraction and loading (see Jentzsch "Ocean Loading", this volume). This situation and its consequences for tidal research are well documented by Harrison (1985, part IV), who also did a good part of the pioneering work towards a better understanding of these problems (see Baker and Lennon, 1973; King and Bilham, 1973; Beaumont and Berger, 1974; Berger and Beaumont, 1975; Harrison, 1976; King et al., 1976; Beavan et al., 1979). The distortions are called cavity, topographic and geological effects in the literature. Sato and Harrison (1990) and Kohl and Levine (1995) made the most recent attempts of interpretation of anomalous tidal strains and tilts, respectively, by taking carefully all the known effects into account. Hart et al. (1996) discuss the problems related to small-scale inhomogeneities in the context of tidal calibration of borehole strainmeters. Actually, it appears to be better to turn around and look for possibilities where the geological effects could tell us something new about the local or regional structure of the earth. This means that in an experiment the geological effects caused by the object under study should dominate, and cavity and topographic effects should be minimized by careful selection of the instrument site. For the interpretation models of the body tide (the global response of the earth to the tidal forcing) and the ocean tide effect are assumed to be much better known than the other distortions and are therefore subtracted from the observations. Cavity effects cannot be avoided by measurements in mines, and they are mostly stronger for instruments with small baselines, therefore the experimental answer is to put small instruments into boreholes and/or to increase the baseline (Zürn et al., 1986; Emter et al., 1989).

After this problem was understood, several attempts were undertaken to deploy arrays of borehole tiltmeters in geologically interesting situations in order to use the spatial (amplitude and phase) variation of observed tidal tilts for the geological studies. However, other researchers tried to find out if it is really possible to get consistent such tidal measurements in places where the rocks were assumed to be very homogeneous and no spatial variation of tidal tilts or strains was expected a-priori. In the following some of this work is referenced. An important theoretical prediction was published by Beaumont and Berger (1974). These authors showed with the use of finite element models, that rather large and easily observable temporal modification of tidal tilt- and strain amplitudes could occur in an earthquake zone, if it becomes dilatant (or its elastic parameters change for other reasons) during the time when crustal stresses built up before an earthquake. In essence this is a positive outcome of the research on the distortion effects.

Three important experiments using relatively small arrays of borehole tiltmeters were designed to test if consistent tidal signals within a supposedly homogeneous volume in the earth's crust would be recorded. Cabaniss (1978) had installed two arrays of three tiltmeters each in New England, USA. For the shallow (18 m depth) array with borehole separations of 100 m he reports agreement of the tidal responses within 2 %, while for the deep array (100 m) with similar spacing instrumental problems obviously contaminated the tidal results. The tidal results reported by Peters and Beaumont (1987) were obtained from three boreholes in the Charlevoix earthquake zone in Québec, Canada. The boreholes with depths of 47, 47 and 110 m were forming a triangle of approximately 80 m sides and were housing Askania tiltmeters. Unfortunately the results from the 110 - m borehole suffer from the fact that the orientation of the tiltmeter at the bottom of this hole was not known with sufficient precision, therefore the orientation had to be deduced from the tidal responses of the two components.

One important but very disappointing result from these three boreholes was the dispersion of the observed tidal admittances (up to 20 % in the amplitude of the M_2 tide), which must be blamed on unexpected strain - tilt coupling by previously unknown heterogeneities and/or fractures in the vicinity of the boreholes.

At the Pinyon Flat Observatory (PFO) in Southern California (Wyatt et al., 1990) an important set of experiments is carried out to determine the significance of observed tilts and strains in the crust not very far from two major faults (one of them the famous San Andreas). Agnew (1981) used the strain tides observed with three laser strainmeters (appr. 700-800 m long) to look for a nonlinear response of the rocks to the earth tides with a negative result. Nonlinear behaviour would express itself by enhanced higher harmonics of the large diurnal and semi-diurnal tides. Such harmonics are visible in tilt and strain observations of high quality from this and other stations (e. g. Peters and Kümpel, 1988), but they are mostly caused by nonlinear loading tides from nearby seas and nonlinear response of the rocks cannot be invoked at the level of precision available. Wyatt and Berger (1980) and Wyatt et al. (1982, 1987, 1988) compared tilts (tides and other frequencies) from several installations (surface long-baseline and shallow and intermediate depth borehole) with each other and found significant discrepancies, which in several cases were interpreted to be due to insufficiently known instrumental properties. Johnson et al. (1994) concentrate on the tidal results from one Askania borehole tiltmeter (this instrument is described by Agnew, 1986) at PFO installed successively in two neighboring boreholes with different depths. Here also it was found that the results differed significantly from each other. Kohl and Levine (1992, 1995) studied the tidal tilts from another set of borehole tiltmeters a few meters horizontally away from the Askania and observed rather large discrepancies between their installations and the Askania tiltmeter. This disturbing result triggered their interesting theoretical investigation of cavity effects in borehole installations and they could correct some of their results to get better agreement. The result of all this research is that tidal amplitudes (and phases) obtained with the same techniques in supposedly homogeneous blocks of the earth's crust at close distance can differ appreciably due to effects from the installation itself or from the immediate neighborhood of the borehole. The results from experiments designed to study a larger heterogeneity of geological interest must therefore be interpreted very cautiously.

Edge et al. (1981) used a borehole tiltmeter array to study the effect of a known batholith in the Lake District of England. They found remarkable coherency between tidal results from the different installations, at least within their uncertainties of about 3 %. A detailed analysis of the tidal results is not available.

Gerstenecker et al. (1986) operated borehole tiltmeters along a profile across a deep-reaching (30 km) fault zone in the Rhenish Massif in Germany to see if this zone would produce anomalous tidal tilts in its vicinity. Fig. 4 shows some of their results, which are the largest tidal tilt anomalies ever reported from borehole installations. In one case the amplitude of the anomaly reaches 100 %

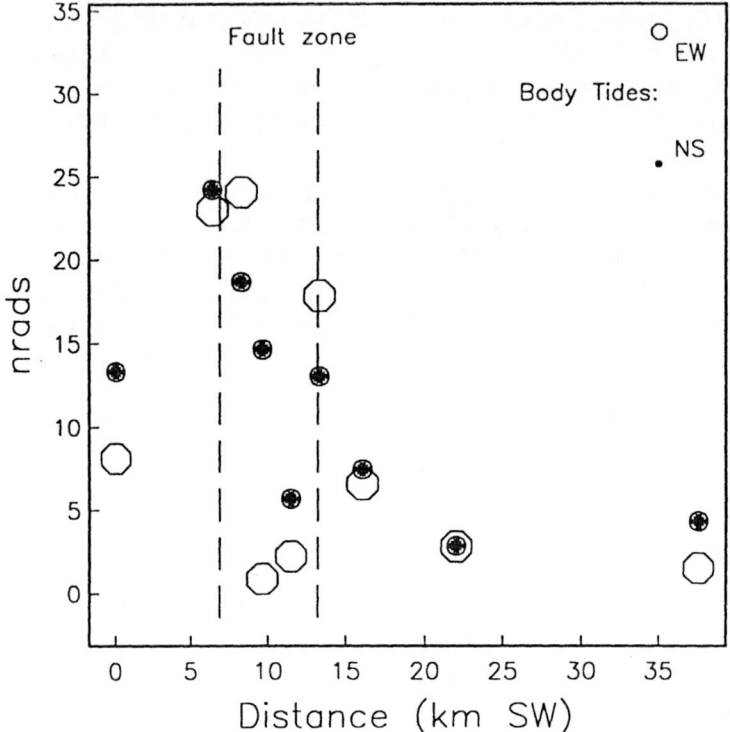

Fig. 4. Amplitude anomalies of tidal NS- and EW-tilts (in nanorads) for M_2, observed along a NE-SW profile across the Hunsrück fault zones in Germany. The anomalies plottet are amplitudes of residual tidal vectors after subtraction of models for the body tide, the ocean loading effect and the topography. These amplitudes would be zero if no other physical mechanism is involved and if corrections were perfect. Body tide amplitudes are given at top right for reference. Note that the anomaly reaches almost 100 % of body tide amplitude for NS - tilt at the station with a distance of 7 km. Vertical dashed lines indicate surface boundaries of fault zone.

of the body tide amplitude. Several models of this fault zone were devised by the experimenters, but due to the large effect these models were rather exotic and somewhat improbable. Of course, and unfortunately, small scale heterogeneities in the immediate neighbourhood of the borehole can easily produce such large effects and therefore one must always worry about a regional interpretation.

Meertens et al. (1989) used the same technique to study the tidal tilts in the Yellowstone, USA, Caldera in order to find information on the structure beneath it. They found effects as large as 50 % of the expected signal from the body tide and ocean loading effects. Their models for a low velocity body underneath the caldera notoriously underpredict the effects. Since the modeling is much constrained by other geophysical data, the authors conclude, that possibly a more complex or more rapid spatial variation of elastic parameters is sensed by the tidal tilts. This result is very similar to that of Gerstenecker et al. (1986) and basically not surprising considering the results obtained in mine installations

and supposedly 'homogeneous' blocks as well as from the many models of heterogeneities studied in that context. In addition, the study by Kohl and Levine
(1995) of installation effects should be taken as a warning. In summary, tilt-
and strain tides are very sensitive to local structure, while for gravity tides large
scale (global) structure dominates (Molodenskii and Kramer, 1980).

Occasionally amplitude variations of tidal strains and tilts with time were
reported in the literature in connection with earthquakes (e. g. Latynina and
Rizaeva, 1976). Some of these claims were later retracted, because instrumental
problems were detected afterwards. Some other claims must be considered du-
bious because the long term stability of the tidal signals at times without quakes
was not demonstrated. Often the effect is seen on one instrument only, which
then could have a very local cause. No effect was seen in the data from several
strain- and tiltmeters at PFO before the 1992 Landers earthquake in California
($M_S = 7.2$) about 70 km from the source (Wyatt, pers. comm. 1995).

A very intensive search for temporal variations of tidal tilts is carried out
with several Askania borehole tiltmeters installed in the North Anatolian fault
zone. The Turkish-German earthquake prediction research program carries out
multidisciplinary studies towards the western end of this very active fault. Westerhaus et al. (1991) and Westerhaus ("Tidal Tilt Modification", this volume)
report about correlations between microseismicity in this area and the tidal response at three stations. However, no big quake has occurred in the region under
study, but the program is still going on.

Tidal tilts and strains have very high potential due to their sensitivity to local
structure and temporal variations of local rock properties. However, extreme care
must be taken in the interpretation of any anomalous results, because this virtue
very easily could turn into a fault, due to unknown small scale structure, and
changes of it, near the instrument. Therefore it is always advisable to obtain
redundant information by means of a local multitude and/or small arrays of
instruments for more certainty in the interpretation.

Acknowledgements: Several colleages and friends have helped to improve my
understanding of tides and tidal instrumentation over many years: the late L.
B. Slichter, D. C. Agnew, T. F. Baker, C. Beaumont, J. C. Harrison, B. Richter, P. A. Rydelek, K. Schüller, H.-G. Wenzel and H. Wilhelm. The *Deutsche
Forschungsgemeinschaft* has financially supported my own work on earth tides
through several grants and the *Alexander von Humboldt - Foundation* made international cooperation possible. All this is gratefully acknowledged.

References

Achilli, V., Baldi, P., Casula., G., Errani, M., Focardi, S., Guerzoni, M., Palmonari,
F., Raguni, G. 1995. A calibration system for superconducting gravimeters. *Bull.
Geod.* **69**: 73 - 80.

Agnew, D. C. 1981. Nonlinearity in Rock: Evidence from Earth Tides. *J. geophys. Res.* **86**: 3969 - 3978.

Agnew, D. C. 1986. Strainmeters and Tiltmeters. *Rev. Geophysics* **24**: 579 - 624.

Agnew, D. C. 1995. Ocean Load Tides at the South Pole: A Validation of Recent Ocean-Tide Models. *Geophys. Res. Lett.* **22**: 3063 - 3066.

Asch, G., Jahr, T., Jentzsch, G., Kiviniemi, A., Kääriäinen, J. 1987. Measurement of gravity tides along the 'Blue Road Geotraverse' in Fennoscandia. *Publ. Finnish Geodetic Inst.* **107**: 1 - 57.

Baker, T. F. 1980a. Tidal gravity in Britain: tidal loading and the spatial distribution of the marine tide. *Geophys. J. R. astr. Soc.* **62**: 249 - 267.

Baker, T. F. 1980b. Tidal tilt at Llanrwst, North Wales: tidal loading and Earth structure. *Geophys. J. R. astr. Soc.* **62**: 269 - 290.

Baker, T. F. 1984. Tidal deformations of the Earth. *Sci. Prog., Oxf.* **69**: 197 - 233.

Baker, T. F., Curtis, D. J., Dodson, A. H. 1996. A new test of Earth tide models in central Europe. *Geophys. Res. Lett.* **23**: 3559 - 3562.

Baker, T. F., Edge, R. J., Jeffries, G. 1989. European tidal gravity: an improvement between observations and models. *Geophys. Res. Lett.* **16**: 1109 - 1112.

Baker, T. F., Lennon, G. W. 1973. Tidal Tilt Anomalies. *Nature* **243**: 75 - 76.

Beaumont, C., Berger, J. 1974. Earthquake Prediction: Modification of Earth Tide Tilts and Strains by Dilatancy. *Geophys. J. R. astr. Soc.* **39**: 111 - 118.

Beavan, J. R., Bilham, R., Emter, D., King, G. C. P. 1979. Observations of Strain Enhancements Across a Fissure. *Deutsche Geod. Komm.* **B 231**: 47 - 58.

Berger, J. 1975. A Note on Thermoelastic Strains and Tilts. *J. geophys. Res.* **80**: 274 - 277.

Berger, J., Beaumont, C. 1975. An Analysis of Tidal Strain Observations from the USA. II. The Inhomogeneous Tide. *Bull. seismol. Soc. Am.* **66**: 1821 - 1846.

Bonatz, M., Melchior, P., Ducarme, B. 1971. Station: Longyearbyen (Spitsbergen) - Mésures faites dans les trois composantes avec six pendules horizontaux VM et trois gravimètres Askania. *Bull. Obs. R. Belg.* **4**: 1 - 110.

Bonatz, M., Schüller, K. 1976. Gravimetrische Erdgezeiten-Station Kerguelen - Parameter der Partialtiden für den Beobachtungszeitraum 1973/74. *Deutsche Geod. Komm.* **B 218**: 1 - 26.

Cabaniss, G. H. 1978. The measurement of long period and secular deformation with deep borehole tiltmeters. In: *Applications of Geodesy to Geodynamics*. pp. 165 - 169. I. Mueller (ed.). Ohio State University Press, Columbus, Ohio, USA.

Dehant, V. 1991. Review of the Earth Tidal Models and Contribution of Earth Tides in Geodynamics. *J. geophys. Res.* **96**: 20235 - 20240.

Dehant, V. 1987. Integration of the Gravitational Motion Equations for an Elliptical Uniformly Rotating Earth with an Inelastic Mantle. *Phys. Earth planet. Inter.* **49**: 242 - 258.

Dehant, V. 1995. Theoretical tidal parameters: state of the art. *Bull. Inf. Marées Terrestres.* **121**: 9027 - 9029.

Dehant, V., Ducarme, B. 1987. Comparison between the Theoretical and Observed Tidal Gravimetric Factors. *Phys. Earth planet. Inter.* **49**: 192 - 212.

Dehant, V., Zschau, J. 1989. The Effect of Mantle Inelasticity on Tidal Gravity: a Comparison between the Spherical and the Elliptical Earth Model. *Geophys. J. Int.* **97**: 549 - 556.

Dittfeld, H.-J., Wenzel, H.-G. 1993. Joint Gravity Tide Recording at Potsdam. In: *Proc. IAG-Symp. 113: Geodesy Phys. Earth.* pp. 186 - 189. Springer, Berlin.

Dziewonski, A. M., Anderson, D. L. 1981. Preliminary reference Earth model. *Phys. Earth planet. Inter.* **25**: 297 - 357.

Edge, R. J., Baker, T. F., Jeffries, G. 1981. Borehole Tilt Measurements: Aperiodic Crustal Tilt in an Aseismic Area. *Tectonophysics* **71**: 97 - 109.

Emter, D., Zürn, W., Mälzer, H. 1989. Underground Measurements at Tidal Sensitivity with a Long Baseline Differential Fluid Pressure. *Deutsche Geod. Komm.* **B 288**: 1 - 74.

Francis, O., Melchior, P. 1996. Tidal Loading in Western Europe: a Test Area. *Geophys. Res. Lett.* **23**: 2251 - 2254.

Friederich, W., Wilhelm, H. 1986. Solar Radiational Effects on Earth Tide Measurements. In: *Proc. 10th Int. Symp. Earth Tides.* pp. 865 - 880. R. Vieira (ed.). Cons. Sup. Invest. Cient., Madrid.

Gerstenecker, C., Zschau, J., Bonatz, M., 1986. Finite Element Modelling of the Hunsrück Tilt Anomalies - A Model Comparison. In: *Proc. 10th Int. Symp. Earth Tides.* pp. 797 - 804. R. Vieira (ed.). Cons. Sup. Invest. Cient., Madrid.

Goodkind, J. M. 1991. The Superconducting Gravimeters: Principle of Operation, Current Performance and Future Prospects. *Cah. Centre Europ. Geodyn. Seism.* **3**: 81 - 90.

Goodkind, J. M., Czipott, P., Mills, A., Murakami, M., Platzman, P., Young, C., Zuckerman, D. 1993. Test of the gravitational inverse-square law at 0.4 m to 1.4 m mass separation. *Phys. Rev. D.* **47**: 1290 - 1297

Harrison, J. C. 1976. Cavity and Topographic Effects in Tilt and Strain Measurements. *J. geophys. Res.* **81**: 319 - 328.

Harrison, J. C. (ed.) 1985. Earth Tides. In: *Benchmark Papers in Geology Series.* pp. 1 - 419. Van Nostrand Reinhold, New York.

Hart, R. H. G., Gladwin, M. T., Gwyther, R. L., Agnew, D. C., Wyatt, F. K. 1996. Tidal calibration of borehole strainmeters: Removing the effects of small-scale inhomogeneity. *J. geophys. Res.* **101**: 25553 - 25571.

Jahr, T., Jentzsch, G., Andersen, N., Remmer, O. 1991. Ocean tidal loading on the shelf areas around Denmark. In: *Proc. 11th Int. Symp. Earth Tides..* pp. 309 -319. J. Kakkuri (ed.). Schweizerbart, Stuttgart.

Jentzsch, G., Ramatschi, M., Madsen, F. 1995. Tidal gravity measurements on Greenland. *Bull. Inf. Marées Terrestres.* **122**: 9239 - 9248.

Johnson, H. O., Wyatt, F., Agnew, D.C., Zürn, W. 1994. Tidal Tilts at Pinyon Flat, California Measured at Depths of 24 and 120 Meters. In: *Proc. 12th Int. Symp. Earth Tides.* pp. 129 - 136. H.-T. Hsu (ed.). Science Press, Beijing.

King, G. C. P., Bilham, R. 1973. Tidal Tilt Measurements in Europe. *Nature* **243**: 74 - 75.

King, G. C. P., Bilham, R. 1976. A Geophysical Wire Strainmeter. *Bull. seism. Soc. Am.* **66**: 2039 - 2047.

King, G. C. P., Zürn, W., Evans, R., Emter, D. 1976. Site Correction for Long Period Seismometers, Tiltmeters and Strainmeters. *Geophys. J. R. astr. Soc.* **44**: 405 - 411.

Knopoff, L., Rydelek, P. A., Zürn, W., Agnew, D. C. 1989. Observations of Load Tides at the South Pole. *Phys. Earth Planet. Inter.* **54**: 33 - 37.

Kohl, M., Levine, J. 1992. Using a Short Baseline Borehole Tiltmeter to Measure the Regional Tilt. *EOS (Trans. Am. geophys. Un.)* **73**: Suppl. 121.

Kohl, M., Levine, J. 1995. Measurement and Interpretation of Tidal Tilts in Southern California. *J. geophys. Res.* **100**: 3929 - 3942.

Latynina, L. A., Rizaeva, S. D. 1976. On Tidal-Strain Variations Before Earthquakes. *Tectonophysics* **31**: 121 - 127.

Llubes, M., Mazzega, P. 1996. The ocean tide gravimetric loading reconsidered. *Geophys. Res. Lett.* **23**: 1481 - 1484.

Melchior, P. 1981. The tides of the planet earth. pp. 1 - 609. Pergamon Press, Oxford.

Melchior, P. 1994. A New Data Bank for Tidal Gravity Measurements (DB 92). *Phys. Earth planet. Inter.* **82**: 125 - 155.

Melchior, P. 1995a. A continuing discussion about the correlation of tidal gravity anomalies and heat flow densities. *Phys. Earth planet. Inter.* **88**: 223 - 256.

Melchior, P. 1995b. The Trends of Earth Tides Research. In: *Proc. 12th Int. Symp. Earth Tides.* pp. 29 - 37. H.-T. Hsu (ed.). Science Press, Beijing.

Meertens, C., Levine, J., Busby, R. 1989. Tilt Observations Using Borehole Tiltmeters: 2. Analysis of Data From Yellowstone National Park. *J. geophys. Res.* **94**: 587 - 601.

Molodenskii, S. M., Kramer, M. V. 1980. The Influence of Large-scale Horizontal Inhomogeneities in the Mantle on Earth Tides. *Bull. Acad. Sci. U.S.S.R., Earth Phys.* **16**: 1 - 11.

Müller, G. 1977. Thermoelastic Deformations of a Half-Space - a Green's Function Approach. *J. Geophys.* **43**: 761 - 770.

Peters, J., Beaumont, C. 1987. Tidal and Secular Tilt from an Earthquake Zone: Thresholds for Detection of Regional Anomalies. *Earth Planet. Sci. Lett.* **84**: 263 - 276.

Peters, J., Kümpel, H.-J. 1988. Non-linear tilt tides from the Charlevoix seismic zone in Quebec. *J. Geophys.* **62**: 128 - 135.

Prothero, W. A. Jr., Goodkind, J. M. 1968. A Superconducting Gravimeter. *Rev. Scient. Instr.* **39**: 1257 - 1262.

Richter, B. 1987. Das supraleitende Gravimeter. *Deutsche Geod. Komm. C* **239**: 1 - 124.

Richter, B. 1991. Calibration of Superconducting Gravimeters. *Cah. Centre Europ. Geodyn. Seism.* **3**: 99 - 107.

Richter, B., Wenzel, H.-G., Zürn, W., Klopping, F. 1995. From Chandler wobble to free oscillations: comparison of cryogenic gravimeters and other instruments in a wide period range. *Phys. Earth planet. Inter.* **91**: 131 - 148.

Richter, B., Wilmes, H., Nowak, I. 1995. The Frankfurt calibration system for relative gravimeters. *Metrologia.* **32**: 217 - 223.

Rydelek, P. A., Knopoff, L. 1982. Long Period Lunar Tides at the South Pole. *J. geophys. Res.* **87**: 3969 - 3973.

Rydelek, P. A., Zürn, W., Hinderer, J. 1991. On Tidal Gravity, Heat Flow and Lateral Heterogeneities. *Phys. Earth planet. Inter.* **68**: 215 - 229.

Sato, T., Harrison, J. C. 1990. Local Effects on Tidal Strain Measurements at Esashi, Japan. *Geophys. J. Int.* **102**: 513 - 526.

Schneider, M. M. 1971. Erste Beobachtungen der Schweregezeiten in der zentralen Antarktis. *Gerl. Beitr. Geoph.* **80**: 491 - 496.

Schwiderski, E. W. 1980. Ocean tides, Part I. global ocean tidal equations; Part II: a hydrodynamical interpolation model. *Mar. Geodesy.* **3**: 161 - 255.

Shibuya, K., Ogawa, F. 1993. Observation and Analysis of the Tidal Gravity Variation at Asuka Station on the Antarctic Ice Sheet. *J. geophys. Res.* **98**: 6677 - 6688.

Timmen, L., Wenzel, H.-G. 1994. Improved Gravimetric Earth Tide Parameters for Station Hannover. *Bull. Inf. Marees Terrestres* **119**: 8834 - 8846.

Torge, W. 1989. *Gravimetry.* pp. 1 - 465. De Gruyter, Berlin.

Varga, P. 1974. Dependence of the Love Numbers upon the Inner Structure of the Earth and Comparison of Theoretical Models with Results of Measurements. *Pure Appl. Geophys.* **112**: 777 - 785.

Varga, P., Denis, C. 1988. A Study of the Variation of Tidal Numbers with Earth Structure. *Geophys. Trans.* **34**: 263 - 282.

Varga, P., Hajosy, A., Csapo, G. 1995. Laboratory Calibration of LaCoste Romberg Type Gravimeters by Using a Heavy Cylindrical Ring. *Geophys. J. Int.* **120**: 745 - 757.

Van Ruymbeke, M. 1989. A Calibration System for Gravimeters Using a Sinusoidal Acceleration Resulting from a Vertical Periodic Movement. *Bull. Geod.* **63**: 223 - 235.

Wahr, J. M. 1981. Body Tides on an Elliptical, Rotating, Elastic and Oceanless Earth. *Geophys. J. R. astr. Soc.* **64**: 677 - 703.

Wahr, J. M., Bergen, Z. 1986. The Effects of Mantle Anelasticity on Nutations, Earth Tides, and Tidal Variation in Rotation Rate. *Geophys. J. R. atr. Soc.* **87**: 633 - 668.

Wang, R. 1991. Tidal Deformations on a Rotating, Spherically Asymmetric, Viscoelastic and Laterally Heterogeneous Earth. pp. 1 - 139. Peter Lang, Frankfurt.

Wang, R. 1994. Effect of Rotation and Ellipticity on Earth Tides. *Geophys. J. Int.* **117**: 562 - 565.

Warburton, R. J., Brinton, E. W. 1995. Recent developments in GWR Instruments' superconducting gravimeters. *Cah. Centre Europ. Geodyn. Seism.* **11**: 23 - 56.

Warburton, R. J., Goodkind, J. M., 1976. Search for Evidence of a Preferred Reference Frame. *Astrophys. J.* **208**: 881 - 886.

Weise, A., Jentzsch, G., Kiviniemi, A., Kääriäinen, J., Ruotsalainen, H. 1995. Tilt measurements in geodynamics - results from the 3-component-station Metsähovi and the clinometric station Lohja / Finland. In: *Proc. 12th Int. Symp. Earth Tides.* pp. 105 - 104. H.-T. Hsu (ed.). Science Press, Beijing.

Wenzel, H.-G., Zürn, W., Baker, T. F. 1991. In-Situ Calibration of LaCoste-Romberg Earth Tide Gravity Meter. *Bull. Inf. Marées Terrestres* **109**: 7849 - 7863.

Westerhaus, M., Welle, W., Büyükköse, N., Zschau, J. 1991. Temporal Variations of Crustal Properties in the Mudurnu Valley, Turkey: an Indication for Regional Effects of Local Asperities. In: *Proc. Int. Conf. Earthquake Predict..* pp. 272 - 281. Strasbourg.

Wilhelm, H. 1978. Upper Mantle Structure and Global Earth Tides. *J. Geophys.* **44**: 435 - 440.

Wilhelm, H., Zürn, W. 1984. Tidal forcing field. *Landolt-Börnstein, Neue Serie V/2a* **2.5.1**: 259 - 279. H. Soffel, K. Fuchs (eds.), Springer, Berlin.

Wyatt, F., Agnew, D.C., Johnson, H. O. 1990. Pinyon Flat Observatory: Comparative Studies and Geophysical Investigations. *Open-file report* **90-380**: 242 - 248. U. S. Geol. Survey.

Wyatt, F., Berger, J. 1980. Investigations of Tilt Measurements Using Shallow Borehole Tiltmeters. *J. geophys. Res.* **85**: 4351 - 4362.

Wyatt, F., Cabaniss, G., Agnew, D. C. 1982. A Comparison of Tiltmeters at Tidal Frequencies. *Geophys. Res. Lett.* **9**: 743 - 746.

Wyatt, F., Levine, J., Agnew, D. C., Zürn, W. 1987. Side - by - Side Tidal Tilt Measurements - Some Disturbing Results. *EOS (Trans. Am. geophys. Un.)* **68**: 1247.

Wyatt, F. K., Morrissey, S.-T., Agnew, D. C. 1988. Shallow Borehole Tilt: A Reprise. *J. geophys. Res.* **93**: 9197 - 9201.

Yanshin, A. L., Melchior, P., Keilis-Borok, V. I., de Becker, M., Ducarme, B., Sadovsky, A. M. 1986. Global distribution of tidal anomalies and an attempt of its geotectonic interpretation. In: *Proc. 10th Int. Symp. Earth Tides.* pp. 731 - 755. R. Vieira (ed.). Cons. Sup. Invest. Cient., Madrid.

Zürn, W., Beaumont, C., Slichter, L. B. 1976. Gravity Tides and Ocean Loading in Southern Alaska. *J. geophys. Res.* **81**: 4923 - 4932.

Zürn, W., Emter, D., Heil, E., Neuberg, J., Grüninger, W. 1986. Comparison of Short- and Long-Baseline Tidal Tilts. In: *Proc. 10th Int. Symp. Earth Tides.* pp. 61 - 69. R. Vieira (ed.). Cons. Sup. Invest. Cient., Madrid.

Zürn, W., Wenzel, H.-G., Laske, G. 1991. High Quality Data from LaCoste-Romberg Gravimeters with Electrostatic Feedback: A Challenge for Superconducting Gravimeters. *Bull. Inf. Marées Terrestres* **110**: 7940 - 7952.

Zürn, W., Wilhelm, H. 1984. Tides of the solid earth. *Landolt-Börnstein, Neue Serie V/2a* **2.5.2.**: 280 - 299. H. Soffel, K. Fuchs (eds.), Springer, Berlin.

The Nearly-Diurnal Free Wobble-Resonance

Walter Zürn

Black Forest Observatory, Heubach 206, D-77709 Wolfach, Germany

Abstract. The theoretical background and observational results for the nearly-diurnal free wobble, one of the rotational eigenmodes of the earth, are presented. Especially the evidence for a shift in eigenfrequency with respect to the value computed for an earth in hydrostatic equilibrium is presented and possible implications are discussed.

1 Introduction

The spatial variation of gravimetric factors for earth tides at present does with high probability not contribute to our knowledge about the internal structure of the earth beyond the knowledge acquired by seismological tools. For different reasons, the global tidal response cannot be extracted with sufficient accuracy from tilt and strain tide observations either (see Zürn "Earth Tide Observations", this volume). In contrast, there is evidence that the variation of the observed response with frequency in the diurnal tidal band tells us something about the interior of our planet which seismology cannot provide. This variation with frequency is caused by resonant behaviour of the earth in the vicinity of 1 cycle/day due to the existence of a normal mode of the rotating earth. Fortunately this resonance is sensitive to the tidal forcing for diurnal tides and the properties of the resonance are such, that its eigenfrequency is located within this tidal band. In the literature this mode is called the 'Nearly Diurnal Free Wobble' (NDFW), the 'Free Core Nutation' (FCN) and the 'Core Resonance', among other names (see Melchior (1980) and Chao (1985), who discuss the rather confusing terminology in the literature from different points of view). It is excited, when the rotation axes of mantle and outer core are slightly misaligned. In that case restoring forces are set up at the core-mantle boundary (CMB), which try to realign the two axes. The major restoring torques by inertial coupling are due to CMB ellipticity. Because the earth is a fast-spinning gyro, the reaction is a damped wobble of the instantaneous rotation axis around the figure axis and a nutation in space of the rotation axis around the axis of total angular momentum. Both motions are two aspects of the same mode (Lambeck, 1988). The possible existence of this oscillation was recognized very early theoretically by Hopkins (1839), Hough (1895) and Sludskii (1896). This mode is described by its angular eigenfrequency σ_{NDFW} or the corresponding period $T_{FCN} = \frac{2 \cdot \pi}{\sigma_{FCN}}$ of the associated free core nutation, and its quality factor Q_{NDFW}. The (angular) eigenfrequencies of the wobble and the nutation are related by the equation:

$$\frac{\sigma_{NDFW}}{2 \cdot \pi} = \nu_{NDFW} = \frac{1}{T_{FCN}} - \frac{\Omega}{2 \cdot \pi} \tag{1}$$

where Ω is the mean angular velocity of the Earth and both σ_{NDFW} and σ_{FCN} are negative quantities indicating retrograde motions.

2   Theoretical Background

The theoretical treatment of the NDFW has a long history and there are many research papers and textbooks dealing with this oscillation (e. g. Munk and MacDonald, 1960; Rochester et al., 1974; Toomre, 1974; Hinderer et al. 1982; Smith, 1977; Sasao et al., 1977, 1980; Wahr, 1981; Lambeck, 1980, 1988), therefore this subject is treated only very briefly here.

For a rigid body Euler's equation (Munk and MacDonald, 1960) describes the rate of change (dot above a symbol represents time derivative) of the total angular momentum $\mathbf{H}$ due to the total external torque $\mathbf{L}$ in inertial space (bold symbols denote vectors and tensors):

$$\dot{\mathbf{H}} = \mathbf{L} \tag{2}$$

In a reference frame rotating with angular velocity ω this equation is modified:

$$\dot{\mathbf{H}} + \omega \times \mathbf{H} = \mathbf{L} \tag{3}$$

where $\times$ denotes the vector cross product. The angular momentum can be written as:

$$\mathbf{H} = \mathbf{I} \cdot \omega \tag{4}$$

where $\mathbf{I}$ is the inertia tensor of the rigid body. If the body is deformable, the last equation must be modified, by adding the angular momentum h due to the deformation:

$$\mathbf{H} = \mathbf{I} \cdot \omega + \mathbf{h} \tag{5}$$

If this is introduced into the Euler equation, we obtain the Liouville equations of motion of the deformable body:

$$\frac{d}{dt}[\mathbf{I} \cdot \omega + \mathbf{h}] + \omega \times [\mathbf{I} \cdot \omega + \mathbf{h}] = \mathbf{L} \tag{6}$$

For a simple earth model consisting of the mantle and the fluid core with a spheroidal CMB we can write the Liouville equations separately for these two parts, but then we must add to the equations coupling torques $\mathbf{N}$ which the mantle and core exert on each other. These torques cancel when the earth as a whole is considered (superscripts m and c refer to the whole earth and the core, respectively):

$$\dot{\mathbf{H}}^c + \omega \times \mathbf{H}^c = \mathbf{L}^c + \mathbf{N} \tag{7}$$

and

$$\dot{\mathbf{H}}^m + \omega \times \mathbf{H}^m = \mathbf{L}^m - \mathbf{N} \tag{8}$$

where $\mathbf{L} = \mathbf{L}^m + \mathbf{L}^c$ is the external torque as before and $\mathbf{N}$ is the sum of all the coupling torques acting between mantle and core. The most important torque arises due to pressure on the CMB by the fluid of the core, when the fluid core rotates around an axis which is not aligned with the axis of symmetry of the ellipsoidal CMB (normally it is assumed that this axis coincides with the figure axis of the mantle). Toomre (1974, p. 337) describes an illustrative analogy for this torque with a marble rolling frictionless inside a spheroidal container along the equator. When the container is tilted with respect to the orbital plane, this plane will regress in space. This torque would vanish for a spherical CMB and is commonly called pressure or inertial coupling. It tries to restore rotation of the core around this axis of symmetry. This coupling was studied very early by Hopkins (1839), Hough (1895), Sludskii (1896) and Poincaré (1910). However, $\mathbf{N}$ will also comprise topographic, electromagnetic, viscous and gravitational coupling terms.

Hinderer et al. (1982) treated this problem again and included elasticity of the mantle (the instantaneous change of CMB ellipticity due to the pressure field must be taken into account in that case) and visco-magnetic coupling. The equations are linearized, only small deviations from the rigid rotation with angular velocity Ω around the figure axis of the axially symmetric earth are allowed. These authors obtain the following differential equations for the equatorial components of the angular velocity of the whole earth and the core using a complex representation of the two components of the rotation vector in the equatorial plane:

$$\dot\omega(1+\alpha\frac{k}{k_s})-i\Omega\omega\alpha(1-\frac{k}{k_s})+(\dot\omega^c+i\Omega\omega^c)(\frac{A^c}{A}-\frac{\alpha k_1}{k_s}) = \frac{3\alpha k}{k_s a^2}(\frac{\dot W}{\Omega}+iW)-\frac{3i\alpha W}{a^2}(9)$$

and

$$\dot\omega(1+\frac{q_o h^c}{2})+\dot\omega^c(1-\frac{q_o h_1^c}{2})+i\Omega\omega^c(1+\alpha^c+K'-iK) = \frac{3q_o h^c}{2a^2}\frac{\dot W}{\Omega} \qquad (10)$$

where $i = +\sqrt{-1}$, $\omega = \omega_1 + i\omega_2$, $\omega^c = \omega_1^c + i\omega_2^c$. Subscript 1 refers to an axis towards the meridian of Greenwich, subscript 2 to an axis $90°$ east of it. $W = W_{21} + i\tilde{W}_{21}$ represents the proper components of the tesseral part of the tidal potential of degree 2 and order 1. The quantities k and h represent various Love numbers for potential perturbations and displacements as defined in Neuberg et al. (1987). A, C and $\alpha = (C - A)/A$ are the Earth's mean equatorial and polar moment of inertia and dynamic ellipticity, A^c, C^c and $\alpha^c = (C^c - A^c)/A^c$ the corresponding quantities for the core (the inner core is not solid in this model), Ω is the mean axial angular velocity. K and K' are dimensionless viscomagnetic coupling constants defined by Hinderer et al. (1982). Dickey (1995) reports the following values for the major moments of inertia of the earth: $A^m = 7.0999 \cdot 10^{37}$ $kg.m^2$, $A^c = 0.9117 \cdot 10^{37}$ $kg.m^2$, $C^m - A^m = 0.02377 \cdot 10^{37}$ $kg.m^2$ and $C^c - A^c = 0.002328 \cdot 10^{37}$ $kg.m^2$. $q_o = \Omega^2 a/g$ is the ratio of centrifugal force to gravity at the surface (equator). Due to the tidal deformation by the tesseral tidal forces the inertia tensor changes with time and the ellipticity of the CMB is also modified, these changes are incorporated in the above equations with the help of the Love number formalism.

The eigenfrequencies of this model can be found by setting W = 0, i. e. no external torques are applied to the planet. There are two solutions of the system of equations:

$$\sigma_{CW} = \frac{A}{A^m}\alpha(1 - \frac{k}{k_s})\Omega \tag{11}$$

and

$$\sigma_{NDFW} = -\Omega[1 + \frac{A}{A^m}(\alpha^c - q_o\frac{h_1^c}{2} + K' - iK)] \tag{12}$$

The first of these solutions represents the well known Chandler wobble for this simplistic earth model, which is treated in detail in the article by Plag ("Chandler Wobble", this volume). The observed eigenfrequency for the Chandler wobble is appreciably modified relative to this formula due to effects from the oceans and mantle anelasticity (Smith and Dahlen, 1981). The second is the nearly diurnal free wobble: all terms after the leading 1 are smaller by two orders of magnitude. The minus sign indicates retrograde motion of the instantaneous rotation axis and the figure axis around each other. The deviation from the diurnal frequency is controlled by the dynamic ellipticity α^c of the core. The term with the "Love" number h_1^c describes the change in the dynamic ellipticity caused by the instantaneous elastic deformation of the CMB due to the wobble, which was estimated by Sasao et al. (1980) to be about 25 % of α^c for seismologically constrained earth models. Note that for an inelastic mantle h_1^c is a complex quantity. The magnitudes of the visco-magnetic contributions described by K and K' and any other effects not accounted for in the above equations are subject to severe uncertainties due to present lack of knowledge.

The associated nutation of the instantaneous rotation axis in space is also retrograde and its period T_{FCN} is given by eq. (1). The motion as a whole can be pictured by a body-fixed cone rolling without slipping inside a space-fixed cone, this is the Poinsot-representation (Fig. 1; e. g. Rochester et al. 1974, Toomre 1974, Lambeck 1988) of this motion.

The symmetry axes of the body-fixed and space-fixed cones are the figure axis of the earth and the angular momentum vector, respectively. These two directions and the vector of instantaneous rotation remain in one plane in space, which revolves slowly around the angular momentum vector, which for zero external torque would stay fixed in inertial space. The sense of this revolution is retrograde, i. e. opposite to the earth' rotation. The geometrical conditions above are equivalent to the angular amplitude relationship between wobble and nutation given by:

$$\frac{\alpha_{NDFW}}{\alpha_{FCN}} = \frac{A\alpha^c}{A^m} \tag{13}$$

i. e. the amplitude of the nutation is about 400 times larger than the amplitude of the wobble. This point was stressed by Toomre (1974) and Rochester et al. (1974)

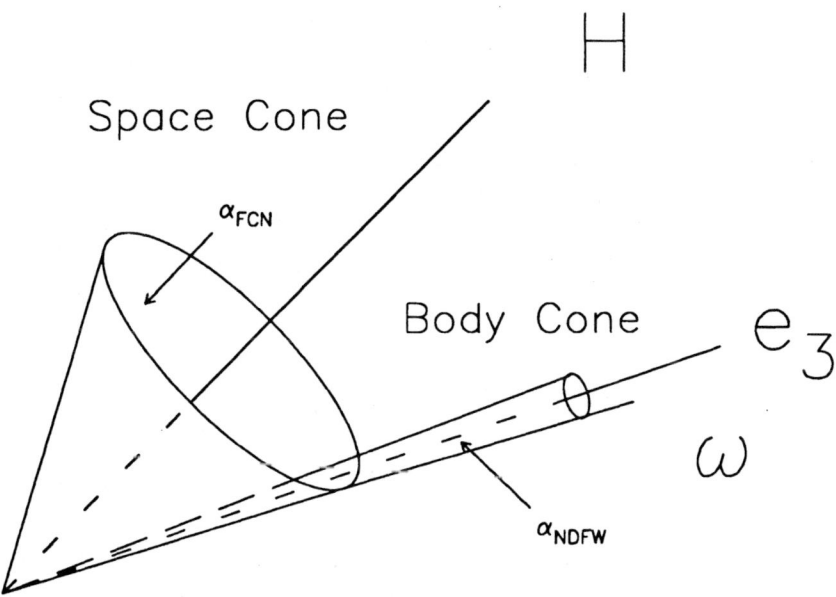

Fig. 1. Poinsot-representation of the NDFW/FCN rotational free mode of the earth. e_3, **H** and ω are the figure axis, the total angular momentum and rotation vectors, respectively. All three vectors remain in the same plane which rotates slowly around **H** in a retrograde sense, i. e. opposite to the earth's rotation.

in reviews of the attempts to observe the wobble. Note that the wobble amplitude α_{NDFW} in Fig. 1 is highly exaggerated. The instantaneous rotation vector ω outlines one circle (the herpolhode) on the invariable plane (perpendicular to the total angular momentum vector **H**) and another circle (the polhode) on the Poinsot- or energy-ellipsoid (defined by the surface of constant energy in ω-space). These two circles are the traces of the tip of the instantaneous rotation vector on the two cones of the Poinsot-representation and represent intersections of the body and space cone with the Poinsot-ellipsoid and the invariable plane, respectively.

With the equations above one also obtains the rotational response of the earth's mantle and core to the tesseral tidal forcing $W = W_0 \cdot exp(i\sigma t)$ at angular frequency σ. The equatorial component of rotation of the whole earth becomes $\omega = \omega_0 \cdot exp(i\sigma t)$:

$$\omega_0 = \frac{A[(\sigma + \Omega)(1 - A^c q_0 h^c/(2\alpha A)) + \Omega(\alpha^c - q_0 h_1^c/2)]3\alpha W_0}{A^m(\sigma - \sigma_{NDFW})\Omega a^2} \tag{14}$$

and for the core $\omega^c = \omega_0^c \cdot exp(i\sigma t)$:

$$\omega_0^c = \frac{A(1 - q_0 h^c/(2\alpha))3\alpha W_0}{A^m(\sigma - \sigma_{NDFW})a^2} \tag{15}$$

The response of the earth in forced motion becomes resonant in the diurnal tidal frequency band because of the presence of this eigenmode. The response of the earth to the tidal forcing in this band in the vicinity of σ_{NDFW} should therefore show a strong frequency dependence. For the special case of tidal gravity observations, Hinderer et al. (1991) derive the following equation for the gravimetric factor δ (the magnitude of the complex admittance of the earth with respect to the tesseral tidal forces) as a function of angular frequency σ:

$$\delta = (1 + h - 3/2k) + \frac{A}{A^m} \cdot \Omega \cdot \frac{(h_1 - 3/2k_1)(\alpha - q_0 h^c/2)}{\sigma - \sigma_{NDFW}} \tag{16}$$

h, k are the Love numbers (e. g. Wang "Solid Earth Response", this volume), which by the first term define the response far away from the resonance frequency. The second term represents the contribution from the NDFW to the response. If ω_{NDFW} would be a real quantity, i. e. if the mode would not be damped by some physical mechanism, this second term and the response would become infinite at $\sigma = \sigma_{NDFW}$. Clearly, there must be such mechanisms and this can be accounted for in the above equations by making σ_{NDFW} a complex quantity. Note that this definition of δ neglects the changes in centrifugal forces accompanying the tides due to the associated forced nutations as described explicitly by Schwahn (1995) and taken into account by Wahr (1981) and Mathews et al. (1995) among others. This effect gives rise to additional forces, but also modifies h and k in the above definition. Similar equations hold for all observable physical quantities influenced by the reaction of the earth to tidal forces. Neuberg and Zürn (1986) showed that this statement is valid even in the presence of local elastic effects (see Westerhaus " Tidal Tilt Modification", this volume). Wahr (1981) gave a different, but equivalent formulation for the frequency dependence of the tidal observables (including the gravimetric factor and Love numbers) in the diurnal band, for example:

$$\delta = \delta_{O_1} + \delta_1 \cdot \frac{\sigma - \sigma_{O_1}}{\sigma_{NDFW} - \sigma} \tag{17}$$

(note that δ_1 is a negative quantity). In this formulation the tide O_1 is taken as a reference, because it is far enough away from the resonance in frequency. Goodkind (1983) showed that the resonance term for a complex eigenfrequency (i. e. finite damping) as a complex function of frequency describes a circle in the complex plane with the real driving frequency as a parameter. The origin then represents the contribution to the response from the resonance at the frequency of the O_1 tide. The resonance term is also equivalent to the response of the classical harmonic oscillator (e. g. Goodkind, 1983; Neuberg et al., 1987).

For hydrostatic flattening of the CMB the period of the nutation in space T_{FCN} was computed for seismologically constrained earth models for the first time by Jeffreys and Vicente (1957) and Molodenskii (1961). Sasao et al. (1980) and Wahr (1981) calculated T_{FCN} again for much improved earth models taking the instantaneous deformation of the CMB into account and obtained 467 and 460 sidereal days, respectively, for their best constrained models. Anelasticity of the mantle results in relaxation of the elastic modulii which in turn lengthens the period (due to an increased instantaneously produced change in CMB ellipticity) by a few days (Wahr and Bergen, 1986; Dehant, 1988). Wahr and de Vries (1989, 1990) considered the possibility of nonhydrostatic structure (i. e. density) in the outer core and its effect on α^c. Such structure could arise from two sources: CMB topography on one hand and lateral density variations in the lower mantle on the other hand. Wu and Wahr (1997) estimated numerically the effects of non-hydrostatic CMB-topography and of the resonances in the outer core on earth's rotation, using a model with a rigid mantle and a homogeneous and incompressible fluid core. They conclude that the resonance strength for the FCN can possibly be affected to observable levels by some non-elliptical components of CMB-topography,, while it is very unlikely, that the sharp resonances in the outer core have an effect on the observable tidal and nutation amplitudes, unless their frequencies are extremely close to ω_{NDFW}. Both properties of the earth, non-hydrostatic CMB-topography and core-mode frequencies are not well known at present.

The existence of the solid inner core leads to two additional eigenmodes, an inner core wobble with eigenfrequency far outside the diurnal band and a prograde free inner core nutation with its associated wobble. The latter has its eigenfrequency also in the diurnal tidal band. This was theoretically treated by De Vries and Wahr (1991), Mathews et al. (1991a, b) and Dehant et al. (1993) and found to have very little influence on the earth tide response and a small, but possibly observable influence on the nutation amplitudes.

3 Observations

The seismic-gravitational free oscillations of the earth are excited by large earthquakes and decay afterwards with their proper damping until the next large earthquake strikes. Seismic observations can in principle (difficulties arise from the density of the mode spectrum) easily be used to extract the properties of these modes, because the spectrum of the excitation is very simple (Masters and Widmer, 1995). For the Chandler wobble the situation is more complicated: the excitation mechanisms are not completely understood and therefore the spectrum of the excitation is not very well known either. The determination of the mode properties is therefore difficult, this is especially the case for the damping constant (see Plag "Chandler Wobble", this volume). Obviously the excitation possesses a reasonably smooth (but not necessarily constant) spectrum across the resonance such that this eigenmode can clearly be seen in the observations. For the NDFW-FCN eigenmode we have two fundamentally different observation methods. First one can try a direct observation of the mode in nutation or wobble data (just like in the case of the Chandler wobble) with unknown and obviously much weaker excitation processes. Secondly an indirect method is provided by the periodic tidal forcing with frequencies very close to the resonance, in this case the excitation spectrum has no energy at all at the eigenfrequency.

In the textbook by Munk and MacDonald (1960) the observational differences between the two aspects of the FCN-NDFW mode are explained: wobble motion is obtained by measuring variations of latitude, nutations in turn cause variations in the declinations of the stars. As Rochester et al. (1974) and Toomre (1974) point out, the observation of the FCN-NDFW eigenmode should be easier in form of the nutation because the wobble amplitude is more than 400 times smaller. Earlier claims for direct observation of the mode are discussed in these two papers. Unambiguous direct observation of the FCN with an amplitude of 174 μas (0.84 nrads) was achieved by Herring and Dong (1994) in an analysis of eight years of very long baseline interferometry (VLBI) data. These authors indicate, that there is more than enough power in the P_{21} atmospheric pressure field to excite the mode to this amplitude level. Jiang and Smylie (1995) also claim the detection of a retrograde nutation with a period of 431 ± 20 solar days in the nutation data obtained with VLBI.

The detection of the resonance in periodically forced motions of the earth has a much longer history. In many earth tide measurements (for separation of the $P_1 - S_1 - K_1$ group of constituents a record length of at least one year is necessary) a strong deviation of the responses of K_1 and ψ_1 from the static response at O_1 was conspicuous (Melchior, 1981). Some of the best early results were obtained by Blum et al. (1973) in tilt, by Abours and Lecolazet (1979) in gravity and by Levine (1978) in strain. Fig. 2 shows a more recent example of gravity tide observations compared to a model of the resonant behaviour.

The first attempts to extract information about the earth were made by Lecolazet (1983), who correlated temporal changes of diurnal tidal gravimetric factors with changes in the length-of-day, and Goodkind (1983), who determined a Q_{NDFW} of about 800 using data from his superconducting gravimeter record. While the apparent correlation found by the former author disappeared with new data, the work by the latter stimulated the earth tide community to study the structure of the resonance in the improved data becoming available in the following years, mostly due to the availability of superconducting gravimeters (e. g. Zürn et al., 1991; Richter et al., 1995).

Measurements of the amplitudes of the forced nutations of the earth using the VLBI - techniques led Gwinn et al. (1986) to suggest a change in the parameters of the FCN from the theoretical values in order to account for these anomalies. While the quality factor was reasonably high, the free period T_{FCN} had to be shifted down from the theoretical value for a hydrostatically prestressed earth by about 30 sidereal days to fit the observations. Recently Haas and Schuh (1996) determined the frequency dependence of the Love numbers in the diurnal

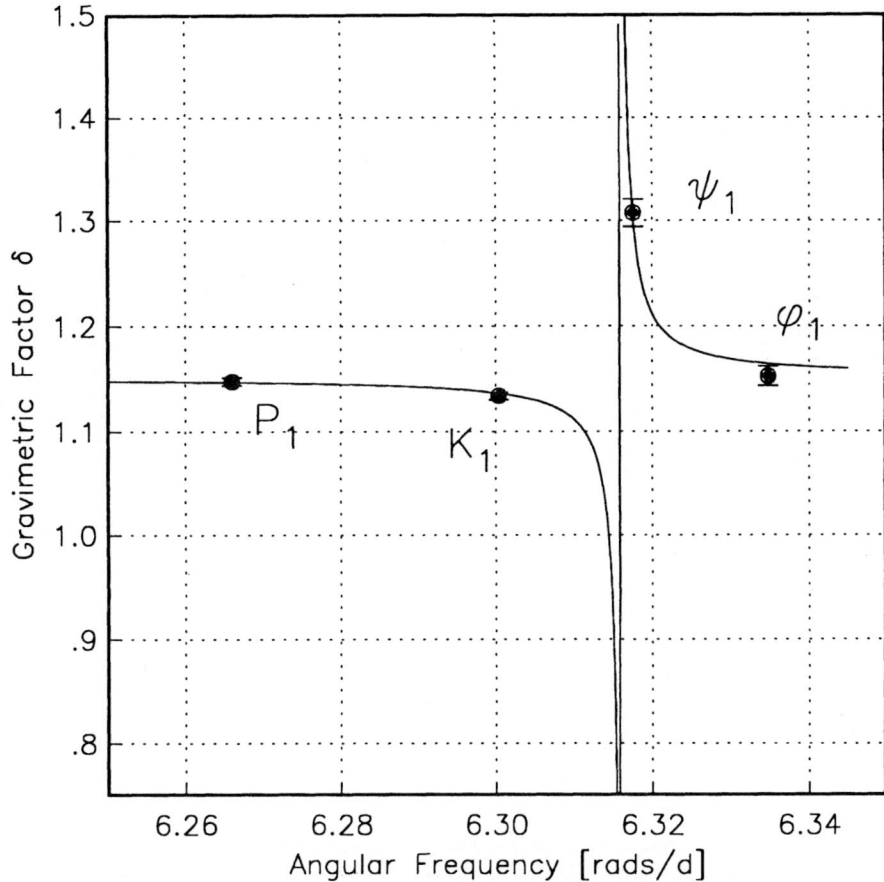

Fig. 2. Observed gravimetric factors δ for diurnal tides obtained from a digital spring-gravimeter record at BFO ($48.33°N, 8.33°E$, 587 m elevation). The record was 13 months long starting November 23, 1988. Error bars comprise a 0.3% calibration uncertainty and 3 standard deviations as given by the tidal analysis program ETERNA 3.20 (Wenzel, 1994). The smooth curve is the theoretical response for $\sigma_{NDFW} = 6.3160$ rads/d, $Q_{NDFW} = 10^6$ and resonance strength $-0.0005 + i \cdot 0.0002$.

frequency band directly from VLBI data and obtained for the period T_{FCN} 420 ± 20 sidereal days.

Independently of the VLBI observations Zürn et al. (1986), Neuberg et al. (1987) and Richter and Zürn (1988) found a very similar frequency shift when the resonance parameters were retrieved from gravimetric factors from high quality tidal gravity data recorded in central Europe. Fig. 3 shows the quality factors and free core nutation periods obtained by these authors and compares them to theoretical estimates and later observations. This is a nice example for mutual corroboration of unexpected results.

Levine et al. (1986) also noted in a gravity tide analysis from data recorded

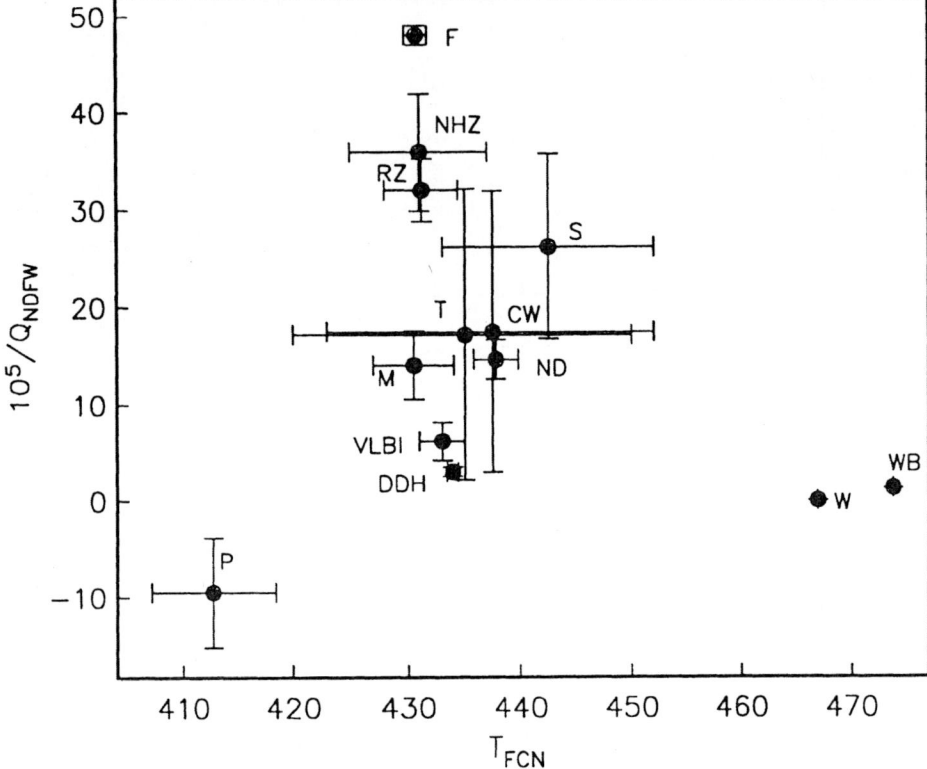

Fig. 3. Estimated parameters ($10^5/Q$ vs. T_{FCN} in sidereal days) of the NDFW/FCN eigenmode of the Earth from different authors: W - theory for hydrostatic, seismologically constrained earth models (Wahr, 1981), WB - effect of mantle anelasticity included (Wahr and Bergen, 1986); VLBI - interpretation of VLBI nutation measurements (Gwinn et al., 1986); results from gravity tide measurements (Neuberg et al., 1987 (NHZ), Richter and Zürn, 1988 (RZ), Sato et al., 1994 (T), Cummins and Wahr, 1993 (CW), Merriam, 1994 (M), Florsch et al. 1994 (F), Neumeyer and Dittfeld, 1997 (ND)); S - result from tidal strain measurements (Sato, 1991); stacks of different types of observations (Defraigne et al., 1994, 1995 (DDH), Polzer et al. 1996 (P)). Note that all results for eigenperiod differ significantly from the predictions. The uncertainty estimates differ methodically from one result to the other.

in the USA, that the resonance could be located at higher frequencies, without being as specific as the above authors. More corroboration for the frequency shift was provided by Sato (1991) studying strain tides in Japan, by Sato et al. (1993) through the study of gravity tides in Japan using records from three superconducting gravimeters, by Cummins and Wahr (1993) through a sparse global stack of data from the IDA - network (Agnew et al., 1986), by Merriam (1994) in a study of the data from the Canadian superconducting gravimeter, by Florsch et al. (1994) and Neumeyer and Dittfeld (1997) with analyses of data from the superconducting gravimeters near Strasbourg, France and at Potsdam, Germany, respectively. These results are also shown in Figure 3. The retrieved eigenperiods agree all nicely within their uncertainties, while the quality factors do not. However, there are several problems inherent in all these analyses. The uncertainties are certainly underestimated due to correlations between the parameters and due to systematic errors in the ocean loading corrections and the modeling of the gravity effects of the atmosphere. Especially the latter problem is an extreme complication, because the effects of the atmosphere are complex right at those tides which must be used for the resonance analysis (e. g. Merriam, 1994). Defraigne et al. (1994) stacked results from VLBI nutation and earth tide gravity measurements together and solved simultaneously for the NDFW-parameters. Their results support previous findings but with extremely small uncertainties. In an analysis of high quality gravity and strain tide data from the Schiltach observatory in southwestern Germany, Polzer et al. (1996) find T_{FCN} periods much lower than the rest of the tidal and VLBI researchers, about 410 sidereal days. Also, if not prohibited by the inversion method, their Q_{NDFW} adopts large negative values. While physics requests that Q_{NDFW} must be a positive quantity, it is clear that for the numerical results very large negative values are possible and need not be rejected for the physical reasons. They simply show that the damping is extremely low and cannot be determined with confidence from the tidal response (which was demonstrated by unpublished synthetic tests). Polzer et al. s (1996) analysis also shows that the effect of the uncertainties in the tidal admittances, basically due to the finite (albeit large) signal-to-noise ratio in the diurnal band has been underestimated in the previous analyses and dominates the error budget for the estimation of NDFW-parameters, at least for the results from earth tides, because of the high sensitivity of the low amplitude ψ_1 tide. For these reasons the discrepancies between the quality factors Q of the individual estimates should not be taken too seriously. A global (sparse) network of superconducting gravimeters, with some help of very few LaCoste-Romberg gravimeters with electrostatic feedback (Zürn et al., 1991), is desperately needed to solidify the present results on the eigenperiod, Q and resonance strengths of this free mode of the earth.

Gwinn et al. (1986) suggested that the significant shift of the eigenperiod to shorter values is caused by a decreased polar axis of the CMB by a about 1/2 kilometer. Several physical mechanisms to explain this shift and the observed Q value were discussed by Neuberg et al. (1990): their conclusion supports the conjecture of Gwinn et al. (1986). More discussion of possible physical causes for the frequency shift is provided by Yoder and Ivins (1988), Wahr and de Vries (1989), De Vries and Wahr (1991) and Mathews et al. (1991a,b). Lumb et al. (1995) and Forte et al. (1995) discuss the possibility that the excess ellipticity is dynamically supported by mantle convection. A decrease of the polar axis of the CMB between 250 and 350 meters suffices to explain the observation. Seismological methods can hardly reach this magnitude of resolution at the CMB, because the highest observable frequencies of body waves have wavelengths there at least an order of magnitude larger and the splitting of seismic free oscillations due to small ellipticity changes cannot easily be separated from other influences. From this point of view it is not surprising that a resonance whose existence is caused by CMB ellipticity also has the highest sensitivity to the shape of this internal boundary. The study of the NDFW, however, provides only information on the ellipticity of the CMB and not about CMB - topography in general (i. e. the P_2^0-term of a spherical harmonic expansion only). From Figure 3 it is obvious, that all modern results indicate an increased frequency of the NDFW. The presently accepted value is the one from the nutation measurements. Hinderer et al. (1991) verified that the observed real part of the resonance strength in tidal gravity is consistent with present seismological earth models. An interpretation of the Q factor has to await more precise estimates, the same is true for the imaginary part of the resonance strength.

Acknowledgements: The input of Jacques Hinderer, Strasbourg, and Helmut Wilhelm, Karlsruhe, over many years was very beneficial to my own work on the NDFW. Gudrun Polzer provided new results from her dissertation and made several helpful suggestions. The work was financially supported by the *Deutsche Forschungsgemeinschaft* through grants Wi 687/1 and We 1653/1. All this is gratefully acknowledged.

References

Abours, S., Lecolazet, R. 1979. New results about the dynamical effects of the liquid outer core as observed at Strasbourg. In: *Proc. 8th Int. Symp. Earth Tides.* pp. 689 - 697. Bonatz, M. and Melchior, P. (eds.), Bonn.

Agnew, D. C., Berger, J., Farrell, W. E., Gilbert, J. F., Masters, G., Miller., D. 1986. Project IDA: A Decade in Review. *EOS - Trans. Am. geophys. Un.* **67**: 203 - 211.

Blum, P. A., Hatzfeld, D., Wittlinger, G. 1973. Résultats expérimentaux sur la fréquence de résonance due à l'effet dynamique du noyau liquide. *C. R. Acad. Sci. Paris* **277**, Ser. B: 241 - 244.

Chao, B. F. 1985. As the World Turns. *EOS Trans. Am. Geophys. Union* **66**: 766 - 770.

Cummins, P. R., Wahr, J. M. 1993. A Study of the Earth's Free Core Nutation Using IDA Gravity Data. *J. geophys. Res.* **98**: 2091 - 2104.

Defraigne, P., Dehant, V., Hinderer, J. 1994. Stacking gravity tide measurements and nutation observations in order to determine the complex eigenfrequency of the nearly diurnal free wobble. *J. geophys. Res.* **99**: 9203 - 9213, and Correction **100**: 2041 - 2042.

Dehant, V. 1988. Nutations and inelasticity of the Earth. In: *The Earth's Rotation and Reference Frames for Geodesy and Geodynamics..* pp. 323 - 329. A. Babcock, G. Wilkins (eds.). Kluwer, Dordrecht, Holland.

Dehant, V., Hinderer, J., Legros, H., Lefftz, M. 1993. Analytical approach to the computation of the Earth, the outer core and the inner core rotational motions. *Phys. Earth planet. Inter.* **76**: 259 - 282.

De Vries, D., Wahr, J. M. 1991. The Effects of the Solid Inner Core and Nonhydrostatic Structure on the Earth's Forced Nutations and Earth Tides. *J. geophys. Res.* **96**: 8275 - 8293.

Dickey, J. O. 1995. Earth Rotation. In: *Global Earth Physics - A Handbook of Physical Constants.* pp. 356 - 368. T. J. Ahrens (ed.). Am. Geophys. Union, Washington.

Florsch, N., Chambat, F., Hinderer, J., Legros, H. 1994. A simple method to retrieve the complex eigenfrequency of the Earth's nearly diurnal-free wobble; application to the Strasbourg superconducting gravimeter data. *Geophys. J. Int.* **116**: 53 - 63.

Forte, A. M., Mitrovica, J. X., Woodward, R. L. 1995. Seismic-geodynamic determination of the origin of excess ellipticity of the core-mantle boundary. *Geophys. Res. Lett.* **22**: 1013 - 1016.

Goodkind, J. M. 1983. Q of the nearly diurnal free wobble. In: Proc. 9th Int. Symp. Earth Tides.. pp. 569 - 575. J. T. Kuo (ed.). Schweizerbart, Stuttgart.

Gwinn, C., Herring, T. A., Shapiro, I. I. 1986. Geodesy by Radio Interferometry: Studies of the Forced Nutations of the Earth. 2. Interpretation. *J. geophys. Res.* **91**: 4755 - 4765.

Haas, R., Schuh, H. 1996. Determination of frequency dependent Love and Shida numbers from VLBI data. *Geophys. Res. Lett.* **23**: 1509 - 1512

Herring, T. A., Dong, D. 1994. Measurement of diurnal and semidiurnal rotational variations and tidal parameters of Earth. *J. geophys. Res.* **99**: 18051 - 18071.

Hinderer, J., Legros, H., Amalvict, M. 1982. A search for Chandler and nearly diurnal free wobbles using Liouville equations. *Geophys. J. R. astr. Soc.* **71**: 303 - 332.

Hinderer, J., Zürn, W., Legros, H. 1991. Interpretation of the Strength of the NDFW Resonance from Stacked Gravity Tide Observations. In: *Proc. 11th Int. Symp. Earth Tides.* pp. 549 - 555. J. Kakkuri (ed.). Schweizerbart, Stuttgart.

Hopkins, W. 1839. Researches in physical geology. *Phil. Trans. R. Soc. London* **129**: 381 - 423.

Hough, S. S. 1895. The oscillations of a rotating ellipsoidal shell containing fluid. *Phil. Trans. R. Soc. London* **186**: 469 - 506.

Jeffreys, H., Vicente, R. O. 1957. The theory of nutation and the variation of latitude. *Mon. Not. R. astr. Soc.* **117**: 162 - 173.

Jiang, X., Smylie, D. E. 1995. A search for free core nutation modes in VLBI nutation observations. *Phys. Earth planet. Inter.* **90**: 91 - 100.

Lambeck, K. 1980. *The Earth's Variable Rotation.* 446 pp. Cambridge University Press, Cambridge, England.

Lambeck, K. 1988. *Geophysical Geodesy: The Slow Deformations of the Earth.* 718 p. Clarendon Press, Oxford.

Lecolazet, R. 1983. Correlation between Diurnal Gravity Tides and the Earth's Rotation Rate. In: Proc. 9th Int. Symp. Earth Tides.. pp. 527 - 530. J. T. Kuo (ed.). Schweizerbart, Stuttgart.

Levine, J. 1978. Strain-tide spectroscopy. *Geophys. J. Int.* **54**: 27 - 41

Levine, J., Harrison, J. C., Dewhurst, W. 1986. Gravity Tide Measurements with a Feedback Gravimeter. *J. geophys. Res.* **91**: 12835 - 12841.

Lumb, L. I., Jarvis, G. T., Aldridge, K. D., DeLandro-Clarke, W. 1995. The period of the free core nutation: towards a dynamical basis for an 'extra-flattening' of the core-mantle boundary. *Phys. Earth planet. Inter.* **90**: 255 - 271.

Masters, T. G., Widmer, R. 1995. Free Oscillations: Frequencies and Attenuations. In: *Global Earth Physics - A Handbook of Physical Constants.* pp. 104 - 125. T. J. Ahrens (ed.). Am. Geophys. Union, Washington.

Mathews, P. M., Buffett , B. A., Herring, T. A., Shapiro, I. I. 1991a. Forced Nutations of the Earth: Influence of Inner Core Dynamics 1. Theory. *J. geophys. Res.* **96**: 8219 - 8242.

Mathews, P. M., Buffett , B. A., Herring, T. A., Shapiro, I. I. 1991b. Forced Nutations of the Earth: Influence of Inner Core Dynamics 2. Numerical Results and Comparisons. *J. geophys. Res.* **96**: 8243 - 8257.

Mathews, P. M., Buffett, B. A., Shapiro, I. I. 1995. Love numbers for diurnal tides: Relation to wobble admittances and resonance expansions. *J. geophys. Res.* **100**: 9935 - 9948.

Melchior, P. 1980. For a clear terminology in the polar motion investigations. In: *Nutations and the Earth's Rotation.* pp. 17 - 21. E. P. Fedorov, M. L. Smith and P. L. Bender (eds.), Intern. Astron. Union, Kiev.

Melchior, P. 1981. *The tides of the planet earth.* pp. 1 - 609. Pergamon Press, Oxford.

Merriam, J. 1994. The Free Core Nutation Resonance in Gravity. *Geophys. J. Int.* **119**: 369 - 380.

Molodenskii, M. S. 1961. The theory of nutation and diurnal Earth Tides. *Comm. Obs. R. Belg.* **188**: 25 - 56.

Munk, W. H., MacDonald, G. J. F. 1960. *The rotation of the Earth.* Cambridge Univ. Press

Neuberg, J., Hinderer, J., Zürn, W. 1987. Stacking Gravity Tide Observations in Central Europe for the Retrieval of the Complex Eigenfrequency of the Nearly Diurnal Free Wobble. *Geophys. J. R. astr. Soc.* **91**: 853 - 868

Neuberg, J., Hinderer, J., Zürn, W. 1990. On the Complex Eigenfrequency of the "Nearly Diurnal Free Wobble" and its Geophysical Interpretation. In: *Variations in Earth's Rotation, Geophys. Monogr.* **59**: 11 - 16. D. D. McCarthy, W. E Carter (eds.). AGU and IUGG, Washington, D. C..

Neuberg, J., Zürn, W. 1986. Investigation of the Nearly Diurnal Resonance Using Gravity, Tilt and Strain Data Simultaneously. In: *Proc. 10th Int. Symp. Earth Tides.* pp. 305 - 311. R. Vieira (ed.). Cons. Sup. Invest. Scient., Madrid.

Neumeyer, J., Dittfeld, H.-J. 1997. Results of three years observation with a superconducting gravimeter at the GeoForschungsZentrum Potsdam. *J. Geodesy.* **71**: 97 - 102.

Poincaré, H. 1910. Sur la précession des corps déformables. *Bull. Astr.* **27**: 321 - 356.

Polzer, G., Zürn, W., Wenzel, H.-G. 1996. NDFW Analysis of Gravity, Strain and Tilt Data from BFO. *Bull. Inf. Marées Terrestres* **125**: 9514 - 9545.

Richter, B., Wenzel, H.-G., Zürn, W., Klopping, F. 1995. From Chandler wobble to free oscillations: comparison of cryogenic gravimeters and other instruments in a wide period range. *Phys. Earth planet. Inter.* **91**: 131 - 148.

Richter, B., Zürn, W. 1988. Chandler Effect and the Nearly Diurnal Free Wobble as Determined from Observations with a Superconducting Gravimeter.In: *The Earth's Rotation and Reference Frames for Geodesy and Geodynamics..* pp. 309 - 315. A. Babcock, G. Wilkins (eds.). Kluwer, Dordrecht, Holland.

Rochester, M. G., Jensen, O. G., Smylie, D. E. 1974. A Search for the Earth's 'Nearly Diurnal Free Wobble'. *Geophys. J. R. astr. Soc.* **38**: 349 - 363.

Sasao, T., Okamoto, I., Sakai, S. 1977. Dissipative core-mantle coupling and nutational motion of the Earth. *Publs. astr. Soc. Japan* **29**: 83 - 105.

Sasao., T., Okubo, S., Saito, M. 1980. A Simple Theory on the Dynamical Effects of a Stratified Fluid Core upon Nutational Motion of the Earth. In: *Proc. IAU Symp.: Nutation and the Earth's Rotation.* **78**: pp. 165 - 183. E. P. Fedorov, M. L. Smith, P. L. Bender (eds.). Reidel, Dordrecht, Holland.

Sato, T. 1991. Fluid Core Resonance Measured by Quartz Tube Extensometers at Esashi Earth Tide Station. In: *Proc. 11th Int. Symp. Earth Tides..* pp. 573 - 582. J. Kakkuri (ed.). Schweizerbart, Stuttgart.

Sato, T., Tamura, Y., Higashi, T., Takemoto, S., Nakagawa, I., Morimoto, N., Fukuda, Y., Segawa, J., Seama, N. 1994. Resonance Parameters of Nearly Diurnal Free Core Nutation Measured from Three Superconducting Gravimeters in Japan. *J. Geomagn. Geoelectr.* **46**: 571 - 586.

Schwahn, W. 1995 Effects of inertial forces due to the forced nutation on the gravimetric factor in the diurnal range. *Bull. Inf. Marées Terrestres* **121**: 9036 - 9042.

Sludskii, F. 1896. De la rotation de la Terre supposée fluide a son interieur. *Bull. Soc. Natur. Moscou* **9**: 285 - 318.

Smith, M. L. 1977. Wobble and nutation of the earth. *Geophys. J. R. astr. Soc.* **50**: 103 - 140.

Smith, M. L., Dahlen, F. A. 1981. The period and Q of the Chandler Wobble. *Geophys. J. R. astr. Soc.* **64**: 223 - 281.

Toomre, A. 1974. On the 'Nearly Diurnal Wobble' of the Earth. *Geophys. J. R. astr. Soc.* **38**: 335 - 348.

Wahr, J. M. 1981. Body Tides on an Elliptical, Rotating, Elastic and Oceanless Earth. *Geophys. J. R. astr. Soc.* **64**: 677 - 703.

Wahr, J. M., Bergen, Z. 1986. The Effects of Mantle Anelasticity on Nutations, Earth Tides, and Tidal Variation in Rotation Rate. *Geophys. J. R. astr. Soc.* **87**: 633 - 668.

Wahr, J., de Vries, D. 1989. The possibility of lateral structure inside the core and its implications for nutation and Earth tide observations. *Geophys. J. Int.* **99**: 511 - 519.

Wahr, J. M., de Vries, D. 1990. The Earth's Forced Nutations: Geophysical Implications. In: *Variations in Earth's Rotation, Geophys. Monogr.* **59**: 79 - 84. D. D. McCarthy, W. E Carter (eds.). AGU and IUGG, Washington, D. C..

Wenzel, H.-G. 1994. Earth tide analysis package ETERNA 3.0. *Bull. Inf. Marées Terrestres* **118**: 8719–8721.

Wu, X., Wahr, J. M. 1997. Effects of non-hydrostatic core-mantle boundary topography and core dynamics on Earth rotation. *Geophys. J. Int.* **128**: 18 - 42.

Yoder, C. F., Ivins, E. R. 1988. On the ellipticity of the core-mantle boundary from earth nutations and gravity. In: *The Earth's Rotation and Reference Frames for Geodesy and Geodynamics..* pp. 316 - 325. A. Babcock, G. Wilkins (eds.). Kluwer, Dordrecht, Holland.

Zürn, W., Rydelek, P. A., Richter, B. 1986. The Core-resonance Effect in the Record from the Superconducting Gravimeter at Bad Homburg. In: *Proc. 10th Int. Symp. Earth Tides..* pp. 141 - 147. R. Vieira (ed.). Cons. Sup. Invest. Cient., Madrid.

Zürn, W., Wenzel, H.-G., Laske, G. 1991. High quality data from LaCoste-Romberg gravimeters with electrostatic feedback: a challenge for superconducting gravimeters. *Bull. Inf. Marées Terrestres* **110**: 7940 - 7952.

Ocean Tides

and Related Phenomena

Ocean Tides

Wilfried Zahel

Institut für Meereskunde, Universität Hamburg
Troplowitzstrasse 7, D-22529 Hamburg, Germany

Abstract. According to their spatial scales and to the generation mechanisms applying, tidal phenomena in the sea are presented together with hydrodynamic models explaining their existence and appearance. The astronomical tide generating forces, to which the tidal variations of the ocean state variables can finally be traced, have planetary scale and therefore can directly excite tidal oscillations in the open ocean. Applying models of schematic ocean basins and of the real global ocean elucidate the dependence of theses forced oscillations on the free oscillation properties of ocean basins and on physical processes. Introducing data information into hydrodynamic-numerical tide models of the global ocean by employing a data assimilation procedure turns out to considerably contribute to the computation of realistic tidal fields and of realistic dependent geophysical quantities. The dynamics governing co-oscillating tides in shelf and adjacent sea areas are discussed subsequently. Having in view analytical wave solutions to the tidal equations, in many cases allows to recognize the importance of specific waves for the formation of tidal regimes. The increasing importance of frictional effects in shelf and adjacent sea areas becomes particularly apparent in current field properties, and quasi-resonantly amplified diurnal and semidiurnal tidal waves originating from the open ocean make clear the decisive role these areas are playing in the tidal energy budget. Proceeding to tidal phenomena characterized by further reduced spatial scales, the existence of over- and compound tides with in parts remarkable energy contents is assigned to nonlinear interactions between tidal waves in shallow water areas. The energy transfer from the astronomical tides to over- and compound tides becomes obvious when applying hydrodynamic-numerical models which include such areas as well as the shelf edge and slope. Also internal tidal motions cannot directly be excited, instead corresponding variations of currents and stratifications with often considerable amplitudes are due to energy transfer from the barotropic tides referred to above.

1   Introduction

The effects of astronomical tidal forces on the ocean are recognizable by corresponding time variations of the oceanic state variables, i.e., of the current velocity components, of pressure, temperature, salinity and density. Careful analysis of sufficiently long time series of measured state variables exhibits by means of the excited astronomical and possibly shallow water tidal periods to what extent the fields of motion, pressure and mass are tide determined in the area of investigation. Traditional tidal measurements of high practical importance have been performed in nearly all coastal regions of the earth using tide gauges. On the basis of the periodicities derived from measurements over time intervals of up to many years, in particular sea-surface elevations can accurately be predicted, on the premises that relevant topographic and non-tidal sea state conditions do not change. Other than tide determined time variations of elevation, those of current velocity, temperature and salinity many times are strongly dependent upon the density stratification, its variation and its influence on interactive processes in the sea. Therefore, tidal signals which are detected in measurements of current velocity, temperature and salinity in many cases are not strictly periodic or are even intermittent.

Tide tables containing tidal height predictions for so-called Standard Ports in all continents and tidal harmonic constants of a great number of so-called Secondary Ports are annually issued, e.g., the Admiralty Tide Tables (ATT) by the Hydrographer of the Navy in London. Additionally, information on pelagic tidal elevations at selected positions has been made available by the operation of deep-sea pressure instruments (Smithson, 1992). Operating current meters at appropriate sites yields data also making possible predictions of tidal currents. So the ATT also include such predictions for a number of waterways and harmonic constants of current velocity components.

Local measurements of tides allowing to predict tidal heights and currents are obviously important for many practical purposes. Information on tidal variations of sea state variables concerning their interdependence and their complete coverage of the ocean can only be obtained by applying hydrodynamical models. Measurements are needed in this connection for the verification of model results and for compensating model deficiencies. Figure 1 depicts two tidal elevation curves of essentially different shapes from two sites of the same oceanic region, viz., the South China Sea. These curves represent the prediction of tides which are predominantly diurnal and semidiurnal, respectively. Precisely, for each of the sites two curves are given, one basing on measurements and one resulting from applying a tide model. Hydrodynamic models yielding realistic results are the basis for understanding the generation of ocean tides in the open ocean areas and for explaining the various tidal phenomena occurring due to co-oscillation in shelf and adjacent sea regions. The free oscillation behavior of ocean basins determines the ocean tides as forced oscillations. Hence studying free oscillation properties as a function of dynamical and topographic ocean parameters is important.

1.1   Open Ocean Tides.

The Dynamical Equations. The spectrum of oscillations occurring in the ocean covers a wide frequency range reaching far beyond the band of forced oscillations generated by the major harmonics of the astronomical tidal potential. Modifying the general system of hydrodynamic equations in order to make them properly describe essentially waves forced by the astronomical tidal potential, leads to

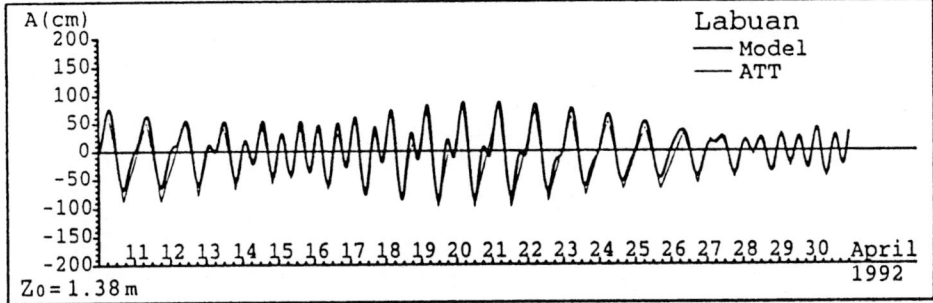

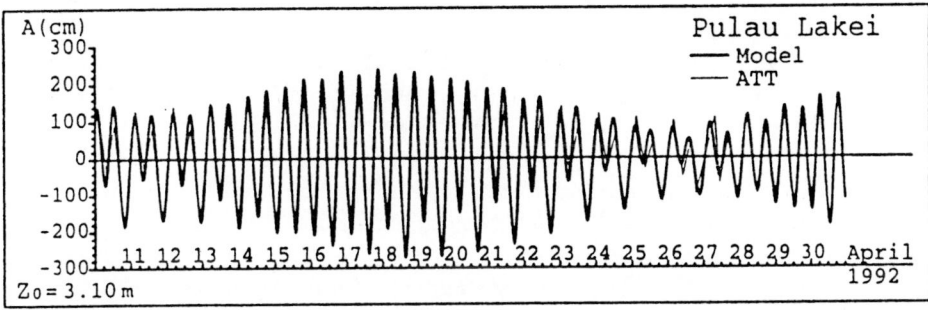

Fig. 1. Predicted tidal elevations at Labuan (5°17′N, 115°15′E) and at Pulai Lakei (1°45′N, 110°30′E). From Stawarz (1994).

the following system of integro-differential equations (Zahel, 1978; Marchuk and Kagan, 1986).

$$\frac{\partial}{\partial t}\mathbf{v} + \mathbf{f} \times \mathbf{v} + \mathbf{F} + r\,|\mathbf{v}|\,\mathbf{v}/H + g\nabla\zeta -$$

$$- \nabla \iint_S \zeta\,(t, \lambda', \varphi')\,G\,(\lambda, \varphi, \lambda', \varphi')\,R_e^2 \cos\,(\varphi')\,d\lambda'd\varphi' = (1 + k_{\bar{m}} - h_{\bar{m}})\,g\nabla\bar{\zeta} \tag{1}$$

$$\frac{\partial}{\partial t}\zeta + \nabla\cdot(H\mathbf{v}) = 0 \tag{2}$$

where ζ, $\mathbf{v}$ denote sea-surface elevation relative to the moving sea bottom and the depth-averaged current velocity vector, respectively, H the instantaneous water depth, g the surface gravity of a spherical earth, r the coefficient of quadratic bottom friction, $\mathbf{f} = 2\Omega\sin\varphi\,\mathbf{z}$ the vector of Coriolis acceleration, $k_{\bar{m}}$ and $h_{\bar{m}}$ Love numbers, and $\bar{\zeta}$ the total uplift of equipotential surface of gravity by the $\bar{m}$th-degree tide-generating potential. With the actual deformation δ of the ocean bottom, the geocentric sea-surface elevation ζ_0 is composed of $\zeta_0 = \zeta + \delta$. $\mathbf{F}$ denotes the vector defined by second order eddy viscosity terms, viz., $(F_\lambda, F_\varphi) = (-A_h\Delta u, -A_h\Delta v)$. S is the surface of the globe with radius R_e and G the Green's function for loading and self-attraction

$$G\left(\lambda,\varphi,\lambda',\varphi'\right)=$$
$$1/\left(4\pi\right)\sum_{n}\left(1+k'_{n}-h'_{n}\right)\alpha_{n}\sum_{m}P^{m}_{n}\left(\sin\varphi\right)P^{m}_{n}\left(\sin\varphi'\right)\cos\left(m\left(\lambda'-\lambda\right)\right) \qquad (3)$$

with normalized associated Legendre functions P^{m}_{n}, loading Love numbers k'_{n}, h'_{n} and the normalized density ratio $\alpha_{n}=(3/2n+1)\left(\rho_{o}/\rho_{e}\right)$. These equations reflect, among other things, that in the case of tidal oscillations forced by the astronomical potential the effect of density stratification can be ignored. Internal tides which are indirectly generated by barotropic tides are governed by a specific system of partial differential equations arising from the general system of hydrodynamic equations (see below). In shallow water areas, in addition to the nonlinear terms having already been considered in (1) and (2), the advective terms in the equations of motion (1) may become important (see below).

Treating (1) and (2) together with homogeneous conditions at the coastal boundaries, viz., vanishing normal component of current velocity and the no-slip condition, has become possible such that co-oscillating regimes in adjacent seas and in extended shelf areas can be resolved in global tidal models. Evaluating the tidal energy balance of appropriate numerical models allows to estimate the role the different ocean areas are playing with respect to the input, distribution and dissipation of tidal energy. From the oceanic free oscillation properties it is easily recognized that the large scale behavior of long waves in the open ocean may allow the planetary scale astronomical tidal potential to efficiently excite oceanic tidal oscillation systems, provided that the resonance properties of the ocean basins are adequate. Thus, investigating the barotropic free oscillation behavior of the ocean basins is basic for understanding the response of the ocean to the astronomical tidal potential. Although this response is strictly linear in the deep open ocean, numerical solutions to the linearized equations (1) and (2) used in homogeneous form are rare (Platzman et al., 1981; Gotlib and Kagan,1982; Gaviño, 1984) as compared with those to the inhomogeneous equations. Moreover, the former solutions include additional simplifications concerning the consideration of dynamical constituents. Nevertheless, the computed free oscillations are suitable for demonstrating the near semidiurnal and diurnal free oscillations systems of oceanic scale the excitation of which determines the principal basin dominating tidal regimes.

Oscillations in Schematic Ocean Basins. Before addressing to the free and forced basin wide barotropic oscillations in the real ocean, reference shall be made to the oscillations in an equatorial hemispherical ocean basin with a realistic mean ocean depth. The solutions of both, the free and the forced oscillation problem of this schematic ocean basin, can readily be obtained by expanding the dependent variables in spherical harmonics and solving the resulting linear algebraic equations for the expansion coefficients. The dynamics considered may also include linear friction and the full loading and self-attraction effect (Zahel, 1991). An overview of classical frictionless tidal solutions for schematic ocean basins is given by Doodson (1964). Longuet-Higgins and Pond (1970) treat the problem of frictionless free oscillations in a hemispherical basin bounded by meridians, while Webb (1980) also considers linear friction in his investigation of free and forced hemispherical oscillation systems. The equations being the basis of the hemispherical ocean model are obtained from (1) and (2) by referring to a hemispherical ocean centered at the equator and having constant depth h, by using a linear friction term $R^* \mathbf{v}$ and by writing the horizontal velocity vector as

$$\mathbf{v} = \frac{\partial}{\partial t} \left(\nabla \Phi + \nabla \Psi \times \mathbf{z} \right) \tag{4}$$

with

$$\Phi = \sum_{r=1}^{\infty} p_r \Phi_r \exp\left(-i\omega t\right) , \ \Psi = \sum_{r=1}^{\infty} p_{-r} \Psi_r \exp\left(-i\omega t\right)$$

$$\Phi_r = a_n^m P_n^m \left(\sin \varphi\right) \cos \left(m\lambda\right) , \ \Psi_r = a_n^m P_n^m \left(\sin \varphi\right) \sin \left(m\lambda\right)$$

where the suffixes r are associated with pairs of suffixes (n, m) and where the validity of $\nabla \Phi_r \cdot \mathbf{n} = 0$, $\Psi_r = 0$ in $\lambda = 0, \pi$ guarantees exact fulfillment of the boundary condition $\mathbf{v} \cdot \mathbf{n} = 0$. The equations for the coefficients p_r and p_{-r} are obtained by successively multiplying (1) with the functions $\nabla \Phi_r$ and $\nabla \Psi_r \times \mathbf{z}$, considering (2) and by integrating over the hemispherical domain.

$$-\omega^2 p_r - i R^* \omega p_r - 2i\omega \Omega \mu_r^{-1} \sum_{s=-\infty}^{\infty} \beta_{r,s} p_s + gh\mu_r \left(1 - \alpha_r'/2\right) p_r$$

$$- \frac{1}{2} gh \sum_{r'=1}^{\infty} \mu_{r'} q\left(r, r'\right) p_{r'} = g\gamma_2 \bar{\zeta}_r \tag{5}$$

$$-\omega^2 p_{-r} - i R^* \omega p_{-r} - 2i\omega \Omega \nu_r^{-1} \sum_{s=-\infty}^{\infty} \beta_{-r,s} p_s = 0 \tag{6}$$

where

$$q\left(r, r'\right) = R_e^4 a_{\bar{n}}^{\bar{m}} a_{n'}^{m'} \times$$

$$\times \sum_{n} \alpha_n' \sum_{\substack{m-m'=\text{odd} \\ m-\bar{m}=\text{odd}}}^{m} \left(2m a_n^m\right)^2 \mathrm{I}\binom{m\,\bar{m}}{n\,\bar{n}} \mathrm{I}\binom{m\,m'}{n\,n'} / \left(\left(m^2 - \bar{m}^2\right)\left(m^2 - m'^2\right)\right)$$

$$\mathrm{I}\binom{m\,m'}{n\,n'} = \int_{-1}^{+1} P_n^m \left(\mu\right) P_{n'}^{m'} \left(\mu\right) d\mu , \ \bar{\zeta}_r = \iint_B \Phi_r \bar{\zeta} dA$$

$$\nu_r = \mu_r = n\left(n+1\right)/R_e^2 , \ \alpha_r' = \left(1 + k_n' - h_n'\right) \alpha_n$$

and

$$\beta_{r,s} = -\iint_B \mathbf{z} \sin \varphi \nabla \Phi_r \times \nabla \Phi_s dA, \ \beta_{r,-s} = \iint_B \sin \varphi \nabla \Phi_r \cdot \nabla \Psi_s dA,$$

$$\beta_{-r,s} = -\iint_B \sin \varphi \nabla \Psi_r \cdot \nabla \Phi_s dA, \ \beta_{-r,-s} = -\iint_B \mathbf{z} \sin \varphi \nabla \Psi_r \times \nabla \Psi_s dA$$

with the $\beta_{\pm r, \pm s}$ referred to as gyroscopic coefficients. With suitable factors a_n^m the systems $\{\ \Phi_r\ \}$ and $\{\ \Psi_r\ \}$ are orthonormal in the hemispherical domain B. $\bar{\zeta}$ denotes the oceanic equilibrium tide on a rigid earth. Equation (2) yields the expansion of the sea-surface elevation relative to the moving sea bottom

$$\zeta = h \sum_{r=1}^{\infty} p_r \mu_r \Phi_r \tag{7}$$

Free oscillations of the hemispherical ocean have been computed by using the values $R^* = 9.26 \cdot 10^{-6}\,\mathrm{s}^{-1}$, $h = 4420\,\mathrm{m}$, $\gamma_2 = 0.69$ and loading Love numbers based on a Gutenberg-Bullen earth model (Zahel, 1980). The depth used is typical of open ocean areas, and the value assigned to the friction parameter corresponds to an estimated realistic energy decay time of 30 hours. Table 1 shows computed real eigenperiods T and imaginary parts of eigenfrequencies ω_I for the two types of possible hemispherical free oscillations, i.e., those which are symmetric and those which are antisymmetric with respect to the equator. Values are opposed to each other which belong to the equations (5) and (6) used in homogeneous form when either considering or neglecting (omitting the terms including α'_r as factor) the loading and self-attraction effect (LSA). There do exist for each, the class of symmetric and that of antisymmetric oscillations, a fastest rotational and a slowest gravitational mode. The first slowest gravitational modes determine the resonant response to the semidiurnal and to the diurnal astronomical tidal potentials, respectively. The significant dependence of the free oscillations on considering or neglecting LSA becomes apparent in the semidiurnal, i.e., symmetric, and in the diurnal, i.e., antisymmetric, hemispherical tidal oscillation systems. Figure 2 demonstrates this dependence by the near-resonant L_2-tide. Applying realistic numerical ocean tide models (see below) leads to significant modifications of the computed semidiurnal and diurnal tidal oscillation systems when considering the full LSA-effect, and this holds true for models with quite different spatial resolutions (Zahel, 1980; Grawunder, 1996). The most obvious of these modifications basically appear in the hemispherical model already. To these tidal LSA-effects in the ocean belongs the delay of approximately 30 degrees the poleward propagating tidal waves experience at the east sides of the Pacific and of the Atlantic.

Oscillations in the Real Global Ocean. The simplicity of the basin geometry allows the exact computation of the free and forced oscillation properties of the hemispherical ocean where all of the most important dynamical constitutents can be taken into account. Due to the complicated bottom and coastline topography the situation becomes different concerning the computation of the free oscillation properties of the real global ocean. The resulting computational expense up to the present excludes an exact resolution of the spectrum of the free oscillations in the relevant time interval. Nevertheless, the results obtained by Gaviño (1984), although referring to a frictionless global model with large scale topographic features resolved only, are appropriate for studying tidal resonance effects. Meanwhile (Gaviño, pers. comm.) the spectrum in the time interval between 6 h and 108 h has completely been resolved for that model. The various

Table 1. Eigenperiods of an equatorial hemispherical ocean

CLSA		LSA neglected	
T, s	ω_I, 10^{-6} s^{-1}	T, s	ω_I, 10^{-6} s^{-1}
Symmetric oscillations			
Fastest rotational eigenoscillation			
384348	-7.9628	380894	-7.9605
Slowest gravitational eigenoscillations			
162728	-4.2875	153806	-4.2231
84273	-4.4669	80960	-4.4390
59452	-5.1319	57644	-5.1284
52989	-6.1891	51476	-6.1098
48870	-5.5802	47541	-5.5263
42762	-5.0242	41635	-5.0057
40753	-5.2533	39733	-5.2205
36054	-5.9685	35322	-5.9419
35010	-5.0128	34204	-4.9790
34219	-5.1182	33455	-5.0859
31779	-5.6005	31145	-5.5650
Antisymmetric oscillations			
Slowest gravitational eigenoscillations			
110937	-6.2428	107145	-6.1237
74672	-5.3368	71992	-5.2776
54628	-5.0022	52999	-4.9707
43821	-5.6154	42778	-5.6286
41553	-5.8475	40561	-5.7437

T and ω_I denoting real eigenperiods and imaginary parts of eigenfrequencies, respectively. CLSA meaning full loading and self-attraction considered. The ocean depth and the friction used are $h = 4420$ m and $R^* = 9.26 \cdot 10^{-6}$ s^{-1}, respectively.

free oscillations differ widely with respect to energy partition, relative energy contents of ocean basins and field of elevation and current patterns. Figure 3 shows computed free oscillations with their distributions of sea-surface elevation ζ and of energy transport vectors $\mathbf{J} = gh\overline{\mathbf{v}\zeta_0}$, with their periods in hours and their relative attributions of energy to different ocean regions, viz., to the Arctic Ocean, North Atlantic, South Atlantic, Indic, South Pacific and North Pacific (from left to right in Fig. 3). The free gravitational oscillations with nearly equipartitioned potential and kinetic energies and periods of 24.549 h and 12.195 h, respectively, play, as will be explained below when referring to recent global tidal solutions, an important role in the oceanic response to diurnal and semidiurnal tide-generating forces, respectively.

The global ocean tide models having been used essentially differ by the extent

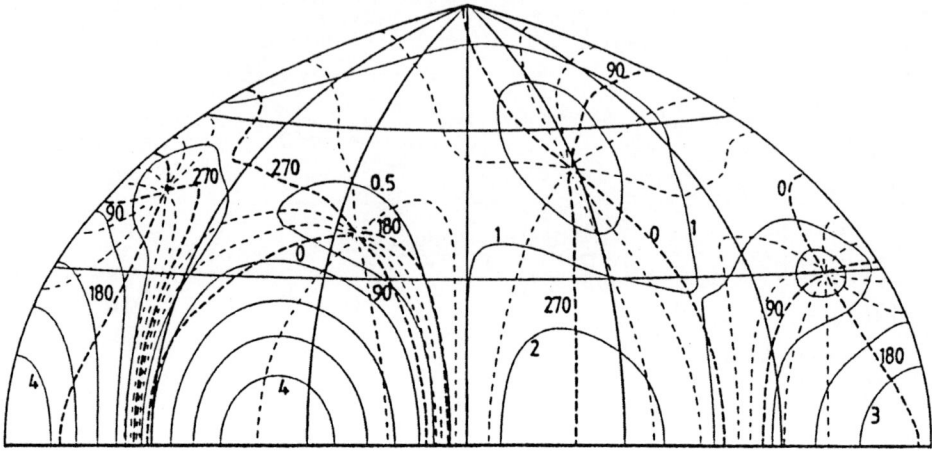

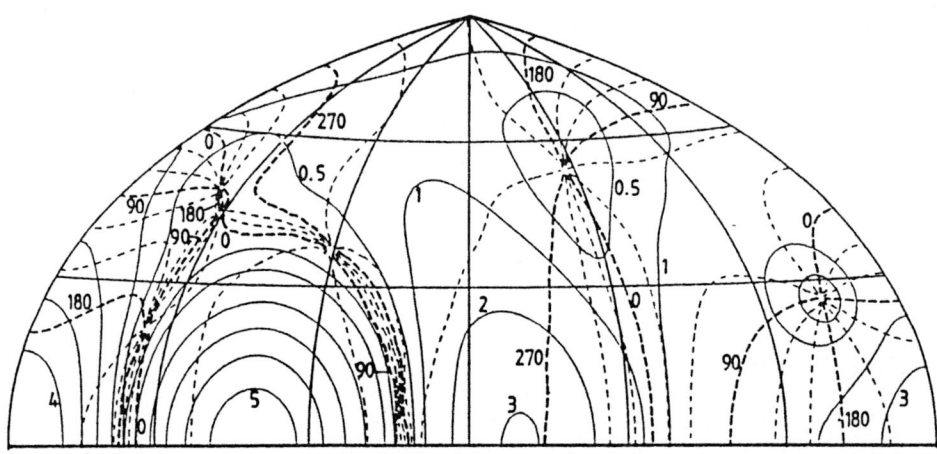

Fig. 2. L_2-tide of a hemispherical ocean model when neglecting (above) and considering (below) LSA. Cotidal lines are dashed, with phases in degrees referred to meridian passage at the western boundary. Coamplitude lines are solid, with amplitudes in units of the maximum equilibrium tide amplitude.

they have been made dependent on measured tidal sea-surface elevations and by the energy dissipation mechanisms included. Free ocean tide models are defined by (1) and (2) making use of homogeneous boundary conditions, exclusively. Thus, these models are completely independent of ocean tide measurements. Semi-empirical models are characterized by prescribing measured tidal elevations at the coastal boundaries where, consequently, an inhomogeneous condition has to be fulfilled. An overview of early solutions and of the principal numerical methods applied is given by Hendershott (1977). Free as well as semi-empirical earlier models yielded many basin dominating features of tidal oscillation which

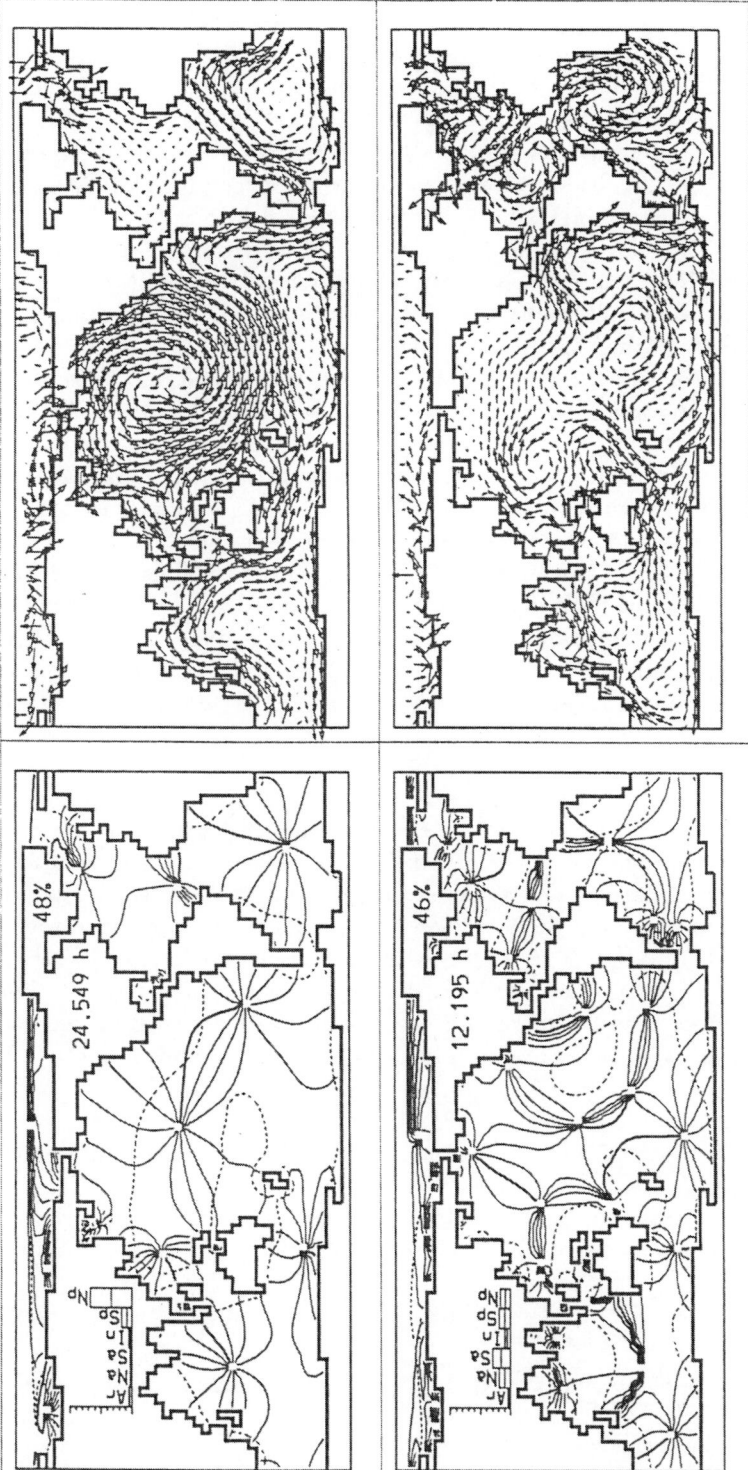

Fig. 3. Cotidal and corange lines (left) and energy transport vectors (right) for a diurnal and a semidiurnal free oscillation in the world ocean. Proportion of potential energy to total energy rates in per cent. From Gaviño (personal communication).

meanwhile have turned out as realistic in view of pelagic conventional (Smithson, 1992) and satellite measurements having recently become available. Free models have proved particularly valuable for estimating the role dynamical constituents are playing for the formation of tidal wave properties, e.g., the tidal effect of LSA in the ocean (Accad and Pekeris, 1978; Zahel, 1978). Particularly extended use of coastal and island data was made by Schwiderski (1983) with a modified semi-empirical approach aiming at the computation of semidiurnal, diurnal and long-period tidal elevation fields being as realistic as possible.

In view of the advent of an increased number of data from deep-sea pressure instruments and from satellite altimetry, global ocean tide models have been developed which allow influencing the model results by data information taking into account that neither the tidal model nor the data are free from errors (Egbert et al., 1995; Zahel, 1991, 1995a). Data are assimilated into these models by minimizing a functional defined by the squares of the dynamical and of the data residuals. By this procedure a balance is created between the approximate fulfillment of the dynamical equations and the approximate adoption of the assimilated data by the corresponding model variables. This balance is controlled by a priori dynamical and data error covariances. In the approach taken by Zahel (1991, 1995a) firstly the tidal elevation is eliminated from the linearized equation (1) using (2) in linearized form. By this way the computational expense is reduced and mass conservation is guaranteed also when assimilating data. Considering the homogeneous boundary conditions the resulting finite-difference-equations take the form of a linear algebraic system with a uniquely determined solution:

$$\mathbf{A}\mathbf{x} = \mathbf{b} \tag{8}$$

A factor $\exp(-i\omega t)$ having been introduced corresponding to the tidal constituent under investigation, the vector of unknowns $\mathbf{x}$ is made up by the horizontal current velocity components at the grid points. The latter are arranged according to a Richardson lattice, i.e., zonal and meridional components are assigned to different points encircling the elevation grid points zonally and meridionally, respectively (Zahel, 1980). The vector $\mathbf{b}$ is defined by the astronomical tidal potential and the matrix $\mathbf{A}$ results from the dynamics and is non-sparse due to the LSA effect. Equation (8) represents the classical free global tidal model for a tidal constituent with frequency ω. The data equations are given by

$$\mathbf{D}\mathbf{x} = \mathbf{d} \tag{9}$$

With observations taken from m positions, only m rows of the matrix $\mathbf{D}$ and of the vector $\mathbf{d}$ contain non-zero entries. When tidal elevations are assimilated, (2) defines the data equations, while tidal loading gravity data are related to the tidal current velocity vectors $\mathbf{v}$ (defining $\mathbf{x}$) by

$$i\omega \hat{g}_{\mathrm{L}} = \rho \iint_S \nabla \cdot (h\hat{\mathbf{v}}) \, \tilde{G} R_{\mathrm{e}}^2 \cos(\varphi') \, d\lambda' d\varphi' \tag{10}$$

with $g_{\mathrm{L}} = \hat{g}_{\mathrm{L}} \exp(-i\omega t)$ denoting loading gravity and $\tilde{G}$ denoting the corresponding Green's function (Scherneck, 1990). The minimization functional used is defined by

$$J(\mathbf{x}) = (\mathbf{Ax} - \mathbf{b})^H \mathbf{C}^{-1}(\mathbf{Ax} - \mathbf{b}) + (\mathbf{Dx} - \mathbf{d})^H \mathbf{S}^{-1}(\mathbf{Dx} - \mathbf{d}) \tag{11}$$

where $\mathbf{C}$ and $\mathbf{S}$ denote dynamical and data error covariance matrices, respectively. Usually assuming the data errors as uncorrelated, $\mathbf{S}$ takes diagonal form. With the lower triangular matrix $\mathbf{R}$ resulting from the Cholesky decomposition $\mathbf{C} = \mathbf{RR}^H$, the least squares solution to the system of equations made up by

$$\mathbf{R}^{-1}\mathbf{Ax} = \mathbf{R}^{-1}\mathbf{b}$$
$$\mathbf{S}^{-1}\mathbf{Dx} = \mathbf{S}^{-1}\mathbf{d} \tag{12}$$

is searched for. Applying the method of conjugate gradients for obtaining this solution only requires performing matrix multiplications and solving linear algebraic equations with square lower and upper triangular matrices, respectively.

Figures 4 and 5 show diurnal O_1 and semidiurnal M_2 tidal elevation maps, respectively. These maps result from the application of a global model with assimilating pelagic elevation data from 45 positions and Topex/Poseidon data (Eanes and Bettadpur, 1995) from 1415 uniformly distributed positions as described above. The tidal elevation patterns shown only slightly deviate from those published in Zahel (1995a) having been constrained by pelagic elevation data from 85 positions and by loading gravity data from 14 positions. However, it turns out that the agreement of the solutions with independent elevation data increases with the growing number of assimilated data. The appearance of the tidal regimes is determined by the distribution of amphidromic systems, their senses of rotation, shapes and extensions as well as by the anti-nodes and their strengths. Owing to the assimilation of data from a great number of uniformly distributed positions, these model generated tidal regimes can be regarded as highly realistic. The characteristic features of the latter are determined, of course, by the free oscillation behavior of the ocean basins. Simply inspecting the main features of the tidal oscillation patterns and comparing them with those of the energetic free oscillations gives an idea of near-resonant excitations occurring. So, e.g., the near resonant O_1-tide oscillation system in the Pacific (Fig. 4) with a left-hand rotating amphidrome north of the equator and a right-hand rotating amphidrome with far weaker energy transport south of the equator clearly resembles the 24.549 h–oscillation (Fig. 3) in view of the well comparing oscillation patterns and energy attributions. Wave propagation to the west along the Antarctic is recognized to take place (Fig. 3 and Fig. 4) by means of a Kelvin wave with the elevation amplitude exponentially increasing to the left toward the continental coast. The semidiurnal tidal solution (Fig. 5) exhibits in the near resonant Atlantic as well as in the Pacific showing a well developed oscillation system good agreement with the main features of the 12.195 h–free oscillation (Fig. 3). To these features belong the specific left-hand rotating amphidrome in the South Atlantic and the northeast–southwest quasi-standing wave pattern in the Pacific with amphidromes corresponding to each other and rotating in same senses.

A complete description of the tidal oscillation systems also requires referring to the tidal current velocities (Fig. 6) which, however, are scarcely known from observations in the open ocean, by this excluding the opposition of these computed fields to correponding observations. The main diurnal and semidiurnal tidal oscillation features had already been yielded by classical free models. However, due to inadequate energy dissipation mechanisms and to other model deficiencies agreement with observations proved to be insufficient in detail. The imperfectness of global tidal models now having been compensated to a large extent by data information has to be attributed, among other things, to insufficient resolution of near coastal areas, to neglecting the interaction of the barotropic tide with density stratification and with the non-tidal field of motion, and generally to the incomplete consideration of sub-grid-scale effects.

The computed global tidal fields of elevation and current allow to establish, e.g., the tidal angular momentum and energy budgets and to compute the effects of the tidal mass redistribution in the ocean on solid earth deformations and on variations of gravity. The relation for computing the solid earth deformation is part of (1) and is given below in connection with the tidal energy equation. The loading gravity variations are evaluated using (10) as has been done in Zahel (1995b). The tidal angular momentum budget is of particular interest for studies of variations of earth rotation.

The Tidal Energy Budget. Computing the values which the constituents of the tidal energy equation take when being averaged over the period of the respective tidal constituent, is of special interest for understanding the mechanisms of generating, distributing and dissipating tidal energy in the ocean. This topic was discussed in some detail by Schwiderski (1985) and in connection with loading and self-attraction effects by Zahel (1980). The time averaged tidal energy equation being consistent with (1) and (2) writes

$$
\overline{\rho r \left(u^2 + v^2\right)^{3/2}} + \overline{\rho H \left(F_\lambda u + F_\varphi v\right)} + \overline{\rho \nabla \cdot \left(g H u \zeta_0, g H v \zeta_0\right)} =
$$
$$
\overline{\rho \left(1 + k_{\bar{m}}\right) \nabla \cdot \left(g H u \bar{\zeta}, g H v \bar{\zeta}\right)} + \overline{\rho \left(1 + k_{\bar{m}}\right) g \bar{\zeta} \partial \zeta / \partial t}
$$
$$
+ \overline{\rho \nabla \cdot \left(H u \Phi^*, H v \Phi^*\right)} + \overline{\rho \Phi^* \partial \zeta / \partial t} + \overline{\rho g \zeta \partial \left(\delta_0 + \delta^*\right) / \partial t} \qquad (13)
$$

with

$$
\Phi^* = g \sum_n \left(1 + k_n'\right) \alpha_n \zeta_n, \quad \delta^* = \sum_n h_n' \alpha_n \zeta_n, \quad \delta_0 = h_{\bar{m}} \bar{\zeta}
$$

and thus Φ^*, $\delta = \delta_0 + \delta^*$ denoting the potential due to LSA and the actual deformation of the ocean bottom, respectively. The first two terms on the left-hand side of (13) indicate energy dissipation by bottom friction and eddy viscosity, respectively, the third term represents divergence of the energy transport. The terms on the right-hand side give the works done by the astronomical potential (first and second), by the secondary potential due to LSA (third and fourth), and by the moving ocean bottom. When integrating (13) over the global ocean only the first and the second term on the left-hand side and the second and the fifth on the right-hand side yield non-zero contributions. When data are assimilated (1) is not exactly fulfilled, leaving a non-zero residual vector $\mathbf{p} = (p_\lambda, p_\varphi)$ on

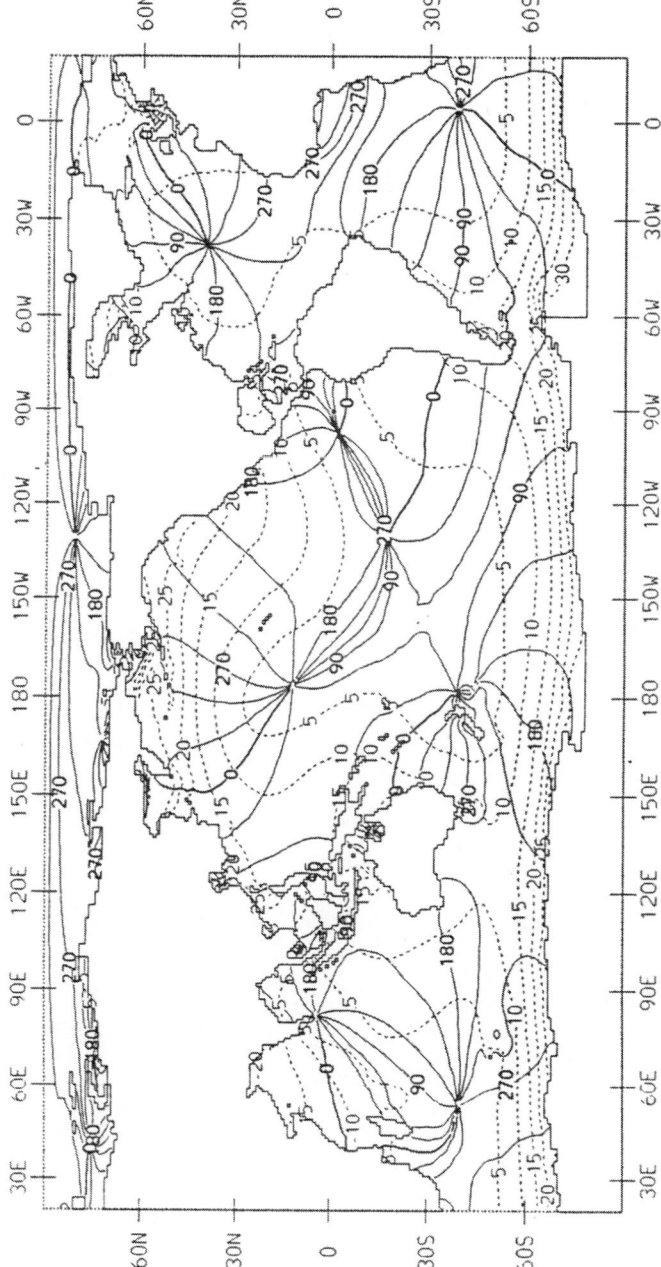

Fig. 4. The 1° World Ocean tide model, O_1-tide elevation. Cotidal lines solid with Greenwich phases in degrees (in steps of 30°). Coamplitude lines dashed with amplitudes in centimeters (5, 10, 15, 20, 25, 30).

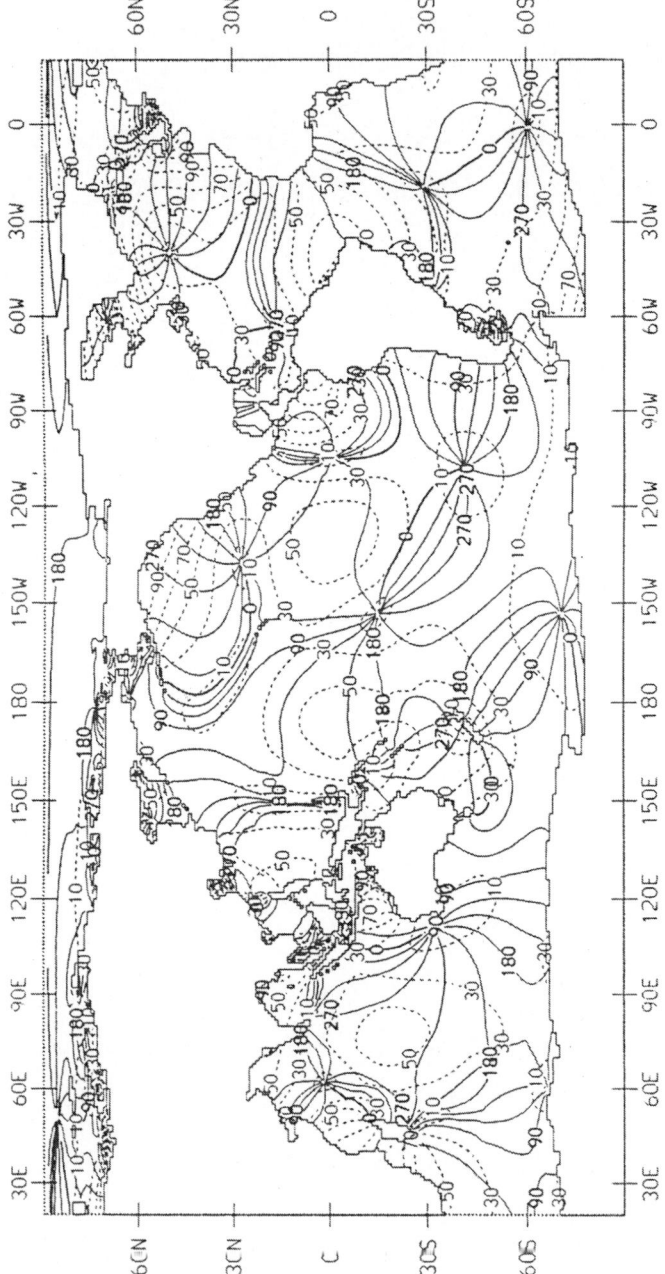

Fig. 5. The 1° World Ocean tide model, M$_2$-tide elevation. Cotidal lines solid with Greenwich phases in degrees (in steps of 30°). Coamplitude lines dashed with amplitudes in centimeters (10, 30, 50, 70, 90, 110, 130).

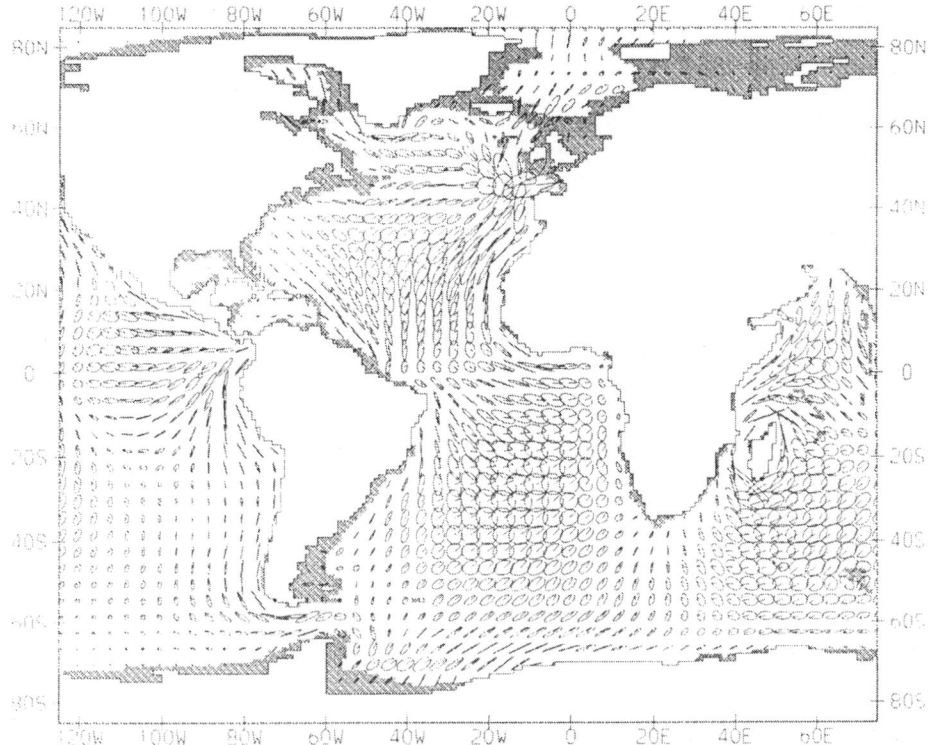

Fig. 6. The 1° World Ocean tide model, M_2-tide ellipses of volume transport in the Atlantic. Ocean areas with water depths smaller than 1000 m shaded.

the right-hand side of (1) which yields the contribution $\overline{\rho H \left(p_\lambda u + p_\varphi v \right)}$ on the right-hand side of (13). These additional terms in the equations of momentum and energy represent implicit corrections of them induced by assimilated data information. The spatial distributions of the work done by the astronomical tidal potential yields for all partial tides positive contributions when being integrated over the main parts of the global ocean. Figure 7 shows the corresponding spatial pattern for the M_2-tide. Moreover, it is recognized that this energy input predominantly takes place in the interior of the ocean basins, where areas of both positive and negative contributions have basin scale order of magnitude. Inspection of the spatial patterns of the other terms in (13) shows the following simple picture which constitutes a good approximation to the exact computed energy balance. In the open ocean the divergence of energy flux and the tidal work terms essentially balance each other. The transport of energy from the open ocean into the coastal regions is brought about by the energy flux term. Finally, energy dissipation is concentrated on the shelf regions, where the tidal current velocities have maximum magnitudes being one to two orders greater than the global rms main half-axis, amounting, e.g., to 116.8 cm s^{-1} and to 4.5 cm s^{-1}, respectively, for the M_2-tide in the global model (Fig. 6). Consequently, the friction terms in (13) yield considerable spatially integrated contributions for these regions only. Deviations from the simple balance described above may primarily take place in shelf regions concerning the role of tidal work.

In Tab. 2 integrated quantities are put together characterizing the overall response of the ocean to the major semidiurnal and diurnal constituents of the astronomical tidal potential. The given rates result from the application of a global model with assimilating data as explained above. When being related to

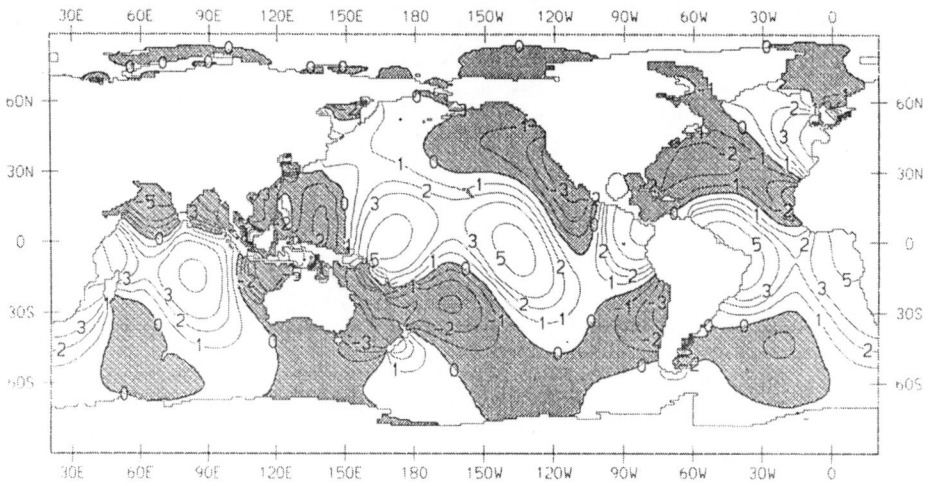

Fig. 7. The 1° World Ocean tide model, M_2 tidal work done by the astronomical forces and the moving sea bottom. Work done per unit area in $10^{-2}\,\mathrm{W\,m^{-2}}$.

the rms amplitudes of the corresponding equilibrium tides, the rates of energy contents, the magnitudes of rms current velocity main half-axes, and the elevation amplitudes indicate that the semidiurnal tides are altogether closer to resonance than the diurnal tides. The rates of tidal power, being equal to those of tidal dissipation, confirm the order of frictional decay time of tidal energy to amount to about 30 h (33 h for semidiurnal and 26 h for diurnal tides on an average), as assumed in the hemispherical model. As can be taken from Tab. 1, the slowest free gravitational oscillations have periods of 45.2 h (symmetric) and 30.2 h (antisymmetric). In the real global ocean the maximum period of free gravitational oscillations approximately takes the value 112 h (Marchuk and Kagan, 1986). Quasi-resonant excitations of free oscillations are thus essential for the development of the global semidiurnal and diurnal tidal regimes. In contrast to this, tidal constituents the periods of which are long as compared to 112 h, e.g., the fortnightly Mf, the monthly Mm and the semiannual Ssa, are characterized in the ocean by regimes becoming increasingly indistinguishable from the corresponding equilibrium tides (Seiler, 1989). Assuming a rigid earth, the space-time dependence of the latter is given by $\bar{\zeta} = \frac{a}{2}\left(1 - 3\sin^2\varphi\right)\cos\left(\omega t\right)$ with amplitudes a of 4.2 cm, 2.2 cm and 1.9 cm, respectively. Linear superposition of tidal currents and elevations, taking into account the actual phase relationships of the astronomical tidal potentials, yields the total tidal current vectors and elevations in the open ocean and generally where shallow water effects are negligible. Interferences of M_2 and S_2 and of K_1 and O_1 have periods of 14.77 days and 13.66 days, respectively.

The semidiurnal and diurnal ages are defined by the time differences between maximum superposition of tidal elevations in the sea and maximum amplitude of potential superposition. The times of the maximum superpositions of the tidal potentials are given by full and new moon and by the northern and southern maximum declinations of the moon, respectively. The rates t_a of the tidal ages result from the phase and frequency differences of the relevant partial tides at the corresponding position

$$t_a = \frac{\psi(M_2) - \psi(S_2)}{\omega(M_2) - \omega(S_2)} \tag{14}$$

For ψ the local or the Greenwich phase may be inserted. Replacing M_2 by O_1 and S_2 by K_1 yields the diurnal age. From the global tidal computations referred to above, the following information on tidal ages is obtained. Positive and negative values occur, but on an average the semidiurnal age amounts to 18.2 h and the diurnal age to 10.4 h. In 65.5 % (semidiurnal) and in 59.0 % (diurnal) of the ocean area the rates of the ages range between 0 and 2 hours. These average delays of the maximum superposition of tidal elevations in the sea are attributed to the effect of friction (Garrett and Munk, 1971).

Table 2. Integrated quantities of computed tidal fields

	M_2	S_2	N_2	K_1	O_1	P_1
Potential energy [10^{16} J]	11.52	1.80	0.56	1.41	0.70	0.15
Kinetic energy [10^{16} J]	16.46	2.87	0.72	1.71	0.89	0.19
Tidal power [10^{11} W]	23.53	3.79	1.10	3.24	1.76	0.35
Elevation rms amplitude [cm]	40.91	16.27	8.60	15.18	11.03	5.00
Equilibrium tide[a] rms amplitude [cm]	12.22	5.69	2.34	7.61	5.07	2.36
Current velocity rms main half-axis [cm/s]	4.54	1.67	0.98	1.87	1.39	0.62
Q-factor	16.7	17.9	16.0	7.0	7.5	7.0

[a] For comparison

1.2   Co-oscillating Tides.

Tides in Adjacent Seas. If shelf areas and adjacent seas are sufficiently extended, i.e., if they have a spatial scale of order $\sqrt{ghT}/4$ at least (Webb, 1976),
they may develop separate eigenoscillation systems which, however, cannot directly be excited by the astronomical tidal potential having planetary scale.
Instead these systems may be excited by the tidal wave having been generated
in the open ocean. The tidal regimes by this way coming into existence are called
co-oscillating tides. When being close to resonance such tidal regimes are characterized by strong currents and elevation amplitudes of up to several meters.
To these regions developing strong co-oscillating tides belong the Patagonian
Shelf and the North Sea (Bundesamt für Seeschiffahrt und Hydrographie: Ge-
zeitentafeln; Davies and Furnes, 1980) concerning semidiurnal constituents, and
the North Siberian Shelf (Kowalik and Proshutinsky, 1994) and the Indonesian
Waters concerning semidiurnal as well as diurnal constituents. Co-oscillating tides play an important role in the global tidal energy budget because they nearly
completely include the tidal energy losses as mentioned above. Co-oscillating
tides can be described by the system of equations

$$\frac{\partial}{\partial t}\mathbf{v} + \mathbf{v}\nabla\mathbf{v} + 2\Omega \times \mathbf{v} + \mathbf{F} - \frac{\partial}{\partial z}A_v\frac{\partial\mathbf{v}}{\partial z} = -\frac{1}{\rho_0}\nabla p - \frac{\rho}{\rho_0}\mathbf{z} \tag{15}$$

$$\nabla \cdot \mathbf{v} = 0 \tag{16}$$

$$\frac{\partial\zeta}{\partial t} + \nabla_h \cdot (H\mathbf{v}_h) = 0 \tag{17}$$

In the case of barotropic tides $\rho = \rho_0$ is assumed, and the third of the equations
of motion (15) for the vertical current velocity component is reduced to the hydrostatic equation. For baroclinic tidal motions ρ denotes the perturbation density, in the third equation of (15) only the linear constituent of the inertial term,
the pressure and the buoyancy terms are usually retained, and the following equation is additionally used with the buoyancy frequency $N = (-g/\rho_0\,\partial\rho_0/\partial z)^{1/2}$:

$$\frac{\partial\rho}{\partial t} - wN^2\rho_0/g = 0 \tag{18}$$

Also in the case of density stratification playing no role, a vertical dependence
of horizontal current velocity becomes apparent close to and in the frictional
bottom boundary layer, the thickness of which is of the order of 20 m, whence,
other than in the open ocean, in shelf areas in a considerable part of the water
column the current velocity vector changes with depth. Figure 8 shows current
velocity vectors as having been observed in the German Bight (Mittelstaedt
et al., 1983) where semidiurnal tides are dominating. The vectors reflect typical
properties of depth dependence, e.g., the currents near the bottom reaching their
maximum magnitudes before the surface current in most cases. On the basis of
the tidal equations and of assumptions on the turbulent bottom boundary layer,
Kagan (1974) has derived for shelf areas characteristic vertical dependencies of
tidal current ellipses related to surface properties and to water depth.

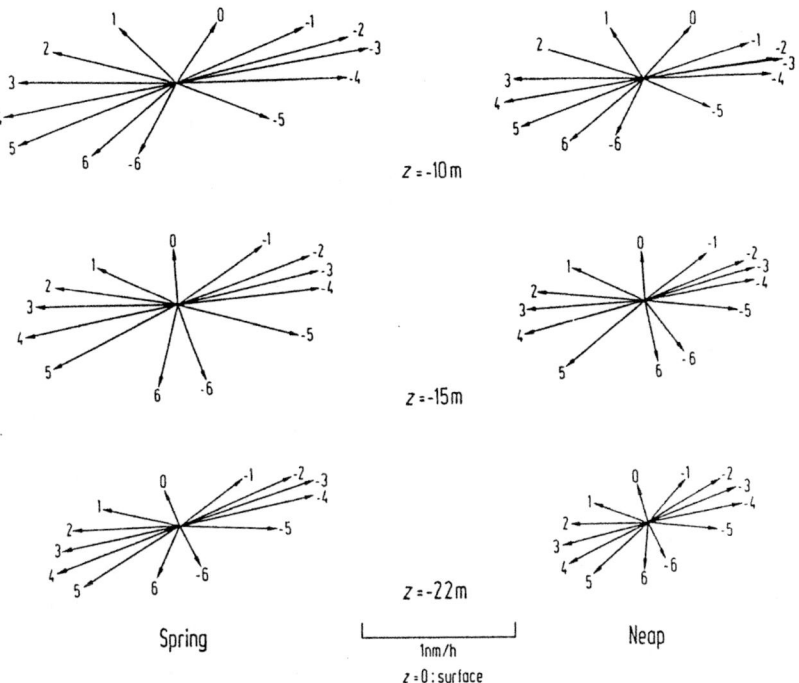

Fig. 8. Time dependence of tidal current velocity vectors (-22 m: close to bottom) at 53°48′N, 6°20′E. Integer numbers at the vectors denote the time difference in hours as against high-water in Helgoland. After Mittelstaedt et al. (1983).

Application of numerical modeling has yielded increasingly reliable results for the tidal fields in the shelf and adjacent sea regions with remarkably developed co-oscillating tides. The reliability of the results also depends on the tidal wave being prescribed at the open boundaries separating the high resolution model area from the open ocean area. The Indonesian Waters are particularly interesting showing due to the complex topography complicated co-oscillating semidiurnal and diurnal tidal regimes, which in some parts are resonantly excited by the Indian and Pacific Ocean tide. Figures 9 and 10 display amplitudes and phases of K_1-elevation as obtained by a hydrodynamic-numerical model with $10' \times 10'$ spatial resolution (Stawarz, 1994). Maximum elevation amplitudes appear in the shelf regions of the South China Sea and the Arafura Sea, where their ratios to the corresponding maximum equilibrium tide amplitudes amount to 9 for the K_1-tide and to 7 for the M_2-tide, approximately, both values clearly indicating strong resonance behavior of the co-oscillating tides. Mihardja (1991) obtains by applying a hydrodynamic-numerical model a total M_2-tide dissipation rate of 0.214 TW for this region, where 98 % of the M_2-tide energy supplied to the Indonesian Waters is transported through the line between the north coast of Australia and the Sunda Island Chain. This means (see Tab. 2) that nearly

10 % of the global M_2-tide energy dissipation rate has to be attributed to this region.

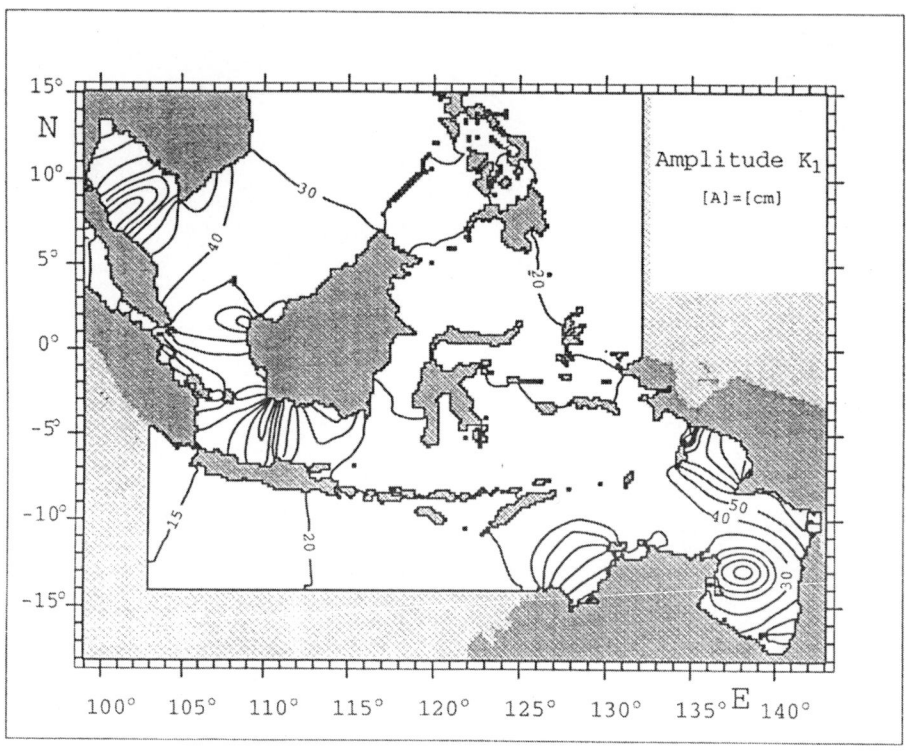

Fig. 9. Computed K_1-tide elevation. Coamplitude lines with amplitudes in centimeters $(5, 10, 15, 20, 30, 40, \ldots)$. From Stawarz (1994).

Analytical Wave Solutions. Other than in real ocean regions, in simply shaped basins or on shelfs infinitely extending along a straight continental coast the tidal wave behavior can be described by analytical functions. Wherever the real ocean topography comes close to such simple geometry, the corresponding analytical wave solutions might have a meaning for the real ocean tide. The wave solutions fulfill, e.g., the simplified equations (15) and (17) describing long wave behavior with f=const. and without resolving the vertical dependence of current velocity. The depth distribution h has to be chosen properly allowing the existence of analytical solutions

$$\frac{\partial u}{\partial t} + R^* u - fv + g\frac{\partial \zeta}{\partial x} = 0$$

$$\frac{\partial v}{\partial t} + R^* v + fu + g\frac{\partial \zeta}{\partial y} = 0$$

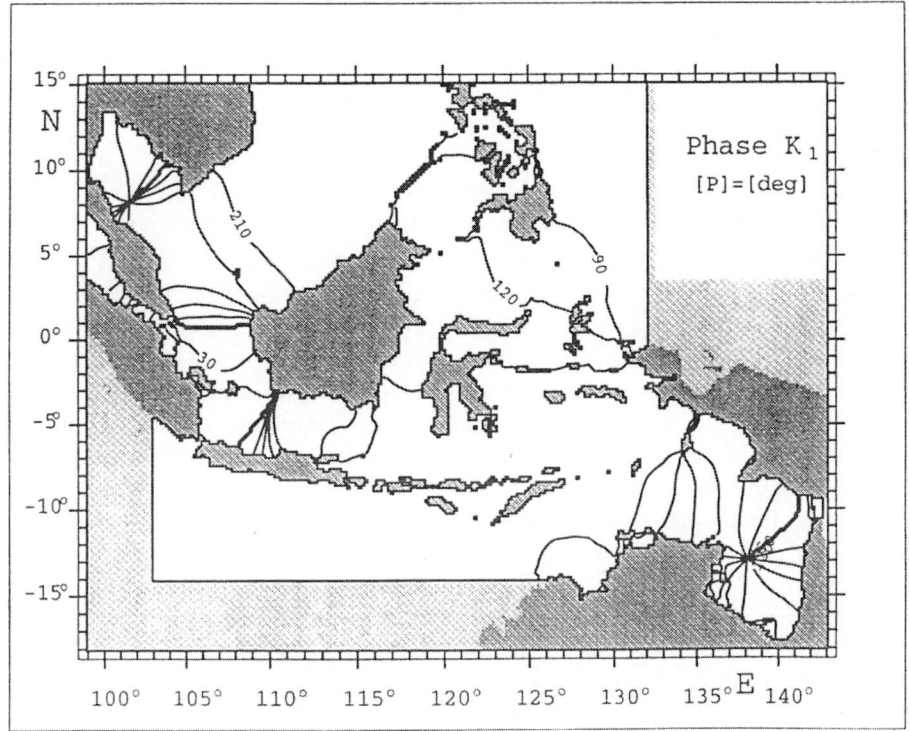

Fig. 10. Computed K_1-tide elevation. Cotidal lines with Greenwich phases in steps of 30°. From Stawarz (1994).

$$\frac{\partial \zeta}{\partial t} + \frac{\partial (hu)}{\partial x} + \frac{\partial (hv)}{\partial y} = 0 \tag{19}$$

So the tidal response of a semi-enclosed rectangular basin with constant depth to a wave prescribed at the open side can be described by linear superposition of two Kelvin waves and an infinite number of second kind Poincaré waves, where these waves are modified by friction (Röber, 1970). Neglecting frictional influence, for constant depth Kelvin waves are characterized by zero current velocity perpendicular to the direction of propagation, the phase velocity $c = \sqrt{gh}$ and by transverse exponential decay of amplitudes:

$$\zeta = \exp\left(-\frac{fy}{\sqrt{gh}}\right) \cos\left(\omega\left(t - \frac{x}{\sqrt{gh}}\right)\right)$$
$$u = \frac{g}{\sqrt{gh}} \zeta \tag{20}$$

Poincaré waves in a channel of width l $(-l/2 \leqq y \leqq l/2)$ propagate with the phase velocity

$$c_n = \left(gh + \frac{f^2}{\kappa_n^2} + \left(\frac{n\pi}{l}\right)^2 \frac{gh}{\kappa_n^2}\right)^{1/2} \tag{21}$$

For $n = 1, 3, 5, \ldots$ the elevation is given by

$$\zeta = a \left(\cos \frac{n\pi}{l}y + \frac{\omega n\pi}{l f \kappa_n} \sin \frac{n\pi}{l}y \right) \cos\left(\omega t - \kappa_n x\right) \tag{22}$$

The expressions for u and v directly result from (19) yielding non-degenerate tidal current ellipses, generally. Corresponding solutions are obtained for even values of n. The progressive Poincaré waves can be understood as having been generated by the reflection of plane Sverdrup waves with $c = \left(gh + f^2/\kappa^2\right)^{1/2}$ and right-hand rotating current ellipses for positive f. As second kind standing Poincaré waves exponentially decaying, e.g., from the closed end of the rectangular basin towards the open end, they require fulfillment of $\omega^2 < f^2 + gh\left(n\pi/l\right)^2 = \omega_n^2$. These latter waves allow constructing together with two Kelvin waves a solution with zero current velocity normal to the rigid boundaries and taking prescribed values at the open boundary, e.g., values of the current velocity component normal to latter. The reflection of a semidiurnal Kelvin wave in a semi-enclosed rectangular basin with the dimensions of the North Sea yields a simple symmetric oscillation pattern which roughly resembles the M_2-tide regime of this adjacent sea. The left-hand rotating amphidromes are located on the central axis half a Kelvin wave length apart with alternating currents and left-hand rotating strong current velocity vectors near the reflecting closed end. If, due to the channel width l being sufficiently large, the tidal period T is less than $2\pi/\omega_1$ the Kelvin wave reflection is no more complete as in the above example of Taylor (1920), but at least one progressive Poincaré wave is involved. This wave might even dominate an asymmetric oscillation pattern with currents which are no more alternating distant from the closed end of the basin (Brown, 1973). Tidal amplitude and phase patterns of many close-to-rectilinear gulfs have successfully been explained by interference of simple wave solutions. This applies, e.g., to the Adriatic Sea and to the Gulf of California (Hendershott and Speranza, 1971).

Simple solutions of progressive Kelvin and Poincaré type waves are also appropriate to give an idea of tidal wave phenomena being governed by the transition from the open ocean to the continental shelf and by the continental boundary assumed to be straight and infinitely extended. Solutions representing free waves travelling parallel to the coast are given for different normal-to-coast depth profiles by Mysak (1980). For a topography defined by the constant depths $\tilde{h}_1$ on the shelf $(-L \leq x \leq 0)$ and $\tilde{h}_2$ off the shelf $(0 < x)$ the dimensionless solution

$$\zeta_{1,2} = \left(\cos\left(\alpha_{1,2}\,x\right) + b_{1,2}\sin\left(\alpha_{1,2}\,x\right)\right)\cos\left(\beta y - \omega t\right) \tag{23}$$

is valid fulfilling the conditions of continuity of elevation and of normal-to-shore transport at the shelf step, and the adiabatic condition at the coast if

$$\omega^2 - 1 = h_1\left(\alpha_1^2 + \beta^2\right)$$
$$\omega^2 - 1 = \alpha_2^2 + \beta^2$$
$$b_1 = \frac{\beta \cot \alpha_1 L - \omega \alpha_1}{\beta + \omega \alpha_1 \cot \alpha_1 L}$$

$$b_2 = \frac{1}{\omega\alpha_2}\left(\beta - \frac{h_1\left(\beta^2 + \omega^2\alpha_1^2\right)}{\beta + \omega\alpha_1\cot\alpha_1 L}\right) \tag{24}$$

At this, depths and elevations have been made dimensionless by $\tilde{h}_2$, velocities by $C = \sqrt{g\tilde{h}_2}$, time and frequency by f^{-1} and f, respectively, distances by Cf^{-1} and wavenumbers by $C^{-1}f$. The frequency ω and the wavenumber β are real, and ω is taken positive. Obviously, a continuum of normal-to-shore trigonometric Poincaré waves with real α_1 and α_2 results if $\omega^2 - 1 - \beta^2 > 0$. Edge waves result for $\omega^2 - 1 < \beta^2 < h_1^{-1}\left(\omega^2 - 1\right)$ being trigonometric ($\alpha_1 = $ real) on the shelf and exponential ($i\alpha_2 = $ positive real) off the shelf, and finally $h_1\beta^2 > \omega^2 - 1$ yields completely exponential shelf and edge waves with $i\alpha_1$ and $i\alpha_2$ both being positive real. In each of the cases $\zeta_{1,2}$ turns out to be real. The edge and shelf waves being exponential off the shelf are discrete due to the requirement that they decay towards the open sea, whence the following relationship has to be fulfilled:

$$\frac{\tan\alpha_1 L}{\alpha_1} = \frac{\omega\left(\beta + i\omega\alpha_2\right)}{\left(\omega^2 - 1\right)\left(\omega^2 - h_1\beta^2\right) - \beta\left(\beta + i\omega\alpha_2\right)} \tag{25}$$

Every $\beta < 0$ a frequency ω with $(0 < \omega < 1)$ is uniquely assigned to fulfilling (24) and (25). The first two equations of (24) and equation (25) constitute the dispersion relation for the discrete waves, generally. The above wave with $\omega < 1$ represents a wave trapped against the shelf edge travelling on the northern hemisphere $(f > 0)$ with the continental coast to the right. The existence of only the gravest shelf mode is due to the assumption of a step shelf. Generally an infinite number of shelf waves with $0 < \omega < 1$ exists.

The necessary condition $\omega^2 < f^2$ for shelf waves, given in dimensional quantities, is fulfilled for diurnal tides already in middle latitudes. Cartwright et al. (1980) detected shelf wave behavior of the northward propagating diurnal tide wave at the St. Kilda shelf off Northwest Scotland, viz., maximum elevation at the shelf edge and tidal current velocity vectors with negative rotation on the shelf and positive rotation off the shelf.

For negative β and tidal frequencies the lowest edge wave mode represents the generalization of the flat bottom Kelvin wave described above. It was shown by Munk et al. (1970) (see also Zahel, 1986) that shelf and deep-sea tidal measurements off the Californian coast can be fitted by introducing a step shelf approximation and by superposing free waves of the above types and a forced wave. A free Kelvin like edge wave travelling northwards along the coast and a free Poincaré like wave travelling southwards along the coast dominate the M_2-tide regime there, which is characterized by a left-hand rotating amphidrome (see Fig. 5). The elevation amplitudes at the coast amount to 62.2 cm (Kelvin wave) and to 18.6 cm (Poincaré wave) (Zahel, 1986).

Shallow Water Tides. In the shallow water parts of the shelf regions the non-linear constituents of (15) and (17), including the parameterization of bottom friction (see (1)), lead to specific alterations of tidal elevations and currents.

Moreover, they can give rise to considerable interactions between the tidal and non-tidal fields of motion. Analyzing time series of elevation and current velocity obtained by measurements or by applying tidal models taking into account the nonlinear terms, generally yields overtides and compound tides for shallow water positions. The latter shallow water tidal constituents have frequencies which are an exact multiple of the frequencies of the astronomical tidal constituents (overtides) or which are equal to linear integral superpositions of the frequencies of two or more astronomical tidal constituents (compound tides). In Tab. 3, basing on five year's measured time series, harmonic constants of Cuxhaven located at the mouth of the river Elbe into the German Bight are put together with the angular frequencies of the constituents given. From 53 constituents with elevation amplitudes of more than 0.2 cm only the astronomical constituents with the largest amplitudes are listed together with overtides and compound tides related to the former and being important at this site.

Table 3. Elevation harmonic constants of Cuxhaven[a] (after Zahel (1986))

Tide	Frequency $°/h$	Amplitude cm	Phase[b] degrees
Astronomical			
O_1	13.9430356	9.3	262.7
K_1	15.0410686	6.5	52.0
N_2	28.4397295	21.1	334.5
M_2	28.9841042	134.4	1.4
S_2	30.0000000	34.4	70.7
K_2	30.0821373	10.2	67.9
Shallow water			
$2SM_2$	31.0158958	3.1	297.1
MO_3	42.9271398	1.6	171.9
MN_4	57.4238338	3.8	224.1
M_4	57.9682085	11.4	248.1
MS_4	58.9841042	7.3	312.7
MK_4	59.0662415	2.1	308.8
$2MN_6$	86.4079380	3.4	75.2
M_6	86.9523127	6.6	106.0
$2MS_6$	87.9682085	6.6	170.0
$2MK_6$	88.0503458	1.9	171.1

[a] Located at 53°53′N, 8°43′E
[b] Referred to local meridian

Overtide and compound tide generation can be attributed to the individual nonlinear terms, i.e., to advection $v\nabla v$, nonlinear continuity $\nabla \cdot (\zeta v)$ and to quadratic bottom friction $\frac{r}{h+\zeta}|\mathbf{v}|\,\mathbf{v}\ \left(\approx\frac{r}{h}|\mathbf{v}|\,\mathbf{v}-\frac{r}{h^2}\zeta\,|\mathbf{v}|\,\mathbf{v}\right)$, by the application of numerical models and of Fourier decomposition technique (Le Provost, 1991; Walters and Werner, 1991). The role the individual nonlinear terms are playing clearly depends upon the topography of the area of investigation. However, some general statements being approximately valid are possible. Second order approximation yields that shallow water continuity and to a smaller extent advection generate even overtides and compound tides as well as mean sea level changes and half monthly and monthly modulations. High-frequency odd harmonics, like M_6, $2MN_6$ and $2MS_6$, and also semidiurnal odd harmonics, e.g. $2SM_2$, are due to the efficiency of bottom friction. Particularly in estuaries also even harmonics can result from the action of bottom friction. Asymmetrical effects due to shallow water continuity and advection as, e.g., represented by M_4, can be exemplified by the distortion of a progressive wave connected with the plane wave phase velocity $c_0=\sqrt{gh}$ changed to $c=c_0\left(3\left(1+\zeta/h\right)^{1/2}-2\right)$. Figure 11 shows tidal elevation curves typically distorted by shallow water effects. The mean spring and neap curves have been obtained from measurements and reflect the contributions of all relevant tidal constituents, whereas the third curve results from the application of a nonlinear M_2-tide model. Particularly striking and significant is that the times between low and high water and between high and low water are clearly smaller and greater than half of the basic tidal period, respectively. Considering shallow water continuity and advection, simple dependencies of elevation amplitudes and phases of overtides and compound tides on those of the involved astronomical tides have been derived by Dronkers (1964). For quarter-diurnal constituents these relationships are given by

$$Z_{aa}/Z_{ab}=Z_a^2/2Z_aZ_b$$
$$\psi_{aa}-\psi_{ab}=\psi_a-\psi_b \tag{26}$$

At this, $Z_{aa},Z_{ab},\psi_{aa},\psi_{ab}$ denote elevation amplitude Z and phase ψ of quarter-diurnal constituents with frequencies 2a (aa) and a+b (ab), while Z_a,Z_b,ψ_a,ψ_b denote amplitudes and phases of astronomical semidiurnal constituents with frequencies a and b, respectively. For extended shallow water regions these relationships prove approximately fulfilled, as has been demonstrated by numerical model application (Le Provost, 1991).

The co-oscillating tides in the Bohai Sea, approximately extending between 118°E and 121°E and between 37°N and 41°N with a mean depth of 18 m, have been computed by Huang (1995) using a spatial three-dimensional barotropic finite-difference-model. The semidiurnal tides are dominating, but also the K_1- and O_1-tide regimes are well developed. The M_2-tide (Figs. 12 and 13) is near-resonant and essentially appears as a reflected Kelvin wave in the northern part with the left-hand rotating amphidrome being a quarter wave length distant from the northern end and a progressive Sverdrup wave in the central basin with right-hand rotating tidal current ellipses. The predominant shallow water tides in the Bohai Sea are M_4 and MS_4. The M_4-tide oscillation pattern (Fig. 14) shows the reduced spatial scale with many amphidromes developing. In the northern part again a reflected Kelvin wave pattern can be identified, the width of the basin

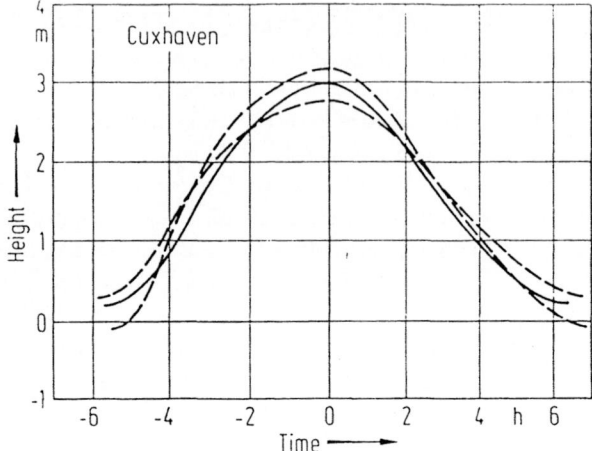

Fig. 11. Mean spring and neap curves (broken lines) for Cuxhaven (see Tab. 3). Heights refer to Chart Datum. Tidal curve (solid line) as obtained by a M_2-tide model.

being not large enough to allow a first kind Poincaré wave to arise.

1.3   Internal Tides.

From (15), (16) and (18) basic free internal wave behavior can be derived (Wunsch, 1975). Taking the linearized equations as a basis and assuming a flat ocean bottom at the depth $z = -h$ one obtains a vertical dependence of the vertical current velocity component in the form

$$W(z) = A_n \sin \alpha_n N (z + h) \tag{27}$$

At this, N is taken as constant and α_n is required to fulfill the n-th-mode relationship

$$\tan \alpha_n N h = N/g\alpha_n^2 \tag{28}$$

Defining $h_n = 1/g\alpha_n$, the horizontally dependent parts of the horizontal current velocity components and the pressure fulfill the tidal equations for barotropic motions with h_n playing the role of depth for the n-th-mode. Equation (28) yields for $n > 1$, i.e., for internal modes, approximately

$$1/\alpha_n = c_n = Nh/n\pi \tag{29}$$

With the realistic values $N = 1.7 \cdot 10^{-3}\,\mathrm{s}^{-1}$, $h = 5000\,\mathrm{m}$ a wavelength of the order of 200 km results for the first mode of diurnal tidal waves. For an interfacial long wave, sometimes being a good approximation to wave propagation at pycnoclines, a similar order of magnitude is obtained from its dispersion relation when applying tidal frequencies:

$$c = \left(\frac{\Delta\rho}{\rho} g \, \frac{h'(h - h')}{h} \right)^{1/2} \tag{30}$$

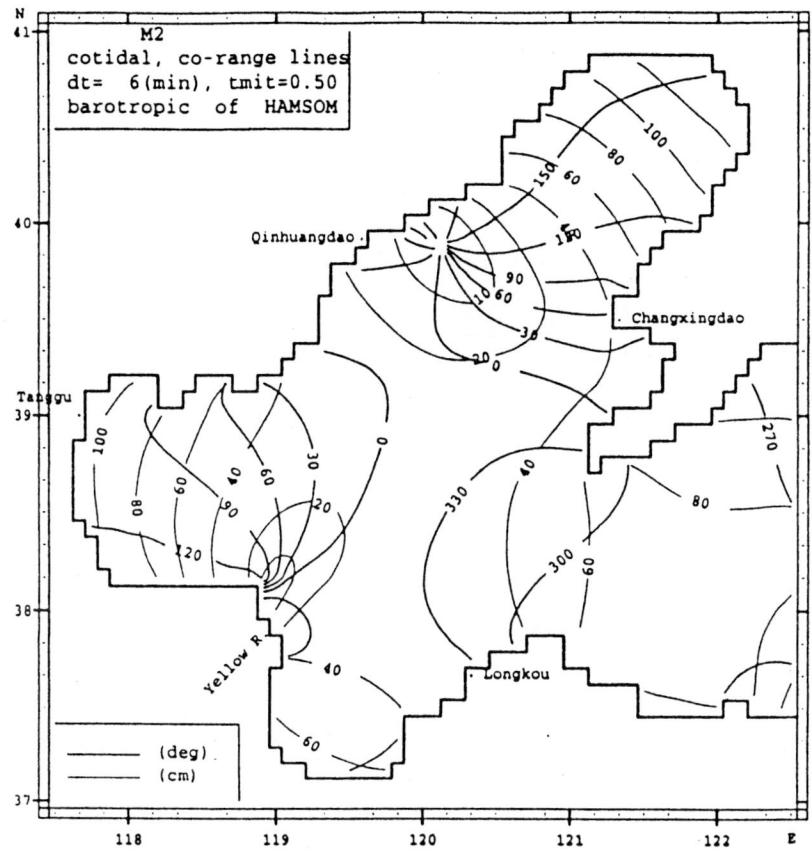

Fig. 12. Computed cotidal and coamplitude charts of the M_2-tide in the Bohai Sea. Phases referred to zonal meridian. From Huang (1995).

At this, h' denotes the thickness of the upper layer and $\Delta\rho$ the density difference between the lower and upper layer. Thus due to mismatch of the spatial scales of the tide generating potential and the possible response in the sea, a direct generation of internal tides by the external tidal forces is not possible. An accepted explanation for the existence of internal tides is that they represent the response to the forcing of isopycnal surfaces over ocean depth variations, e.g., the continental shelf slope, by the barotropic tide. To the order of the approximation (29), obviously (see (27)) internal waves are not connected with a moving free surface, and in fact the surface elevation is hardly influenced by internal tides. Subsurface measurements of current velocity and temperature often show internal tidal signals clearly. The vertical displacements of the isotherms, as measured by Meincke (1978) at $0°\,01'\,S$, $28°\,52'\,W$ in a region of the mid-Atlantic ridge with strong semidiurnal barotropic tides (Fig. 5), include distinct semidiurnal tidal signals. As can be taken from Fig. 15, the first-mode like pattern shows maximum amplitudes for the $15°C$–isotherm of the order of $20\,m$.

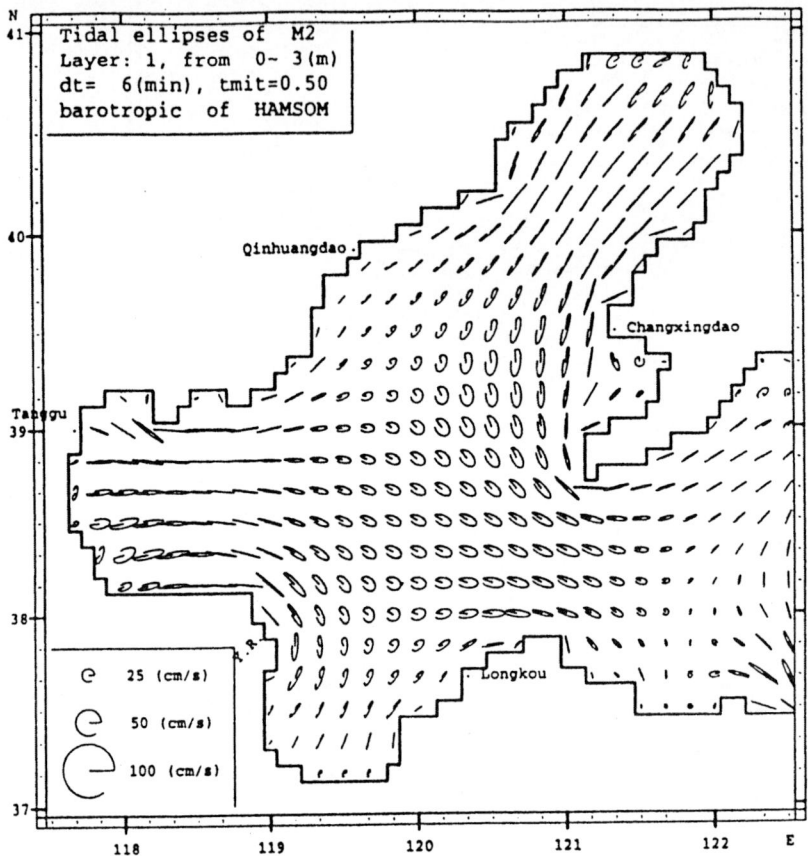

Fig. 13. Computed M_2-tide current ellipses in the surface layer of the Bohai Sea. From Huang (1995).

Although being generated by the barotropic tide, the internal tide is very variable (Sandstrom, 1991). This property is attributed to stratification changes and to the internal tides interacting with the non-tidal field of motion being particularly energetic in the continental slope region. Internal tidal wave behavior, viz., phase velocities being small as compared to barotropic long wave speeds and comparatively large amplitude-wavelength ratios, leads to the formation of internal bores and of internal solitary waves. The tidal energy lost due to internal wave breaking, shear instability and increased bottom friction is regarded to represent a non-negligible energy sink in the global tidal energy budget.

References

Accad, Y., Pekeris, C.L. 1978. Solution of the tidal equations for the M_2 and S_2 tides in the world ocean alone. *Phil. Trans. R. Soc. London A.* **290**: 235 - 266.

Brown, P.J. 1973. Kelvin-wave reflection in a semi-infinite canal. *J. Mar. Res.* **31**: 1 - 10.

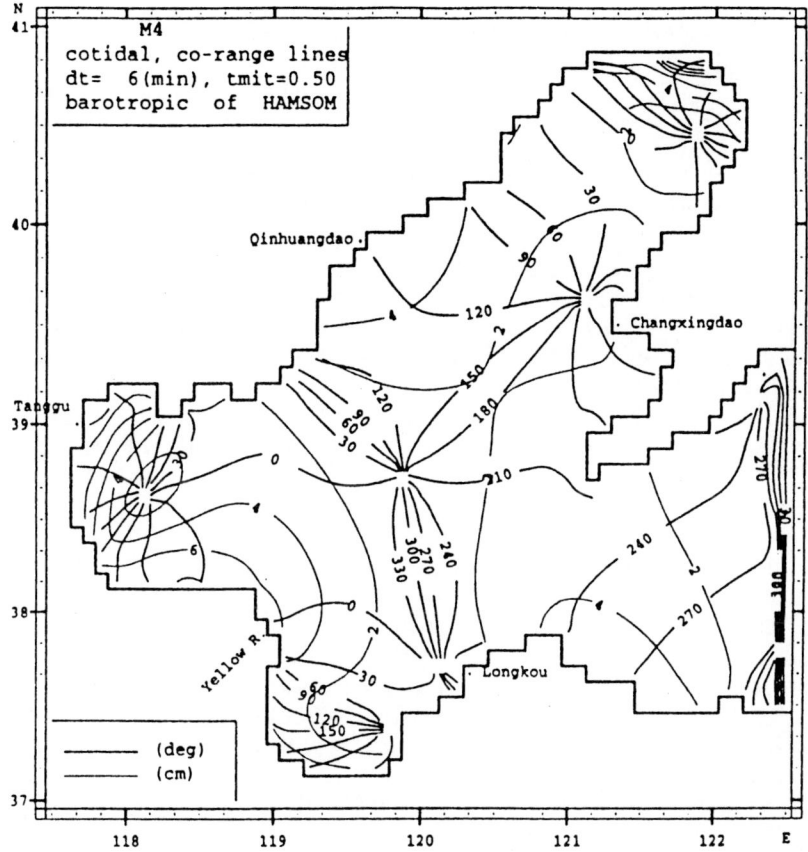

Fig. 14. Computed cotidal and coamplitude charts of the M₄-tide in the Bohai Sea. Phases referred to zonal meridian. From Huang (1995).

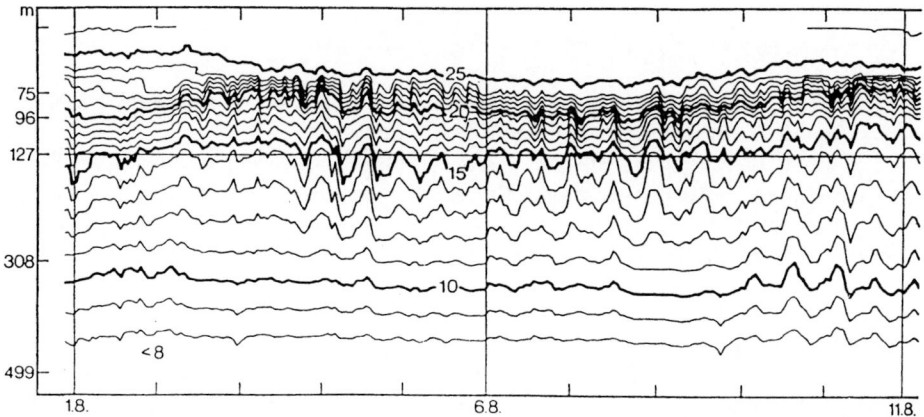

Fig. 15. Isotherms from moored instruments at position 00°01′S, 29°52′W between 1.8.1974 and 11.8.1974. Temperatures are given in °C. After Meincke, personal communication.

Cartwright, D.E., Huthnance, J.M., Spencer, R. and Vassie, J.M. 1980. On the St. Kilda shelf tidal regime. *Deep-Sea Res.* **27**: 61 - 70.

Davies, A.M., Furnes, G.K. 1980. Observed and Computed M_2 Tidal Currents in the North Sea. *J. Phys. Oceanogr.* **10**: 237 - 257.

Doodson, A.T. 1964. Oceanic tides. *Adv. Geophys.* **5**: 117 - 152.

Dronkers, J.J. 1964. Tidal computation in rivers and coastal waters. pp. 1 - 299. North Holland Publ. Comp.

Eanes, R., Bettadpur, S. 1995. The CSR3.0 Global Ocean Tide Model. *Center Space Research, Techn. Memorandum* CSR-TM-95-06.

Egbert, G.D., Bennett, A.F., Foreman, M.G.G. 1995. TOPEX/POSEIDON tides estimated using a global inverse model. *J. geophys. Res.* **99**: 821 - 24, 852.

Garrett, C.J.R., Munk, W.H. 1971. The age of the tide and the Q of the oceans. *Deep-Sea Res.* **18**: 493 - 504.

Gaviño, J.H. 1984. On the calculation of resonance oscillations of a World Ocean's finite difference model by means of the Lanczos method. *Mitt. Inst. Meereskd. Univ. Hamburg* 27: 1 - 78.

Gotlib, V.Y., Kagan, B.A. 1982. Numerical simulation of the tides in the world ocean: 3. A solution to the spectral problem. *Dt. hydrogr. Z.* **35**: 45 - 58.

Grawunder, D. 1996. Assimilation von Daten in ein numerisches Modell der Gezeiten des Weltozeans. Dissertation Univ. Hamburg.

Hendershott, M.C., Speranza, A. 1971. Co-oscillating tides in long narrow bays; the Taylor problem revisited. *Deep-Sea Res.* **18**: 959 - 980.

Hendershott, M.C. 1977. Numerical models of ocean tides. The Sea: Ideas and Observations on Progress in the Study of the Seas **6**: 47 - 95. John Wiley, New York.

Huang, D. 1995. Modelling studies of barotropic and baroclinic dynamics in the Bohai Sea. *Berichte ZMK Univ. Hamburg.* **17**: 1 - 26.

Kagan, B.A. 1974. Hydrodynamic models of tidal motions at sea. *Techn. Transl. DMAAC-TC-2028 St. Louis* 1 - 260.

Kowalik, Z., Proshutinsky, A.Y. 1994. The Arctic Ocean Tides. Geophys. Monograph **85**. Am. Geophys. Union. pp. 137 - 158.

Le Provost, C. 1991. Generation of overtides and compound tides (Review). In: *Tidal Hydrodynamics*. pp. 269 - 295. John Wiley, New York.

Longuet-Higgins, M.S., Pond, G.S. 1970. The free oscillations of fluid on a hemisphere bounded by meridians of longitude. *Phil. Trans. Roy. Soc. London A.* **266**: 193 - 223.

Marchuk, G.I., Kagan, B.A. 1986. Dynamics of ocean tides. In: *Oceanographic Sciences Library.* **3**: 1 - 327. Kluwer Academic Publ.

Meincke, J. 1978. Measurements of currents and stratification during the GATE Equatorial Experiment - Data Report. *Meteor. Forsch.-Ergebnisse A.* **20**: 81 - 100.

Mihardja, D.K. 1991. Energy and momentum budget of the tides in Indonesian Waters. *Berichte ZMK Univ. Hamburg.* **14**: 1 - 183.

Mittelstaedt, E., Lange, W., Brockmann, C., Soetje, K.C. 1983. Die Strömungen in der Deutschen Bucht. *Dtsch. Hydrograph. Inst., Hamburg.* **2347**: 1 - 141.

Munk, W., Snodgrass, F., Wimbush, M. 1970. Tides Off Shore: Transition from California Coastal to Deep-Sea Waters. *Geophys. Fluid Dynamics.* **1**: 161 - 235.

Mysak, L.A. 1980. Topographically trapped waves. *Ann. Rev. Fluid Mech.* **12**: 45 - 76.

Platzman, G.W., Curtis, G.A., Hansen, K.S. and Slater, R.D. 1981. Normal modes of the World Ocean. Part II: Description of modes in the period range 8 to 80 hours. *J. Phys. Oceanogr.* **11**: 579 - 603.

Röber, K. 1970. Analytische und numerische Lösungen für Mitschwingungsgezeiten in einem Rechteckbecken konstanter Tiefe unter Berücksichtigung von Bodenreibung, Corioliskraft und horizontalem Austausch. *Mitt. Inst. Meereskd. Univ. Hamburg.* **16**: 1 - 119.

Sandstrom, H. 1991. The origin of internal tides (A revisit). In: *Tidal Hydrodynamics.* pp. 437 - 447. John Wiley, New York.

Schwiderski, E.W. 1983. Atlas of ocean tidal charts and maps, I, The semidiurnal principal lunar tide M_2. *Mar. Geod.* **6**: 219 - 265.

Schwiderski, E.W. 1985. On tidal friction and the deceleration of the Earth's rotation and Moon's revolution. *Mar. Geod.* **9**: 399 - 450.

Scherneck, H.-G. 1990. Loading Green's function for a continental shield with a Q-structure for the mantle and density constraints from the geoid. *Bull. Inform. Marées Terr.* **108**: 7775 - 7792.

Seiler, U. 1989. An investigation to the tides of the world ocean and their instantaneous angular momentum budgets. *Mitt. Inst. Meereskd. Univ. Hamburg.* **29**: 1 - 104.

Smithson, M.J. 1992. Pelagic tidal constants 3. *Publ. Sci. Int. Assoc. Phys. Sci. Oceans.* **35**: 1 - 191.

Stawarz, Z.M. 1994. Bestimmung von Geoidkorrekturen der Südostasiatischen Gewässer aus simulierten und altimetrischen Meeresoberflächenhöhen. Diplomarbeit Univ. Hamburg 1 - 104.

Taylor, G.I. 1920. Tidal oscillations in gulfs and rectangular basins. *Proc. London Math. Soc.* **20**: 148 - 181.

Walters, R.A., Werner, F.E. 1991. Nonlinear generation of overtides, compound tides, and residuals. *Tidal Hydrodynamics.* pp. 297 - 320. John Wiley, New York.

Webb, D.J. 1980. A model of continental-shelf resonances. *Deep-Sea Res.* **23**: 1 - 15.

Webb, D.J. 1980. Tides and tidal friction in a hemispherical ocean centered at the equator. *Geophys. J. Roy. astron. Soc.* **61**: 573 - 600.

Wunsch, C. 1975. Internal tides in the ocean. *Rev. Geophys. Space Phys.* **13**: 167 - 182.

Zahel, W. 1978. The influence of solid earth deformations on semi-diurnal and diurnal oceanic tides. In: Tidal friction and Earth's rotation I. pp. 98 - 124. Springer, Heidelberg.

Zahel, W. 1986. Mathematical modelling of global interaction between ocean tides and earth tides. *Phys. Earth Planet. Inter.* **21**: 202 - 217.

Zahel, W. 1986. Astronomical tides. In: *Landoldt-Börnstein, New Series V/3c.* pp. 83 - 134. Springer, Heidelberg.

Zahel, W. 1990. The influence of ocean and solid earth parameters on oceanic eigenoscillations, tides and tidal dissipation. Geophysical Monograph **59**: 33 - 41. Am. Geophys. Union.

Zahel, W. 1991. Modeling ocean tides with and without assimilating data. *J. geophys. Res.* **96** 20379 - 20391.

Zahel, W. 1995a. Assimilating ocean tide determined data into global tidal models. *J. Mar. Systems.* **6**: 3 - 13.

Zahel, W. 1995b. The influence of assimilated data in global ocean and loading tide modeling. In: *Proc. 12th Int. Symp. on Earth Tides.* pp. 413 - 422. H.-T. Hsu (ed.). Science Press Beijing.

Earth Tides and Ocean Tidal Loading

Gerhard Jentzsch

Friedrich-Schiller-Universität Jena, Institut für Geowissenschaften, Burgweg 11, D-07749 Jena, Germany

Abstract. The tidal deformation of the solid earth is superimposed by the loading effect caused by the ocean tides. There are two contributions: the attraction of the moving water masses and the deformation of the crust due to the water load. Thus, the elasticity of the local crust-mantle structure controls the response. The distribution of the continents and the topography of the ocean bottom avoid a uniform response of the ocean to the tidal forces. Therefore, the tidal loading signal is generally not in phase with the body tides.

The different transfer functions for gravity, tilt, and strain result in different loading effects as well: The loading signal in gravity is close to the ocean mostly smaller than the body tide, but it can still be observed at great distances from the oceans. On the other hand, in tilt and strain the loading signal may surmount the body tide close to the coast, but it vanishes below the noise level far from the coast. Therefore, tidal gravity provides an independent tool to constrain ocean tidal models, whereas tidal tilt and strain can be used to determine local crustal structures.

The vertical displacement due to ocean tidal loading may reach values of more than a decimeter close to the coasts, and it must be taken into account for high precision earth monitoring. Models for the correction of the body tides exist, but since there is still no global ocean tidal model available that also fits all the shelf tides further gravity measurements are needed.

1 Introduction: Earth tides and ocean tides

The variations of the surface of the oceans due to tidal forces related to moon and sun are well known. First descriptions are more than two thousand years old. In contrast, the knowledge about the deformation of the solid earth caused by the same forces is much younger: This was proved at the end of the last century. Ekman (1993) tells the story of the discovery of the relations between the tidal forces, the marine tide and the tidal deformation of the solid earth. Referring to old original publications he covers the development from antique time to the year 1950, whereas the latest developments are compiled by Jentzsch (1986).

George Howard Darwin (1882) analysed tide gauge data from fourteen harbours distributed all over the globe. He concentrated on the long-period tides Mm and Mf (periods of about 13.66 and 27.55 days, resp.), because these show nearly no resonance compared to the diurnal and semidiurnal tides. The ratio of the observed amplitudes and the theoretical amplitudes derived from the tidal potential was about two thirds and not unity. This could only be explained by the ocean bottom undergoing vertical displacements with these tidal periods, thus diminishing the ocean tidal amplitudes. From this indirect prove of the tidal deformation of the solid earth the first estimation of global elasticity was possible.

Baker & Lennon (1976) compared the results of tidal gravity observations throughout Europe. They found, that the amplitude factor δ and the phase difference α varied between 1.13 and 1.31, and - 4.9° and + 1.1°, resp. For a standard solid earth the amplitude factor should be around 1.16 and the phase difference 0.0°.

This scatter of the tidal parameters is due to the effect of ocean loading: The movement of the water masses causes changes in the earth's gravity field, and the load of the water results in additional vertical displacements. Both phenomena have tidal periods, but they are not necessarily in phase with the body tides, because the marine tides are restricted by coast lines and the topography of the ocean bottom. Therefore, we observe resonance tides in shelf areas and amphidromes (zero tidal amplitudes) at distinct areas in the open ocean. Thus, the signal of the load tide is a vector which is added to the theoretical tidal vector (comp. Fig. 1.1). This load vector contains information both on the distribution of the marine tide and the regional elastic properties of crust and upper mantle (comp. Baker, 1978).

The difference between theory and observation is called the residual vector, because in principle it contains further additional effects, too. This is of special importance to tilt and strain tides, which are not discussed here (comp. Harrison, 1976).

2   Deformation due to ocean loading

The following discussion of the computation of the deformation caused by surface loads is orientated on the basic article of Farrell (1972). Further explanation is provided by Harrison (1985). We have to distinguish three steps:

(1) *Computation of load Love numbers for a given layered earth:*
     In contrast to the conventional Love numbers that describe the static deformations of an elastic earth due to tidal forces the load Love numbers are related to the boundary condition of a surface mass load. They are functions of the radius r and the degree n. In the loading case it is not sufficient to discuss low degrees only: Farrell (1972) computes the load Love numbers up to degree n = 10,000.

(2) *Green's functions:*
     The Green's functions are formed by proper weighted sums of the load Love numbers. They depend both on the radius r and the angular distance Θ we are only interested in the dependence on Θ at the surface r = a.

(3) *Computation of the loading effect:*
     The ocean tidal chart of a specific tidal constituent is tabulated for a grid with individual amplitudes and phases for each cell. The convolution of all

cells with the Green's function gives the loading effect in gravity, tilt and strain.

The shape of the finite sized load of each cell is assumed as a disc. Therefore, so-called disc factors are needed if the distance is smaller than ten times the radius of the disk (Farrell, 1973).

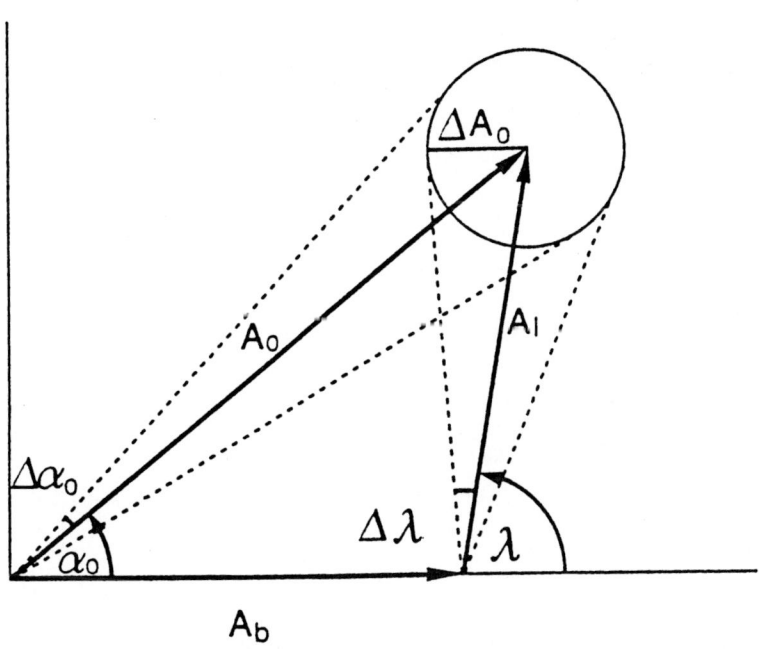

Figure 1.1. Load vector as difference between theory and observation:

A_b : theoretical tidal amplitude multiplied by the amplitude factor for a standard earth model

A_0 : observed amplitude

α_0 : phase difference between theory and observation

A_l, λ : amplitude and phase of the load vector

$\triangle A_0, \triangle \alpha_0, \triangle \lambda$: errors

2.1 Computation of load Love numbers for a given layered earth

The basic solution is usually derived for a point load on a nongravitating, elastic half-space (Lamé parameters $\lambda, \mu = $ const.). This boundary-value problem is associated with Boussinesque (1885). The solution can be extended later to our case of a spherical, layered and gravitating earth. The elastic equilibrium equation for the displacement vector $\vec{s}$ is

$$\sigma \nabla \nabla \cdot \vec{s} - \mu \nabla \times \nabla \times \vec{s} = 0 \tag{1}$$

with $\sigma = \lambda + 2\mu$. The problem is axially symmetric, and the solution should satisfy the condition of a stress-free surface except at the point load. Thus,

$$\vec{s} = u(z,r)\vec{e}_z + v(z,r)\vec{e}_r \tag{2}$$

With $\eta = \lambda + \mu$, $R^2 = r^2 + z^2$ (r: distance, z: depth) we find the solutions for a unit point force pressing vertically on the surface of an elastic half-space:

$$
\begin{aligned}
u(z,r) &= -\frac{1}{4\pi\mu R}\left(\frac{\sigma}{\eta} + \frac{z^2}{R^2}\right) \\
v(z,r) &= -\frac{1}{4\pi\eta r}\left(1 + \frac{z}{R} + \frac{\eta r^2 z}{\mu R^3}\right)
\end{aligned}
\tag{3}
$$

The multiplication with a disc factor allows to adjust for cylindrical or elliptical loads. This was necessary to reduce computing time and improve convergence. Today, due to available computer power we use a finer grid especially near the coast close to the observation point. Therefore, we do not use disc factors any more.

The attraction of the load follows from Poisson's equation for the gravitational potential perturbation:

$$\nabla^2 \Phi = -4\pi G \rho \nabla \cdot \vec{s} \tag{4}$$

where Φ = disturbing potential, ρ = density, G = gravitational constant. At the deformed surface this results in the acceleration

$$\vec{a} = -\nabla \Phi(0^+, r) \tag{5}$$

with the vertical component $a_z^{\Phi} = (G\rho)/(2\mu r)$. The load causes two more components: The attraction of the load on an unperturbed earth has the horizontal and vertical components a_r^m and a_z^m. The vertical displacement u causes an acceleration change of $a_z^u = 2(\text{g}u)/a$ (vertical gradient). The tilt of the deformed surface causes a horizontal acceleration $a_r^u = -\text{g}\,du/dr$. To the first order $a_z^m = 0$ and $a_r^{\Phi} = 0$. Typical values are:

$$a_z^{\Phi} = -0.29\, a_z^u \quad \text{and} \quad a_r^m = 0.175\, a_r^u.$$

This means that the free-air gradient is reduced by about a third, and the horizontal acceleration due to tilt is six times larger than the direct attraction of the mass load.

For the stratified half-space we have to consider

$$\lambda = \lambda(z), \qquad \mu = \mu(z)$$

Now, analytical solutions are only available for special cases. The Green's functions are found by numerical methods. Farrell (1972) discusses different approaches for plane-layered models.

In the case of *spherical models* the displacement $\vec{s}$ and the potential disturbance Φ caused by the surface loads are expanded in spherical harmonics:

$$\vec{s} = \sum_{n=0}^{\infty} \left(U_n(r) P_n(cos\Theta) \vec{e}_r + V_n(r) \frac{\partial P_n(cos\Theta)}{\partial \Theta} \vec{e}_\Theta \right) \qquad (6)$$

$$\Phi = \sum_{n=0}^{\infty} \Phi_n(r) P_n(cos\Theta) \qquad (7)$$

Since there is no longitude dependence only the particular Legendre order $m = 0$ occurs.

Within the earth the displacement decreases with $(r/a)^n$, which means that for $n > 10$ the core is of no influence. Loads of degree $n = 0$ and $n = 1$ have a special meaning: P_o describes a uniformly distributed load. Since the total load of the ocean must be constant (conservation of mass) P_o should be equal zero. But Agnew (1983) could show that this condition is not sufficient: Mass conservation has to be performed by a special normalisation which is realised by adding a constant tide such that at any time the sum over all oceans is zero. The case $n = 1$ denotes the shift of the center of the earth: This should be impossible, but with regard to the load case it can be shown that this restriction is only valid for both the load and the solid earth, but there is a proportional shift of the center of the earth with regard to the redistribution of the load.

The static deformation of the earth as an elastic sphere is caused by a perturbing potential Φ with

$$\Phi = \Phi_1 + \Phi_2 \qquad (8)$$

where Φ_1: potential of the earth's distorted density field
$\quad\quad\Phi_2$: potential of the applied mass load

Then the Love numbers referred to the potential $\Phi_{1,n}$ and the displacements U_n, V_n (axially symmetric force) are defined as

$$\begin{bmatrix} U_n(r) \\ V_n(r) \\ \Phi_{1,n} \end{bmatrix} = \Phi_{2,n}(r) \begin{bmatrix} \frac{h_n(r)}{g} \\ \frac{l_n(r)}{g} \\ k_n(r) \end{bmatrix} \qquad (9)$$

Here, r is the radius of the earth with $r = a$ as the surface. The displacements U_n, V_n and the potentials are transformed quantities. The transformed potential of the point load is $\Phi_{2,n}(r)$ with

$$\Phi_{2,n}(r) = \frac{ag}{m_e}.$$

There are two different boundary conditions from the force fields that lead to different Love numbers:

(1)   tidal potential (no normal stress at the surface): h_n, l_n, k_n
(2)   potential of the mass load (normal stress at the surface): h_n', l_n', k_n'

The superscript prime denotes the load Love numbers. Both sets of Love numbers describe the response of the earth and for the incompressible earth a relation can be given (Munk & MacDonald, 1960):

$$h_n' = -\frac{2}{3}(n-1)h_n, \qquad k_n' = -\frac{2}{3}(n-1)k_n \qquad (10)$$

The negative sign expresses that the displacement is opposite to the load.

Whereas for the tidal potential the degrees n = 2 and n = 3 are mainly important (we consider the degree n = 4 only for the analysis of high resolution data) Farrell (1972) showed that for loading computations the load Love numbers up to degree 10,000 are needed. Tab. 2.1 contains these load Love numbers for the PREM reference earth model (Dziewonski & Anderson, 1981; comp. Farrell, 1972, for the Gutenberg-Bullen earth model). If the distance Φ from the load is small, the terms with high n are dominating. Then $P_n(cos\Phi)$ can be replaced by Bessel functions, and the sums in equ. (6) and (7) are asymptotically equivalent to the integrals

$$\vec{s} = \int_0^\infty \left\{ \frac{ah_n'}{m_e} J_0 \left[(n+\frac{1}{2})\Theta \right] \vec{e}_r - \frac{al_n'}{m_e}(n+\frac{1}{2})J_1 \left[(n+\frac{1}{2})\Theta \right] \vec{e}_\Theta \right\} dn$$

$$\Phi_1 = \int_0^\infty \left\{ \frac{agk_n'}{m_e} J_0 \left[(n+\frac{1}{2})\Theta \right] \right\} dn \qquad (11)$$

Here, r = a, and the variables U_n, V_n and $\Phi_{1,n}$ are expressed as functions of the load Love numbers and the potential of the point mass $\Phi_{2,n}$. Thus, close to the load a layered earth model responds like a half-space with the properties of the uppermost layer. For high degrees this allows for the comparison of the numerical and the analytical solutions.

2.2   Computation of Green's functions

The Green's functions depend both on the radius r and the angular distance Θ. They are formed by proper weighted sums of the load Love numbers and the spherical harmonics, and they consist of two displacement components and the gravitational potential perturbation. Since we are interested in observable effects, i. e. horizontal and vertical accelerations, the tilt, and the strain tensor, we define six Green's functions in three groups:

  - horizontal and vertical displacement
  - horizontal and vertical acceleration
  - elements of the strain tensor

The surface *vertical displacement* is

$$u(\Theta) = \frac{a}{m_e} \sum_{n=0}^{\infty} h'_n P_n(\cos\Theta) \tag{12}$$

Table 2.1. Load Love numbers for the PREM reference earth model computed up to degree n = 10,000 (comp. Farrell, 1972, tab. A2).

n	$-h_n$	nl_n	$-nk_n$
1	.290	.113	0
2	1.006	.457	.648
3	1.049	.212	.602
4	1.047	.235	.540
5	1.078	.231	.524
6	1.134	.231	.540
7	1.202	.239	.572
8	1.274	.252	.610
9	1.344	.267	.650
10	1.412	.283	.689
18	1.860	.430	.962
32	2.314	.643	1.242
56	2.656	.814	1.404
100	2.937	.899	1.454
180	3.319	.895	1.505
325	4.008	.927	1.761
550	4.923	1.191	2.253
1,000	5.850	1.657	2.833
1,800	6.155	1.860	3.038
3,000	6.175	1.875	3.054
10,000	6.177	1.875	3.056
∞	6.239	1.893	3.072

Taking into account that for large n the h'_n, nl'_n, nk'_n become constant

$$h'_n \rightarrow h'_\infty, \qquad nl'_n \rightarrow l'_\infty, \qquad nk'_n \rightarrow k'_\infty$$

we can write using the asymptotic value for h'_n

$$u(\Theta) = \frac{ah'_\infty}{m_e} \sum_{n=0}^{\infty} P_n(\cos\Theta) + \frac{a}{m_e} \sum_{n=0}^{\infty} (h'_n - h'_\infty) P_n(\cos\Theta) \tag{13}$$

where $(h'_n - h'_\infty) = 0$ for n > N. Using the exact solution for the first term we find

$$u(\Theta) = \frac{ah'_\infty}{2m_e \sin(\frac{\Theta}{2})} + \frac{a}{m_e} \sum_{n=0}^{\infty}(h'_n - h'_\infty)P_n(cos\Theta) \tag{14}$$

The *horizontal displacement* is given by

$$v(\Theta) = \frac{a}{m_e} \sum_{n=1}^{\infty} l'_n \frac{\partial P_n(cos\Theta)}{\partial\Theta} \tag{15}$$

(There is no horizontal displacement for n = 0). As above we find:

$$v(\Theta) = \frac{al'_\infty}{m_e} \sum_{n=1}^{\infty}\frac{1}{n}\frac{\partial P_n(cos\Theta)}{\partial\Theta} + \frac{a}{m_e}\sum_{n=1}^{\infty}(nl'_n - l'_\infty)\frac{1}{n}\frac{\partial P_n(cos\Theta)}{\partial\Theta} \tag{16}$$

Again the first term is exactly known, and we can write:

$$v(\Theta) = \frac{al'_\infty}{m_e}\left(-\frac{cos(\frac{\Theta}{2})\left[1 + 2sin(\frac{\Theta}{2})\right]}{2sin(\frac{\Theta}{2})\left[1 + sin(\frac{\Theta}{2})\right]}\right)$$
$$+\frac{a}{m_e}\sum_{n=1}^{\infty}(nl'_n - l'_\infty)\frac{1}{n}\frac{\partial P_n(cos\Theta)}{\partial\Theta} \tag{17}$$

For the *vertical and horizontal acceleration* we first have to note:

- For both tilt and gravity changes the direct attraction of the mass load is important. They have to be added to the acceleration of the perturbed density field proportional to k'_n.
- An additional vertical acceleration is caused from moving through the gradient of the unperturbed field (proportional to h'_n).
- An additional horizontal acceleration is caused by the tilt of the deformed surface (proportional to h'_n).

Hence we find for gravity and tilt applying a unit load and with N as the number of the last term of the expansion (usually N = 10,000):

$$g_s(\Theta) = \frac{g}{m_e}\sum_{n=0}^{N}\{n + 2h'_n - (n+1)k'_n\}P_n(cos\Theta) \tag{18}$$

(gravity, positive upward)

$$g_n(\Theta) = -\frac{1}{m_e} \sum_{n=0}^{N} \{1 + k'_n - h'_n\} \frac{\partial P_n(cos\Theta)}{\partial \Theta} \tag{19}$$

(tilt)

The first terms in each bracket are the Newtonian attraction which has the exact sum (superscript N):

$$g_s^N(\Theta) = -\frac{g}{4m_e \, sin(\frac{\Theta}{2})} \qquad \text{(vertical)} \tag{20}$$

$$g_n^N(\Theta) = \frac{cos(\frac{\Theta}{2})}{4m_e \, sin^2(\frac{\Theta}{2})} \qquad \text{(horizontal)} \tag{21}$$

Thus, the elastic accelerations due to the displacement are (superscript E):

$$g_s^E(\Theta) = g_s(\Theta) - g_s^N(\Theta) \qquad \text{(vertical)} \tag{22}$$
$$g_n^E(\Theta) = g_n(\Theta) - g_n^N(\Theta) \qquad \text{(horizontal)} \tag{23}$$

The four nonzero elements of the *strain tensor* are (u and v being the radial and tangential displacements at the surface):

$$
\varepsilon_{rr} = \frac{\partial u}{\partial r} \qquad 2\varepsilon_{r\Theta} = \frac{1}{a}\frac{\partial u}{\partial \Theta} + \frac{\partial v}{\partial r} - \frac{v}{a}
$$
$$
\varepsilon_{\Theta\Theta} = \frac{1}{a}\frac{\partial v}{\partial \Theta} + \frac{u}{a} \qquad \varepsilon_{\lambda\lambda} = \frac{u}{a} + cot\Theta\frac{v}{a} \tag{24}
$$

Introducing the Love numbers and taking into account that at the free surface τ_{rr} is a δ-function (point load) and $\tau_{r\Theta} = 0$ we find for the two radial derivatives at the surface

$$\frac{\partial u}{\partial r} = -\frac{2\lambda}{\sigma a}u + \frac{\lambda}{\sigma m_e} \sum_{n=0}^{\infty} n(n+1)l'_n P_n(cos\Theta) \tag{25}$$

$$\frac{\partial v}{\partial r} = -\frac{1}{m_e} \sum_{n=0}^{\infty} h'_n \frac{\partial P_n(cos\Theta)}{\partial \Theta} + \frac{v}{a} \tag{26}$$

Comparing equations (24) and (26) we see that $\varepsilon_{r\Theta} = 0$. The strain component ε_{rr} is given by equ. (25), and $\varepsilon_{\Theta\Theta}$ is

$$\varepsilon_{\Theta\Theta} = \frac{u}{a} + \frac{1}{m_e} \sum_{n=0}^{\infty} l'_n \frac{\partial^2 P_n(cos\Theta)}{\partial \Theta^2} \tag{27}$$

At the stress-free surface $\tau_{rr}(\Theta) = 0$ for $\Theta > 0$; hence for all $\Theta > 0$ we have:

$$\varepsilon_{rr} = -\frac{\lambda(a)}{\sigma(a)}(\varepsilon_{\Theta\Theta} + \varepsilon_{\lambda\lambda}) \tag{28}$$

with $\lambda(a)$ und $\sigma(a) = \lambda(a) + 2\mu(a)$ as the surface parameters of the material. Thus, ε_{rr} is a simple combination of the surface areal strain and the Lamé parameters of the top layer.

The Green's functions are usually normalized by the attraction. Figs. 2.2 a/b provide the gravity and tilt normalised Green's functions for the PREM reference earth model with different crustal depths as a function of the angular distance Θ. It is obvious that the oceanic crust can take less elastic deformation than the thicker continental crust. The depth of the Moho controls the result until nearly 1.0° corresponding to about 100 kilometers. With regard to the assumed continental Moho depth of 40 km we can point out that, as a rule of thumb, the penetrating depth is about half the distance to the load.

Farrell (1972) computed the Green's functions for a Gutenberg-Bullen standard earth model. In the appendix A.1/2 the Green's functions for the PREM model are given for different Moho depths (oceanic and continental crusts). The comparison of these tables to Farrell's reveals that the difference is not big (comp. Jahr, 1989; see also Baker, 1978, who discussed the requirements for the measurement to check different mantle models by ocean loading).

Goad (1980) improved Farrell's algorithm by using integrated Green's functions. This technique simplifies the computation and improves the convergence of infinite series. It explains the disk factors Farrell had to use to overcome these problems. He points out that this method also includes the height of the station above sea level (which is crucial for stations close to the coast); but this is incorporated into Farrell's process, too.

Up to now only earth models have been taken into account that consist of elastic shells: Farrell's algorithm allowed the earth to be compressible, layered, with a solid inner core and a fluid outer core. But there where several attempts to extend the theory (and the experiments) to realistic earth models: Zschau (1977, 1978) calculated phase shifts of the ocean tide loading effects due to low-viscosity layers in the interior of the earth, and Pagiatakis (1990) extended the theory to a self-gravitating, anisotropic viscoelastic and rotating earth.

2.3   Computation of the loading effect

The loading effect is computed by the convolution of the Green's function and the ocean tidal model (Farrell, 1973):

$$L_t(\vec{r}) = \rho \int \int_{oceans} G(\vec{r} - \vec{r}') H(\vec{r}') dA \tag{29}$$

Here,

$G(\vec{r} - \vec{r'})$	-	Green's function
ρ	-	density of the sea water
$H(\vec{r'})$	-	amplitudes (and phases) of the specific marine tide with the position vector
$\vec{r}$	-	position vector of the observation point

In practical computation the integration over the whole ocean has to be replaced by a sum over all ocean cells. These have to be chosen properly, or they are already given by the numerical model of the marine tide. The necessary steps are:

(1) The numerical model is sorted into individual areas (North- / Southatlantic, different parts of the Pacific and Indian Oceans, Arctic oceans, shelf areas). If the model has a 1° x 1° grid we have about 40,000 cells (e.g. Schwiderski, 1979c; 1980).

(2) Since this grid provides an insufficient approximation to the real coast lines a local shelf model has to be derived to replace the local cells of the global model. Here, a resolution down to 1' x 1' is useful. The tidal amplitudes and phases of the global model are then used to compute the values for the fine mesh. If available a local model is applied or derived from local tide gauge data.

(3) The whole model must conserve mass. It may be useful to conserve the mass of attached shelf seas separately.

(4) For the computation of the load Love numbers an earth model with the local crust of the observation point nearest to the coast must be used.

(5) Disk factors are computed, if necessary.

(6) Convolution over different areas of the oceans to provide separate load vectors in order to control especially the effect of the local shelf. The sum of these vectors forms the total load.

Typical values for the ratio of the amplitudes of load tides versus body tides are compiled in tab. 2.2. The declining properties of the different Green's functions are also given.

We observe typical amplitudes for a main constituent of the gravity tides in the order of e.g. 30 μGal[1]. That means that on the basis of a signal-to-noise ratio of about 1,000 the loading effect can be monitored world wide. Furthermore, comparing the observation with the computed load vector the resolution allows to check the fit of global models of the marine tide. This is also true for tilt, although the s/n-ratio is mostly smaller by one order of magnitude. But the load effect is much bigger. Especially close to the sea local crust/mantle structures can be derived from tilt residuals (comp. results for the Irish Sea; Baker, 1980b). But it should be pointed out here, that tilt and strain measurements are often affected by local geological inhomogeneities, like cavities, faults.

---

[1] The common unit μGal is used here. The conversion to the equivalent SI-unit is: $1\,\mu\text{Gal} \equiv 10\,\text{nm/s}^2$

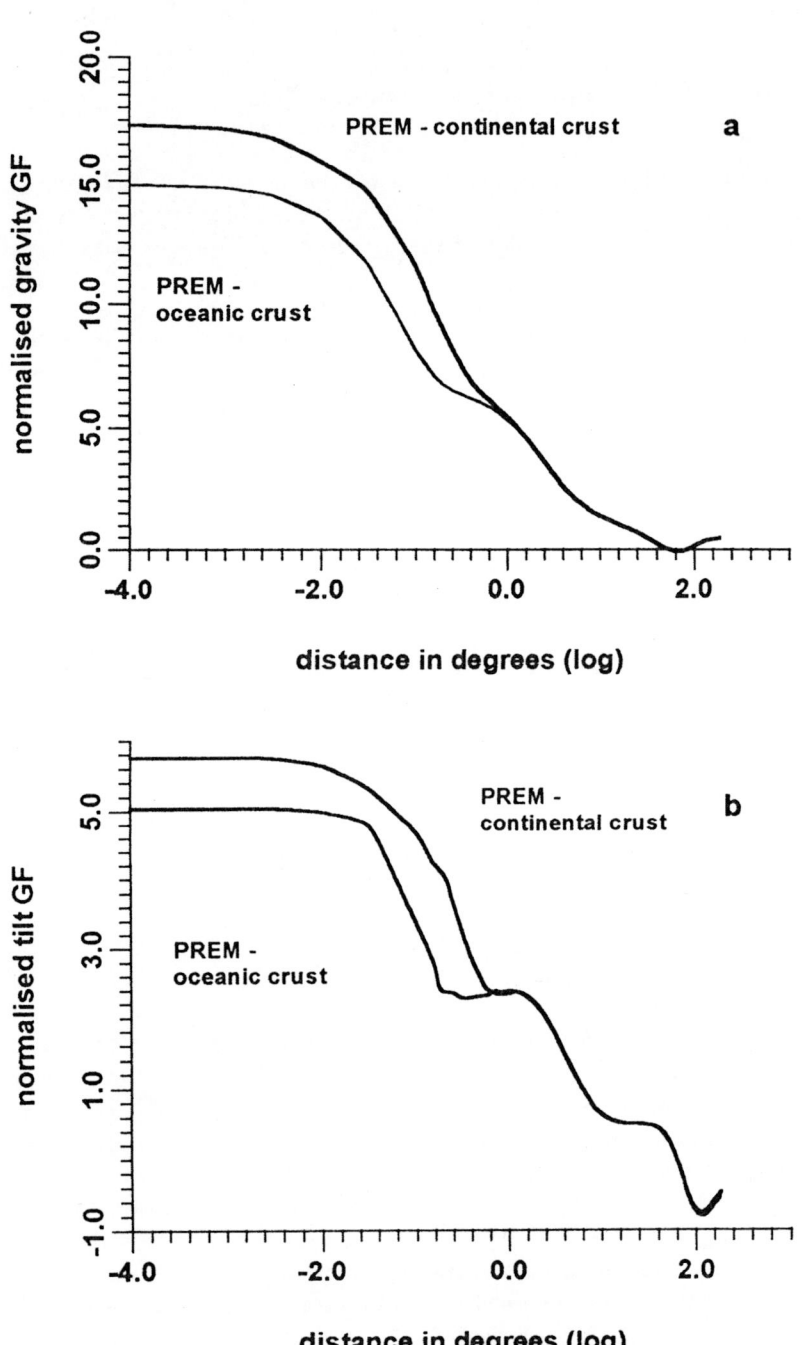

Figure 2.2. Normalised Green's functions for gravity (a) and tilt (b) for the PREM reference earth model (continental and oceanic crust).

Table. 2.2. Ratio of load tides to body tides in per cent compared to the properties of the Green's function close to the coast (approximately).

	coast	continent	Green's function
gravity	10	2	r^{-1}
tilt	1000	20	r^{-2}
strain	100	20	r^{-2}

3   Modeling ocean tides

The modeling of global ocean tides is treated by Zahel ("Ocean Tides", this volume). Here, only some points are discussed which are related to the interpretation of the observed load vector by loading computations. Tidal models are provided for the main tidal constituents, e.g. the diurnal tides O1, P1, K1, and the semidiurnal tides M2, S2, N2, and K2. Because of resonance in some areas like bights or estuaries higher harmonics can be important (e.g. M4).

The first global models were derived by extrapolation of tidal constants computed from tide gauges on islands and at coastal stations. Later, hydrodynamic algorithms were used to investigate the distribution of the respective constituents. Two different approaches can be distinguished: The purely hydrodynamic computation without any observations as constraints, and the so-called hydrodynamic interpolation. The latter method applied by Schwiderski (1979a/b) is based on tidal constants which are interpolated using a hydrodynamic formalism.

Today, due to the availability of high resolution satellite altimeter data ocean tidal models are derived from this huge amount of surface data (comp. e.g.: Cartwright, 1991; TOPEX/ POSEIDON: Andersen, 1994; Eanes, 1994).

The amplitudes of the marine tides depend on the body tides, too. This was already found by Darwin (1982; see also Hendershott, 1972; Schwiderski, 1979a). The vertical displacement of the ocean bottom through body tides and ocean tidal loading perturbs the equilibrium of the marine tide.

The computation of global tidal models is treated elsewhere in this volume (Zahel "Ocean tides"); for the interpretation and application of loading computations it is of interest here, how useful empirical shelf models can be. In many cases the interpolation of local tidal constants is the only way to provide local models and to combine them with the global data set. Especially for complicated coastal structures like the Norwegian fjords it is not feasible to derive hydrodynamic models.

Experiences show that the errors of such local models can be estimated quite well. Here not only the effect of the grid size but also the correct position of the mass center of the cell are of interest (Büker et al., 1993; Jentzsch, 1985), as well as the distribution of the amplitudes and phases in shelf areas (Asch et al., 1986; 1987; Baker, 1980a; Jahr, 1989; Jentzsch, 1986; Jahr et al., 1991).

In many cases we concentrate on the discussion of the two main constituents of the tidal spectrum, M2 (semidiurnal) and O1 (diurnal), because for these waves the residual vectors have the highest signal-to-noise ratios. But for tidal corrections of the whole spectrum the computations have to be carried out for all available constituents. To test the effects of the waves with very small amplitudes for which no models exist it is the only possibility to assume the same resonance pattern of the existing model closest to this frequency and to apply a proper amplitude ratio.

4   Some examples

The investigation of oceanic loading has in principle three different goals which in some cases are connected (comp. Baker, 1978). The first task is the correction of geodetical and geophysical measurements of high precision. Close to the sea the loading tides have to be taken into account for repeated gravity surveys (e.g. microgravimetry along the Fennoscandian land uplift lines). Even precise levelling can be affected, and the same holds true for installations like VLBI antennas. Since only large scale effects can be corrected, local effects must be monitored.

The second goal concerns the test of different ocean tidal models by comparison of observed and computed loading tides. Here we have the advantage that tidal gravimetry (and with some reservation tidal tilt) offers a method independent from direct water level monitoring.

Finally, the evaluation of the loading signal with a known ocean tidal input allows the investigation of the local crust-mantle structure.

Some examples are outlined in the following:

(1) Correction of geodetical and geophysical measurements of high precision

In connection with the gravity tidal profile along the 'Blue Road Geotraverse' in Fennoscandia (Asch et al., 1986; 1987; Jentzsch, 1986) computations concerning the improvement of tidal corrections of repeated gravity surveys along the land-uplift lines were carried out. It turned out that close to the coast the effect of loading tides must be corrected because they sum up to about $10\,\mu$Gal, which surmounts the achieved errors of repeated gravity measurements ($< 10\,\mu$Gal; comp. Jentzsch, 1981; Weise, 1986).

During the recent years these corrections have received special attention, e.g. in connection with climate research ('Global Change'): The resolution of distance measurements (satellite-laser-ranging) and radar-altimetry is now clearly below the decimeter level. To benefit from this achievement the position of the satellite or air plane must be well known accordingly using kinematic GPS linked to reference stations (fiducial sites).

For continental reference stations the correction of the dynamic contribution to the local coordinates caused by the deformation due to earth tides is unproblematic. But, since many reference stations are close to the oceans (Greenland, Arctic, Antarctic) the contribution of ocean tidal loading cannot be neglected. Especially at high latitudes an automatic correction is impossible because of lacking or unprecise models.

Thus, the monitoring of the Greenland ice cap by air borne methods (GAP - Greenland Aerogeophysics Project; Brozena, 1992) requires tidal gravity measurements at the stations. Therefore, in summer 1993 a tidal gravimeter started to record at the fiducial site Scoresbysund / Eastern Greenland for a one year period (Engen et al., 1993). These measurements were continued in Narsarsuaq / Southern Greenland (summer 1994 to 1995) and Thule in the north-west (1995 - 1996). The measurements are continued in Godhavn / Western Greenland. In parallel to this cooperation with the Geodetic Division of the Office for National Survey and Cadastre, the Geo-Research Center, Potsdam, cooperates with the Norwegian Institute for Surveying and Mapping; they performed similar measurements in Tromsø / Northern Norway (October 1994 to April 1996) and continued in Stavanger / Western Norway.

These observations show that the observed tidal residuals for the main tidal constituents cannot be explained by computations based on existing ocean tidal models (Jentzsch et al., 1995). As an example the results of the tidal analyses from Scoresbysund are presented in tab. 4.1.

For a standard earth model the tidal parameters (amplitude factor and phase shift) provide values around 1.16 and 0°. Tab. 4.2 contains the derived residuals with the errors corresponding to the noise level of the Fourier-spectrum. These results can be compared to the computational results for two different ocean tidal models (tab. 4.3; global models of Schwiderski; same models, but North Atlantic replaced by Flather's models). It is obvious that the observed residuals cannot be explained by the computations. The differences summed up over the whole spectrum and converted into height changes arrive at ± 5 cm. This bias will affect repeated measurements if global loading models are applied.

The different loading amplitudes for M2 and O1 are due to the fact that in the North Atlantic the amplitude of O1 is smaller compared to M2. This means, that the effects of distant oceans surmount local loading. Therefore, we find a better fit of Schwiderski's global O1 model compared to the combination with Flather's, although Flather's grid covers a great part of the Scoresbysund. But, since it does not fit well in that area, the whole result is deteriorated.

In case of constituent M2 it is just opposite: Since the amplitudes in the North Atlantic are bigger the regional / local contributions are bigger as well. Therefore, the computed load vector (SCH) does not fit as well as the vector of the combined model (SCH/FL). But still the result is not at all satisfying. This points to the necessity to model the local shelf areas separately.

Table 4.1. Tidal analyses results for the record from Scoresbysund / Eastern Greenland (gravimeter LCR ET-18; analysed period July 31, 1993 until February 2, 1994; coordinates: 70.485° N, 21.953° W, elevation 64.8 m); errors are based on the noise level of the spectrum.

Constituent	Amplitude [μGal]	Signal/ noise ration	Amplitude factor	Phase shift
O1	21.751 ± 0.077	282	1.111 ± 0.004	1.63 ± 0.20
M2	7.962 ± 0.025	314	0.950 ± 0.003	- 9.45 ± 0.18

(2) Test of ocean tidal models

From Table 4.2 / 3 we see that it is possible to test ocean tidal models with the help of land-based tidal gravimetry. Baker (1980a) already demonstrated this for the shelf tides around Great Britain using spatially distributed stations. In this way he could derive misfits of the model of Flather (1976).

Similar investigations were carried out for the Norwegian shelf around the polar circle ('Blue Road Geotraverse'; Asch et al., 1986; 1987; Jentzsch, 1986) Jentzsch, 1986; Asch et al., 1986; 1987) and the eastern part of the North Sea / western Baltic Sea (Jahr, 1989; Jahr et al., 1991). For the German Bight it turned out that the combination of Flather's model with the local model of the German Hydrographic Service (Deutsches Hydrographisches Institut, Hamburg) provided the best fit if all models conserved water.

In the case of the Norwegian shelf it could be shown that even tide gauges on islands outside the fjords were still strongly affected by the resonance caused by the coast. This means that the tidal constants are not representative for the whole area. The variation of the phases for the shelf tides by - 15° and - 20°, resp. (phase lead), reduced the amplitudes of the rest vector which cannot be explained by the loading model.

On a global scale Melchior et al. (1981) carried out world-wide tidal gravity profile measurements to map the ocean load signal and to compare the tidal residuals with standard ocean tide maps.

This data bank was used by Llubbes and Mazzega (1996) to test ten ocean tidal models for M2 and O1, including very recent ones, on a global scale. For the load residues they could show that the standard deviations of the differences between observed and calculated gravity effects are mainly well below 0.5 . There is no model that performs best in every part of the globe. Because of lacking satellite data, there are still problems with the modeling in polar areas.

Table 4.2. Tidal residuals for constituents O1 and M2 observed at Scoresby-sund / Eastern Greenland (from Jentzsch et al., 1995).

Constituent	Amplitude [μGal]	Error [μGal]	Phase [°]	Error [°]
O1	1.146	0.077	147.25	3.85
M2	2.281	0.025	-145.03	0.64

New regional studies were also carried out: Agnew (1995) used the observations of the gravity tide at the South Pole to test the modeling of the ocean tides south of 66S. He found that the tides beneath the ice shelves have to be included to improve the fit. The South Pole is a unique point where the body tides for gravity vanish, and thus, the loading signal can be observed directly. Francis and Melchior (1996) pointed out that Agnew (1995) used for obvious reasons only one station, and they presented their own results for Southwestern Europe as a "test area". They found that there, the loading effect varies even over short distances significantly, which means that also the validation of the models needs more stations. As another example Melchior et al. (1996) present results from Southeast Asia. Regarding O1 they found that Schwiderski's (1980) model fits well, which is in agreement with Llubbes and Mazzega (1996). Unfortunately, they did not yet apply other, new models.

In contrast to satellite altimetry earth tides are not used to derive the models, but they are used to constrain and validate these models. Therefore Agnew's (1995) statement, that satellite altimetry has 'outstripped' earth tides to validate ocean tides is misleading (comp. also the comment of Francis and Melchior (1996).

(3) Evaluation of local crust / mantle - structures

Although gravity is not as sensitive to crustal structure as tilt it could be shown that the final residual vector for the Norwegian stations of the 'Blue Road Geotraverse' (fig. 4.1) was especially small if a normal crust without any mountain root was applied (30 km rather than 50 km). Fig. 4.2 shows the Green's functions for both models: The differences between 1 km and 100 km are due to the penetrating depth of the method equivalent to about half the distance from the load.

A similar result was found by Zürn et al. (1976) at the southern coast of Alaska: The gravity residual could be explained up to 80 % by local loading, whereas the effect of the subducting oceanic plate is much too weak to model the remaining difference.

Table 4.3. Computed load vectors for the models of Schwiderski (global, 1° x 1°) - **SCH** - and a combined model Schwiderski (global) and Flather (Northatlantic ocean) - **SCH/FL** -; Newtonian attraction and elastic deformation are given separately.

constituent	Newtonian [μGal] [°]	elastic [μGal] [°]	total [μGal] [°]
O1 (SCH)	0.111 -169.8	0.494 140.1	0.571 149.1
O1 (SCH/FL)	0.042 -58.1	0.183 121.0	0.141 120.7
M2 (SCH)	0.083 -4.4	0.187 -42.6	0.257 -31.2
M2 (SCH/FL)	0.559 -81.6	0.826 -105.8	1.355 -96.1

Baker (1980b) applied this investigation also to the results obtained with the tiltmeters in the mine of Llanrwst / Wales. This work was originally intended to investigate the effect of the cavity by 'strain-tilt coupling'. These are tidal tilts induced by tidal strain around geological inhomogeneities or cavities (comp. Harrison, 1976). He could show that the loading signal has to be separated with an error below 1 to 2 % in order to enable the improvement of local crust-mantle structures. This was not possible because strain induced local effects were too dominant.

Tilt measurements in boreholes are much more suitable for this purpose. This could be shown by Zschau (1976) with the results from three neighbouring boreholes near Kiel.

5   Recent developments

To check ocean tidal models and to consider realistic earth models are the main issues of actual developments. Regarding the earth structure we try to combine the results from different disciplines of research, e.g. travel times, free oscillations, earth tides and ocean tidal loading as well as nutation and tidally-induced variations in the earth's rotation with long period dynamics of the earth.

Of special interest is to detect non-linearities in the tidal signal and to investigate the viscoelastic behaviour of the earth.

Thus, for the equations of motion Pagiatakis (1990)adopted not only a standard linear solid type rheology but also a grain-boundary relaxation model for the dissipation mechanism within the earth. In this way viscosity and Q profiles enter the process. He can prove that the complex load Love numbers h'_n, l'_n, k'_n (calculated up to degree 10,000) are affected by a few per cent by the rotation of the earth, as well as by the anisotropy in the upper mantle. There is even a weak latitude dependence for degrees $n \leq 4$. And at fortnightly periods, the viscoelastic load numbers are 1 to 2 % larger than the corresponding elastic values. These findings will have to be taken into account for future studies.

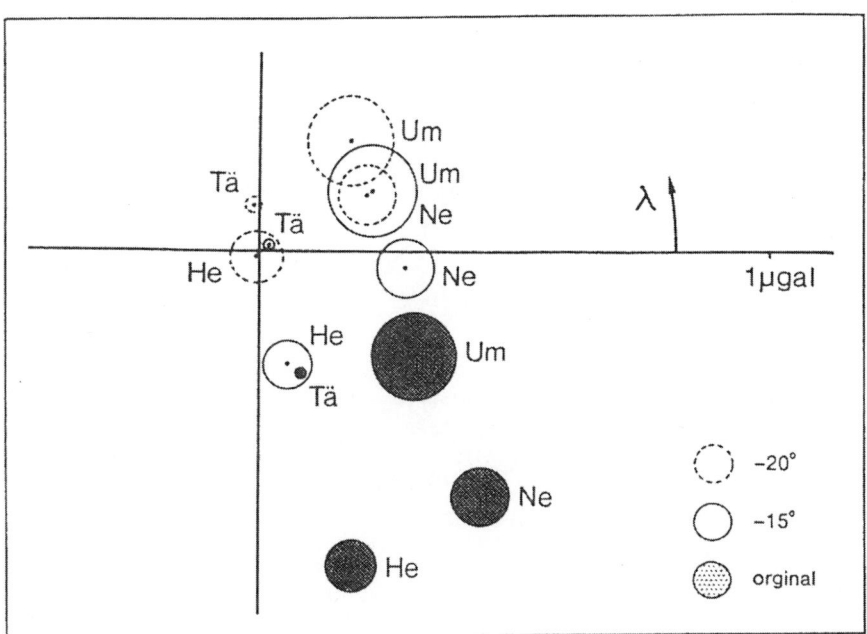

Figure 4.1. Local rest vectors for constituent M2 at four stations along the 'Blue Road' with increasing distances from the coast: Three shelf models with different phases are used; the stations are in or near the villages Ne - Nesna, He - Hemnesberget, Um - Umbukta (Norway) and Tä - Tärnaby (Sweden); normal crust (Jentzsch, 1986).

Baker et al. (1996) emphasize that the new improved ocean tide models allow a corresponding improvement of loading computations. Together with the progress made with tidal gravity measurements more comprehensive tests of body tide models are now possible. Even constraints on the anelasticity of the earth can be derived at tidal frequencies.

On the basis of PREM Pagiatakis (1990) used an anisotropic earth model with lateral isotropy with the vertical as symmetry axis. With his anlytical process he is able to extend his method also to more complicated types of anisotropy. But to investigate complex structures like subducting plates (Zürn et al., 1976) numerical methods (finite element modeling) have to be applied rather than analytical ones.

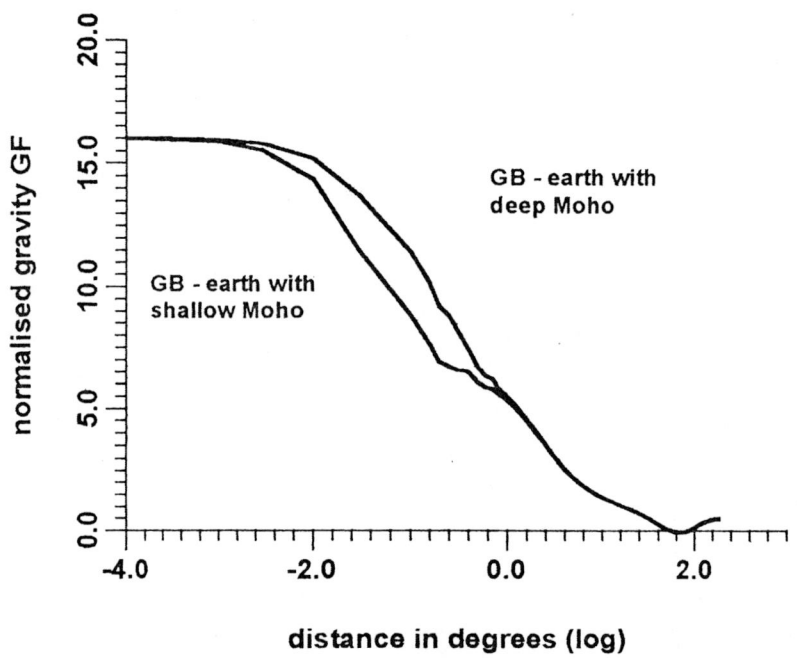

Figure 4.2. Green's functions for two extremely different crust-mantle models below the Norwegian Caledonides close to the polar circle (from Jentzsch, 1986).

References

Agnew, D.C. 1983. Conservation of mass in tidal loading computations. *Geophys. J. R. astr. Soc.* **72**: 321 - 325.

Agnew, D. C. 1995. Ocean-load tides at the South Pole: A validation of recent ocean-tide models. *Geophys. Res. Lett..* **22**: 3063 - 3066.

Andersen, O.B. 1994. Global ocean tides from ERS-1 and TOPEX/ POSEIDON altimetry. *J. geophys. Res.* TOPEX/POSEIDON special issue 2 (in press).

Asch, G., T. Jahr, G. Jentzsch, A. Kiviniemi, J. Kääriäinen, H.-P. Plag and W. Thiel 1986. Loading tides along the 'Blue Road Geotraverse'. In: *Proc. 10th Int. Symp. Earth Tides.* pp. 707 - 717. R. Vieira (ed.). Consejo Sup. Invest. Cient., Madrid.

Asch, G., T. Jahr, G. Jentzsch, A. Kiviniemi and J. Kääriäinen. 1987. Measurement of gravity tides along the 'Blue Road Geotraverse' in Fennoscandia. *Publ. Finnish Geodetic Inst.* **107**: 1 - 57.

Baker, T.F. 1978. What can earth tide measurements tell us about ocean tides or earth structure? In: *Proc. 9th GEOP Conf. Dept. of Geodetic Science.* **280**, Ohio State Univ.

Baker, T.F. 1980a. Tidal gravity in Britain; tidal loading and the spatial distribution of the marine tide. *Geophys. J. R. astr. Soc.* **62**: 249 - 267.

Baker, T.F. 1980b. Tidal tilt at Llanrwst, North Wales: tidal loading and Earth structure. *Geophys. J. R. astr. Soc.* **62**: 269 - 290.

Baker, T. F., D. J. Curtis and A. H. Dodson 1996. A new test of earth tide models in central Europe. *Geophys. Res. Lett..* **23**: 3559 - 3562.

Baker, T.F. and G.W. Lennon 1976. Spatial coherency and tidal tilts. In:*Proc. 7th Int. Symp. Earth Tides.* pp. 479 - 493. G. Szàdeczky-Kardoss (ed.)., Schweizerbart, Stuttgart.

Boussinesque, J. 1885. *Application des Potentiels à l'Étude de l'Équilibre et du Movement des Solid Élastiques.* 508 pp. Gauthiers-Villars, Paris.

Brozena, P. 1992. The Greenland Aerogeophysics Project: Airborne gravity, topographic and magnetic mapping of an entire continent. 6th Int. Symp. on Satellite Pos., Columbus, OH, March 1992.

Büker, F., G. Jentzsch and S. Koa. 1995. Ocean tidal loading and topographic effects on tilt at Blå Sjø, Southern Norway. In: *Proc. 12th Int. Symp. Earth Tides.* pp. 475 - 486. H.-T. Hsu (ed.). Science Press, Beijing.

Cartwright, D.E. 1991. Detection of tides from artificial satellites (Review). In: Tide Hydrodynamics. pp. 547 - 568. Parker, B.B. (ed.). Wiley, New York.

Darwin, G.H. 1882. A numerical estimate of the rigidity of the earth. *Nature.***27**.

Dziewonski, A.M. and D.L. Anderson. 1981. Preliminary reference Earth model. *Phys. Earth planet. Int.* **25**: 297 - 356.

Eanes, R.J. 1994. Diurnal and semidiurnal tides from TOPEX/ POSEIDON altimetry. *EOS (Trans. Am. geophys. Union).* **75**: 108.

Ekman, M. 1993. A concise history of the theories of tides, precession-nutation and polar motion (from antiquity to 1950). *Surv. Geophys..***14**: 585 - 617.

Engen, B., G. Jentzsch, F. Madsen and P. Schwintzer. 1995. NATURE: North Atlantic Tides Under Research - Tidal gravity measurements around the Norwegian Greenland Sea -. In:*Proc. 12th Int. Symp. Earth Tides.* pp 451 - 461. H.-T. Hsu (ed.). Science Press, Beijing.

Farrell, W.E. 1972. Deformation of the earth by surface loads. *Rev. Geophys. Space Phys.***10**: 761 - 797.

Farrell, W.E. 1973. Earth tides, ocean tides and tidal loading. *Phil. Trans. Roy. Soc. London* A **274**: 253 - 259.

Flather, R.A. 1976. A tidal model of the north-west European continental shelf. *Mem. Soc. R. Sci. Liège* **10**: 141-164.

Francis, O. and P. Melchior 1996. Tidal loading in south western Europe: a test area. *Geophys. Res. Lett..* **23**:2251 - 2254.

Goad, C.C. 1980. Gravimetric tidal loading from integrated Green's functions. *J. Geophys. Res..***85**: 2679 - 2683.

Harrison, J.C. 1976. Cavity and topographic effects in tilt and strain measurement. *J. geophys. Res.*. **81**: 319 - 328.

Harrison, J.C. 1985. Earth tides. *Benchmark Papers in Geology Series*. Van Nostrand Reinhold Comp., New York, 419pp.

Hendershott, M.C. 1972. The effects of solid earth deformation on global ocean tides. *Geophys. J.R. astr. Soc.*.**29**: 389 - 402.

Jahr, T. 1989. Gezeitengravimetrie in Dänemark. *Berl. Geowiss. Abh.* **B 16**: 1 - 137.

Jahr, T., G. Jentzsch, N. Andersen and O. Remmer. 1991. Ocean tidal loading on the shelf areas around Denmark. In: *Proc. 11th Int. Symp. Earth Tides*. pp. 309 - 319. J. Kakkuri (ed.). Schweizerbart, Stuttgart.

Jentzsch, G. 1981. Improved tidal corrections for high precision gravity surveys. *Earth Evol. Scs.*. **1**: 89 - 91.

Jentzsch, G. 1985. The influence of the grid structure on the results of loading calculations. *Bull. Inf. Marées Terr.*. **94**: 6382 - 6386.

Jentzsch, G. 1986. Auflastgezeiten in Fennoskandien. *Berl. Geowiss. Abh.* **B 13**: 1 - 184.

Jentzsch, G., M. Ramatschi and F. Madsen. 1995. Tidal gravity measurements on Greenland. *Bull. Inf. Marées Terrestres* **122**: 9239 - 9248.

Llubbes, M. and P. Mazzega 1996. The ocean tide gravimetric loading reconsidered. *Geophys. Res. Lett.*. **23**: 1481 - 1484.

Longman, I.M. 1962. A Green's function for determining the deformation of the earth under surface mass loads, 1, theory. *J. geophys. Res.* **67**: 845 - 850.

Longman, I.M. 1963. A Green's function for determining the deformation of the earth under surface mass loads, 2, computations and numerical results. *J. Geophys. Res.***68**: 485 - 496.

Melchior, P., M. Moens, B. Ducarme and M. van Ruymbeke. 1981. Tidal loading along profile Europe - East Africa - South Asia - Australia and Pacific Ocean. *Phys. Earth planet. Inter.* **25**: 71 - 106.

Munk, W.H. and G.J.F. MacDonald. 1960. The rotation of the earth. Cambridge Univ. Press, New York, 323p.

Pagiatagis, S.P. 1990. The response of a realistic earth to ocean tide loading. *Geophys. J. Int.*.**103**: 541 - 560.

Schwiderski, E.W. 1979a. Ocean tides, part I: Global tidal equations. *Marine Geodesy*. **3**: 161.

Schwiderski, E.W. 1979b. Ocean tides, part II: A hydrographical interpolation model. *Marine Geodesy*.**3**: 219.

Schwiderski, E.W. 1979c. Global ocean tides II: The semidiurnal principal lunar tide (M2). Atlas of tidal charts and maps, NSWC, Dahlgren.

Schwiderski, E.W. 1980. Global ocean tides V. The diurnal principle lunar tide (O1). Atlas of tidal charts and maps, NSWC, Dahlgren.

Weise, A. 1986. Prediction of tidal corrections by collocation of tidal residuals. In:*Proc. 10th Int. Symp. Earth Tides*. pp. 521 - 529. R. Vieira (ed.). Cons. Sup. Invest. Cient., Madrid.

Zschau, J. 1976. Tidal sea load tilt of the crust and its application to the study of crustal and upper mantle structures. *Geophys. J. R. astr. Soc.* **44**: 577 - 593.

Zschau, J. 1977. Phase shifts of tidal sea load deformations of the earth's surface due to low viscosity layers in the interior. in: *Proc. 8th Int. Symp. Earth Tides*. pp. 372 - 398. Bonatz, M. and P. Melchior (ed.). Bonn.

Zschau, J. 1978. Tidal friction in the solid earth: loading tides versus body tides. In: Tidal Friction and the Earth's Rotation. pp 62 - 94. P. Brosche, J. Sündermann (ed.). Springer, Berlin.

Zürn, W., C. Beaumont and L.B. Slichter. 1976. Gravity tides and ocean loading in southern Alaska. *J. geophys. Res.*. 81: 4923 - 4932.

Appendix

A.1 Gravity Green's functions for the model PREM / continent / ocean

CONTINENTAL CRUST

distance [°]	radial displacem. $\cdot 10^{12}$ $(a\Theta)$	tangential displacem. $\cdot 10^{12}$ $(a\Theta)$	GF vert. elastic $\cdot 10^{18}$ $(a\Theta)$	GF hor. elastic $\cdot 10^{12}$ $(a\Theta)^2$	strain $r\Theta\Theta$ $\cdot 10^{12}$ $(a\Theta)^2$
.0001	-39.061	-13.106	-90.380	39.093	13.103
.0010	-38.703	-13.102	-89.576	39.092	13.106
.0030	-37.919	-13.072	-87.817	39.067	13.134
.0100	-35.555	-12.778	-82.532	38.310	13.327
.0300	-32.726	-12.097	-76.392	36.181	12.155
.1000	-25.562	-10.731	-60.311	31.759	11.859
.1600	-21.396	-9.523	-50.767	28.583	13.761
.2000	-19.651	-8.805	-46.669	27.589	12.693
.2500	-17.977	-8.045	-42.615	24.851	11.302
.3000	-16.811	-7.464	-39.675	22.593	10.385
.4000	-15.390	-6.710	-35.862	19.318	8.774
.5000	-14.637	-6.323	-33.635	17.390	7.484
.6000	-14.179	-6.149	-32.172	16.282	6.623
.7000	-13.842	-6.092	-31.078	15.994	6.107
.8000	-13.536	-6.081	-30.140	15.829	5.906
.9000	-13.228	-6.078	-29.269	15.893	5.932
1.0000	-12.909	-6.063	-28.428	15.989	6.088
1.2000	-12.245	-5.969	-26.805	16.080	6.527
1.6000	-10.931	-5.579	-23.842	15.695	7.153
2.0000	-9.761	-5.084	-21.312	14.862	7.259
2.5000	-8.545	-4.478	-18.713	13.617	6.966
3.0000	-7.572	-3.948	₊16.625	12.366	6.477
4.0000	-6.176	-3.129	-13.560	10.148	5.410
5.0000	-5.273	-2.562	-11.493	8.381	4.450
6.0000	-4.679	-2.169	-10.062	7.016	3.623
7.0000	-4.279	-1.900	-9.042	5.989	2.922
8.0000	-4.001	-1.718	-8.289	5.237	2.343
9.0000	-3.800	-1.597	-7.712	4.696	1.878
10.0000	-3.647	-1.518	-7.249	4.317	1.517
12.0000	-3.413	-1.433	-6.531	3.858	1.041
16.0000	-3.038	-1.375	-5.480	3.496	.671
20.0000	-2.664	-1.337	-4.639	3.378	.625
25.0000	-2.143	-1.268	-3.711	3.343	.758
30.0000	-1.559	-1.167	-2.858	3.312	1.060
40.0000	-.301	-.875	-1.379	2.926	1.998
50.0000	.863	-.540	-.300	1.872	2.874
60.0000	1.707	-.284	.275	.323	3.150
70.0000	2.125	-.200	.387	-1.340	2.694
80.0000	2.103	-.311	.145	-2.745	1.645
90.0000	1.690	-.583	-.315	-3.720	.271
100.0000	.961	-.945	-.875	-4.200	-1.183
110.0000	.000	-1.317	-1.446	-4.248	-2.540
120.0000	-1.108	-1.622	-1.971	-3.970	-3.701
130.0000	-2.286	-1.799	-2.427	-3.456	-4.656
140.0000	-3.462	-1.806	-2.808	-2.830	-5.430
150.0000	-4.566	-1.619	-3.123	-2.159	-6.079
160.0000	-5.533	-1.240	-3.389	-1.481	-6.688
170.0000	-6.304	-.688	-3.622	-.780	-7.347
180.0000	-6.820	.000	-3.833	.000	-8.140

OCEANIC CRUST

distance [°]	radial displacem. $\cdot 10^{12}$ ($a\Theta$)	tangential displacem. $\cdot 10^{12}$ ($a\Theta$)	GF vert. elastic $\cdot 10^{18}$ ($a\Theta$)	GF hor. elastic $\cdot 10^{12}$ ($a\Theta$)^2	strain $\epsilon_{\Theta\Theta}$ $\cdot 10^{12}$ ($a\Theta$)^2
.0001	-34.175	-11.428	-77.695	34.202	11.426
.0010	-33.872	-11.426	-77.034	34.201	11.427
.0030	-33.204	-11.411	-75.576	34.191	11.441
.0100	-31.008	-11.252	-70.827	33.907	11.566
.0300	-26.333	-10.468	-61.106	32.597	11.692
.1000	-18.034	-7.585	-42.411	22.793	9.777
.1600	-15.920	-6.557	-36.925	18.477	8.714
.2000	-15.402	-6.205	-35.185	16.064	7.587
.2500	-15.085	-6.017	-33.974	16.023	6.480
.3000	-14.941	-5.969	-33.264	15.558	5.919
.4000	-14.731	-6.012	-32.291	15.514	5.698
.5000	-14.501	-6.050	-31.503	15.689	5.817
.6000	-14.227	-6.048	-30.743	15.811	6.041
.7000	-13.927	-6.014	-30.003	16.205	6.168
.8000	-13.607	-5.958	-29.263	16.087	6.250
.9000	-13.273	-5.886	-28.523	16.186	6.414
1.0000	-12.930	-5.801	-27.786	16.202	6.446
1.2000	-12.241	-5.603	-26.335	16.149	6.622
1.6000	-10.926	-5.148	-23.621	15.639	6.722
2.0000	-9.768	-4.683	-21.241	14.802	6.602
2.5000	-8.564	-4.151	-18.739	13.589	6.266
3.0000	-7.597	-3.690	-16.694	12.371	5.841
4.0000	-6.199	-2.969	-13.646	10.188	4.953
5.0000	-5.290	-2.459	-11.569	8.425	4.135
6.0000	-4.690	-2.098	-10.123	7.051	3.405
7.0000	-4.287	-1.848	-9.093	6.012	2.770
8.0000	-4.007	-1.677	-8.334	5.253	2.235
9.0000	-3.806	-1.562	-7.751	4.711	1.800
10.0000	-3.651	-1.486	-7.285	4.327	1.458
12.0000	-3.417	-1.405	-6.563	3.853	.999
16.0000	-3.048	-1.351	-5.523	3.497	.630
20.0000	-2.676	-1.316	-4.689	3.365	.586
25.0000	-2.160	-1.250	-3.774	3.334	.718
30.0000	-1.583	-1.151	-2.940	3.302	1.013
40.0000	-.334	-.867	-1.484	2.929	1.939
50.0000	.822	-.538	-.428	1.892	2.800
60.0000	1.663	-.289	.142	.362	3.078
70.0000	2.081	-.208	.257	-1.264	2.628
80.0000	2.065	-.321	.032	-2.646	1.597
90.0000	1.660	-.593	-.401	-3.588	.244
100.0000	.944	-.953	-.924	-4.045	-1.185
110.0000	-.001	-1.323	-1.450	-4.069	-2.511
120.0000	-1.094	-1.626	-1.926	-3.767	-3.646
130.0000	-2.254	-1.801	-2.328	-3.253	-4.561
140.0000	-3.411	-1.806	-2.655	-2.631	-5.290
150.0000	-4.498	-1.619	-2.919	-1.982	-5.894
160.0000	-5.450	-1.240	-3.140	-1.342	-6.458
170.0000	-6.209	-.687	-3.337	-.701	-7.073
180.0000	-6.718	.000	-3.525	.000	-7.829

A.2 Normalized Green's functions for gravity and tilt versus distance

CONTINENTAL CRUST

RAD	DEG	LOG(DEG)	KM	GRAVITY	TILT
.000002	.000	-4.00	.01	17.27	5.76
.000017	.001	-3.00	.11	17.12	5.76
.000052	.003	-2.52	.33	16.78	5.75
.000175	.010	-2.00	1.11	15.77	5.64
.000524	.030	-1.52	3.34	14.60	5.33
.001745	.100	-1.00	11.12	11.52	4.68
.002793	.160	-.80	17.79	9.70	4.21
.003491	.200	-.70	22.24	8.92	4.06
.004363	.250	-.60	27.80	8.14	3.66
.005236	.300	-.52	33.36	7.58	3.33
.006981	.400	-.40	44.48	6.85	2.85
.008727	.500	-.30	55.60	6.43	2.56
.010472	.600	-.22	66.71	6.15	2.40
.012217	.700	-.15	77.83	5.94	2.36
.013963	.800	-.10	88.95	5.76	2.33
.015708	.900	-.05	100.07	5.59	2.34
.017453	1.000	.00	111.19	5.43	2.35
.020944	1.200	.08	133.43	5.12	2.37
.027925	1.600	.20	177.90	4.56	2.31
.034907	2.000	.30	222.38	4.07	2.19
.043633	2.500	.40	277.98	3.58	2.01
.052360	3.000	.48	333.57	3.18	1.82
.069813	4.000	.60	444.76	2.59	1.49
.087266	5.000	.70	555.95	2.20	1.23
.104720	6.000	.78	667.14	1.92	1.03
.122173	7.000	.85	778.33	1.73	.88
.139626	8.000	.90	889.52	1.58	.77
.157080	9.000	.95	1000.71	1.47	.69
.174533	10.000	1.00	1111.90	1.38	.64
.209440	12.000	1.08	1334.28	1.25	.57
.279253	16.000	1.20	1779.04	1.04	.52
.349066	20.000	1.30	2223.80	.88	.50
.436332	25.000	1.40	2779.75	.70	.50
.523599	30.000	1.48	3335.70	.54	.49
.698132	40.000	1.60	4447.60	.26	.44
.872665	50.000	1.70	5559.50	.06	.29
1.047198	60.000	1.78	6671.40	-.05	.05
1.221730	70.000	1.85	7783.30	-.07	-.21
1.396263	80.000	1.90	8895.20	-.03	-.45
1.570796	90.000	1.95	10007.10	.05	-.63
1.745329	100.000	2.00	11119.00	.15	-.74
1.919862	110.000	2.04	12230.90	.24	-.79
2.094395	120.000	2.08	13342.80	.31	-.80
2.268928	130.000	2.11	14454.70	.37	-.77
2.443461	140.000	2.15	15566.60	.41	-.72
2.617994	150.000	2.18	16678.50	.44	-.67
2.792527	160.000	2.20	17790.40	.46	-.62
2.967060	170.000	2.23	18902.30	.46	-.59
3.141593	180.000	2.26	20014.20	.47	.00

OCEANIC CRUST

RAD	DEG	LOG(DEG)	KM	GRAVITY	TILT
.000002	.000	-4.00	.01	14.85	5.04
.000017	.001	-3.00	.11	14.72	5.04
.000052	.003	-2.52	.33	14.44	5.04
.000175	.010	-2.00	1.11	13.53	4.99
.000524	.030	-1.52	3.34	11.68	4.80
.001745	.100	-1.00	11.12	8.10	3.36
.002793	.160	-.80	17.79	7.06	2.72
.003491	.200	-.70	22.24	6.72	2.37
.004363	.250	-.60	27.80	6.49	2.36
.005236	.300	-.52	33.36	6.36	2.29
.006981	.400	-.40	44.48	6.17	2.29
.008727	.500	-.30	55.60	6.02	2.31
.010472	.600	-.22	66.71	5.87	2.33
.012217	.700	-.15	77.83	5.73	2.39
.013963	.800	-.10	88.95	5.59	2.37
.015708	.900	-.05	100.07	5.45	2.38
.017453	1.000	.00	111.19	5.31	2.39
.020944	1.200	.08	133.43	5.03	2.38
.027925	1.600	.20	177.90	4.51	2.30
.034907	2.000	.30	222.38	4.06	2.18
.043633	2.500	.40	277.98	3.58	2.00
.052360	3.000	.48	333.57	3.19	1.82
.069813	4.000	.60	444.76	2.61	1.50
.087266	5.000	.70	555.95	2.21	1.24
.104720	6.000	.78	667.14	1.93	1.04
.122173	7.000	.85	778.33	1.74	.89
.139626	8.000	.90	889.52	1.59	.77
.157080	9.000	.95	1000.71	1.48	.69
.174533	10.000	1.00	1111.90	1.39	.64
.209440	12.000	1.08	1334.28	1.25	.57
.279253	16.000	1.20	1779.04	1.05	.52
.349066	20.000	1.30	2223.80	.89	.50
.436332	25.000	1.40	2779.75	.72	.50
.523599	30.000	1.48	3335.70	.56	.49
.698132	40.000	1.60	4447.60	.28	.44
.872665	50.000	1.70	5559.50	.08	.29
1.047198	60.000	1.78	6671.40	-.03	.06
1.221730	70.000	1.85	7783.30	-.05	-.20
1.396263	80.000	1.90	8895.20	-.01	-.43
1.570796	90.000	1.95	10007.10	.07	-.61
1.745329	100.000	2.00	11119.00	.16	-.71
1.919862	110.000	2.04	12230.90	.24	-.76
2.094395	120.000	2.08	13342.80	.30	-.76
2.268928	130.000	2.11	14454.70	.36	-.72
2.443461	140.000	2.15	15566.60	.39	-.67
2.617994	150.000	2.18	16678.50	.41	-.61
2.792527	160.000	2.20	17790.40	.42	-.57
2.967060	170.000	2.23	18902.30	.43	-.53
3.141593	180.000	2.26	20014.20	.43	.00

Ocean Tides and Earth Rotation

Johannes Wünsch

Sternwarte Sonneberg, Sternwartestr. 32, D-96515 Sonneberg, Germany

Abstract. Periodic influences of ocean tides on Earth rotation are described in the first part of this paper. UT1 changes (Universal Time) as well as a small polar motion are affected. The second part concerns tidal friction by the ocean tides. Observational evidence and hydrodynamic theory are treated. Tidal friction leads to the secular deceleration of the Earth's rotation and to the evolution of the lunar orbit.

1 Periodic changes of Earth rotation by ocean tides

This new field of work comprises *periodic* variations of Earth rotation (not secular braking by tidal friction). The following scientists from Bonn and Hamburg are cooperating in this effort: P. Brosche, J. Wünsch (Sternwarte Univ. Bonn), U. Seiler, J. Sündermann (Institut für Meereskunde Univ. Hamburg), J. Campbell and H. Schuh (Geodätisches Institut Univ. Bonn). The periodic variations of the angular momentum of the oceans are obtained by hydrodynamic tidal models (see Zahel "Ocean Tides", this volume). Based on the conservation of angular momentum the consequences for the solid Earth can be calculated in variations of length-of-day and in polar motion. The strongest partial tides M_2, S_2, N_2, K_1, O_1, P_1, Mf(13.66 d), Mf'(13.63 d) and Mm were treated. In the rotation angle of the Earth $\Delta UT1$ (Universal Time) a typical amplitude of 30 μs (M_2, O_1) is to be expected. The sum of all amplitudes has an order of magnitude of 0.1 ms, i.e. 4.6 cm at the Earth's equator. For ocean tides both tensor of inertia changes $\Delta\Theta_{ij}$ and relative angular momentum J_{rel} (by tidal currents) have to be considered in Earth rotation. In the case of UT1 only the oceans are active in the semidiurnal and diurnal tidal bands; there is no solid Earth contribution because of the symmetry of the spherical harmonics involved.

The hydrodynamic computations were performed in Hamburg by Seiler (1989, 1990, 1991). A 1° × 1° grid and the Schwiderski topography are used. Tidal loading and self-attraction are treated according to Accad and Pekeris (1978). The Seiler model does not use empirical boundary conditions because it is a purely theoretical model. One obtains tidal maps and global values of the components of J_{rel} and J_Θ, accurate to about ±10% in amplitude and ±10° in phase. These values are written as harmonic waves:

$$A \cos(\sigma t - \phi) \tag{1}$$

with amplitude A and phase angle ϕ. In the z component we have

$$\Delta\omega = \Delta J_{SE}/C_m = -\Delta J_{OC}/C_m \tag{2}$$

if exchange with the atmosphere is excluded and the periodic transfer to the Moon is neglected (J = angular momentum, SE = solid Earth, OC = ocean, C_m = moment of inertia of the Earth's mantle). From $\Delta\omega$ we easily obtain the change in length-of-day, i.e. in the rotational period ΔLOD. However, the angle $\Delta UT1$, which is observed, is given by an integration in time from $\Delta\omega$:

$$\Delta UT1 = \tfrac{86400s}{2\pi} \int \Delta\omega dt. \tag{3}$$

1.1   VLBI data analysis

The existence of UT1 variations due to ocean tides has been pointed out by Brosche (1982). Brosche et al. (1991) carried out a VLBI analysis (Very Long Baseline Interferometry) aimed at the empirical verification of the UT1 terms described above. They used geodetic VLBI experiments, especially the so-called IRIS-A experiments (International Radio Interferometric Surveying; A for Atlantic) from December, 1985, to December, 1986. These sessions of 24 h duration took place every 5 days. Three parabolic antennas in the USA plus the station in Wettzell, Germany, participated. The long East-West-extension is favourable for measuring UT1. In one of the analyses, 48 IRIS-A experiments were used and the Earth rotation parameter UT1 was solved for every 3 h during a single 24 h session. This was done with the software CALC/SOLVE. Thus, one obtains a time series of UT1 values. The theoretical oceanic UT1 variations were verified in two ways:

a) By computing the cross-correlation function between the empirical UT1 results and the corresponding theoretical time series for a range of assumed time lags τ. This shows a clear maximum near a time lag $\tau = 0$ h.
b) By a Fourier fit using the method of least squares for the known periods of the tidal constituents.

The results are significant for M_2, O_1 and K_1. See also Wünsch and Bußhoff (1992).

In the long period tides Mf, Mm the model according to Brosche et al. (1989) yields both in-phase corrections and small out-of-phase parts in addition to the commonly used UT1 model of Yoder et al. (1981) which is based on an equilibrium ocean. The fitted parameter is the Love number k. The results of Hefty and Capitaine (1990) showed good agreement of the daily Earth rotation data with Brosche et al. (1989).

In the meantime, other VLBI authors have also solved for the short periodic oceanic variations and have confirmed them in their data analyses. The theoretical model for UT1 fits very well but not completely. Thus, the most recent ocean tide models with data assimilation will be of interest. Herring (1993) and Sovers et al. (1993) have also studied the polar motion caused by ocean tides (see below) with VLBI and obtain verification of its existence and order of magnitude. The mathematically elegant ocean tide model by Dickman (1993) which uses spherical harmonics yields amplitudes which are very small. The subdiurnal Earth rotation variations have also been confirmed by SLR and GPS campaigns.

1.2   Short periodic polar motion caused by ocean tides

The ocean tides are also causing a very small periodic polar motion. This can be computed by inserting the equatorial components of angular momentum into linearised equations of gyroscopic motion (Liouville equations, cf. Munk and MacDonald, 1960). The small quantities $m_1 := \omega_1/\Omega$ and m_2 are considered. With the complex quantities $m := m_1 + im_2$ etc. the Liouville equations are:

$$dm/dt = i\sigma_0(m - \psi); \tag{4}$$

where σ_0 is that angular frequency which corresponds to the observed Chandler period and $\psi = \psi_1 + i\psi_2$ is the excitation function which is a linear combination of J_{rel}, J_Θ and their time derivatives. The angular momentum components are given by Seiler (1991). Each tidal constituent can be treated separately (angular velocity σ_j). The solution is obtained by quadrature. It consists of prograde and retrograde parts, proportional to $e^{i\sigma t}$ and $e^{-i\sigma t}$. However, there are two problems to be taken into account:

a) The equations must contain the FCN resonance (free core nutation, a normal mode of the mantle + core system; $\sigma_{FCN} \approx -1.0023$ cycles per sidereal day; see Zürn "Nearly-Diurnal Resonance", this volume); this especially concerns the diurnal tides.

b) At these short periods it is not the ω axis (instantaneous axis of rotation) which is observed but the so-called Celestial Ephemeris Pole CEP (see e.g. Gross, 1992).

Each tidal constituent gives rise to an ellipse of polar motion. The superposition of all the constituents is a complicated curve with beats between diurnal and semidiurnal tides. The astronomical argument of each tide is needed to find the correct starting point. For the 'strongest' tides the order of magnitude is 0.5 milliarcseconds, i.e. 1.5 cm on the Earth's surface (Gross, 1993; Brosche and Wünsch, 1994). At present the VLBI accuracy is close to ±0.2 mas over 24 h. For the long period tides observers report a strong fortnightly period in polar motion data obtained by VLBI (Schuh, 1990) whereas ocean models show only rather small amplitudes (0.1 mas). This requires further study.

Furthermore, the Seiler model also predicts short period variations of the **centre of mass** of the solid Earth as a counter-motion to the centre of mass of the oceans. The tidal constituents M_2, S_2, K_1, O_1 dominate with an order of magnitude of 5 mm in X, Y, Z (ellipses) (Brosche and Wünsch, 1993). These can be checked by SLR or GPS.

2   Secular evolution due to tidal friction

An equilibrium tide on the Earth caused by the Moon gives rise to a prolate spheroid with its long axis pointing to the Moon. But this orientation is delayed: by internal friction (in a solid planet) or by bottom friction (in the case of the Earth's oceans). As the angular velocity of the Earth's rotation Ω is greater than the angular velocity of the Moon's orbit n the tidal 'bulge' is dragged along a little bit by the Earth. The tidal bulge will be turned by an angle δ. This phase angle has a meaning only as a schematic mean in the presence of ocean tides (irregular ocean basins). The tidal bulge causes a non-central perturbing force and also a torque acting on the Moon. By this, the Moon gains orbital angular momentum and slowly recedes from the Earth. Simultaneously, the rotational angular momentum and the Ω of the Earth are decreasing. At each moment, Kepler's third law is valid for the lunar orbit:

$$n^2 a^3 = G(M + m); \tag{5}$$

where $a(t)$ = semi-major axis of the lunar orbit, G = constant of gravitation, M and m = masses of Earth and Moon.

2.1 Observations

Observations give the integrated 'clock reading' ΔT of Earth's rotation. The changes in length-of-day LOD(T) and angular velocity $\Omega(T)$ are obtained by differentiation. We have:

$$\Delta T := TT - UT1; \quad \Delta LOD/LOD = d(\Delta T)/dt; \quad \Omega = 2\pi/LOD; \tag{6}$$

TT = Terrestrial Time is supposed to be an ideal timelike argument (earlier definitions are ET = Ephemeris Time, corresponding to planetary ephemerides and TDT). ΔT corresponds to a turn of the Earth around its axis of rotation. As a function of time it looks like a perturbed parabola. The secular, parabolic part mainly originates from tidal friction, whereas the fluctuations over decades are explained by core-mantle coupling. UT1 (Universal Time) is defined from sidereal time Θ today, while it was intended as mean solar time in the past. It contains all irregularities of the Earth as a clock (timekeeper). (In modern applications one puts $dUT1 := UT1 - TAI$, i.e. with the opposite sign.) During the time interval ΔT the celestial bodies Moon, Sun, Mercury, Venus etc. move by an observable amount $\Delta \lambda_i$ in ecliptic longitude.

Lunar Laser Ranging (LLR) is an extremely accurate method for the measurement of the lunar orbit. Since 1969, four retroreflectors have been positioned on the Moon. Observatories in Texas, Hawaii and in France (CERGA) regularly measure the distances from the Earth. The accuracy of distances has been increased to ± 3 cm. The most recent results for the secular rate of change of a and n are (Williams et al., 1993):

$$\dot{a} = +3.7cm/yr; \quad \dot{n} = -26.0 \pm 1.0''/cy^2; \tag{7}$$

This is the non-conservative part of $\dot{n}$. There is also a known conservative part which arises from the secular perturbations of the planets on the lunar orbit. It is in fact long-periodic. The analysis of LLR data requires: a numerical integration of the lunar orbit, the libration of the Moon (e.g. Lambeck, 1988), the tides of the Moon's body, reflector coordinates, observatory coordinates, Earth rotation parameters like UT0 etc. (UT0 is not corrected for polar motion, whereas UT1 is.) A recent LLR analysis is given by Müller et al. (1991) with a test of certain aspects of General Relativity Theory.

The theoretical formula for $\dot{n}$ is:

$$\dot{n} = -9k_2\delta \; n^2 \tfrac{m}{M}(R/a)^5(1 + 27e^2/2)\cos i_E; \tag{8}$$

(e = eccentricity of the lunar orbit, i_E = inclination of the lunar orbit relative to the Earth's equator). Putting $k_2 = 0.30$ leads to $\delta = 2.35°$. More exactly, the factor $k_2\delta\cos i_E$ splits up into two different functions which contain phase angles ϵ_{22} (caused by semidiurnal tides) and ϵ_{21} (by diurnal tides). These could be determined from LLR observations spanning more than 18.6 years.

Positions of the Moon can be obtained from **occultations of stars by the Moon.** Each observation of an ingress or egress of a star at the lunar limb yields an information about the mean longitude L of the Moon, which is also called s. Since 1955.5 atomic time has been available for an independent time determination. With earlier data a suitable linear combination of lunar observations with those of Sun, Mercury or Venus has to be formed in order to separate the variable Earth rotation and the gradual increase of the lunar orbit. Then a value of $\Delta T(T)$ can be deduced from every occultation observation. Lunar occultations have been recorded since about 1620 and in a greater number since about 1800. The following by-products have been derived in a series of papers by Morrison et al.: certain corrections to the star catalogues FK4, FK5; corrections to the obliquity of the ecliptic ϵ and its time rate of change; bounds for a hypothetical time dependence of the gravitational constant G (Van Flandern). Jordi et al. (1994) have considered monthly means of ΔT in data since 1925. This can be compared with El Niño events. **Transits of Mercury** in front of the solar disk allow an astrometric measurement complementary to lunar occultations. They take place about 14 times per century.

Historical **solar eclipses** have been recorded for a long time, for example since 1000 B.C. Tidal friction causes a typical change of 15° within two millennia in the geographical longitude of the totality path. One needs historic reports (chronicles) which state the year and place of observation. In some cases not all the information required is given. Hellenistic authors, babylonian clay tablets, chinese chronicles and arabic sources have been used by various authors. The place of observation of a total solar eclipse determines a linear combination of $\dot{n}$ and $\dot{\Omega}$ (e.g. Lambeck, 1980). In the most recent work the value of $\dot{n}$ is taken from modern methods (LLR) as shown above. The order of magnitude of $\dot{\Omega}$ is $\approx -1000''/cy^2$.

In the 1920ies Fotheringham published astrometric discussions of the hellenistic solar eclipses. R. R. Newton has written several books about the topic. Using arabic observations Stephenson and Morrison (1984) found a kink in the curve of $\Delta LOD(T)$ around 1000 AD, i.e. not one straight line throughout. Dalmau (1993) gives a novel philological discussion of the arabic observations and does not confirm this kink. He considers a time change of the moment of inertia of the Earth $\dot{C}$ caused by postglacial rebound. His value of $\dot{C}$ is obtained from the LAGEOS value of $\dot{J}_2$. Stephenson and Morrison give a synopsis in their most recent publication (1995).

Growth rhythms are extending even farther into the past, into geologic time. Bivalves and corals are useful as well as stromatolites. Their growth increments per solar day are periodically modulated within one or one half synodic month and by the seasons. Thus, by counting growth increments one can determine the number of days per synodic month and the number of days per year, i.e. angular velocities. Brosche, Sündermann and collaborators have computed theoretically the energy dissipation and the torque for schematic paleooceans. Recently, Williams (1990) has obtained the following result from tidally induced **sedimentation rhythms** in the Elatina formation (Australia): for the epoch ≈ 650 Ma BP: 400 ± 7 solar days per year; 13.1 ± 0.1 months per year (present: 12.37); 30.5 ± 0.5 solar days per month. (Note that: $13.1 \cdot 30.5 = 400$.) From these results he derives a distance of the Moon of 58.28 ± 0.30 Earth radii (present: 60.27).

2.2   Oceanic tidal friction

If certain assumptions are used, three quantities are sufficient to describe the system Earth-Moon under the influence of tidal friction: rotational energy of the Earth, energy and magnitude of the angular momentum of the lunar orbit. The conservation of total angular momentum can be employed. When writing down hydrodynamic energy balances care has to be taken whether the rotating coordinate system (Ω) or the inertial system is used. Two components of the equation of motion and the equation of continuity are valid for the vertically integrated variables ζ (tidal elevation), u and v (components of velocity). The yielding of the solid Earth is accounted for by the Love factor $\gamma = 1+k-h$. Each tidal constituent is treated separately. A periodic solution is computed with the boundary condition that the normal component of velocity at the coast vanishes. Oceanic friction is usually considered as quadratic bottom friction acceleration B_λ, B_ϕ:

$$(B_\lambda, B_\phi) = r|\mathbf{v}|\mathbf{v}/(h + \zeta); \tag{9}$$

where $r=0.003$ is a friction coefficient from experience, h is the undisturbed water depth and $\mathbf{v} = (u, v)$ is a two component vector. Some authors also use a turbulent eddy viscosity. A purely harmonic bottom friction would give zero as a mean over time and thus no secular torque would appear!

The net torque L is given by (Brosche, 1981):

$$L = \int\int \rho R \cos\phi \, \zeta K_\lambda dt dF; \tag{10}$$

(K_λ = zonal tidal acceleration, ϕ = latitude, ρ = density of sea water, R = radius of a spherical Earth). This leads to about $-5 \cdot 10^{16}$ Nm for the M_2 tide. The corresponding energy dissipation $\dot{E}$ is about $-4 \cdot 10^{12}$ W.

To write down the rate of change of the orbital elements of the lunar orbit a coordinate transformation and series expansion according to Kaula (e.g. Lambeck, 1980) can be used. Each tidal constituent is assigned a phase angle ϵ_{lmpq}. This shows, among other things, that the N_2 tide contributes the greatest part to the increase of the eccentricity of the lunar orbit. ϵ_{lmpq} is simply connected to the coefficients of the spherical harmonic expansion of the tidal elevation ζ: C_{mn}^{+} [cm], ϵ_{mn}^{+} [°]. These two quantities can be determined from the orbits of artificial satellites.

From the hydrodynamic equations of motion detailed balances for the z component of angular momentum and for the energy can be derived by multiplication with the radius vector and with the velocity vector, respectively. This is then to be integrated over the ocean surface and over one tidal period (Sündermann and Brosche, 1984). This balance permits a discussion of the various parts.

3   Long time integrations of the Earth-Moon system

Many authors have reached the conclusion that as 'recently' as 1 to $2 \cdot 10^9$ years ago, the Moon was very close to the Earth, although there is no geological evidence for this (the so-called Gerstenkorn event). The simplest case of secular orbital evolution (only Earth and Moon; instantaneously a circular orbit; rotational axis of the Earth perpendicular to the lunar orbit plane) is described by the differential equation for the distance r

$$\dot{r} = const \; r^{-5.5} \tag{11}$$

which can be integrated analytically. About 25 long-time integrations have been published in the literature with various assumptions about the phase angle δ. There are also differences depending on whether the lunar orbit is treated as a circle or as an ellipse, and whether the solar influence is considered. Daily and monthly variations are removed by averaging. The use of δ would be more adequate for a viscoelastic planet. The existence of oceans on the Earth leads to resonance behaviour. Brosche and Hövel (1982) made oceanographic calculations and they found a change in the tidal friction torque by a factor of 2 within $30 \cdot 10^6$ years if a continental drift is introduced. Hansen (1982) integrated tidal equations on a schematic Earth with one continent surrounded by shallow sea and deep ocean. The resonance behaviour of ocean tides was not so pronounced in the past, he wrote, and thus the tidal friction torque not as big as at present. He came to the conclusion that the Moon has never been closer than 225000 km to Earth which supports the hypothesis of binary accretion. Similar results are described by Webb (1982).

A long time integration gives the temporal evolution of

$$a(t), \; n(t), \; \Omega(t), \; \epsilon(t), \; i_E(t), \; e(t) \; etc. \tag{12}$$

In the far future Ω and n will become synchronous (at a distance of 86.4 Earth radii). Then, one month will be as long as one day = 47 present days. The remaining tidal friction due to the Sun will then cause the Moon to spiral inwards.

However, the Sun will expand to become a red giant star before this happens. The other direction of time (the far past) leads to the question of the origin of the Moon. For a long time, the three hypotheses: fission, capture or binary planet have been discussed. At present the hypothesis of a 'giant impact' is favoured by many planetologists: the collision of a Mars-sized body with the proto-Earth and the formation of the Moon from the resulting debris.

Acknowledgements: Thanks are due to P. Brosche, J. Campbell, H. Schuh and H.-J. Tucholke for reading a draft of this paper.

References

Accad, Y., Pekeris, C. L. 1978. Solution of the tidal equations for the M_2 and S_2 tides in the world ocean from a knowledge of the tidal potential alone. *Phil. Trans. Roy. Soc. London A.* **290**: 235 - 266.

Brosche P.: 1981, Gezeitenreibung im Erde-Mond-System, *Mitt. Astron. Ges.* **51**: 81 - 100.

Brosche, P. 1982. Oceanic tides and the rotation of the Earth.In: *Sun and Planetary System.* pp 179 - 184. W. Fricke, G. Teleki (eds.). Dordrecht, Holland.

Brosche P., Hövel W. 1982. *Naturwiss..* **69**: 241

Brosche P., Seiler U., Sündermann J., Wünsch J. 1989. Periodic changes in Earth's rotation due to oceanic tides. *Astron. Astrophys..* **220**: 318 - 320.

Brosche P., Wünsch J., Campbell J., Schuh H. 1991. Ocean tide effects in universal time detected by VLBI. *Astron. Astrophys.* **245**: 676 - 682.

Brosche P., Wünsch J. 1993. Variations of the solid Earth's center of mass due to oceanic tides. *Astron. Nachr..* **314**: 87 - 90.

Brosche P., Wünsch J. 1994. On the rotational angular momentum of the oceans and the corresponding polar motion. *Astron. Nachr..* **315**: 181 - 188.

Dalmau W. 1993. Bestimmung des säkularen Verhaltens der Erdrotation. *Veröff. Inst. Geschich-te der Arab.-Islam. Wiss..* **A 6**: Frankfurt a. M.

Dickman S. R. 1993. Dynamic ocean-tide effects on Earth's rotation. *Geophys. J. Int..* **112**: 448 - 470.

Gross R. S. 1992. Correspondence between theory and observations of polar motion. *Geophys. J. Int..* **109**: 162 - 170.

Gross R. S. 1993. The effect of ocean tides on the Earth's rotation as predicted by the results of an ocean tide model. *Geophys. Res. Lett..* **20**: 293 - 296.

Hansen K. S. 1982. Secular effects of oceanic tidal dissipation on the Moon's orbit and the Earth's rotation. *Rev. Geophys. Space Phys..* **20**: 457 - 480.

Hefty J., Capitaine N. 1990. The fortnightly and monthly zonal tides in the Earth's rotation from 1962 to 1988. *Geophys. J. Int..* **103**: 219 - 231.

Herring T. 1993. Diurnal and semidiurnal variations in Earth rotation. *Adv. Space Res..* **13**: 281 - 290.

Jordi C., Morrison L. V., Rosen R. D., Salstein D. A., Rossello G. 1994. Fluctuations in the Earth's rotation since 1830 from high-resolution astronomical data. *Geophys. J. Int..* **117**: 811 - 818.

Lambeck K. 1980. The Earth's variable rotation. Cambridge University Press, Cambridge, pp. 449.

Lambeck K. 1988. Geophysical Geodesy. Oxford University Press, Oxford, pp. 718.

Munk W. H., MacDonald G. J. F. 1960. The rotation of the Earth. Cambridge University Press, Cambridge, pp. 323.

Müller J., Schneider M., Soffel M., Ruder H. 1991. Testing Einstein's theory of gravity by analysing Lunar Laser ranging data. *Astrophys. J.*. **382**: L101 - L103.

Schuh H. 1990. In: Earth's Rotation from Eons to days. p. 1 - 12. P. Brosche, J. Sündermann (eds.). Springer-Verlag, Berlin.

Seiler U. 1989. An investigation to the tides of the world ocean and their instantaneous angular momentum budget. *Mitt. Inst. f. Meereskunde, Univ. Hamburg.* **29**: 1 - 101.

Seiler U. 1990. In: Earth's Rotation from Eons to days. p. 81 - 94. P. Brosche, J. Sündermann (eds.). Springer-Verlag, Berlin.

Seiler U. 1991. Periodic changes of the angular momentum budget due to the tides of the world ocean. *J. geophys. Res.*. **96**: 10287 - 10300.

Sovers O. J., Jacobs C. S., Gross R. S. 1993. Measuring rapid ocean tidal orientation variations with very long baseline interferometry. *J. geophys. Res.*. **98**: 19959 - 19971.

Stephenson F. R., Morrison L. V. 1984. Long-term changes in the rotation of the Earth: 700 BC to AD 1980. *Phil. Trans. R. Soc. Lond.*. **A 313**: 47 - 70.

Stephenson F. R., Morrison L. V. 1995. Long-term fluctuations in the Earth's rotation: 700 BC to AD 1990. *Phil. Trans. R. Soc. Lond.*. **A 351**: 165 - 202.

Sündermann J., Brosche P.. 1984. Tidal friction and dynamics of the Earth-Moon-system. In: *Landolt-Börnstein, neue Serie.* **V2a**: 299 - 310. K. Fuchs, H. Soffel (eds.). Springer-Verlag, Berlin,

Webb D. J. 1982. In: Tidal friction and the Earth's rotation II. p. 210 - 221. P. Brosche, J. Sündermann (eds.). Springer-Verlag, Berlin.

Williams G. E. 1990. Tidal rhythmites. *J. Phys. Earth.* **38**: 475 - 491.

Williams J. G., Newhall X X, Dickey J. O. 1993. In: Contributions of space geodesy to geodynamics: Earth Dynamics. p. 83 - 88. D. E. Smith, D. L. Turcotte (eds.). Geodynamics series Vol. 24. AGU, Washington D. C..

Wünsch J., Bußhoff J. 1992. Improved observations of periodic UT1 variations caused by ocean tides. *Astron. Astrophys.*. **266**: 588 - 591.

Yoder C. F., Williams J. G., Parke M. E. 1981. Tidal variations of Earth rotation. *J. geophys. Res.*. **86**: 881 - 891.

Chandler Wobble and Pole Tide in Relation to Interannual Atmosphere-Ocean Dynamics

Hans-Peter Plag

Institut für Geophysik, Christian-Albrechts-Universität zu Kiel, Olshausenstr. 40, D-24118 Kiel, Germany

Abstract. Since the discovery of the Chandler wobble in polar motion more than a century ago, the cause of the wobble remained obscure. As long as one assumes the observed wobble to be a free damped mode of the rotating Earth, a reoccurring excitation of the wobble has to be assumed likewise. Neither the cause and mechanism of this excitation nor the damping of the wobble have satisfactorily been explained. Furthermore, the analyses of polar motion data under the above assumption lead to contradictory results, namely (1) a multi-frequency or a single frequency wobble, (2) an amplitude-dependent frequency, (3) a large diversity of Q-values.

A detailed study of polar motion, oceanographic, and meteorological data gave rise to the hypothesis that the observed wobble in fact is a forced oscillation, with a slightly variable forcing frequency. The consequences of this hypothesis are discussed. A possible forcing mechanism is found in a large-scale, quasi-periodic variation in air pressure within the Chandler band. This fourteen–to–sixteen months atmospheric fluctuation is responsible for most of the oceanic pole tide hitherto attributed to the Chandler wobble, and it is the most prominent candidate for forcing the observed wobble.

Regarding the observed Chandler wobble as a forced resonant phenomenon and not as a purely free wobble raises the question of the true Chandler period and the wobble Q. However, the determination of both, period and Q, is strongly limited by the amount of available data and the still unknown amplitude of the forcing function.

1 Introduction

For a body rotating about its axis of the main moment of inertia a suitable excitation may result in a displacement of the figure axis with respect to the rotation axis. The analysis of the equation of rotation of a rigid body with rotational symmetry reveals that for such a displacement exactly one eigenmode exists, i.e., after excitation, the figure axis oscillates around the rotation axis with this oscillation commonly being denoted as wobble. The path of the instantaneous rotation axis $p(t)$ at the surface of the body with respect to an arbitrary point p_0 close to the figure axis, i.e., the path $\hat{p}(t) = p(t) - p_0$ is usually considered as polar motion (PM).

Of course, the rotation axis may also move in space. For the Earth's rotation, these motions are known as precession and nutation (see, e.g. Munk and MacDonald 1960). Here the motion of the rotation axis in space will not be considered and the discussion will be restricted to PM.

Already Euler had shown that the period of the free wobble depends on the angular velocity and the main moments of inertia of the rotating body. Taking the Earth as a rigid body, Euler found the resulting eigenperiod to be 305 days or approximately 10 months. For an observer at the Earth's surface, PM results in latitude variations and in variations of the siderial time. In the nineteenth century, when latitude observations became available with sufficient precision, Euler's results stimulated an extensive search for the wobble in these observations.

A peculiar controversy started at the end of the last century between S. C. Chandler and several theoreticians. Chandler, who analyzed observations of PM without having any preconceptions concerning the period of the free Eulerian nutation found this period to be about 427 days (approx. 14 months) instead of the expected Eulerian period of 305 days or about ten months (Chandler, 1891). Furthermore, his results indicated a variable wobble period (Chandler 1892, 1893), and in 1902 he published his 'inverse relation between angular velocity and radius' (Chandler, 1902). Theoreticians including Newcomb rejected these results when first published defending the 'laws of dynamics' (see Mulholland and Carter, 1982, for a more detailed discussion of this controversy). The period of the wobble was accepted only after Newcomb's – partly wrong – explanation, but the latter results continuously are a matter of discussion.

Today, a century later, the controversy still is not resolved, and the interpretation of the observed Chandler wobble (CW) continues to depend mainly on the various preconceptions of the different authors. Analyzing the most advanced of the long polar PM series, both, the annual and Chandler wobbles turn out to have variable amplitudes (Chandler, 1902), and, while the instantaneous frequency and phase of the annual wobble (AW) are nearly fixed, the respective parameters of the CW exhibit considerable temporal fluctuations (Guinot, 1982). The relationship between the instantaneous CW amplitude and frequency already detected by Chandler (1902) has been confirmed for the subsequent data, too (Vondràk, 1985). On the other hand, the spectra of the long series exhibit multiple Chandler peaks with at least two distinct main peaks. Depending on the author's preconceptions, most of these aspects of the observed CW were attributed to data errors or inhomogeneities (Lambeck, 1980; Lenhardt and Groten, 1985), to a multiple-frequency CW (Ooe, 1978; Dickman, 1981; Chao, 1983), a frequency-modulated wobble (Carter, 1981), non-linear effects (Vondràk, 1985), or multiple excitations (Plag, 1988). However, up to now, besides dismissing the observations as being erroneous, no physically sound explanation of the observed characteristics or the proposed nature of the CW has been given.

Since the discovery of the CW, there has been an ongoing discussion concerning the origin of this wobble. The CW observed in the polar motion data with a period of nearly fourteen months commonly has been considered as the

damped free rotational mode of the rotating, viscoelastic Earth with oceans and
continents, corresponding to the ten months Eulerian period of a rigid, oceanless
Earth. Due to the damping, the wobble should die out after a certain period of
time elapsed since the last excitation. Since observations tell us, that the wobble
is not dying out but rather exhibiting increasing amplitudes at times, it has
been assumed that there are sources frequently exciting the wobble anew. The
problem of the CW excitation has been attacked in a large number of papers,
and earthquakes (Smylie and Mansinha, 1971; Kanamori, 1976; Mansinha et al.,
1979; Souriau, 1986; Gross, 1986; Maddox, 1988; Preisig, 1992) and the atmosphere (Wilson and Haubrich, 1976a, 1976b; Merriam, 1982; Barnes et al., 1983;
Hide, 1984) have been considered as possible causes for the excitation of the
wobble (see Runcorn et al., 1988, for a comprehensive review). However, according to our current knowledge, the displacements due to earthquakes and the
associated changes in the moment of inertia are an order of magnitude too small
to account for the observed wobble amplitude. Using the polar motion equations
for a viscoelastic Earth with non-global oceans, (Pejovič and Vondràk, 1991)
could show that a large part of the wobble may be due to atmospheric forcing.
Nevertheless, there remain considerable discrepancies between model predictions
and the observed polar motion.

In the last two decades, several long series of polar motion became available.
For the present study, which depends strongly on the temporal variations of the
observed wobbles, it is of importance to select the most homogeneous of the
available series. Therefore, in Appendix A the different data sets are compared
with each other, using the geometric properties of the pole path to discuss data
homogeneity and the smoothness of the path. Based on the material compiled in
this appendix, it is concluded that the series provided by Yumi and Yokoyama
(1980; denoted as YY) are the most homogeneous ones.

In the following section, a short introduction to the theory of PM is given,
which is used to simulate PM forced by atmospheric excitation. In section 3,
the – largely well known – characteristics of the CW as determined from the
YY-series are discussed. These characteristics are essential for the hypothesis
concerning the nature of the observed wobble put forward in section 4. Based
on this new hypothesis, sections 5 and 6 introduce the oceanic and atmospheric
phenomena which potentially may force the CW in PM. In section 7, the equations for PM of a simple Earth model are used to predict PM for atmospheric
excitations constructed from observations. The characteristics of the CW in the
predicted PM are then compared to those of the observed CW. The main conclusions are summarized and discussed in section 8. Finally, in an appendix the
available different PM series are critically assessed with respect to their aptitude
for studies of the characteristics of PM at interannual time scales.

2   Theory of polar motion

In general, there are two principle ways of modeling polar motion due to angular
momentum exchange between the Earth's mantle and the core, atmosphere and

oceans: (1) based on the momentum balance, deformations of the Earth due to the momentum and pressure forcing on the mantle are calculated and the mean vorticity of the surface deformations is used to determine polar motion, and (2) based on the angular momentum balance equations for the rotation vector are derived and solved for given excitations due to, for example, the atmosphere (for a more detailed description of the advantages and problems of these approaches see, for example Lambeck, 1988).

Only the second approach will be considered here. The angular momentum balance is given as

$$\frac{d}{dt}\mathbf{H} = \mathbf{L} \tag{1}$$

with $\mathbf{H}$ the total angular momentum of the body, and $\mathbf{L}$ the external torque in an inertial frame of reference. Transforming this balance into a reference frame rotating with $\boldsymbol{\Omega}(t)$ results in

$$\mathbf{H} = \boldsymbol{\Theta} \cdot \boldsymbol{\Omega} + \mathbf{h} \tag{2}$$

where $\mathbf{h}$ is the relative angular momentum of the body, and $\boldsymbol{\Theta}$ is the (time-dependent) inertia tensor defined as

$$\boldsymbol{\Theta}(t) = \int_{V(t)} \rho(\mathbf{x}, t)(\mathbf{x}^2\mathbf{I} - \mathbf{x} \otimes \mathbf{x})dV \tag{3}$$

($\mathbf{x} \otimes \mathbf{x}$ results in a tensor with components $c_{ij} = x_i x_j$). In a reference frame rotating with angular velocity $\boldsymbol{\Omega}$ relative to the inertial frame, the angular momentum balance is written as

$$\boldsymbol{\Omega} \times (\boldsymbol{\Theta} \cdot \boldsymbol{\Omega} + \mathbf{h}) + \dot{\mathbf{h}} + \frac{d}{dt}(\boldsymbol{\Theta} \cdot \boldsymbol{\Omega}) = \mathbf{L} \tag{4}$$

This equation is non-linear in $\boldsymbol{\Omega}$ and describes the global rotation of an arbitrary body. In general, the external torque $\mathbf{L}$, the total angular momentum $\mathbf{H}$, the relative angular momentum $\mathbf{h}$ and the inertia tensor $\boldsymbol{\Theta}$ are all time-dependent parameters.

In literature, different special cases are used for $\boldsymbol{\Omega}$, and the case $\boldsymbol{\Omega} = \boldsymbol{\Omega}_0 =$ const. is emphasized here (see e.g. Smith, 1977). Selecting a uniformly rotating reference frame (sometimes called the nutation system) has several advantages but also disadvantages, and the latter motivate the use of a reference frame fixed to the rotating body.

For a nearly constant angular speed, a perturbation approach with

$$\boldsymbol{\Omega}(t) = \Omega_0(\mathbf{e}_z + \mathbf{m}(t)) = (m_1(t), m_2(t), 1 + m_3(t))^t \Omega_0 \tag{5}$$
$$\mathbf{H}(t) = \boldsymbol{\Theta}(t) \cdot \boldsymbol{\Omega}(t) + \mathbf{h}(t) \tag{6}$$
$$\boldsymbol{\Theta}(t) = \boldsymbol{\Theta}_0 + \mathbf{c}(t) \tag{7}$$

where m_i, h_i and c_{ij} are small in first order, can be used to rewrite the system of equations (4). The coordinate axes are aligned with the main axes of the moment of inertia tensor, i.e., this tensor is given as

$$\boldsymbol{\Theta}_0 = \begin{pmatrix} A & 0 & 0 \\ 0 & B & 0 \\ 0 & 0 & C \end{pmatrix} \tag{8}$$

Dropping all terms of order higher than one in system (4) leads to the Euler-Liouville-equations (ELE)

$$\boldsymbol{\Omega}_0 \times \mathbf{h} + \boldsymbol{\Omega}_0 \boldsymbol{\Omega}_0 \times (\boldsymbol{\Theta}_0 \cdot \mathbf{m}) + \boldsymbol{\Omega}_0 \mathbf{m} \times (\boldsymbol{\Theta}_0 \cdot \boldsymbol{\Omega}_0) + \tag{9}$$
$$\boldsymbol{\Omega}_0 \times (\mathbf{c} \cdot \boldsymbol{\Omega}_0) + \dot{\mathbf{h}} + \boldsymbol{\Omega}_0 \boldsymbol{\Theta}_0 \cdot \dot{\mathbf{m}} + \dot{\mathbf{c}} \cdot \boldsymbol{\Omega}_0 = \mathbf{L}$$

where the dotted quantities indicate their derivatives with respect to time. Assuming rotational symmetry (i.e., $A = B$), introducing the complex quantities

$$m = m_1 + im_2 \tag{10}$$
$$c = c_{13} + ic_{23}$$
$$h = h_1 + ih_2$$
$$L = L_1 + iL_2$$

and defining the excitation functions

$$\Psi^{PM} = \frac{-1}{\Omega_0^2(C - A)}(\Omega_0 \dot{c} + i\Omega_0^2 c + \dot{h} + i\Omega_0 h - L) \tag{11}$$

$$\Psi^{LOD} = \frac{-1}{\Omega_0 C}(\Omega_0 c_{33} + h_3 - L_3) \tag{12}$$

results in the final form of the linear equations for PM and length-of-day (LOD) changes, i.e.

$$\frac{\dot{m}}{\sigma_r} - im = \Psi^{PM} \tag{13}$$

$$\dot{m}_3 = \dot{\Psi}^{LOD} \tag{14}$$

where the wobble frequency σ_r is given by

$$\sigma_r = \Omega \frac{(C - A)}{A} \tag{15}$$

For a rigid body with $\dot{\boldsymbol{\Theta}} = 0$ and in a body-fixed reference frame (i.e. $h = 0$), the excitation functions (11) and (12) in the absence of external torque $\mathbf{L}$ reduce to

$$\Psi^{PM} = \frac{-1}{\Omega_0^2(C - A)}(\Omega_0 \dot{c} + i\Omega_0^2 c) \tag{16}$$

$$\Psi^{LOD} = \frac{-1}{C}c_{33} \tag{17}$$

The ELE given in (13) and (14) show that in the linearized case PM and LOD changes are decoupled. For polar motion, the equations describe a linear oscillator, while LOD changes can be computed directly by integration of the respective excitation function over time.

Strictly speaking, the ELE describe the rotation of the Earth without a core. Is, in general, the body separable into layers which may rotate with respect to each other, then for each layer the angular momentum balance has to be considered, i.e.

$$\boldsymbol{\Omega} \times \sum_{i=1}^{n} \mathbf{H^i} + \frac{d}{dt} \sum_{i=1}^{n} \mathbf{H^i} = \mathbf{L} \tag{18}$$

$$\boldsymbol{\Omega} \times \mathbf{H^i} + \frac{d}{dt} \mathbf{H^i} = \mathbf{K^i} , \ i = 1, \ldots, n-1$$

where $\mathbf{H^i}$ is the angular momentum of the i-th layer and $\mathbf{K^i}$ the torque acting on that layer, which depends on the coupling between the layers. It should be mentioned here, that for a simple Earth model with a mantle and a fluid core, the expression for σ_r has to be changed to

$$\sigma_r = \Omega_0 \frac{C_M - A_M}{A_M}, \tag{19}$$

where the index M indicates that the moments of inertia for the mantle alone have to be used.

For the Earth system, a variety of geophysical effects contributes to the excitation functions through changes of the inertia tensor or changes of the relative angular momentum. Examples are the (visco)-elastic deformations of the Earth due to changes in the Earth's rotation, influences of the oceans, atmosphere, terrestrial hydrosphere and cryosphere. The deformations of the Earth result in changes of the inertia tensor. These deformations may result from internal mass movements (earthquakes, convection) or can be due to exogenic forces such as tides or surface loads. The viscoelastic properties of the mantle as well as the various couplings to the other system constituents (inner and outer core, ocean, atmosphere) may influence the wobble period to a large extent. Thus, the elastic response of the Earth's mantle to PM changes the wobble period from 305 days to 445 days (see, e.g. Lambeck, 1980). Other important effects affecting the wobble period are due to the ocean's response to PM (shortening) and the viscous response of the mantle to PM (lengthening). The latter effect also leads to a damping of the wobble. Finally, it should be mentioned that additional layers in the system not only affect the period of the CW but also give rise to additional wobbles (see Zürn "Nearly-Diurnal Resonance", this volume).

The influence of the atmosphere (and likewise hydrosphere and cryosphere) on the rotation of the Earth could be described by the torque between atmosphere and the solid Earth. However, the torque is difficult to model as it not only depends on the relative wind speed but also on, for example, the surface roughness, vegetation, and topography. Therefore, in most PM model studies, the total angular momentum of the Earth system is considered to be constant (i.e. no external torque, thus excluding body and ocean tides), and in that case, changes in angular momentum of the atmosphere have to be compensated by changes in the other layers. The present-day changes in angular momentum of the atmosphere can be calculated from observational data. It is generally assumed, that the angular momentum exchange between atmosphere and ocean is small compared to that between atmosphere and solid Earth. Separating the angular momentum into an atmospheric part $\mathbf{H}_A$, and a solid Earth part $\mathbf{H}_E$ and doing the same for the excitation functions leads to

$$\mathbf{H} = \mathbf{H}_E + \mathbf{H}_A \tag{20}$$
$$\Psi^{PM} = \Psi_E^{PM} + \Psi_A^{PM} \tag{21}$$
$$\Psi^{LOD} = \Psi_E^{LOD} + \Psi_A^{LOD} \tag{22}$$

Barnes et al. (1983) derive approximations for Ψ_A, which express these excitation functions in terms of the wind and surface pressure fields. However, they combine the angular momentum change of the atmosphere and the deformations of the Earth due to atmospheric loading to "Atmospheric Angular Momentum Functions (AAMF)", which is not desirable since the latter are depending on the Earth model while the first is not.

In section 7 the ELE for a layered Earth are used to simulate the PM due to atmospheric excitation for the last 100 years. These model predictions will be compared to observations in order to see whether the predicted PM exhibits the same properties as those of the observed PM described in the next section.

3   Properties of the observed Chandler wobble

The spectrum of PM has been discussed in a number of studies (e.g. Carter, 1981; Dickman, 1981; Lambeck, 1980; Chao, 1983; Lenhardt and Groten, 1985; Gross, 1985; Plag, 1988), and the basic features will be summarized. The most dominant peaks in the spectrum are those due to CW and AW (Fig. 1). The only peak in the retrograde part of the spectrum is the small peak in the amplitudes at the annual frequency, which, however, does not contribute significantly to the variance, as is seen clearly in the variance spectrum. In the prograde part, the annual and Chandler peaks provide the maximum contribution to the variance, while the contribution of the interdecadal peaks again is negligible. In the prograde Chandler band, there are several peaks present. Depending on the approach, the variations in this band have been modeled by one (Lenhardt and Groten,1985), two (Colombo and Shapiro, 1968), three (Carter, 1981), or four harmonic constituents (Gaposchkin, 1972). Using autoregressive processes, Chao (1983) confirmed the results of Gaposchkin, while Plag (1988) found five harmonic constituents to be qualified for modeling the Chandler band. The values of Q determined in various studies of the YY series also cover a wide range between 20 and 600 (see Tab. 1), and Chao (1983) even finds two of the constituents to be characterized by negative Q values. Taking up an idea of Colombo and Shapiro (1968), Chao explains the multiple peak structure with the existence of non-elastic layers in the Earth (such as the hydrosphere, asthenosphere and outer core) and their coupling with the (visco)-elastic spheres of the Earth.

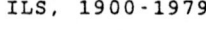

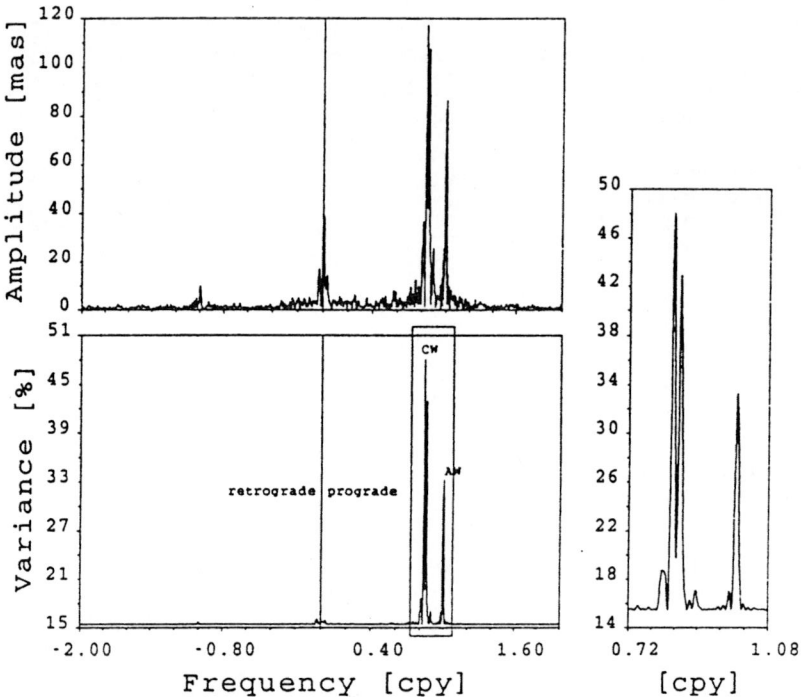

Fig. 1: Spectrum of polar motion.
Lower diagram: variance spectrum, upper diagram: corresponding amplitude spectrum. The smaller diagram to the right is enlarging the prograde Chandler to annual band of the variance spectrum. Amplitudes are given in milli-seconds of arc (mas). The variance spectrum of the ILS/IPMS series is calculated by using a circular motion with variable frequency as base function. The spectrum is defined according to an extended version (Plag, 1988) of the simple variance spectrum (Vanicek, 1970): Let $\mathbf{X} = \mathbf{x}(t_i)$, $i = 1, \ldots, N$ be a not necessarily equidistant vector time series, with the variance $v(\mathbf{X})$ being defined in the usual way, i.e. $v(\mathbf{X}) = 1/(N-1) \sum_{i=1}^{N} \|\mathbf{x}(t_i) - <\mathbf{X}>\|$, where $<\mathbf{X}>$ is the vector of the arithmetic mean values of all $\mathbf{x}(t_i)$, $i = 1, \ldots, N$. For a frequency-dependent base function $\mathbf{Y}_\omega = \mathbf{y}_\omega(t, \mathbf{p}(\omega))$ which is linear in $\mathbf{p}$, $\mathbf{p}(\omega)$ may be calculated for any value of ω, using a least squares fit of $\mathbf{Y}$ to $\mathbf{X}$. With the residual $\mathbf{R}_\omega = \mathbf{X} - \mathbf{Y}_\omega$, the variance spectrum $V\{\mathbf{X}\}$ is defined as $V\{\mathbf{X}\}(\omega) = (1 - v(\mathbf{R}_\omega)/v(\mathbf{X}))$. Here, $V\{\mathbf{X}\}$ is given in per cent. The spectra shown use the base function $\mathbf{y}_\omega(t, \mathbf{p}(\omega)) = (a\cos(\omega t) + b\sin(\omega t) + c_x + d_x t, -a\sin(\omega t) + b\cos(\omega t) + c_y + d_y t)$ with $\mathbf{p}(\omega) = (a(\omega), b(\omega), c_x(\omega), d_x(\omega), c_y(\omega), d_y(\omega))$, which is a harmonic rotation with additional trends in both components. $V\{\mathbf{X}\}$ may be estimated for any value of ω within the interval $[1/(2(t_N - t_1)), 1/\overline{\delta t}]$, where $\overline{\delta t}$ denotes the mean sampling interval.

Table 1. Some selected previous results for CW parameters.

Reference	Main conclusions
Colombo and Shapiro (1968)	Double peak structure from ILS data
Gaposchkin (1972)	Four peaks in the Chandler band
Guinot (1972)	Temporal variations in period, amplitude, and phase
Currie (1974)	One broad peak in ILS/IPMS data, $T_w = 432.95 \pm 1.02$ d, $Q_w = 36 \pm 10$
Graber (1976)	One peak with $T_w = 430.8$ d, $Q_w = 600$ from 15 yrs of IPMS data
Ooe (1978)	One peak with $T_0 = 0.8400 \pm 0.0039$ cpy and $50 < Q_w < 300$
Wilson and Vicente (1980)	Best estimates of $T_0 = 0.843$ cpy and $Q_w = 170$
Carter (1981)	Temporal variability attributed to frequency modulation due to solid Earth/ocean interaction
Okubo (1982)	Stable Chandler period, $50 < Q_w < 100$
Chao (1983)	Two major and two minor constituents in the YY series, Q ranging from -1930 to $+700$
Vondràk (1985)	Non-linear relationship between frequency and amplitude
Lenhardt and Groten (1985)	Several different models for the Chandler peak, Q is estimated to be as low as 24
Wilson and Vicente (1990)	CW period of 433.0 ± 1.1 days and $Q = 179$ with a range of 74 to 789
Kuehne et al. (1996)	CW frequency/period of 0.831 ± 0.004 cpy/439.5 ± 1.2 days
Furuya and Chao (1996)	CW period of 433.7 ± 1.8 days and $Q = 49$ with a range from 35 to 100

As was mentioned in the introduction, already Chandler noticed the variable amplitude of both, CW and AW, and a variable CW period. Contrary to what Chandler found from the early data, the YY series clearly demonstrate that the AW is far more stable in time than the CW (Fig. 2). However, the probably most controversial feature of the CW in the time is the large shift of the instantaneous phase of this wobble between 1925 and 1945 (Guinot, 1972; Vondràk, 1985). This phase shift has been attributed to the double or multiple peak structure in the CW band, and, vice-versa, the multi-peaks have been attributed to the phase shift (see e.g. Dickman, 1981). In the latter case, the phase shift would be due to a hitherto unmodeled excitation process.

Already in 1902, Chandler postulated an inverse relationship between amplitude and period. This notion was taken up again by Carter (1981), who found a linear increase in the period for a decreasing CW amplitude, which he attributed tentatively to a non-equilibrium pole tide. Vondràk (1985) determined the ins-

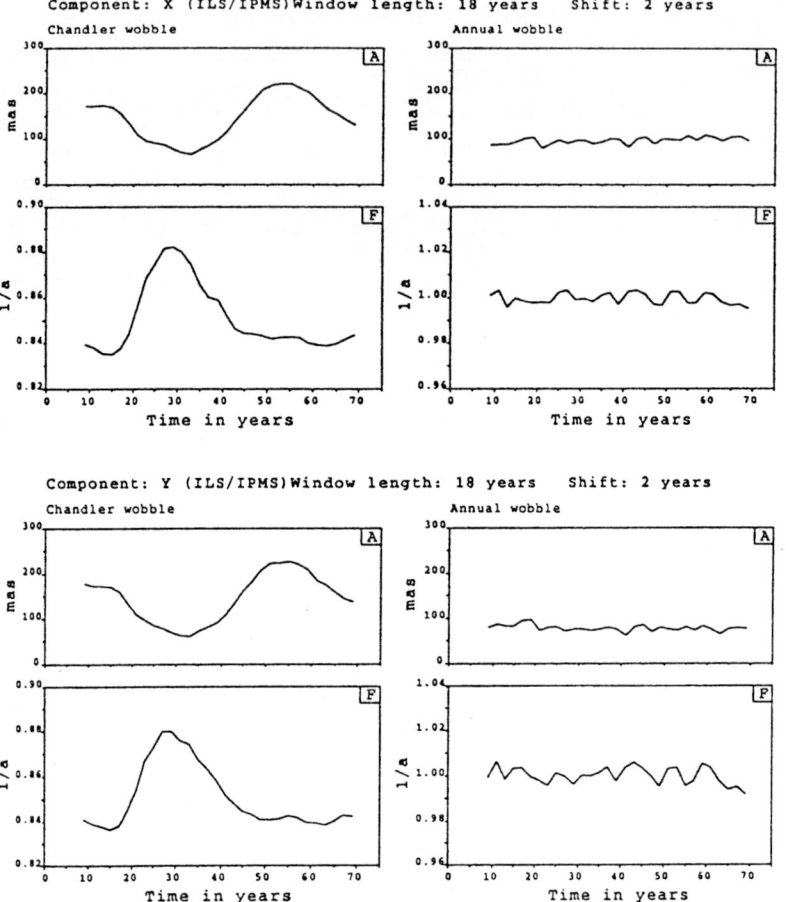

Fig. 2. Instantaneous amplitudes and frequencies of the wobbles in polar motion. The instantaneous CW and AW parameters are separately calculated for X- and Y-components from a fit of a model function to a moving data window. The model function comprises an AW, a CW, and the frequencies are iteratively determined. The window length is 18 years. Compare also to Plag (1988), who used a window length of 25 years, and to Vondràk (1985), who kept the annual frequency constant. A is amplitude, F frequency of the wobble.

tantaneous amplitude, period and phase of the CW (which are essentially the same as in Fig. 2. Fitting several functional relationships to amplitudes and periods, he obtained the best fit for the reciprocal dependence

$$f = 0.816 + \frac{0.0037}{a} \tag{23}$$

where f is the frequency in cpy, and a the total amplitude of the PM in arcseconds.

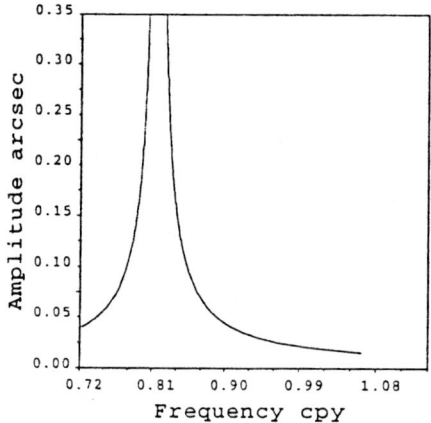

Fig. 3. Chandler wobble amplitude as function of frequency. The functional relationship shown is the inversion of the relation $f = 0.816 + 0.0037/a$, which expresses the CW frequency f as a function of the CW amplitude a (f in cycles per year, a in seconds of arc). Vondràk (1985) found this relation to fit the pairs of instantaneous amplitude and frequencies best. Note, that Vondràk claimed non-linear effects as a cause for the dependency of the free wobble's frequency on amplitude. The inverted relation shown here, however, is strikingly similar to a resonant response to a forcing with variable frequency.

In all the discussions during the last thirty years, the AW and the CW have been considered to be a forced motion and a free mode, respectively. Specifically, the observed CW is treated as a continuously or at least frequently excited, damped harmonic oscillation. This notion leads, of course, to discuss the period of the CW as function of amplitude, a notion which had already been dismissed by Newcomb (1891) as contradicting the 'laws of dynamics'. Nevertheless, such a dependence may result from some strong non-linear effects in the rotational dynamics of the real Earth. However, up to now, no evidence has been provided for non-linearity of the solid Earth's rotation. Therefore, one may ask for the consequences of dropping the notion that the observed CW includes a significant free damped contribution. In this case, the observed CW would be a forced motion, too. Particularly in the presence of a nearby resonance frequency (i.e. the Chandler frequency), we could expect the instantaneous amplitude of the wobble to be a pronounced function of the forcing frequency. Therefore, the best-fit functional given by Vondràk (1985) has been inverted and plotted in Fig. 3. The resemblance of the curve to a resonance curve is striking. Thus, the amplitude-frequency relation of the observed CW is in agreement with a resonance curve of a damped forced oscillation.

4   A new hypothesis: the observed Chandler wobble is a resonant forced oscillation

The theoretical obstacles in explaining an amplitude-dependent Chandler period, which would indicate a highly non-linear system have been discussed in the previous section. In the presence of a resonance frequency, it is more obvious to interpret the results presented in the previous section as a frequency-dependent amplitude, instead. Thus, the observed CW may well be a forced motion instead of a freely decaying wobble. To state this hypothesis explicitly: *'The observed quasi-periodic oscillation in the pole position with a mean period of approximately 14 months is a forced, quasi-periodic motion close to a resonance period.'*

If this hypothesis is correct, it would require a quasi-periodic forcing mechanism to maintain the observed wobble. The discussion of possible candidates and forcing mechanisms will be postponed until the next sections. Here, it will first be tested whether the hypothesis is capable of explaining the observed features of PM in the 14 months period band. Forced oscillation of a damped harmonic oscillator are described by the (one-dimensional) differential equation

$$\ddot{z} + \omega_0 Q^{-1}\dot{z} + \omega_0^2 z = A_0 e^{i\omega t} \tag{24}$$

The stationary part of the solution is given by

$$z_s(t) = A(\omega)e^{i\omega t} \tag{25}$$

with the frequency-dependent amplitude $A(\omega)$,

$$A(\omega) = \frac{A_0(\omega)}{\sqrt{(\omega_0^2 - \omega^2)^2 + \omega^2 \omega_0^2 Q^{-2}}} \tag{26}$$

where $\omega_0 = 2\pi/T_0$ and T_0 is the resonance period. For a known forcing $A_0(\omega)$, the parameters T_0 and Q of the free wobble can be estimated by fitting the base function (26) to the instantaneous CW amplitude as function of the instantaneous CW frequency, i.e. to $A_{CW}(\omega)$. However, no knowledge of the amplitude A_0 of the quasi-periodic forcing is yet available. Therefore, $A_0(\omega)$ is assumed to be constant, and the reasonability of the assumption will be discussed below.

$A_{CW}(\omega)$ is determined by iteratively fitting a base function

$$f(t) = a_1 \sin(\omega_1 t) + b_1 \cos(\omega_1 t) + a_2 \sin(\omega_2 t) + b_2 \cos(\omega_2 t) + c \tag{27}$$

to a moving data window, using the start values of 2π yr^{-1} and $2\pi/1.19$ yr^{-1} for ω_1 and ω_2, respectively. Vondràk (1985) used a similar procedure assuming, however, the annual frequency ω_1 to be constant. In sampling of $A_{CW}(\omega)$, there is a serious trade off between the window length used to get one sample and the number of uncorrelated samples obtainable. To avoid a correlation between different samples, only non-overlapping windows should be used. If a small window length is selected, the number of samples increases, but the uncertainties for each of the samples increase as well, and if the window length becomes less than the beat period between CW and AW, the fit of eq. (27) becomes numerically instable. To assess the effect of the window length on the resonance parameters obtained from the fit of eq. (26) to $A_{CW}(\omega)$ window lengths from 6 to 25 years have been used and, in addition to eq. (27), also a base function of a harmonic circular rotation has been analyzed (see Fig. 1). Using this latter base function, both the prograde and retrograde CW have been investigated. Moreover, since eq. (26) is non-linear in T_0 and Q, these parameters have to be determined iteratively, and the results may depend on the start values used. Therefore, a set of different start values have been used, too.

As expected from the discussion of the spectrum of the PM series given above, the retrograde wobble is highly variably in time and the pairs of instantaneous amplitudes and frequencies appear to be random for all window lengths. On the other hand, the prograde wobble clearly displays the amplitude-frequency relation shown in the previous section. Therefore, only the prograde wobble will be considered in the discussion below.

In Fig. 2 the instantaneous amplitudes and frequencies are exemplarily shown for a window length of 18 years. The respective resonance curves are displayed in Fig. 4 for both the X- and Y-components. The resonance periods and Q values obtained in these fits are roughly 450 days and 100, respectively. The period thus is considerably larger than those generally resulting in analyses of the total PM series (see Table 1). However, the largest periods are calculated for a window length of 18 years, while particularly for larger window lengths the periods are lower, thus tending to the value obtained from a fit of eq. (27) to the total PM series. There is no tendency for the X- and Y-components to produce systematically different values for T_0 or Q. In general, increasing the window length smoothes $A_{CW}(\omega)$, and the fit of the resonance curve results in a larger fraction of the variance of $A_{CW}(\omega)$ being explained. When small start values for Q are employed in the iterative fit of eq. (26) particularly for short window lengths, some fits may result in values for T_0 and Q scattering around 437 days and 30, respectively. However, these fits generally do not approximate the $A_{CW}(\omega)$ samples very well. Averaging over all window lengths and both PM components, the resulting resonance period and Q value are 444 ± 6 days and 85 ± 22, respectively.

A caveat is required here: the assumption of both, stationarity (i.e. a purely forced wobble) and constant amplitude of the quasi-periodic forcing may well bias the T_0- and Q-values given here. There is, however, no way of resolving this problem without knowledge of the excitation mechanism and the amplitude of the forcing.

The AW in PM is mainly attributed to atmospheric forcing (see e.g. Chao and Au, 1991). Therefore, we may use the resonance curve shown in Fig. 4 to derive the fraction of the annual forcing amplitude required to sustain a forced term in the Chandler band, too. In the spectrum, the CW amplitudes are not greater than 1.5 times the annual amplitude. At the annual frequency, the resonance curves determined by the different fits is between 5 and 10 % of the maximum at the resonance period. Thus, in the Chandler band, a forcing with an amplitude of 7.5 to 15 % of the annual forcing amplitude could suffice to maintain the ob-

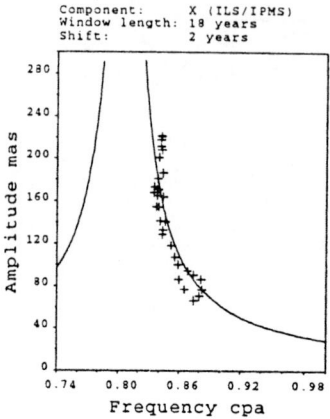
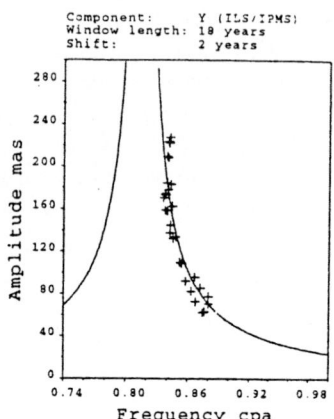

Fig. 4. Resonance curves for polar motion. The resonance curves are shown for frequencies close to the maximum of the resonance. Crosses indicate the pairs of amplitude and frequency values determined in a least-squares fit of eq. (27) to the PM series. The resonance curves eq. (26) are fitted for an assumed "constant amplitude, variable frequency" forcing. The fit results in frequencies (periods) and Q values for the X and Y-components of $\omega_0 = 0.80586$ cpy ($T_0 = 453.2$ days), $Q = 77$ and $\omega_0 = 0.81400$ cpy ($T_0 = 448.7$ days), $Q = 104$, respectively. The resonance frequencies should indicate the "true" (and, of course fixed!) Chandler frequency if the "constant amplitude, variable frequency" assumption holds.

served wobble as a forced term.

Both, in ocean and atmosphere well-known signals exist in the Chandler band, which potentially are candidates for the quasi-periodic forcing mechanism. Therefore, in the next two sections the relation of these signals to the CW and their capability to provide about 15 % of the annual forcing will be discussed.

5   The pole tide

The most prominent oceanographic phenomenon in the CW band is, of course, the so-called pole tide, which is observed in nearly all coastal tide gauge records around the world. Shortly after the discovery of the CW, a fourteen-monthly signal was discovered in tide gauge recordings, too (Christie, 1900). This tide is commonly assumed to result from variations in the Earth's centrifugal potential due to the CW (see e.g. Munk and MacDonald, 1960). If this notion is realistic, then the pole tide would not be able to contribute to an excitation sustaining a forced CW.

As mentioned above, PM changes latitude and thus the centrifugal forces at a point. The potential Φ of the centrifugal force due to PM was given by Schweydar (1916) with

$$\Phi = -\frac{a^2\Omega^2}{2}\sin 2\theta(m_1\cos\lambda + m_2\sin\lambda) \tag{28}$$

where a is the Earth's radius, θ the co-latitude, and λ the (east) longitude. For a spherical Earth, the height ξ of the equilibrium pole tide in a global ocean is then

$$\xi(\theta, \lambda) = \frac{1 + k_2 - h_2}{g} \Phi(\theta, \lambda) \tag{29}$$

where g is the acceleration due to gravity, and k_2 and h_2 are the Love numbers for the changes in potential and the vertical displacement due to a body force of spherical degree 2. For a non-global ocean, mass has to be conserved. In a first order approximation, this can by achieved with the Darwin correction, which results in

$$\xi(\theta, \lambda) = -\frac{1 + k_2 - h_2}{g} \frac{a^2 \Omega^2}{2} \cdot$$
$$\left(\sin 2\theta (m_1 \cos \lambda + m_2 \sin \lambda) - \frac{m_1 a_2^1 + m_2 b_2^1}{5 a_0^0} \right) \tag{30}$$

for the first order approximation of the equilibrium pole tide, where a_2^1, b_2^1 and a_0^0 are the coefficients of the spherical harmonic expansion of the ocean function (see e.g. Munk and MacDonald, 1960). For typical PM amplitudes of 200 mas, the pole tide amplitudes calculated from eq. (30) are of the order of 1 cm and lower.

However, detailed studies of tide gauge records showed that in some regions the observed signal deviates from the equilibrium pole tide predicted from PM observations (Maksimov, 1954; Haubrich and Munk, 1959). In some regions like the Baltic Sea the deviation may reach more than ten times the equilibrium amplitude and large phase shifts, and this has been a puzzling feature of the pole tide for the last 50 years. It has been taken as an indication for a dynamic answer of the ocean to PM forcing. The extent to which the pole tide is a dynamic tide determines the dissipation of the CW energy. Therefore, in the last twenty years, a number of theoretical studies and investigations of observations have attempted to solve this problem (see e.g. Dickman, 1979; Dickman and Steinberg, 1986; Dickman, 1988; Tsimplis et al. 1994 and references therein). Effects of a dynamical response of the ocean to PM forcing derived from hydrodynamical models for non-global oceans were found to be regionally of the order of 50 % of the equilibrium signal (Dickman, 1988) and thus are too small to explain the observed excess.

Naito (1977) was the first to show that globally the instantaneous period, amplitude and phase of the pole tide are temporally variable, but neither the period nor the amplitude of the pole tide are correlated with the instantaneous wobble parameters. Among others, Plag (1988) arrived at the same results; he showed that at the Norwegian coast most of the pole tide is due to an "inverse barometer" response of the sea level to an air pressure signal. His results are supported by Tsimplis et al. (1994), who used a meteorologically forced hydrodynamical tide and surge model of the North Sea to show that the non-equilibrium part of the pole tide in that region is fully accounted for by the meteorological forcing.

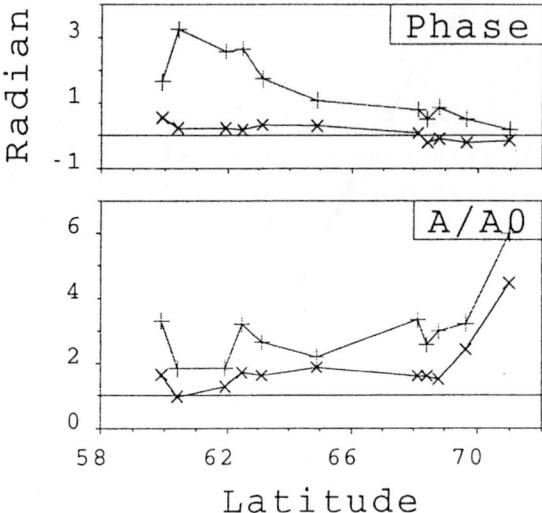

Fig. 5. Transfer function of the equilibrium pole tide. The transfer function is given as an amplitude ratio and a phase difference between the observed signal and the equilibrium pole tide predicted from the observed PM. +: uncorrected monthly mean sea levels, x: after correction of the isostatic air pressure effect. From Plag (1988).

In a recent study of the tide gauge data, Xie and Dickmann (1996) arrived at at similar conclusions. These results suggest that the atmosphere may contribute to PM excitation not only at the annual period but also in the CW band.

6   The fourteen-to-sixteen-months oscillation (FSO)

Maksimov (1954) and Maksimov et al. (1967) described a dominant air pressure signal with a period of about fourteen months which is most pronounced over the North Atlantic in the position of the Islandic low. Bryson and Starr (1977) and Starr (1983) discussed the relation between this meteorological signal and the CW and concluded that the atmospheric pole tide is not due to the CW. Plag (1988) studied the spatial distribution of the air pressure signal along the Norwegian Coast and found the amplitude increasing northwards, a feature contradicting the pattern expected for a CW induced pole tide. Furthermore, in this area, the amplitudes of the 14 months signal are nearly half the annual air pressure variations. He concluded that this atmospheric signal causes the excess of the observed sea-level signal with respect to the equilibrium pole tide at the Norwegian coast. More evidence for this signal in the atmosphere and the oceans to be of internal origin was produced from coupled atmosphere-ocean GCMs which exhibit eigenmodes at 14 months (Hameed and Currie, 1989; Currie and

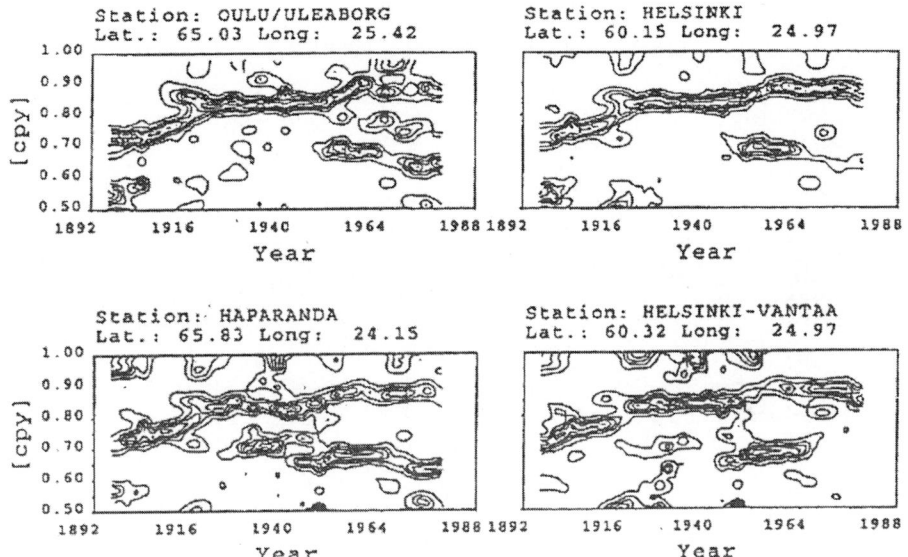

Fig. 6. FSO in sea level and air pressure. Upper row: FSO in tide gauges; lower row: FSO in air pressure. The isolines are lines of constant variance, which are constructed from normalized variance spectra. For the spectra, a window length of 18 years is used. The base function includes an annual constituent to reduce the effect of this constituent on the interannual time scales.

Hameed, 1990). However, the temporal variability of the observed fluctuations, i. e. their quasi-periodicity, is hardly explainable as being due to an eigenmode. The frequency of the signal in air pressure is temporally highly variable (Fig. 6), and these variations are correlated with those of the frequency of the apparent pole tide, while there is no simple correlation with the CW frequency. Thus, the atmospheric signal cannot be explained as purely an atmospheric pole tide.

The period of the signals in sea level and atmospheric pressure is highly variable, which led Plag (1995) to denote the phenomena as the "Fourteen-to-Sixteen-months-Oscillation (FSO)". The high temporal and spatial correlation of the temporal variations in particularly the instantaneous periods, underlines the coupled nature of the FSO in air pressure and sea level. Particularly in the North and Baltic Seas, the FSO masks the pole tide.

It should be mentioned that in other meteorological and oceanic observations, a fourteen months signal has been detected, too. For example, Kikuchi and Naito (1982) describe variations of the sea-surface temperature in the Japan Sea.

Fig. 6 also shows that the frequencies do not vary uniformly in time. There are several intervals, where the FSO displays a nearly constant frequency, and these long stable intervals are interrupted by short intervals of rapid to jump-like variations in the frequency. Based on the oceanic FSO, where this pattern is slightly more pronounced, four steps (Table 2) within the last 90 years can be

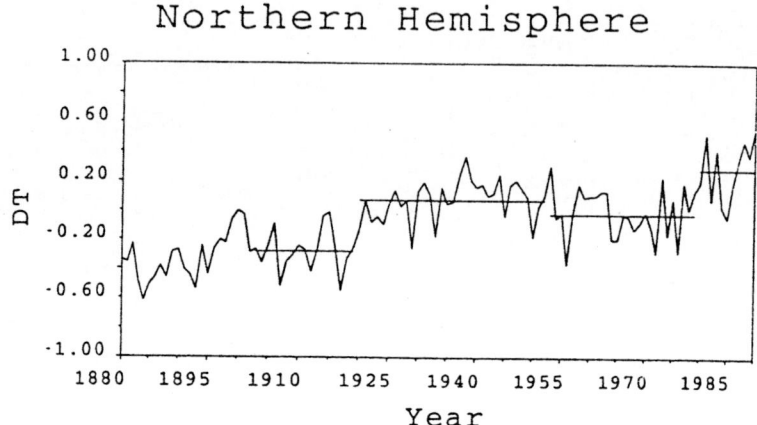

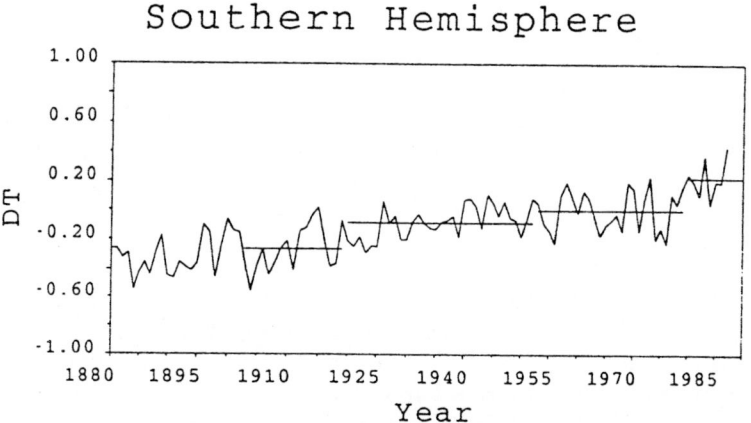

Fig. 7. Segmentation of hemispheric temperature anomalies. DT denotes temperature anomalies (in K), which are taken from Jones et al. (1986) and Jones (1992, pers. comm.). The horizontal lines indicate the mean value of the temperature anomalies within the segments defined by the frequency jumps of the FSO. Note that for the northern hemisphere the temperature steps around 1920 and 1978 are as large as 0.35 K.

identified. Applying this segmentation of the FSO to the northern hemispheric mean temperature record, significant jumps in the mean temperature of the different segments are found (Fig. 7) which break the general temperature increase over the last 100 years into a sequence of a few distinct jumps. No such segmentation is found for the southern hemisphere, where the mean temperature displays a nearly linear increase.

Some of the jumps in the northern hemispheric temperature are well known

Table 2. Time intervals with stable periods of the FSO.

The periods given are mean values determined for the oceanic FSO all suitable tide gauge records in the North and Baltic Seas. The period before the jump around 1902 could not be determined due to lack of data. From Plag (1993).

Interval	Mean Period days	Jump days
1903-1919	480±1	
1920-1955	436±1	−44 ± 1
1956-1978	418±1	−18 ± 1
1979-1991	430	12

among climatologists (particularly the one around 1920, see Ellsaesser et al., 1986, for more references). Thus, jumps in the frequency of the FSO seem to correlate with jumps in hemispheric temperature. On the other hand, Schlesinger and Ramankutty (1994) describe an oscillation of 65-70 years in the (smoothed) mean temperature of the northern hemisphere, which is most pronounced over the North Atlantic (Fig. 8). This oscillation is in agreement with the strongly smoothed jump sequence, and thus it is not clear whether there is a more or less harmonic variation in the hemispheric temperature or rather an indication of rapid transitions between different states of the hemispheric climate system. It is clear, that unraveling this problem is of particular weight for understanding climate variability on interannual to interdecadal time scales, and these time scales are of paramount interest for predicting future anthropogenic climate variations.

The restriction of the phenomenon (i.e. the oscillation of 65-70 years and the segmentation determined from the FSO) to the northern hemisphere may indicate an influence of the land-ocean distribution, which is markedly different on the two hemispheres.

7   Model predictions of polar motion due to atmospheric excitation

Returning to the question of what is forcing the observed CW, a combination of the FSO and the oscillation of 65-70 years described in the previous section will be used to construct an atmospheric forcing function for the ELE given in section 2. Eqs. (18) are set up for an Earth model with a viscoelastic mantle and fluid core, and the excitation functions given by eq. (20) include a first order approximation of an equilibrium ocean. To predict PM due to a given forcing, these equations are integrated over time. However, modeling atmospherically forced PM over time intervals of 10^2 years introduces several problems. The atmospheric and oceanic data base available for the last hundred to two hundred years is not sufficient to be used directly as input for the PM equations. The observational evidence therefore needs to be interpolated somehow. The initial

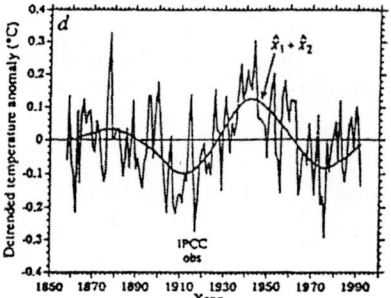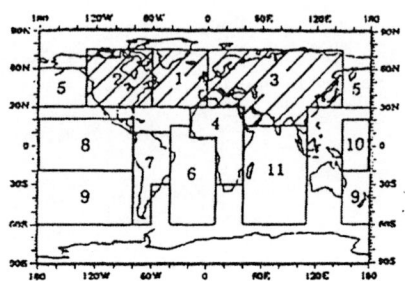

Fig. 8. The 65-70 years oscillation in temperature. Left: temporal variation of the oscillation of 65-70 years in the mean temperature of the northern hemisphere ($\hat{x}_1 + \hat{x}_2$). Right: geographical regions, in which the oscillation is most pronounced (hatched areas). According to Schlesinger and Ramankutty (1994).

conditions excite a free damped CW, which requires a long time before being sufficiently damped. Thus, a considerable time interval is needed to reduce the effect of the initial conditions in the predicted PM series. Therefore, the forcing function needs to be extended before the time interval actually being of interest, i.e. back into the last century.

The coupled atmosphere-ocean FSO loads and deforms the solid Earth and thus, in principle, is capable of forcing PM. Particularly when the instantaneous frequency of the FSO is close to the (still unknown) eigenfrequency of the Earth in the Chandler band, the effect of the FSO on PM would be strong. Therefore, the temporal variations of the FSO frequency may be crucial for PM excitation. The spatial distribution of the FSO in both sea level and air pressure is in agreement with the spatial extent of the oscillation of 65-70 years as given by Schlesinger and Ramankutty (1994). Therefore, the atmospheric forcing function includes a FSO in air pressure with a spatial extent as given in Fig. 8. In addition, an annual forcing term is included in order to determine the amplitude ratio of the predicted AW and CW. Before 1902, when the first detectable frequency shift in the FSO occurred, the excitation consists of a annual constituent and a fixed-frequency FSO.

The PM series predicted for this forcing function (Fig. 9) are analyzed in the same way as the PM observations. The resulting instantaneous amplitudes and frequencies of CW and AW shown in Fig. 10 are similar to those of the observed wobbles (Fig. 2). Moreover, a variety of different forcing functions with and without the FSO were used as input for the model equations, proving that only those including the FSO in the way described above predict a CW with characteristics similar to those of the observed CW. In this case, a FSO in air pressure with an amplitude of only 50 Pa is sufficient to excite the CW with amplitudes of the same order as the observed ones.

The resonance curves fitted to the pairs of instantaneous amplitudes and fre-

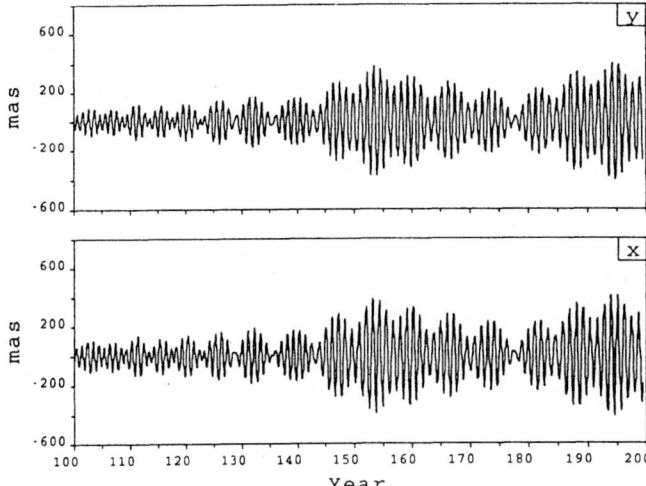

Fig. 9. Predicted polar motion due to atmospheric forcing. Time in years after the start of integration. The excitation is consistent with a start of integration at 1800. The atmospheric excitation includes an annual constituent and a FSO with spatial extent given by the hatched area in Fig. 8 and temporal variations of the period as given in Table 2.

quencies result in values for T_0 and Q of the predicted CW of 437.6 days and 230. It is interesting to note that the free mode period built into the model is 435.0 days, thus, the period determined in the analysis of the predicted PM is slightly longer than the actual free period. The free mode in the model has a Q of 100 whereas the analysis of the predicted forced wobble damping of the model leads to a Q appreciably larger than that. This may be taken as an indication that the T_0 and Q determined for the observed CW are also tending to be slightly larger than the respective parameters of the free wobble of the real Earth.

8 Conclusions

The characteristics of the observed CW are in agreement with those expected for a resonant forced oscillation. Therefore, the hypothesis forwarded here to explain the features of the CW derived from PM observations considers the wobble to be mainly a forced motion instead of a purely free one. Under this hypothesis, the resonance frequency (i.e. the "true" Chandler frequency) is fixed in time, while the forcing frequency is variable, thus producing the characteristics of a resonance curve in the observed CW. Fitting a resonance curve to the instantaneous parameters of the observed CW, the resonance period turns out to be significantly larger than the widely accepted approximately 435 days.

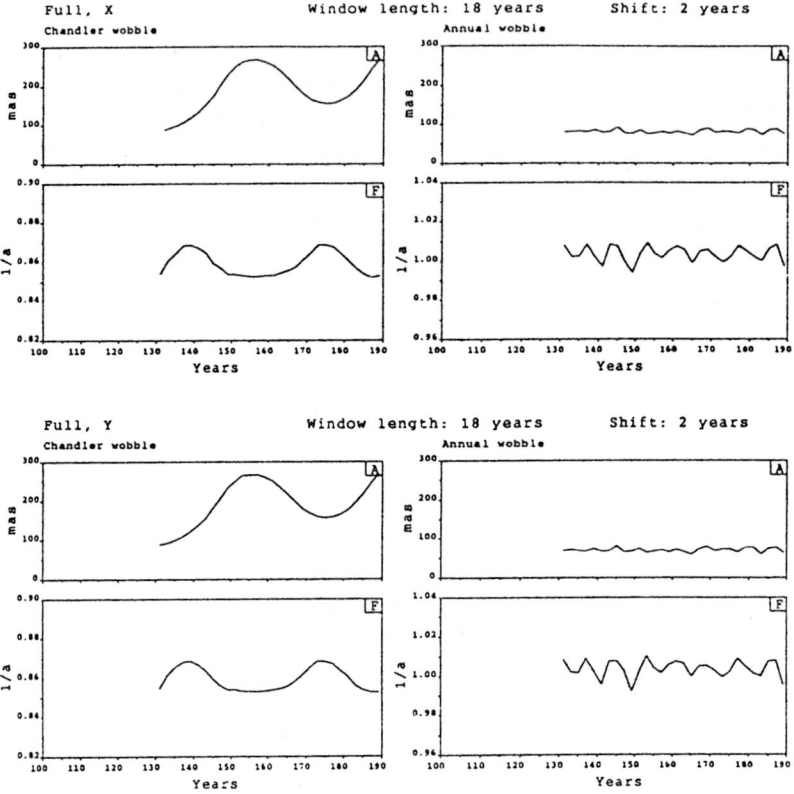

Fig 10. Instantaneous amplitudes and frequencies of predicted annual and Chandler wobbles. For an explanation see Fig. 2.

In a recent study, Kuehne et al. (1996) arrive at a similar conclusion finding the free wobble period to be 439.5 days. In studies of this type, excitation functions are derived in a demodulation of the PM observations and are correlated with other observations such as AAM series. In Kuehne et al. (1996) the determination of the period is based on a least squares fit of excitations functions derived from 9 years of recent PM observations combined with proxy excitations constructed from atmospheric data. The fit is then backed by extensive Monte Carlo tests. The result obtained by Kuehne et al. (1996) appears to be in contrast to that of Furuya and Chao (1996), who base their study on 10.8 years of modern PM observations together with AAM data. Proceeding in essentially the same way as Kuehne et al. (1996) but making fuller use of the AAM observations, the Monte Carlo test leads them to a best estimate of the CW period of 433.7 days. If, however, the observed CW has a strong forced part, as was demonstrated in the previous sections to be in agreement with the characteristics of the observed PM, with the forcing not fully represented by the atmospheric or proxy data, then the methodology applied by Kuehne et al. (1996) and Furuya and Chao

(1996) can be expected to result in different (apparent) CW periods for different time intervals.

The forcing mechanism in the Chandler band may well be the FSO, which is a quasi-periodic oscillation of the coupled atmosphere-ocean system. In all studies of CW excitation, the possible role of the ocean has been neglected, due to insufficient amount of observations. If, as the pieces of evidence discussed here do suggest, the oceans significantly take part in this FSO, the missing contribution for sustaining the observed wobble may stem from the combined effect of the atmosphere and oceans.

Furuya and Chao (1996) refer to a study by Furuya et al. (1996), where quasi-periodic wind signals in the Chandler band are considered as possible excitation of the CW. However, these wind signals are likely to be correlated with the FSO in pressure (see Plag, 1988) and thus may not rule out the pressure forcing as the major forcing of PM in the Chandler band. In fact, as reported in the previous section, in model studies the pressure forcing due to the FSO is found to be sufficient to excite the PM in the Chandler band. Moreover, Volland (1996) rules out wind as a major forcing factor for PM and concludes that only atmospheric pressure loading contributes to atmospheric PM excitation. He finds Rossby-Haurwitz waves in the atmosphere with amplitudes of the order of 50 Pa to be sufficient to excite the PM in the Chandler band. Interestingly, such waves can be excited by suitably heating the atmosphere from below (Volland, 1994, pers. comm.). Ideally, such heating would have a sectorial spatial pattern. Thus, if the FSO is such a Rossby-Haurwitz wave, its restriction to the northern hemisphere alone could be a consequence of the hemispheric differences in the land-ocean distributions, which are nearly sectorial and zonal in the northern and southern hemispheres, respectively.

However, if this turns out to be right, this does not answer the question concerning the ultimate cause, since the origin of the FSO in the atmosphere has not been discussed yet. To understand this signal, the complete interannual meteorological spectrum has to be considered, which exhibits a number of additional quasi-periodic fluctuations such as the quasi-biennial oscillation. Plag (1988) suggested a non-linear response of the atmosphere to the solar forcing as a possible mechanism creating most of the observed interannual signals. The slight modulation of the annual cycle in insolation due to the quasi-periodic solar sunspot cycle, in principle, has the potential of creating a large set of interannual frequencies, with these frequencies coinciding with most of those frequently reported in the interannual part of meteorological or oceanographic spectra. In the non-linear atmosphere-ocean system, the amplitudes of the system's response to these modulation frequencies may be considerable, and ultimately, we may find that they contribute substantially to the interannual spectrum of the atmosphere-ocean system.

As discussed in the previous section, a FSO with amplitude of only 50 Pa is sufficient to excite the PM in the Chandler band. This high sensitivity of the PM to atmospheric forcing requires atmospheric data of high quality as input for polar motion models. However, vanDam et al. (1994) pointed out that, for example, the global atmospheric data set of the National Meteorological Centers (NMC) only poorly represent air pressure at high latitudes with differences between the global data set and station data of up to 3000 Pa. Unfortunately, the FSO has large amplitudes particularly at high latitudes. Therefore, it cannot be expected that the atmospheric excitation of the CW can be modeled on the basis of these data.

The high sensitivity of the CW could, however, be used in the validation process of climate models. Using the air pressure field predicted by climate models for the last 100 years as input for a (sufficiently complex) polar motion model, the predicted PM should exhibit a CW with characteristics comparable to those of the observed wobble. If not, the climate models would not be accounting for the FSO, which may turn out to be a major feature of the atmospheric circulation over the northern hemisphere.

A   Properties of the polar motion data

Most recent papers concerning the interannual part of the spectrum of (PM) are based upon the 80 years record derived from the combined observations of the International Latitude Service (ILS) and the International Polar Motion Service (IPMS). This series starts in 1900 and is derived from optical astrometric observations at a varying number of stations (Yumi and Yokoyama, 1980). A second frequently used but somewhat shorter record starts in 1964 and was, until 1988, supplied by the Bureau International de l'Heure (BIH). Currently, the data set is updated by the International Earth Rotation Service (IERS). This latter record results from a number of different observation techniques such as Satellite Laser Ranging, Lunar Laser Ranging and Very Long Base Line Interferometry (Kolaczek, 1989).

Table 3. Available PM series.

Series given in the lower part of the table are too short to derive the temporal variability of the observed CW.

Abbreviation and Source	Interval	Sampling interval
ILS/IPMS, Yumi and Yokoyama (1980)	1899 - 1979	1 month
GROSS, Gross (1990)	1899 - 1979	1 month
IERS93 (C04)	1854 - 1993	1 month
BIH	1969 - 1985	5 days
VLBI	1981 - 1992	1 to 5 days
IERS/ERP (C01)	1963 - 1993	1 day

Though the Earth rotation parameters (ERP) based on space techniques have a precision about an order of magnitude better than those determined from astrometric methods, the long astrometric series will continue to be important for quite a long time to come. They constitute a unique source of our knowledge of the past PM. On time scales from months to years there are, however, systematic differences between the astrometric PM and those determined from VLBI and SLR, and these differences are still not completely understood (Kolaczek and Hua, 1991).

For our study of the nature of the 14-monthly oscillation in PM, commonly termed the CW, the temporal variability is an important criterion. Therefore, both the length of the series as well as the data homogeneity are important in selecting the data, and the results are heavily dependent on the quality of the PM data used. Therefore, we will first assess the long series to decide which of these is best suited for studies of long-term characteristics of the CW. Besides the ILS/IPMS series mentioned above, there are two more long PM series (see Tab. 3), one of them being a reanalysis of the ILS/IPMS data performed by Gross (1990), the other one being compiled and distributed by the IERS. This latter series is a combination of PM data obtained by different methods of observation covering the period from 1854 to 1993.

The origin of the ILS/IPMS astrometric observations has been reviewed by e.g. Lambeck (1980), who also gives an account of possible sources of data inhomogeneities and errors. In a great effort, Yumi and Yokoyama (1980, denoted by YY) used the original observations to derive the homogeneous series of the ILS/IPMS PM data for the period 1899.0 to 1979.0. In YY, the observed latitude variations $\Delta\phi_i$, where i denotes the station number, are fitted by

$$\Delta\phi_i = x(t)\cos\lambda_i + y(t)\sin\lambda_i + z(t) \tag{31}$$

where ϕ_i and λ_i are the geographical latitude and longitude, respectively, of the i-th station. $x(t)$ and $y(t)$ are the displacements of the pole towards $0°$ and $90°$ W, respectively, relative to an arbitrary origin. $z(t)$ is a station-independent error term introduced by Kimura to increase the overall quality of the fit (see e.g. Lambeck, 1980). The resulting series are given as monthly values of the x- and y-coordinates of the pole position and the Kimura error term. These PM series stimulated a number of analyses (e.g. Dickman, 1981; Daillet, 1981; Chao, 1983; Guinot, 1982), but the analyses did not succeed in clarifying the open questions concerning the properties of the CW.

Vondràk (1985) used a data set comprising the ILS/IPMS data augmented by the more recent BIH and some additional data. His series were not available for the present study, but some of his results will be discussed further below.

Gross (1990) reanalyzed the original ILS/IPMS data to obtain new PM series. The difference between YY's analysis and the Gross analysis lies in the treatment of the error term, i.e. the Kimura term. Gross assumes a station-dependent error and uses

$$\Delta\phi_i = x(t)\cos\lambda_i + y(t)\sin\lambda_i + z_i(t) \tag{32}$$

instead of (31) to fit the observations. In the Gross-analysis, the system of normal equations is underdetermined, requiring additional constraints (see Gross, 1990, for more details).

ILS

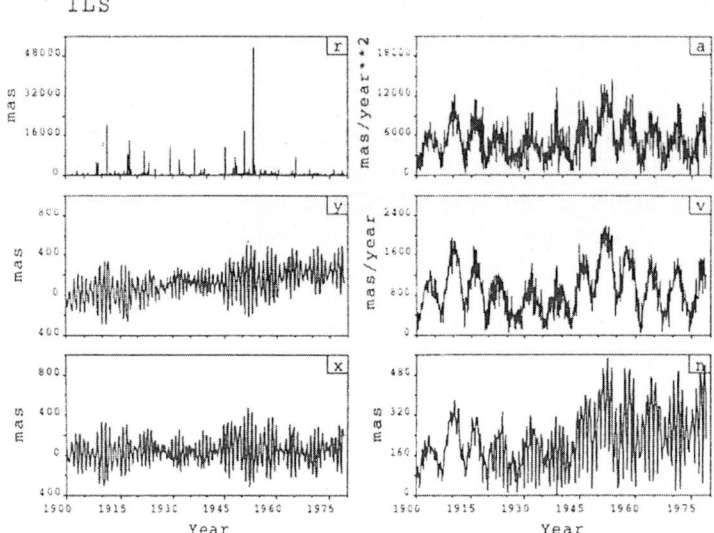

Fig. 11. Characteristics of the ILS/IPMS polar motion series. The pole position is considered as a path in $\mathbf{IR}^2$. The diagrams x and y are the $X-$ and $Y-$coordinates of the pole position in milliarc seconds, respectively, n is the norm, v the velocity, a the acceleration and r the radius determined as described in the text.

The longest PM series covering the interval from 1854 to 1992 are supplied by the IERS. However, these series are a conglomerate of different data sets originating from different observational methods (see below).

To assess the data homogeneity, the pole path is considered as a function in $\mathbf{IR}^2$, i.e.

$$s(t) = \begin{pmatrix} x(t) \\ y(t) \end{pmatrix} \tag{33}$$

From this function, a number of parameter functions are derived, which will be used for the assessment the data quality. Thus norm

$$n(t) = \|s(t)\| = \sqrt{x^2(t) + y^2(t)} \tag{34}$$

the velocity

$$v(t) = \left\| \frac{ds(t)}{dt} \right\| = \sqrt{\left(\frac{dx(t)}{dt} \right)^2 + \left(\frac{dy(t)}{dt} \right)^2} \tag{35}$$

and the acceleration

$$a(t) = \left\| \frac{d^2 s(t)}{dt^2} \right\| = \sqrt{\left(\frac{d^2 x(t)}{dt^2} \right)^2 + \left(\frac{d^2 y(t)}{dt^2} \right)^2} \tag{36}$$

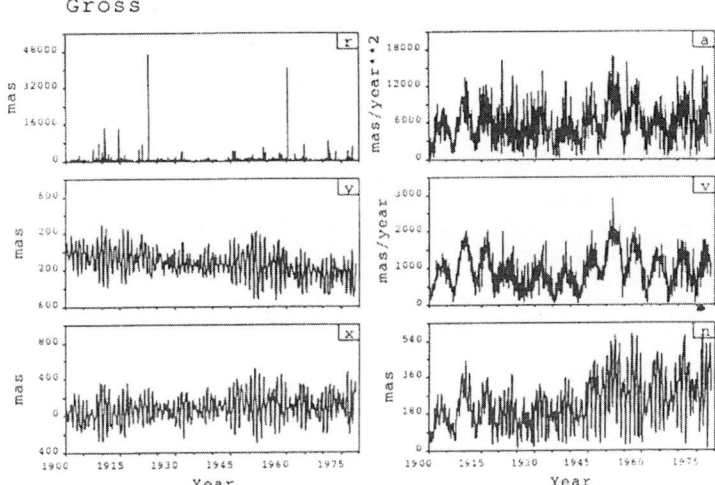

Fig 12. Characteristics of the Gross polar motion series. See Fig. 11 for an explanation. Note that Gross (1990) used a coordinate system with the y-axis pointing towards the 90° E meridian instead of the 90° W as used by the other series. The Gross-series have a higher noise level compared to the YY solution (see text), as is best illustrated by the acceleration and the radius.

are calculated for all series.

The curvature κ of the path could be calculated from

$$\kappa(t) = \frac{\|\mathbf{v}(t) \times \mathbf{a}(t)\|}{v^3(t)} \tag{37}$$

but this formula is inconvenient for numerical calculations and somewhat inaccurate due to the numerical differentiations involved. Therefore, alternatively, the calculation of the radius of curvature

$$r = \kappa^{-1} \tag{38}$$

by fitting a circle to three consecutive pole positions is preferred. All these parameters help to discuss some of the properties inherent in the PM, which are of potential importance in understanding the CW.

Starting with the ILS/IPMS series, Fig. 11 displays the x- and y-coordinates of the pole position together with the parameters derived from the pole path. In the coordinate system used by the ILS/IPMS to represent the pole position the x-axis points toward the longitude $\lambda = 0°$ (Greenwich) while the y-axis points to $\lambda = 90°$ W. In both, the x- and y-coordinate of the pole position, the characteristic beat due to the superposition of AW and CW is the most prominent feature. This superposition also produces the protruding variation of about 5 to 8 years duration in both velocity and acceleration.

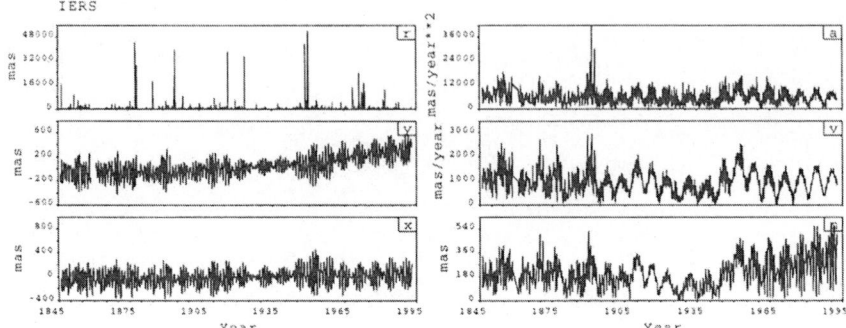

Fig. 13. Characteristics of the IERS monthly polar motion series. See Fig. 11 for an explanation. Note the sudden changes in the characteristics of particularly the acceleration and the radius at about 1895. These changes are taken as evidence for data inhomogeneities. For r the same scale as in Fig. 11 is used, though the maximum values are as large as 10^6 mas.

Especially the norm and the velocity show that the whole record separates into three distinct parts: Between 1900 and about 1920, the beat of CW and AW is characterized by a period of about 6.5 years and a comparatively large amplitude. Within the second part from 1920 to about 1943, the beat amplitude is rather small while the period is almost 8 years. The third part again is characterized by a larger beat amplitude and a smaller beat period.

Leaving Fig. 11 for a moment, the other two long series will be considered. In Fig. 12 the characteristics of the reanalysis of the ILS/IPMS data by Gross (1990) are shown. As was discussed above, the basic difference between the YY-analysis and the Gross-analysis lies in the treatment of the error term (see equations (31) and (32). A high frequency noise is present in the Gross-series, as is most obvious in the acceleration, while the overall characteristics of the pole path are more or less the same for both analyses. There are also some pronounced differences, which are most obvious in the interval 1925–1930 and around 1973, where the characteristic pole path due to CW and AW is almost below the noise level. In the norm of the Gross series, the CW and AW appear right from the start, which implies an origin different from that of the ILS/IPMS series. In the norm, some pairs of adjacent maxima have exchanged their ranks compared to the YY-series (e.g. around 1955 and 1976). The higher noise level of the Gross series might well be due to numerical problems in the solution of the normal equations of the fit of (32) to the observations. Therefore, the original YY series seem to be better representing the actual PM. At a first glance, the IERS monthly series covering the interval from 1854 to 1993 is most attractive due to its length of almost 150 years. However, Fig. 13 reveals severe data inhomogeneities, with the complete record separating into at least three distinct parts. The first part prior to 1895 is characterized by a high noise level and large kinks indicated by very large values of the acceleration and the radius (up to 10^6 mas). The part between 1900 and 1979 is more or less comparable to the YY-series, though there are still some discernible differences. After 1979, the noise in velocity and acceleration decreases strongly, which is characteristical for the strongly filtered BIH data (see e.g. Guinot, 1978).

Due to these inhomogeneities, the IERS series are virtually useless for any long-term studies. It should even be stressed here, that using such inhomogeneous data might result in the detection of apparent phenomena being solely due to artificial effects. As will be discussed elsewhere, the same characteristics described here for the IERS monthly series are found for the IERS daily data (see Tab. 3) supplied for the interval from 1962 to 1993. Here, too, the most protruding properties of the data are the inhomogeneities!

In what follows, this study will be based on the series resulting from the original YY-analysis. Returning to Fig. 11, the norm reveals another interesting fact: during the first part of the record, the motion is mainly about the center of coordinates, and the CW and AW do not appear in the norm. Around 1920, the wobbles appear in the norm, and another change of the norm is discernible around 1943. During each intermediate part, the pattern of the norm is seemingly stable. The feature may be due to both, an increase in the amplitudes of the wobbles and a more or less sudden shift of the center of motion. The latter of these causes introduces a step function rather than a linear trend in the instantaneous center of motion, and this might result in a serious bias of the results of any temporal analysis. Therefore, it is worthwhile trying to exclude the latter cause as being responsible for the changes in the norm. For this purpose, statistical tests for sudden changes within time series may be utilized. The Mann-Kendall-Sneyers sequential test resulting in the so-called C-functions is one of those tests (see e.g. Sneyers, 1975; Kendall and Stuard, 1979) and has been successfully applied to different geophysical time series by e.g. Goossens and Berger (1986; 1987) and Demaree and Nicolis (1990). For the data points $x_1, ..., x_n$ this nonparametric test is based on Mann's rank statistics, which gives, for each sample x_i, the number n_i of preceding samples x_j ($j < i$), with $x_j < x_i$. Under the null hypothesis (i.e. stationary data, no trend), the test statistic

$$t_k = \sum_{i=1}^{k} n_i \tag{39}$$

is normally distributed and the mean and variance of t_k are given by

$$\overline{t_k} = \frac{k(k-1)}{4} \tag{40}$$

and

$$\overline{\delta t_k^2} = \frac{k(k-1)(2k+5)}{72}, \tag{41}$$

respectively. For the normalized variable

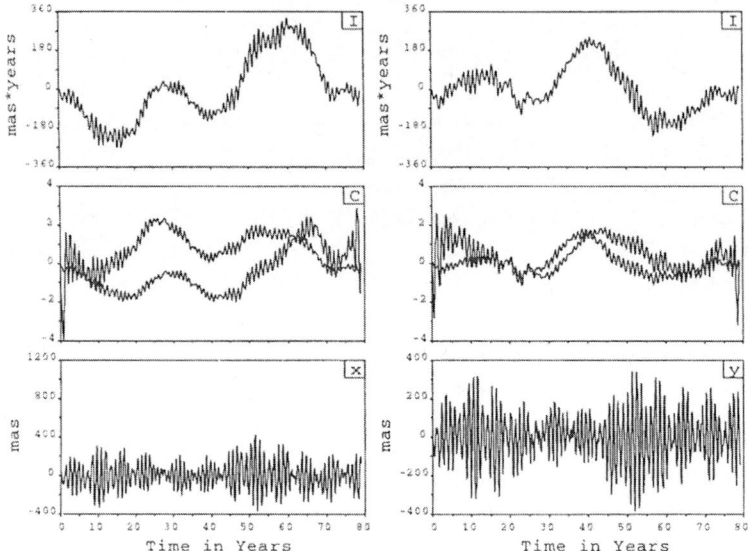

Fig. 14. C-functions of the polar motion components. The lower diagrams display the detrended x- and y-coordinates of the pole position. In the middle diagrams, the C-functions are given for the two coordinates separately. The C-functions are calculated for the detrended data according to Goossens and Berger (1986). Confidence limits for 95 % are at ±2.03. The upper diagrams show the integrals of the two detrended components.

$$u_k = \frac{t_k - \overline{t_k}}{\sqrt{\delta t_k^2}} \tag{42}$$

the probability

$$\alpha = \text{prob}(|u_\alpha| < |u_k|) \tag{43}$$

is used to accept or reject the null hypothesis. For any significance level α_0 of the test, a corresponding level of u_0 can be determined. For $\alpha_0 = 0.05$, the corresponding value of u_0 is 2.03 (Goossens and Berger, 1986). In the sequential version of this test, both forward and backward functions of u_k are calculated (the so-called C-functions), and an intersection of the two functions within the confidence interval may be used to determine the time of the onset of either a trend or a rapid change like a step. The C functions of the detrended x- and y-coordinates of PM (Fig. 14) do not indicate a step where the characteristics of the norm and velocity change. Therefore, it may safely be concluded that these changes are merely due to an increase in the CW amplitude. In Fig. 14, the integrals $I_x(t) = \int_0^t x(s)ds$ and $I_y(t) = \int_0^t y(s)ds$ of the detrended coordinates are plotted. The integral is a rather effective low pass filter. An interdecadal fluc-

BIH

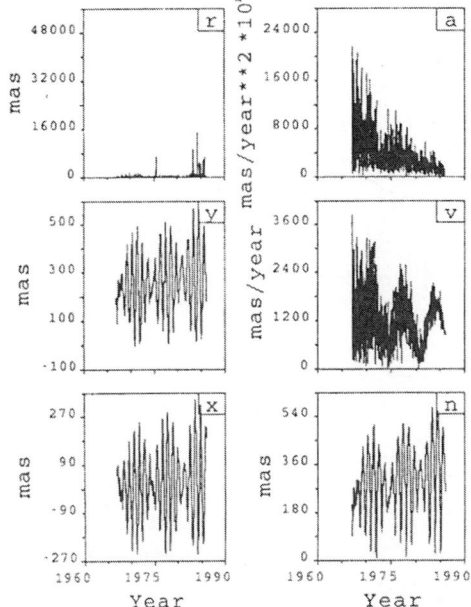

Year

Fig 15. Characteristics of the BIH polar motion data. See Fig. 11 for explanation. Data are the unfiltered BIH series. Note the different scales for the acceleration in the present figure compared to Fig. 11, which may be due to the shorter sampling interval for the BIH series (see Tab. 3). The decrease of the high frequency-noise in both, velocity and acceleration is an indicator of the data quality increasing with time.

tuation clearly emerges from both components, which has been termed as Markowitz wobble (e.g. Dickman, 1981; Vondràk, 1985). Denoting this fluctuation a wobble is still a matter of discussion, and this topic will be taken up again elsewhere.

It is of interest to compare the long time series to the more precise series based on space-geodetic techniques. Therefore, one of the three shorter data sets mentioned in Tab. 3 will be considered, namely the ERP series available from the BIH. The unsmoothed BIH PM-series exhibit features in x, y, and the norm, which are similar to those of the YY series (Fig. 15). The norm indicates that the center of motion is not identical to the origin. Both, velocity and acceleration display large variations in the beginning, which could well be due to larger uncertainties in the measurements. For the velocity, the characteristic beat emerges somewhere after 1970, while for the acceleration no such feature is discernible. Furthermore, the scale for the BIH acceleration is larger than for the ILS acceleration by more than one order of magnitude. This increase in acceleration is most likely due to the shorter sampling interval, which is five days for BIH compared to one month for ILS.

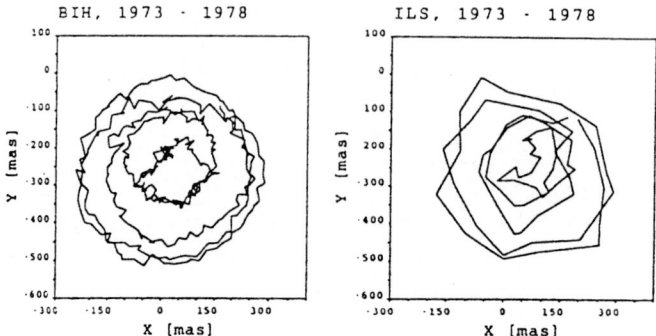

Fig. 16. Comparison of the ILS/IPMS and BIH pole paths. The data have been converted such that the positive y-axis points towards the 90° E meridian.

Comparing directly the pole paths as given by the ILS/IPMS series and BIH series (Fig. 16), the stronger short period noise in the unsmoothed BIH data is obvious, while the long-term pattern is the same for both data sets. There are, however, some protruding differences between the ILS/IPMS and the BIH data, that are not explainable by the high-frequency noise level in the BIH data. These differences indicate the level of uncertainty in the individual ILS/IPMS values, which are higher than the uncertainties ascribed to the individual observations. The source of these systematic differences investigated in more detail by Kolaczek and Hua (1991) remains unknown.

The radius of curvature has not yet been considered. Taking into account the different features for the acceleration, the similarity of this parameter calculated from the different data sets including the BIH data is rather striking. For most of the time, the radius is varying rather smoothly between 100 and 300 mas. However, frequently there are large spikes, amounting to seconds of arc and occasionally radii of as much as 35,000 mas and 90,000 mas may occur for the ILS/IPMS and BIH data, respectively. It is interesting to note that of all available series, the YY data not only exhibit the smoothest acceleration but also the least spikes in the radius. These spikes fix the points of departure from a smooth polar path, and they are related to the kinks already noted by Mansinha and Smylie (1970). They also found a correlation between these kinks and the occurrence of large earthquakes, but these results were questioned on both, observational (Haubrich, 1970) and theoretical basis (Dahlen, 1971; 1973). Recently, Preisig (1992), making use of high precision PM data, again related irregularities of the polar path to large earthquakes. Eventually both, acceleration and radius of the polar path may be used for the search of correlations between geodynamic events and PM. Especially in high precision data such as the IRIS data, these parameters may potentially be of importance, but they will not be discussed here any further.

Summarizing the discussion of the available long series of PM, it is concluded that the ILS/IPMS series are best suited for studying the long-term variability of the CW. Comparing the ILS/IPMS data to the BIH data indicates the individual values to carry uncertainties, which sometimes may be quite large. However, the long-term stability of the series seems to be sufficient enough to study the temporal variability of the CW.

Acknowledgements: The PM data was supplied by the ILS in Mizusawa, the former BIH, Richard Gross and the IERS. The author would like to thank these institutions and persons. Also appreciated are the many discussions with V. Rautenberg, J. Zschau, H.-U. Jüttner and P. Ruf, who carefully read the manuscript and made numerous helpful comments.

References

Barnes, R. T. H., Hide, R., White, A. A., and Wilson, C. A. 1983. Atmospheric angular momentum fluctuation, length-of-day changes and polar motion. *Proc. Royal Soc..* **A 387**: 31 - 73.

Bryson, R. A. and Starr, T. B. 1977. Chandler tides in the atmosphere. *J. Atmos. Sci..* **34**: 1975 - 1986.

Carter, W. E. 1981. Frequency modulation of the Chandlerian component of polar motion. *J. geophys. Res..* **86**: 1653 - 1658.

Chandler, S. C. 1891. On the variation of latitude, II. *Astron. J..* **XI**: 65 - 70.

Chandler, S. C. 1892. On the variation of latitude, VII. *Astron. J..* **XII**: 97 - 101.

Chandler, S. C. 1893. On the variation of latitude, VIII. *Astron. J..* **XIII**: 159 - 162.

Chandler, S. C. 1902. *Astron. J..* **22**: 154.

Chao, B. F. 1983. Autoregressive harmonic analysis of the Earth's polar motion using homogeneous International Latitude Service data. *J. geophys. Res..* **88**: 10299 - 10307.

Chao, B. F. and Au, A. Y. 1991. Atmospheric excitation of the earth's annual wobble: 1980-1988. *J. geophys. Res..* **96**: 6577 - 6582.

Christie, A. S. 1900. The latitude variation tide. *Bull. Phil. Soc. Washington.* **13**: 103 - 122.

Colombo, G. and Shapiro, I. I. 1968. Theoretical model for the Chandler wobble. *Nature.* **217**: 156 - 157.

Currie, R. G. 1974. Period and Q_w of the Chandler Wobble. *Geophys. J. R. astr. Soc..* **38**: 179 - 185.

Currie, R. G. and Hameed, S. 1990. Atmospheric signals at high latitudes in a coupled ocean-atmosphere general circulation model. *Geophys. Res. Lett..* **17**: 945 - 948.

Dahlen, F. A. 1971. The excitation of the Chandler wobble by earthquakes. *Geophys. J. R. astr. Soc..* **25**: 157 - 206.

Dahlen, F. A. 1973. A correction to the excitation of the Chandler wobble by earthquakes. *Geophys. J. R. astr. Soc..* **32**: 203 - 217.

Daillet, S. 1981. Secular variation of the pole tide: correlation with Chandler Wobble ellipticity. *Geophys. J. R. astr. Soc..* **65**: 407 - 421.

Demaree, G. R. and Nicolis, C. 1990. Onset of Sahelian drought viewed as a fluctuation-induced transition . *Q. J. R. Meteorol. Soc..* **116**: 221 - 238.

Dickman, S. R. 1979. Continental drift and true polar wandering. *Geophys. J. R. astr. Soc..* **57**: 41 - 50.

Dickman, S. R. 1981. Investigation of Controversial Polar Motion Features Using Homogeneous International Latitude Service Data. *J. geophys. Res..* **86**: 4904 - 4912.

Dickman, S. R. 1988. The self-consistent dynamic pole tide in non-global oceans. *Geophys. J. Int..* **94**: 519 - 543.

Dickman, S. R. and Steinberg, J. R. 1986. New aspects of the equilibrium pole tide. *Geophys. J. R. astr. Soc..* **86**: 515 - 529.

Ellsaesser, H. W., MacCracken, M. C., Walton, J. J., and Grotch, S. L. 1986. Global climatic trends as revealed by the recorded data. *Rev. Geophys..* **24**: 745 - 792.

Furuya, M. and Chao, B. F. 1996. Estimation of period and *q* of the Chandler wobble. *Geophys. J. Int..* **127**: 693 - 702.

Furuya, M., Hamano, Y., and Naito, I. 1996. Quasi-periodic wind signal as possible excitation of Chandler wobble. *J. geophys. Res..* **101**: 25537 - 25546.

Gaposchkin, E. M. 1972. Analysis of pole positions from 1846–1970. In: *Rotation of the Earth*, pp. 19 -32. P. Melchior, S. Yumi (eds). Reidel, Dordrecht.

Goossens, C. and Berger, A. 1986. Annual and seasonal climatic variations over the northern hemisphere and Europe during the last century. *Ann. Geophys..* **4**: 385 - 400.

Goossens, C. and Berger, A. 1987. How to recognize an abrupt climatic change? In: *Abrupt Climatic Change, Evidence and Implications.* pp. 31 - 47. W. H. Berger and L. Labeyrie (eds.). Reidel, Dordrecht.

Graber, M. A. 1976. Polar motion spectra based upon Doppler, IPMS and BIH data. *Geophys. J. R. astr. Soc..* **46**: 75 - 85.

Gross, R. S. 1985. Signal detection techniques applied to the Chandler Wobble. *J. geophys. Res..* **90**: 10281 - 10290.

Gross, R. S. 1986. The influence of earthquakes on the Chandler wobble during 1977-1983. *Geophys. J. R. astr. Soc..* **85**: 161 - 177.

Gross, R. S. 1990. The secular drift of the rotation pole. In: *Earth Rotation and Coordinate Reference Frames.* pp. 146 - 153. Boucher, C., Wilkins, C.A. (eds.). Springer, New York.

Guinot, B. 1972. The Chandlerian nutation from 1900 to 1970. *Astron. Astrophys..* **19**: 207 - 214.

Guinot, B. 1978. Rotation of the earth and polar motion services. In: *Proc. of the 9th GEOP Conference.*

Guinot, B. 1982. The Chandlerian Nutation from 1900 to 1980. *Geophys. J. R. astr. Soc..* **71**: 295 - 301.

Hameed, S. and Currie, R. G. 1989. Simulation of the 14-Month Chandler Wobble in a Global Climate Model. *Geophys. Res. Lett..* **16**: 247 - 250.

Haubrich, R. A. and Munk, W. 1959. The pole tide. *J. geophys. Res..* **64**: 2373.

Haubrich, R. A. 1970. An examination of the data relating pole motion to earthquakes. In: *Earthquake Displacement Fields and the Rotation of the Earth* . Mansinha, L., Smylie, D.E., Beck, A.E. (eds.). Reidel, Dordrecht.

Hide, R. 1984. Rotation of the atmospheres of the Earth and planets. *Phil. Trans. R. Soc. London A.* **313**: 107 - 121.

Jones, P. D., Raper, S. C. B., Bradley, R. S., Diaz, H. F., Kelly, P. M., and Wigley, T. M. L. 1986. Northern hemispheric surface air temperature variations: 1851-1984. *J. Clim. and App. Met..* **25**: 161 - 179.

Kanamori, H. 1976. Are earthquakes a major cause of the Chandler wobble? *Nature.* **262**: 254 - 255.

Kendall, M. and Stuard, A. 1979. The advanced theory of statistics. Vol. 2 Inference and relationship. Griffin and Co..

Kikuchi, I. and Naito, I. 1982. Sea surface temperature (SST) analyses near the Chandler period. In: *Proc. Int. Latitude Observ. Mizusawa.* **21:** 64 - 70.

Kolaczek, B. 1989. Observational determination of the Earth's rotation. In: Gravity and Low Frequency Geodynamics. pp. 295 - 361. R. Teisseyre (ed.). Elsevier Warszawa.

Kolaczek, B. and Hua, Y. S. 1991. Astronomical Series of Earth rotation parameters. **177:** 121 - 138.

Kuehne, J., Wilson, C. R., and Johnson, S. 1996. Estimates of the Chandler wobble frequency and Q. *J. geophys. Res..* **101:** 13573 - 13579.

Lambeck, K. 1980. The Earth's Variable Rotation: Geophysical Causes and Consequences. Cambridge University Press.

Lambeck, K. 1988. Geophysical Geodesy - The Slow Deformations of the Earth. Oxford Science Publications.

Lenhardt, H. and Groten, E. 1985. Chandler wobble parameters from BIH and ILS data. *Manuscripta Geodaetica.* **10:** 296 - 305.

Maddox, J. 1988. Earthquakes and the Earth's rotation. *Nature.* **332:** 11.

Maksimov, I. V. 1954. On long period tidal phenomena in the sea and in the atmosphere of the earth (in Russian). *Trans. Inst. Okeanol..* **8:** 18 - 40.

Maksimov, I. V., Kraklin, V. P., Sarukhanyan, E. I., and Smirnov, N. P. 1967. Nutational migration of the Iceland Low. *Dokl. Akad. Nauk SSSR.* **177:** 3 - 6.

Mansinha, L. and Smylie, D. E. 1970. Seismic excitation of the Chandler wobble. In: Earthquake Displacement Fields and the Rotation of the Earth. Mansinha, L., Smylie, D.E., Beck, A.E. (eds.). Reidel, Dordrecht.

Mansinha, L., Smylie, D. E., and Chapman, C. H. 1979. Seismic excitation of the Chandler wobble revisited. *Geophys. J. R. astr. Soc..* **59:** 1 - 17.

Merriam, J. B. 1982. Meteorological excitation of the annual polar motion. *Geophys. J. R. astr. Soc..* **70:** 41 - 56.

Mulholland, J. R. and Carter, W. E. 1982. Seth Carlo Chandler and the observational origins of geodynamics. In: High-precision Earth rotation and Earth-Moon dynamics. *Proc. 63rd Colloq. Int. Astr. Union, Grasse, France.* pp. XV–XIX. O. Calame (ed.). Reidel, Dordrecht.

Munk, W. H. and MacDonald, G. J. F. 1960. The Rotation of the Earth. Cambridge University Press, Cambridge.

Naito, I. 1977. Secular variation of the pole tide. *J. Phys. Earth.* **125:** 221 - 231.

Newcomb, S. 1891. *Astron. J..* **11:** 81 - 83.

Okubo, S. 1982. Is the Chandler period variable? *Geophys. J. R. astr. Soc..* **71:** 629 - 646.

Ooe, M. 1978. An optimal complex ARMA model of the Chandler wobble. *Geophys. J. R. astr. Soc..* **53:** 445 - 457.

Pejovič, N. and Vondràk, J. 1991. Polar motion: Observations and atmospheric excitation. Techn. Rep. IUGG Special Study Group 5-98, Bull. 5.

Plag, H. - P. 1988. A regional study of Norwegian coastal long-period sea-level variations and their causes with special emphasis on the Pole Tide. *Berl. Geowiss. Abhandl. Reihe A.* **14:** 1 - 175.

Plag, H. - P. 1993. The "sea level rise" problem: An assessment of methods and data. In: *Proc. Int. Coastal Congr., Kiel 1992.* pp. 714 - 732. P. Lang Verlag, Frankfurt.

Plag, H. - P. 1995. Coastal relative sea level: A valuable indicator of climate variability? In: *Abstr., XXI Gen. Assembly Int. Union. Geodesy Geophys.:* B 317.

Preisig, J. R. 1992. Polar motion, atmospheric angular momentum excitation and earthquakes - correlations and significance. *Geophys. J. Int..* **108**: 161 - 178.

Runcorn, S. K., Wilkins, G. A., Groten, E., Lenhardt, H., Campbell, J., Hide, R., Chao, B. F., Souriau, A., Hinderer, J., Legros, H., LeMouel, J. - L., and Feissel, M. 1988. The excitation of the Chandler Wobble. *Surveys Geophys..* **9**: 419 - 449.

Schlesinger, M. E. and Ramankutty, N. 1994. An oscillation in the global climate system of 65-70 years. *Nature.* **367**: 723 - 726.

Schweydar, W. V. 1916. Die Bewegung der Drehachse der elastischen Erde im Erdkörper und im Raum. *Astron. Nachr..* **203**: 103 - 114.

Smith, M. L. 1977. Wobble and nutation of the earth. *Geophys. J. R. astr. Soc..* **50**: 103 - 140.

Smylie, D. E. and Mansinha, L. 1971. The elasticity theory of dislocations in real earth models and changes in the rotation of the Earth. *Geophys. J. R. astr. Soc..* **23**:, 329 - 354.

Sneyers, R. 1975. Sur l'analyse statistique des séries d'observations. Note Techn. 143, OMM-No. 415, Geneva.

Souriau, A. 1986. Random walk of the Earth's pole related to the Chandler wobble excitation. *Geophys. J. R. astr. Soc..* **86**: 455 - 465.

Starr, T. 1983. On the dynamic atmospheric response to the Chandler wobble forcing. *J. Atmos. Sci..* **40**: 929 - 940.

Tsimplis, M. N., Flather, R. A., and Vassie, J. M. 1994. The North Sea Pole Tide described through a tide-surge numerical model. *Geophys. Res. Lett..* **21**: 449 - 452.

vanDam, T. M., Blewitt, G., and Heflin, M. B. 1994. Atmospheric pressure loading effects on Global Positioning System coordinate determinations. *J. geophys. Res..* **99**: 23939 - 23950.

Vanicek, P. 1970. An analytical technique to minimize noise in a search for lines in the low frequency spectrum. *Observ. Royal Belg., Comm. A.* **96**: 170 - 173.

Volland, H. 1996. Atmosphere and Earth's rotation. *Surveys Geophys..* **17**: 101 - 144.

Vondràk, J. 1985. Long-period behaviour of polar motion between 1900.0 and 1984.0. *Ann. Geophys..* **3**: 351 - 356.

Wilson, C. R. and Haubrich, R. A. 1976a. Atmospheric contribution to the excitation of the Earth's wobble 1901-1970. *Geophys. J. R. astr. Soc..* **46**: 745 - 760.

Wilson, C. R. and Haubrich, R. A. 1976b. Meteorological excitation of the Earth's wobble. *Geophys. J. R. astr. Soc..* **46**: 707 - 743.

Wilson, C. R. and Vicente, R. O. 1980. An analysis of the homogeneous ILS polar motion series. *Geophys. J. R. astr. Soc..* **62**: 605 - 616.

Wilson, C. R. and Vicente, R. O. 1990. Maximum likelihood estimates of polar motion parameters. In: Variations in Earth Rotation. *Geophys. Monographs.* **59**: 151 - 155. D. D. McCarthy and W. E. Carter (eds.). Am. geophys. Union, Washington D. C.

Xie, L. and Dickman, S. R. 1996. Tide gauge analysis of the pole tide in the North Sea. *Geophys. J. Int..* **126**: 863 - 870.

Yumi, S. and Yokoyama, K. 1980. Results of the International Latitude Service in a homogenous system 1899.9 – 1979.0. Techn. Rep., Central Bureau Int. Polar Motion Service, Misuzawa.

Atmospheric Tides

and Related Phenomena

Atmospheric Tides

Hans Volland

Radioastronomical Institute, University of Bonn
Auf dem Hügel 71, 53121 Bonn, Germany

Abstract. Atmospheric tides are diurnal and annual variations as well as higher harmonics of atmospheric parameters like pressure, temperature, or winds. They are generated mainly thermally by the regular solar heat input into the system atmosphere–earth's surface (solar tides). The daily variations of the solar tides have a basic period of one solar day. The semidiurnal component reaches maximum amplitudes of its ground pressure of the order of about 1 hPa, a value just above the meteorological noise. The pressure amplitudes of the zonally averaged seasonal waves with the basic period of one tropical year, however, have amplitudes of the order of 20 hPa and are therefore prominent global–scale atmospheric wave structures. The gravitationally generated lunar semidiurnal atmospheric tides have maximum pressure amplitudes on the ground that are about a factor 20 smaller than those of the solar tides. In order to detect such small a signal, it must be filtered out of the meteorological noise by a statistical analysis spanning several decades. Observations, excitation mechanisms, and the theory of atmospheric tides are reviewed in this chapter.

1 Introduction

Atmospheric tides are global–scale waves excited by regular external forces such as the gravitational tidal force of the moon (gravitational tides) or differential heating of the sun (thermal tides). In contrast to the interactions with solid earth and ocean, gravitational tidal forces of moon and sun have only minor effects on lower atmosphere dynamics. The dominant regular daily and seasonal variations of atmospheric parameters like pressure, temperature, or wind are of thermal origin, depending on differential solar radiation due to the earth's rotation and the geometry of the sun–earth system.

The atmosphere behaves like a huge waveguide closed at the bottom (the earth's surface) and open to space at the top. In such a waveguide, an infinite number of atmospheric wave modes can be excited. Because the waveguide is imperfect, however, only wave modes of lowest degree with large horizontal and vertical scales can develop sufficiently well that they can be filtered out from the meteorological noise. All other wave modes interfere destructively and contribute, in fact, to that noise. The observed atmospheric tides belong to the ultralong or planetary waves with zonal wavenumbers $m < 4$. We define atmospheric tidal waves as those planetary waves which are excited by regular external forces (thermal or gravitational) of sun and moon. These waves produce daily variations of atmospheric parameters with frequencies $\omega = -m\Omega_i$, where Ω_i is $\Omega_S = 7.2722 \times 10^{-5} \text{ s}^{-1}$, the angular frequency of one mean solar day, or $\Omega_L = 7.0259 \times 10^{-5} \text{ s}^{-1}$ the angular frequency of one mean lunar day. The waves are classified by their zonal wavenumbers $m = 1, 2, \cdots$ with $m = 1$: diurnal wave; $m = 2$: semidiurnal wave, etc. The minus sign indicates that these waves are migrating waves propagating to the west with sun and moon, respectively. Seasonal waves with angular frequencies $j\Omega_A$ ($\Omega_A = 1.991 \times 10^{-7} \text{ s}^{-1}$ is the angular frequency of one tropical year; $j = 1$: annual wave; $j = 2$: semiannual wave, etc.), as well as the climatic mean ($j = 0$) are also atmospheric tides according to this definition.

The pressure amplitude of the solar annual wave on the ground, which is of the order of $20\,\text{hPa}$, is a factor of about 20 larger than that of the solar semidiurnal wave (which is the strongest daily wave) although the magnitudes of their heat inputs are nearly the same. This results from a low pass filtering effect of the atmospheric waveguide which suppresses the higher frequencies. The pressure amplitudes of the daily lunar tides near the ground are about 20 smaller than those of the daily thermal solar tides and are thus far below the meteorological noise. At present, only the modes M_2 and O_1 could be detected. All other lunar tides are negligible.

Local and regional oscillations with basic periods of one year and one solar day, respectively, can be generated by internal redistribution of solar insolation within the atmosphere (transport of latent heat), or from the surface (differential surface temperatures). The land–sea wind with surface winds from the water to the coast during sunlit hours and vice versa during the night is an example of a local tidal–like diurnal wave. It is caused by the larger heat capacity of water compared to that of the solid surface. The monsoon wind blowing from the Indian ocean to the Indian subcontinent during summer and vice versa during winter is an example of a regional annual wave. In principle, those waves also belong to the atmospheric tides. However, contrary to the regular migrating tides with well–known horizontal structures, these indirectly driven local and regional tides have, in general, very complicated and almost unpredictable horizontal and vertical structures. We shall therefore deal in the following only with the global–scale atmospheric tides with zonal wavenumbers $m < 4$, directly or indirectly driven by the external forces of sun and moon with frequencies $\Omega_S, \Omega_L, \Omega_A$, and harmonics.

2   Observations

2.1   Solar Daily Tidal Waves

The most important solar daily tidal waves which travel westward with the sun are the diurnal and the semidiurnal components. The annual mean of their pressure amplitudes on the ground can be represented by empirical functions (Siebert, 1961):

$$S_1(p) = 59 \ \cos^3 \phi \ \cos(\tau_S - 5.2 \text{ h}) \qquad \text{Pa}$$
$$S_2(p) = 116 \ \cos^3 \phi \ \cos[2(\tau_S - 9.7 \text{ h})] \ \text{Pa}$$

(1)

with $\tau_S = \Omega_S t + \lambda$ the solar local time , (ϕ, λ) geographic latitude and longitude, and t the Universal Time. Both waves are symmetric with respect to the equator. It seems surprising that the amplitude of the semidiurnal component is larger than that of the diurnal component since one expects a dominant diurnal component of the solar heat input. We shall see in section 5 that the diurnal wave behaves anomally. An antisymmetric terdiurnal wave, which disappears during equinox conditions and reverses its phase from winter to summer, has also been observed (Bartels and Kertz, 1952):

$$S_3(p) = 61 \, \sin \phi \, \cos^3 \phi \, \cos[3(\tau_S - 2.6 \text{ h})] \, \cos \Omega_A t \quad \text{Pa.} \tag{2}$$

All higher harmonics have much smaller amplitudes.

The temperature amplitudes of these waves have magnitudes on the ground of the order of 1 K, and their wind velocities are of the order of 0.5 m/s. These amplitudes increase with altitude. On the ground, the pressure amplitudes of the solar daily tidal waves are small as compared to the irregularly varying interdiurnal pressure fluctuation (meteorological noise), particularly at medium and high latitudes where the noise may be of the order of 10 hPa and larger. However, at upper atmospheric heights above about 85 km, the magnitudes of the solar daily tides increase considerably and become prominent features of the atmosphere dynamics. Temperature wave amplitudes of 150 K and more can be reached above about 200 km altitude.

In addition to the migrating waves which depend on solar local time, there exist global–scale solar diurnal and semidiurnal standing waves that depend on Universal Time. Their pressure amplitudes on the ground can be represented by the following empirical formulas (Chapman and Lindzen, 1970):

$$\begin{aligned} S_{1,0}(p) &= 6.1 \, P_2(\phi) \, \cos(\Omega_S t + 92°) \quad \text{Pa} \\ S_{2,0}(p) &= 8.5 \, P_2(\phi) \, \cos(2\Omega_S t + 28°) \, \text{Pa} \end{aligned} \tag{3}$$

with $P_2(\phi) = 0.5(\sin^2 \phi - 1)$ the Legendre polynominal of second degree. Evidently, their amplitudes are an order of magnitude smaller than those of the migrating waves in (1). They become significant at higher latitudes, however, as can be seen in Fig. 1, where amplitude and phase of the semidiurnal pressure wave on the ground are plotted versus latitude and longitude. The lines of constant phase run almost parallel to the meridians at higher latitudes, indicating standing waves.

The increase in amplitude with height of the solar semidiurnal tidal wave is shown in Fig. 2 where the northwind (wind blowing from the north) is plotted versus altitude between 30 and 60 km height in 37° N latitude in amplitude (left panel) and phase (right panel). The solid lines represent model calculations.

In general, the pressure amplitudes of the individual solar daily tidal wave components of wavenumber m on the ground can be represented mathematically by

$$S_m(p) = \sum C_{s,m}{}'^m \, P_s{}^m(\phi) \, \cos(m\lambda - m'\Omega_S t + \sigma_{s,m}{}'^m) \tag{4}$$

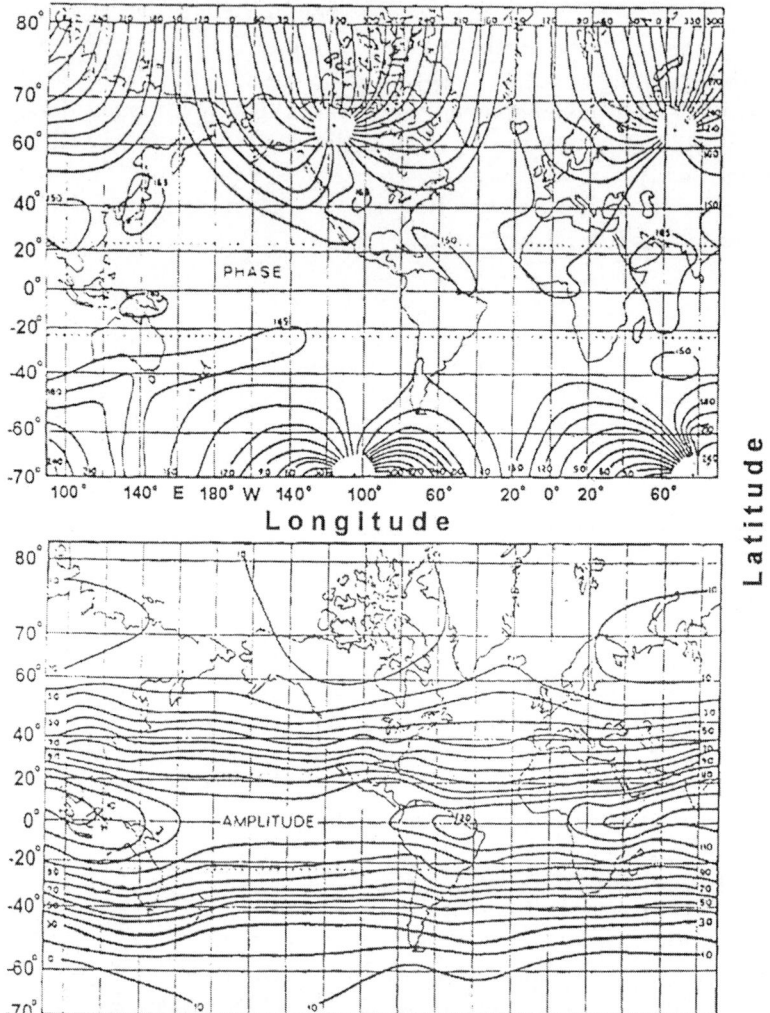

Fig. 1. Pressure amplitude on the ground of the solar semidiurnal tidal wave (sum of S_2 from (1) and $S_{2,0}$ from (3)) as function of latitude and longitude. Lower panel: isolines of amplitude (in units of Pa). Upper panel: phases (in units of degrees) (from Haurwitz, 1956).

with $s \geq m \geq 0$, $C_{s,m}{}^m$ and $\sigma_{s,m}{}^m$ amplitudes und phases, and $P_s^m(\phi)$ latitude dependent spherical functions (in Neumann's representation). Waves with $m' = m$ are travelling toward the east, those with $m' = -m$ are travelling to the west (see (1)). Apart from the westward migrating waves, all other waves owe their existence to orographic and meteorological differences near the ground (e.g., ocean–continent distribution, cloudiness, topography, ice coverage, vegetation, etc.). The standing waves of (3) ($m = 0$) are mainly generated by the ocean–

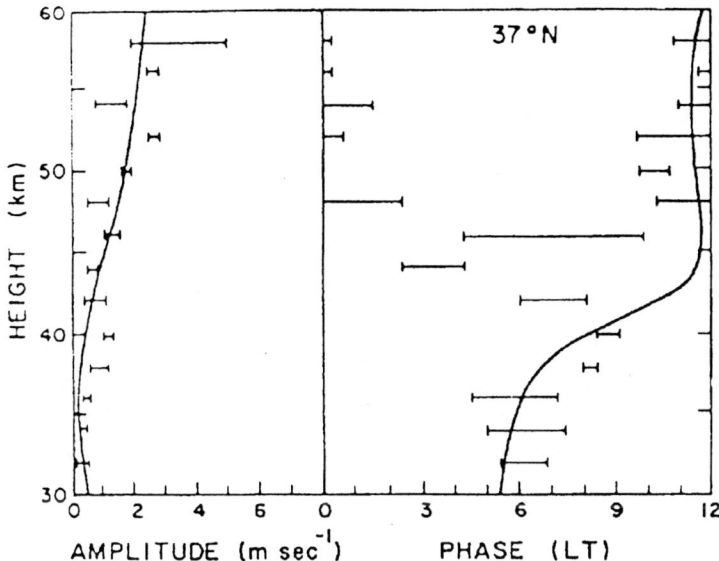

Fig. 2. Observed northwind of the solar semidiurnal tidal wave S_2 plotted as a function of height at 37° latitude in amplitude (left) and phase (right). The horizontal bars are data from Reed (1972). The solid lines are model calculations (from Forbes, 1982).

continent configuration. The complex regional and local distributions of these indirect heat sources can then lead to very complicated wave structures that interfere with each other.

2.2 Lunar Daily Tidal Waves

Lunar gravitationally excited atmospheric daily tidal waves are smaller in amplitude than the solar daily tides by a factor of about 20. The dominant lunar tidal wave is a semidiurnal symmetric wave generated mainly by the component M_2 of the lunar tidal force. Its pressure amplitude on the ground can be represented by (Chapman and Lindzen, 1970)

$$L_2(p) = 5.5 \ \cos^3 \phi \ \cos[2\tau_L + 75°] \ \text{Pa} \tag{5}$$

where τ_L is the lunar local time. The signal–to–noise ratio of this wave is so low that it can be filtered out from the meteorological noise only by long–term statistics of many years. Even this is only possible because the lunar gravitational tidal force is one of the most regularly varying forces in the solar system.

Furthermore, an antisymmetric lunar diurnal wave excited by the O_1–mode of the lunar tidal force with maximum amplitudes of the order of 1–2 Pa has been detected. In the middle atmosphere (ca. 10 to 85 km height), the amplitudes of the lunar tidal waves increase with altitude by a factor of about 10 to 100.

In addition to the direct generation of the atmospheric lunar tides by the lunar tidal force, the oceans have an indirect influence on these waves. The oceanic tides, which reach amplitudes of 1 m and greater, can create atmospheric ground pressure amplitudes comparable to those directly excited by the lunar tidal force (Hollingworth, 1971).

2.3   Zonally Averaged Seasonal Tidal Waves ($m = 0$)

The seasonally dependent solar direct radiation into the atmosphere, as well as infrared reradiation from the earth, generate wind, pressure, and temperature variations as functions of season, latitude, and longitude. The zonally averaged zonal wind ($m = 0$) (the zonal background flow) is plotted in Fig. 3a (upper panel) as function of latitude and height for solstice conditions. Positive numbers indicate isolines of equal westwind velocities (blowing from the west; in units of m/s). Negative numbers indicate eastwinds (blowing from the east). One notices westwinds at medium and higher latitudes within the troposphere, culminating in two bands of westwind jetstreams near about 12 km altitude and ±40° latitude. An eastwind belt is established at lower latitudes with wind velocities on the ground of about 5 m/s near the equator, decreasing with height. Over the ocean, one observes this eastwind belt together with meridional winds blowing toward the equator as the NE– and SE–trade winds. The westwind jetstream in the winter hemisphere is about twice as strong as that in the summer hemisphere.

In the middle atmosphere, there exists an eastwind belt in the summer hemisphere and a westwind belt in the winter hemisphere with wind velocities of the jets of −60 and 80 m/s, respectively, at about 65 km height and ±50° latitude. The wind velocity decreases above that height, but increases again above 120 km altitude.

The associated temperature configuration is shown in Fig. 3b (lower panel). Within the troposphere, the temperature decreases toward the poles. At the tropopause, however, which is the region of the jetstreams, this temperature gradient reverses to produce higher temperatures at the poles than at the equator. Within the middle atmosphere, the equator is warmer than the poles below the height region of the jets, and colder above. This behavior is associated with the well–known thermal wind equation which relates the meridional temperature gradient to the vertical gradient of the zonal wind (e.g., Holton, 1983).

During equinox conditions, the zonal wind configuration is rather symmetric with respect to the equator with westerly jets in both hemispheres, having maximum wind velocities of about 25–30 m/s in the tropopause as well as in the middle atmosphere. The climatic mean zonal background wind has a similar configuration with somewhat smaller wind velocities.

One can separate schematically the zonal wind structure into a climatic mean ($j = 0$), an annual component ($j = 1$), and a semiannual component ($j = 2$):

$$u(\phi, z, t) = u_{0,s}(\phi, z) + [u_{1,s}(\phi, z) + u_{1,a}(\phi, z)]\ \cos \Omega_A t$$
$$+ u_{2,s}(\phi, z)\ \cos 2\Omega_A t \cdots . \tag{6}$$

Here, the indices 's' und 'a' indicate symmetric and antisymmetric wind structures with respect to the equator. The symmetric component dominates within the lower atmosphere (troposphere), i.e. $|u_{0,s} + u_{2,s}| > |u_{1,s} + u_{1,a}|$. Within the middle atmosphere, the reverse is true. The summation shown in (6) without the term $u_{1,s}$ corresponds to direct excitation by solar insolation. However, a symmetric term $u_{1,s}$ is observed within the lower atmosphere. This term very probably comes from the different land–ocean configuration on both hemispheres. Redistribution of the solar heat input by the transport of latent heat within clouds causes the dominance of the symmetric components within the lower atmosphere.

Meridional wind systems complement the zonal wind. Within the troposphere, there are three circulation cells in each hemisphere: first, a Hadley cell at lower latitudes with updraft winds near the equator, cloud development there (intertropical convergence zone), downdraft winds near about $\pm 30°$ latitude with dissolution of the clouds which means low precipitation (desert belts of the earth), and finally closure of the wind circulation on the ground with meridional winds blowing toward the equator (the meridional component of the trade winds); second, a meridional cell at medium latitudes with reversed circulation, the Ferrel cell; and third, a small cell with the same sense of circulation as the Hadley cell at high latitudes. Contrary to the daily tides, the seasonal tides are prominent and rather regular phenomena which significantly determine the seasonal weather.

2.4   Seasonal Tidal Waves with Wavenumbers m > 0

While the seasonal tides of zonal wavenumber $m = 0$ are generated mainly by direct solar insolation within the atmosphere and by infrared reradiation from the ground, the longitude dependent seasonal waves of wavenumbers $m > 0$ are indirectly excited mainly from the ground due to orography or to globally varying surface temperatures. The different heat capacity of oceans and continents causes drastic differences in the surface temperatures at the same latitude. For instance, the sea surface temperature of the North Atlantic ocean near $55°$ latitude is about 40 K warmer than in Siberia in January. In July, this is reversed with higher temperatures in Siberia than on the North Atlantic. The earth's surface thus behaves like a standing heat source to generate large–scale planetary waves with wavenumbers $m > 0$ and frequencies $j\Omega_A$ where $j = 0, 1, 2, \cdots$. Such waves propagate upward.

Figure 4 shows a height–latitude profile of planetary waves with zonal wavenumber $m = 1$ in January in amplitude (left) and phase (right). The amplitude is given in geopotential meters (gpm), a measure of the pressure amplitude. The phase (in degrees) indicates the longitude of maximum amplitude (the pressure high). One notices an increase of gpm with altitude in both hemispheres up to a height of about 10 km. This amplitude continues to increase within the winter hemisphere up to 60 km. In the summer hemisphere, however, it decreases above 10 km height and disappears above about 25 km. This behavior reverses during southern winter.

Within the lower northern hemispheric atmosphere, the amplitude is weaker by about 30 % in July than in January, while within the southern hemisphere

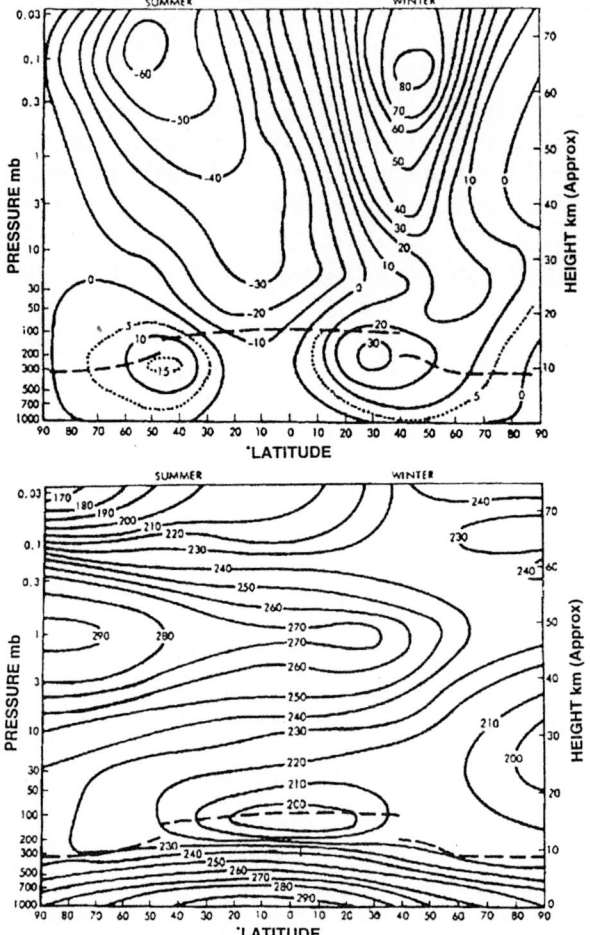

Fig. 3. Upper panel: Isolines of typical zonally averaged zonal winds as a function of latitude and height during solstice conditions. Wind velocities in m/s. Positive numbers: westwinds (winds blowing from the west); negative numbers: eastwinds (winds blowing from the east). The dashed line is the position of the tropopause. Lower panel: same as upper panel except for isolines of temperature (in units of K) (from Murgatroyd, 1969).

not much changes from winter to summer. The behavior of the semiannual wave is similar to the annual wave with the difference that its amplitude is smaller by a factor of about two. The suppression of the planetary waves in the middle atmospheric summer hemisphere with its almost exponential decrease of their amplitudes is related to the seasonally varying propagation conditions of planetary waves at those heights depending on the opposite directions of the two jetstreams in Fig. 3a (Charney and Drazin, 1961). However, the seasonal depen-

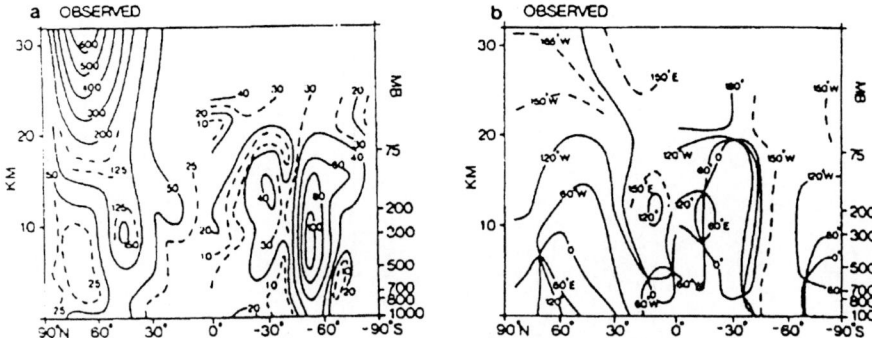

Fig. 4. Latitude–height profile of planetary waves of wavenumber one in January. Left panel: Isolines of amplitude in units of geopotential meters (gpm). Right panel: phases in units of degrees (longitudinal position of the pressure high) (from McAvaney *et al.*, 1978).

dence of the waves within the lower atmosphere is determined by the different orography in both hemispheres which is more pronounced in the northern than in the southern hemisphere.

The pressure amplitude of the planetary waves of wavenumber $m = 1$ within the troposphere behaves like a quasi–standing wave. It can be schematically separated into a symmetric climatic mean $(j = 0)$ and the sum of a symmetric (index 's') and an antisymmetric (index 'a') annual wave $(j = 1)$:

$$p(\phi, z, t) = u_{0,s}(\phi, z) \cos(\lambda + \sigma_{0,s})$$
$$+ [u_{1,s}(\phi, z) + u_{1,a}(\phi, z)] \cos(\lambda + \sigma_{1,s}) \cos \Omega_A t + \cdots \qquad (7)$$

where the two annual waves with amplitudes $u_{1,s}$ und $u_{1,a}$ interfere constructively in the northern hemisphere and destructively in the southern hemisphere. In the northern hemisphere, the superposition of both waves with the climatic mean $(u_{0,s})$ is constructive during winter and destructive during summer. In the southern hemisphere, the climatic mean dominates throughout the year.

3   The Sun as a Thermal Tidal Source

3.1   Heat Balance of the System Earth–Atmosphere

Part of the visible solar light is directly absorbed in the atmosphere, mainly within clouds (about 24 %; see Fig 5). A larger amount (46 %) penetrates down to the surface and heats the ground. Clouds as well as the earth's surface reflect about 30 % of this light back into space (albedo).

The earth radiates almost like a black body with a mean temperature of about 288 K. The amount of this longwave infrared reradiation is about 112 % of

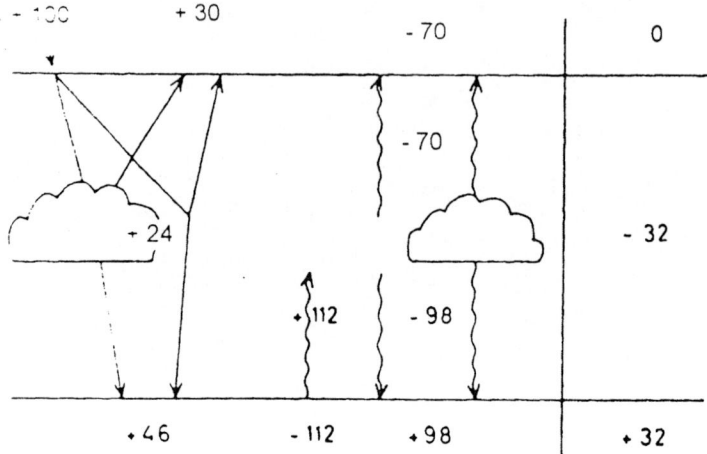

Fig. 5. Radiation balance of the system earth–atmosphere (yearly average in the northern hemisphere). Left: incoming radiation of the visible solar light. Right: infrared reradiation from the earth's surface (from Möller, 1973).

the total visible solar light. However, part of this is absorbed by water vapor in the clouds, or it is reflected at the cloud bottoms and reradiated to the ground (98 %). The region between the lower boundaries of the clouds and the ground is therefore about 30 K warmer than without an atmosphere. This is the well-known greenhouse effect. The total loss of infrared radiation from the surface is therefore only 14 %, while the atmosphere loses 70 % to space. The balance of the surface is positive (+32 %) compared to the same negative amount of the atmosphere. In order to compensate this balance to zero, the surface loses heat by evaporation, particularly over the oceans, where evaporation energy is transfered into latent heat and transported by the clouds. Condensation of water vapor within the clouds adds then condensation energy to the atmospheric gas. Moreover, turbulent convection transports sensible heat from the surface into the atmosphere. Evidently, the ocean circulation plays a significant role for the geographic distribution of latent and sensible heat (e.g., Kiehl, 1994).

An exact determination of the geographical and temporal configuration of the effective indirect solar heat input into the atmosphere to excite tidal waves is extremely difficult because the meteorological conditions like cloud formation, dissolution, and transport change continuously. For global wave structures like the atmospheric tides, however, these local and regional irregularities of the effective heat source are of minor importance because the ultralong waves 'see' only those heat structures which have horizontal scales of at least one quarter of their wavelengths. Clearly, signals at the frequencies Ω_S und Ω_A compete with noise, and line broadening takes place which makes it practically impossible to decompose sidebands like $(\Omega_S \pm \Omega_A)$.

At heights between 20 and 60 km, almost the total band of ultraviolet solar light between 200 and 310 nm is absorbed to maintain the ozone layer. The heat energy transfered to the atmosphere at those heights is thus a second heat source to generate tidal waves. The solar radiation at wavelengths smaller than 200 nm (extreme ultraviolet and X–rays) is totally absorbed within the upper atmosphere above 85 km and acts there like a third source of atmospheric tidal waves. This spectral band of solar radiation comes from higher regions of the sun (chromosphere and corona) where the temperature is much larger than in the photosphere, the source region of the visible light. As a result, the thermal efficiency of solar radiation to excite atmospheric waves increases significantly with altitude. In the thermosphere above 85 km, its value ranges from 1 to 10 W/kg. In the troposphere, this number is down to about 0.01 to 0.1 W/kg. Correspondingly, the generation of tidal waves is much more intensive within the thermosphere than within the troposphere. In contrast to the lower atmosphere, tidal waves of upper atmospheric heights thus represent prominent wave phenomena. On the other hand, the solar ultraviolet radiation is highly variable and thus causes rather irregular fluctuations of these tidal waves (e.g., Volland, 1988).

3.2   Solar Irradiance

The 'Solar Constant', which is the total solar irradiance at the mean distance (see also Wilhelm "Solar Irradiance", this volume) between sun and earth outside the atmosphere, is about $S \simeq 1370 \, W/m^2$. This number does not vary more than about 0.1 % during a time interval of several tens of years (Fröhlich, 1994). A total amount of heat energy averaged over the whole earth, taking account of the albedo of 30 % of

$$\bar{S} = \frac{0.7S}{4} \simeq 240 \ \ W/m^2 \tag{8}$$

is then available within a vertical column of air of unit cross–section. For a 'homogeneous' atmosphere with a constant density of $\bar{\rho} = 1.28 \, kg/m^3$ (the mean density on the ground) and a scale height of $H = 8 \, km$, this amounts to an energy input per mass of

$$\bar{Q} = \frac{S'}{\bar{\rho}H} \simeq 23 \ \ mW/kg \tag{9}$$

3.3   Estimates of 'Equilibrium' Tides

It is possible to estimate the amplitudes of tidal waves which may be excited by heat sources such as given in section 3.2 by using the first law of thermodynamics (energy balance equation) and neglecting dynamics. The increase in internal energy of the atmospheric gas must then equal the external heat energy input:

$$c_v \frac{dT}{dt} \simeq Q \tag{10}$$

with $c_v = 717 \, J/kg/K$ the specific heat at constant volume of the atmospheric gas, T its temperature (in K), t the time (in sec) and Q the external heat source (in W/kg). Applying the gas equation:

$$p = R\rho T \tag{11}$$

($R = 287\,\text{J/kg/K}$ is the gas constant), one arrives at

$$|\Delta p| \simeq \frac{(\gamma - 1)\bar{\rho}\,Q}{\omega} \tag{12}$$

with Δp the pressure amplitude, $\gamma = c_p/c_v = 1.4$ the ratio of the specific heats, and $\bar{\rho} = 1.28\ \text{kg/m}^3$ the mean air density on the ground. The angular frequency of the atmospheric wave, ω, is related to its wave period by $\Delta t = 2\pi/\omega$. For an estimate of the semidiurnal wave, we use the number Q from (9), furthermore $\Delta t = 0.5\,\text{days}$ ($\omega = 2\Omega_S$) and find from (12) a pressure amplitude of $\Delta p \simeq 80\,\text{Pa}$, which is the right order of magnitude for the solar semidiurnal wave in (1). For the solar diurnal wave, it is $\Delta t = 1\,\text{day}$, and the heat source of this wave is expected to be twice as large as that of the semidiurnal wave. Thus, its amplitude should be larger by a factor of about four. From the observations, however, follows an amplitude smaller by a factor of about two as compared to the semidiurnal wave (see (1)) which means a total reduction by a factor of about eight. That apparent discrepancy was a mystery for many decades and was resolved only thirty years ago (see section 5.1).

It follows from (12) that the atmosphere seems to behave like a low–pass filter, suppressing smaller periods. At periods Δt larger than about 20 days, however, turbulent feedback between large–scale and smaller–scale atmospheric waves becomes increasingly important. We can simulate this mutual coupling by an attenuation factor

$$\omega \rightarrow \omega + i\omega_R \tag{13}$$

with $\omega_R \simeq \Omega_S/10$ a so–called Rayleigh friction factor. In the case of the climatic mean, ($\omega = 0$), the heat difference between poles and equator averaged over one year generates a zonally averaged pressure difference between equator and the poles which can be estimated from (12) if one replaces ω by ω_R:

$$|\Delta p| \simeq \frac{(\gamma - 1)\bar{\rho}\,Q}{\omega_R} \tag{14}$$

With Q from (9), one finds a pressure difference of 16 hPa between equator and poles which is, indeed, the right order of the observed magnitude.

In a similar manner, one can estimate the influence of the lunar tidal waves generated by the lunar gravitational tidal force. We use here the horizontal equation of motion, again without dynamics, and arrive at

$$\nabla(\Delta p/\bar{\rho} + \Phi_L) \simeq 0 \tag{15}$$

with ∇ the horizontal gradient operator, and Φ_L the lunar tidal potential. Now, it is $|\Phi_L| \simeq D = 0.75\,Gm_L a^2/r_L^3 = 2.63\,\text{m}^2/\text{s}^2$ with D the Doodson constant, a the earth's radius, m_L and r_L mass and mean distance of the moon, respectively, and G the gravitational constant. From (15) one may estimate the lunar semidiurnal tidal pressure amplitude on the ground:

$$|\Delta p| \simeq \bar{\rho} D \simeq 3 \,\mathrm{Pa} = 0.03 \,\mathrm{hPa} = 30 \,\mu\mathrm{bar} \tag{16}$$

which is again of the right order of magnitude as in (see (5)). Evidently, the reason for such a small amplitude of the lunar atmospheric tides is the air density, which is much smaller than that of water (ratio 1 : 800).

4 Theory

4.1 Basic Equations

The amplitudes of the atmospheric tidal waves within the troposphere are sufficiently small to allow a linearization of the hydrodynamic and thermodynamic equations. However, nonlinear coupling between the seasonal waves with wavenumbers $j = 0, 1$, and 2 within the middle atmosphere during winter can have dramatical consequences (sudden stratospheric warmings; see Labitzke (1981)). Moreover, nonlinear coupling between the solar seasonal and daily tides can take place within the height region between about 60 and 100 km (Forbes, 1984).

The vertical wavelengths of nearly all relevant tidal waves are large compared to the structure of the vertical temperature profile of the background atmosphere. Therefore, it is possible in many theoretical approaches to simulate the real background atmosphere by an isothermal atmosphere with a mean temperature $\bar{T} \simeq 273 \,\mathrm{K}$, a scale height $H \simeq 8 \,\mathrm{km}$, and a background density decreasing exponentially with height z ($\rho_0 = \bar{\rho} \exp(-z/H)$ with $\bar{\rho} = 1.28 \,\mathrm{kg/m^3}$ the mean density on the ground). A more realistic temperature profile appears to be necessary only for more detailed numerical calculations.

In the case of a linearized model, it is suffcient to consider only harmonic waves of angular frequency ω. Each arbitrary wave configuration can then be represented by the superposition of individual harmonic waves (Fourier decomposition). Since the effective heat source is noisy, it is often sufficient to replace in numerical calculations the value of the synodic day and the lunar day, respectively, by the value of the sidereal day ($\Omega_S \simeq \Omega_L \simeq \Omega = 7.2921 \times 10^{-5}\mathrm{s}^{-1}$). In the following, the frequency is normalized according to $\nu = \omega/\Omega$. The temporal and longitudinal dependence of the waves can then be represented as

$$e^{i(m\lambda - \Omega \nu t)} \tag{17}$$

Within the troposphere, the amplitudes of the zonally averaged seasonal tides ($m = 0$) are small compared to the absolute magnitudes of the background atmosphere (ratio between the temperature amplitudes about 1:10). Furthermore, the ratio between the amplitudes of the seasonal tides with $m > 0$ and those of $m = 0$ is also about 1:10. One has, therefore, a kind of hierarchy of waves in the form

Background atmosphere $\Rightarrow$ Zonally averaged wind ($m = 0$) $\Rightarrow$ Waves ($m > 0$). From that follows an influence of the zonally averaged zonal wind on the waves. If this wind (the zonal background flow) can be approximated by the superrotation of the whole atmosphere:

$$u_o = \tilde{U} \cos \phi \tag{18}$$

(with $\tilde{U}$ the zonal wind at the equator), then the total acceleration in the horizontral equations of motion considering (17) becomes

$$\frac{d}{dt} \simeq \frac{\partial}{\partial t} + \frac{u_o}{a \cos \phi} \frac{\partial}{\partial \lambda} = -i\Omega \left(\nu - \frac{m\tilde{U}}{a\Omega} \right) = -i\Omega\nu_D \tag{19}$$

where $\omega_D = \Omega\nu_D$ is the Doppler–shifted frequency measured by an observer on the ground.

Wave dissipation due to turbulent interaction between the large–scale and the medium– and small–scale waves will be taken into account by Rayleigh friction where wave dissipation is assumed to be proportional to the horizontal wave velocity. The proportionally factor is the normalized Rayleigh friction factor $\nu_R = \omega_R/\Omega$ (see (13)). Correspondingly, heat losses due to heat conduction and radiation can be simulated by a normalized Newtonian cooling factor $\nu_N = \omega_N/\Omega$, which is proportional to the temperature amplitude. The normalized frequency ν must now be replaced by the two normalized complex frequencies $\nu_r = \nu + i\nu_R$ and $\nu_h = \nu + i\nu_N$ in the hydrodynamic and thermodynamic equations.

Linearization allows solutions in form of the separation of the variables height, latitude, longitude, and time. We normalize the height coordinate according to $\zeta = z/2H$. The components of a wave mode propagating within an isothermal background atmosphere of constant temperature $\tilde{T}$ and density $\rho_0 = \tilde{\rho} \exp(-2\zeta)$ can then be composed as products of functions of latitude, height, longitude, and time:

$$\left. \begin{array}{ll} \xi = & i\nu_r a\, \xi_n^m(\phi)\, U_n^m(\zeta) \\ \psi = & a\, \psi_n^m(\phi)\, U_n^m(\zeta) \\ L = & ia\Omega\nu_r\, \ell_n^m(\phi)\, U_n^m(\zeta) \\ \Phi = & \theta_n^m(\phi)\, \Phi_n^m(\zeta) \\ w = & \theta_n^m(\phi)\, W_n^m(\zeta) \\ T = & 1/(2R)\, \theta_n^m(\phi)\, d\Phi_n^m(\zeta)/d\zeta \\ Q = & \theta_n^m(\phi)\, Q_n^m(\zeta) \\ M = & \theta_n^m(\phi)\, M_n^m(\zeta) \end{array} \right\} e^{i(m\lambda - \nu\Omega t)} \tag{20}$$

In these equations, $U_n^m = [\Phi_n^m + M_n^m]\Upsilon_n^m/(2a\Omega)$ is the height structure function of the horizonal wind with Υ_n^m a normalization factor, Φ the amplitude of the geopotential [1], w a measure for the vertical wind velocity, T the temperature amplitude, and Q an external heat source. The function θ_n^m is called Hough function. The horizontal wind vector $\mathbf{U}$ is derived from a curl–free (ξ) and a source–free (ψ) function: ($\mathbf{U} = -\nabla\xi + \nabla \times \psi\mathbf{r}$, where $\mathbf{r}$ is the unit vector in radial direction and ∇ and $\nabla\times$ are the horizontal components of the *grad* and the *curl* operators). Explicitly, the horizontal winds may be written

---

[1] Geopotential is the equipotential surface of the gravitational potential $\tilde{\Phi} = \int_0^z g\,dz = \Phi_o + \Phi$ with $\Phi = g\Delta z$ the deviation from its mean value.

$$u = -\frac{1}{a\cos\phi}\frac{\partial\xi}{\partial\lambda} + \frac{1}{a}\frac{\partial\psi}{\partial\phi}; \qquad v = -\frac{1}{a}\frac{\partial\xi}{\partial\phi} - \frac{1}{a\cos\phi}\frac{\partial\psi}{\partial\lambda} \tag{21}$$

where u is the zonal wind (positive to the east), and v is the meridional wind (positive to the nord). In a similar way, the mechanical force can be derived from a curl–free (M) and a source–free (L) component:

$$\mathbf{K} = -\nabla M + \nabla \times L\mathbf{r} \tag{22}$$

With these preconditions, the original system of the hydrodynamic and thermodynamic equations can be reduced to two separated sets of ordinary differential equations (Volland, 1988). The first set, called Laplace's tidal equations, depends only on latitude and longitude (with the longitude dependence according to (17)) and describes the meridional wave structure:

$$[\nu_r^2\nabla^2/2 - m\nu_r/a^2]\xi_n^m - \hat{\nabla}^2\psi_n^m - \nabla^2\theta_n^m/\Upsilon_n^m = 0$$
$$[\nabla^2/2 - m/a^2\nu_r]\,\psi_n^m - \hat{\nabla}^2\xi_n^m + \nabla^2\ell_n^m/2 = 0$$
$$\nabla^2\xi + \epsilon_n^m\theta_n^m/(2a^2\Upsilon_n^m) = 0 \tag{23}$$

with ∇^2 the horizontal Laplace operator, and $\hat{\nabla}^2 = \sin\phi\nabla^2 + \cos\phi\partial/(a^2\partial\phi)$ a modified Laplace operator [2]. The functions $\xi_n^m, \psi_n^m, \theta_n^m$, and ℓ_n^m are dimensionless.

The second system describing the vertical wave structure is a function only of height:

$$dW_n^m/d\zeta - 2W_n^m + iX_n^m(\Phi_n^m + M_n^m) = 0$$
$$d\Phi_n^m/d\zeta + iY(W_n^m - Q_n^m/g) = 0 \tag{24}$$

L, M, Q are external mechanical and thermal energies. Here, the following abbreviations have been used: $X_n^m = \epsilon_n^m\nu_r H/(2\Omega a^2)$ and $Y = 2Rg/(c_p\nu_h\Omega)$, with $a = 6371\,\mathrm{km}$ the earth's radius, $c_p = 1004\,\mathrm{J/kg/K}$ the specific heat at constant pressure, $g = 9.81\,\mathrm{m/s^2}$ the gravitational acceleration, and ϵ_n^m an eigenvalue, called the Lamb parameter. The values ν_r und ν_h are the two complex normalized frequencies (mentioned above), which parameterize the mechanical losses (Rayleigh friction: ν_R) and the thermal losses (Newtonian cooling: ν_N).

4.2 Horizontal Wave Structures

The system of equations (23) is an eigenvalue system with the eigenvalues ϵ_n^m and the eigenfunctions ξ_n^m, ψ_n^m and θ_n^m, which must be solved numerically, in general. A simple analytic solution is possible in the case $|\nu_r| \gg 1$ and $L = 0$:

$$\xi_n^m = P_n^m(\phi)$$
$$\psi_n^m = 0$$
$$\theta_n^m = 2\Upsilon_n^m n(n+1)P_n^m/\epsilon_n^m$$
$$\epsilon_n^m = 4n(n+1)/\nu_r^2 \tag{25}$$

---

[2] Note that in (23) the common factor $\exp(im\lambda)$ must be taken into account in the vector operations.

with $P_n^m(\phi)$ spherical functions in Neumann's normalization. This solution is valid for the upper atmosphere where friction between the ionospheric plasma and the neutral gas is sufficiently large to allow the neglection of the Coriolis force.

A second simple analytic solution holds for the case $\epsilon = 0$, $L = 0$, and $\nu_R = 0$:

$$
\begin{aligned}
\xi_{-n}^m &= 0 \\
\psi_{-n}^m &= P_s^m \\
\theta_{-n}^m &= -\Upsilon_{-n}^m \left[s(s - m + 1) P_{s+1}^m/(s+1) \right. \\
&\quad \left. + (s + 1)(s + m) P_{s-1}^m/s \right]/(2s + 1) \\
\nu &= -2m/[s(s + 1)]
\end{aligned}
\tag{26}
$$

with $s = -n > 0$. The minus sign on the meridional wavenumber n indicates the class of the wave (class 2) (see Fig. 7). These waves, called Rossby–Haurwitz waves, propagate to the west and can exist only at discrete frequencies. Several regional and global wave structures within the troposphere can be represented by such waves. If one introduces the Doppler–shifted frequency ν_D from (19), one notices that these waves migrate to the west relative to the zonally averaged westwind. However, for an observer on the ground, they will appear to propagate to the east if $\tilde{U} > 2a\Omega/[s(s + 1)]$.

Figure 6 shows the Hough functions θ_{-n}^1 of Rossby–Haurwitz waves with zonal wavenumber $m = 1$. The waves concentrate increasingly at higher latitudes with increasing meridional wavenumber n.

Figure 7 shows the eigenvalues ϵ of loss–free waves ($\nu_R = \nu_N = 0$) of wavenumber $m = 1$ as a function of normalized frequency ν. Positive frequencies belong to eastward migrating waves, negative frequencies to westward migrating waves. One notices positive and negative eigenvalues as well as two classes of waves: class 1 waves indicated by positive meridional wavenumbers n, and class 2 waves with negative meridional wavenumbers n. The class 2 waves can only exist at frequencies smaller than two (or periods larger than 12 hours). We shall see later (section 4.3) that there also exists a separation between internal and external waves . This separation takes place at the critical eigenvalue $\epsilon_c \simeq 10$ (horizontal dashed line in fig. 7). The waves in (25) are class 1 waves, and the waves in (26) are class 2 waves with $\epsilon = 0$. In the following, we shall classify the wave modes by the symbol $(m, n; \nu)$. The Rossby–Haurwitz wave from (26) shall be specified by the symbol $(m, n; \nu)^*$.

The diurnal tides $(1, n; -1)$ are westward migrating waves with normalized frequency $\nu = -1$ (symbol 'DT' in Fig. 7). The seasonal tides of wavenumber $m = 1$ can be approximated by the internal class 2 Rossby waves (symbol 'R'). In this case, however, one must replace the frequency on the abscissa by the Doppler–shifted frequency ν_D in (19).

In Fig. 8 (left panel), the Hough functions of the diurnal tides $(1, n; -1)$ are plotted as a function of latitude in the northern hemisphere. A comparison with Fig. 7 shows that wave $(1, -2; -1)$ is a class 2 wave with no zero crossings between the poles. This wave therefore optimally matches the meridional structure of the solar heat input. However, as a wave with a negative eigenvalue, its propagation is suppressed. Figure 8 (right panel) shows the Hough functions of the

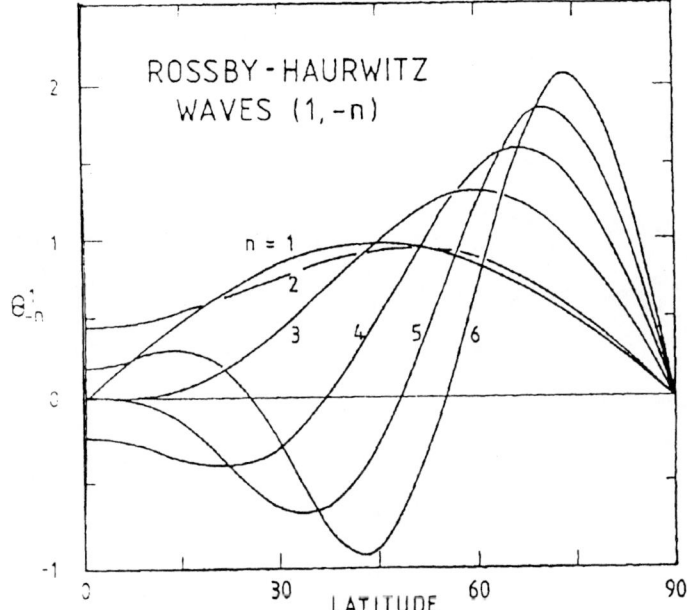

Fig. 6. Hough functions θ_{-n}^{1} of Rossby–Haurwitz waves of zonal wavenumber $m = 1$ as a function of latitude within the northern hemisphere. Waves with even meridional wavenumbers n are symmetric with respect to the equator; waves with odd numbers are antisymmetric.

semidiurnal tides $(2, n; -2)$. These waves belong to the class 1 waves with positive eigenvalues. In particular, wave $(2, 2; -2)$ has a meridional structure which optimally matches the corresponding component of the solar heat input. Eigenvalues and eigenfunctions of the daily tides are tabulated by Chapman and Lindzen (1970).

The eigenvalues for waves with wavenumber $m = 0$ are plotted in Fig. 9 as a function of normalized frequeny ν. Here too, class 1 and class 2 waves, as well as internal and external waves, exist. We shall later identify the zonally averaged seasonal tides as external class 2 waves of wavenumber $m = 0$. One notices in this figure that the eigenvalues of the class 2 waves do not change significantly for $\nu < 1$. Figure 11 shows the corresponding Hough functions of the class 2 waves $(0, n; 0)$ as a function of latitude. Eigenvalues and eigenfunctions of these waves are tabulated by Volland (1988).

4.3   Vertical Wave Structures

For the determination of the vertical wave structures, the set of equations (24) must be solved. Analytical solutions are possible if the terms X_n^{m} and Y are assumed to be constant. These terms contain the loss parameters ν_R and ν_N, the numerical values of which are purely empirically determined. For a simple

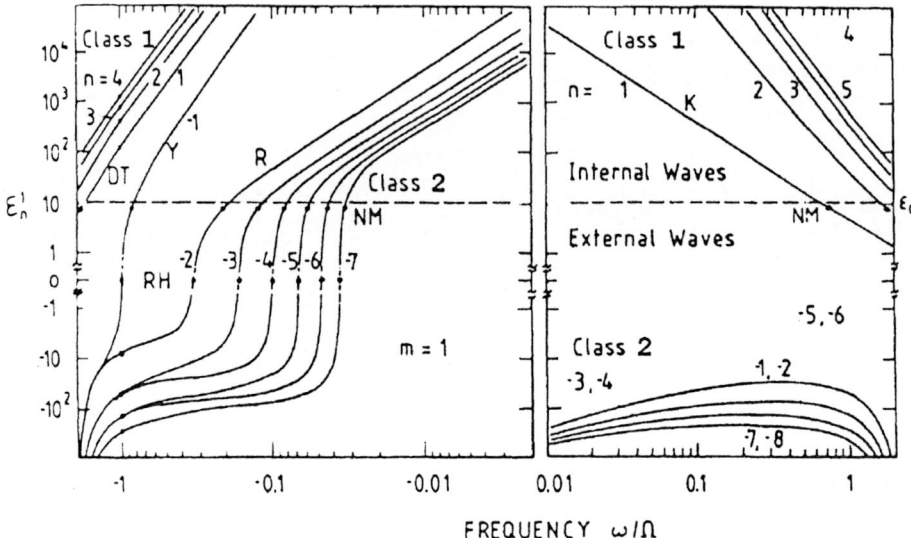

FREQUENCY ω/Ω

Fig. 7. Eigenvalues ϵ_n^1 of the waves $(1, n; \nu)$ with zonal wavenumber $m = 1$ plotted vs. normalized frequency $\nu = \omega/\Omega$. The westward migrating diurnal tidal waves (symbol 'DT') have the frequency $\nu = -1$. The classification of the other waves is as follows: 'RH': Rossby–Haurwitz waves; 'Y': Yanai wave; 'K': Kelvin wave; 'R': Rossby waves; 'NM': normal modes. The horizontal dashed line at the critical eigenvalue ϵ_c separates the internal from the external waves.

estimate, it is sufficient to use constant values. The system (24) is linear and inhomogeneous. Its general solution is the sum of the solution of the homogeneous system plus a particular solution of the inhomogeneous system. For the homogeneous system with $Q = M = 0$, one finds a height structure function

$$F(\zeta) = e^{(1 \mp ik)\zeta} \tag{27}$$

with $k = \sqrt{\epsilon_n^m \nu_r/(\epsilon_c \nu_h) - 1} \times \text{sign}(\omega)$ a vertical wavenumber, and the constant critical eigenvalue $\epsilon_c = (a\Omega)^2 c_p/(RgH) \simeq 10$. Equation (27) consists of two characteristic waves which propagate upward and downward. The expression $\exp(\zeta)$ causes an exponential increase of their amplitudes with height, that results from the linearization of the basic equations. Loss–free waves propagating up into a background atmosphere with almost exponentially decreasing density increase their amplitude, as observations within the middle atmosphere confirm (e.g., Hines, 1960). Such amplification ends when the relative wave amplitudes become sufficiently large so that wave breaking starts.

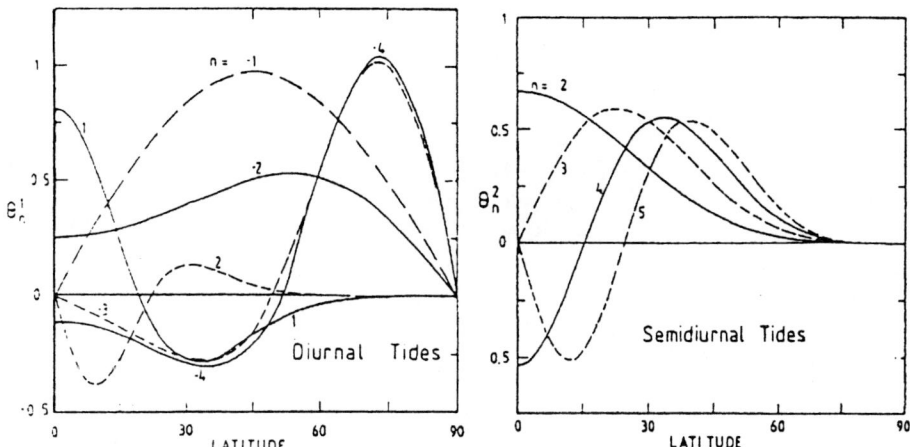

Fig. 8. Hough functions of the daily tides of lowest degrees plotted vs. latitude in the northern hemisphere. Left panel: diurnal tides $(1, n; -1)$; Right panel: semidiurnal tides $(2, n; -2)$. Solid lines: symmetric waves; dashed lines: antisymmetric waves.

In the real atmosphere, the waves are attenuated and lose their energy to the background atmosphere. We simulate in our model wave attenuation by the loss parameters ν_R und ν_N. In the case of a loss–free wave ($\nu_R = \nu_N = 0$), the vertical wavenumber k is real as long as $\epsilon_n{}^m > \epsilon_c$. These waves, with vertical wavelengths $\lambda_z = 4\pi H/k$, are called internal waves (see Fig. 7). For $\epsilon_n{}^m < \epsilon_c$, k is imaginary, and the waves become external or evanescent waves. Their amplitudes now change with height according to $\exp(1 - |k|\zeta)$, and they cannot transport wave energy upward. Their vertical wavelengths become infinite. The positive and negative signs on k in (27), which depend on the sign of the frequency, indicate that the vertical components of their phase and their group velocity have opposite signs. An upward propagating planetary wave therefore progresses downward in phase (Hines, 1960).

For the solution of the inhomogeneous equation (24), we apply a simple vertical profile for the heat source:

$$Q_n{}^m = \tilde{Q}\, e^{(1-\Lambda)\zeta} \tag{28}$$

with Λ a constant. The mechanical force M is assumed to be zero. Furthermore, we use as lower boundary condition $W_n{}^m = i\nu\Omega\Phi_n{}^m/g$ at $\zeta = 0$, corresponding to zero vertical velocity on the ground (Holton, 1983). At the upper boundary ($\zeta \to \infty$), the radiation condition must be fulfilled which demands that no wave from above enters the model atmosphere. With these assumptions, one obtains

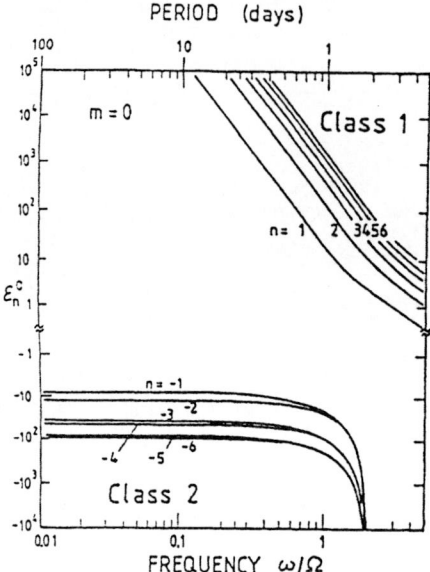

Fig. 9. Eigenvalues ϵ_n^0 of the waves $(0, n; \nu)$ with zonal wavenumber $m = 0$ plotted vs. normalized frequency $\nu = \omega/\Omega$. Waves with positive meridional wavenumbers $(n > 0)$ are class 1 waves; thoses with negative wavenumbers are class 2 waves. Waves with $\epsilon > 10$ are internal waves; those with $\epsilon < 10$ are external waves.

as a solution for the vertical structure functions

$$W_n^m = \frac{(1 + k^2)\tilde{Q}}{g(\Lambda^2 + k^2)}\left[e^{(1-\Lambda)\zeta} + \frac{2\kappa\nu(1 + \Lambda) - \nu_h(1 + k^2)}{(\nu_h(1 - ik) - 2\kappa\nu)(1 + ik)}e^{(1-ik)\zeta}\right]$$

(29)

$$\Phi_n^m = -\frac{2i\kappa(1 + \Lambda)\tilde{Q}}{\nu_h\Omega(\Lambda^2 + k^2)}\left[e^{(1-\Lambda)\zeta} + \frac{2\kappa\nu(1 + \Lambda) - \nu_h(1 + k^2)}{(\nu_h(1 - ik) - 2\kappa\nu)(1 + \Lambda)}e^{(1-ik)\zeta}\right]$$

with $\kappa = R/c_p \simeq 0.29$. It is composed of a particular solution of the inhomogeneous system (24) (first term on the right hand side) and the solution of the homogeneous system (second term on the right hand side) which is an upward propagating characteristic wave (see (27)).

5   Comparison between Theory and Observations

5.1   Daily Tides

In this section, we want to compare observations of tidal events mentioned in section 2 with the theory outlined in the last section. First, we consider daily tidal waves. Their frequencies are large compared with the loss factors ($|\nu| \gg \nu_R, \nu_N$),

Geopotential $(m=0; \omega = 0)$

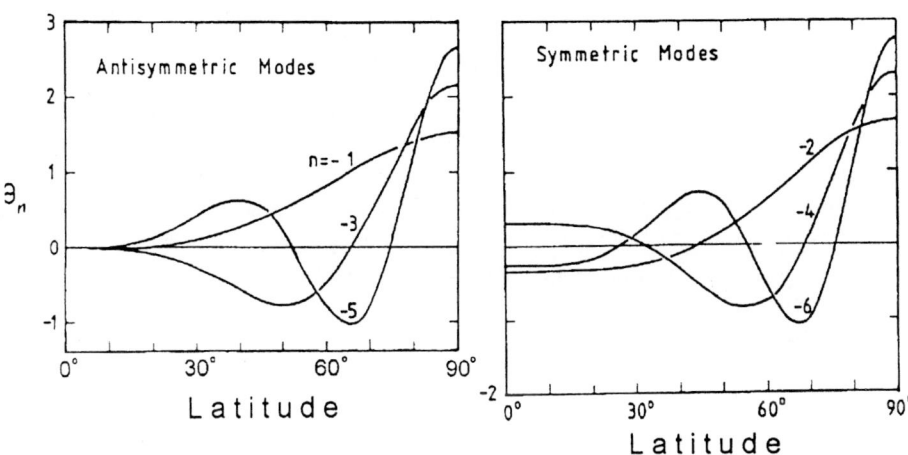

Fig. 10. Hough functions θ_n^0 of the class 2 waves $(0, n; 0)$ $(n < 0)$ plotted vs. latitude in the northern hemisphere.

so that $\nu_r = \nu_h = -m$. According to (29), the vertical structure of their pressure amplitudes becomes

$$p(\zeta) = \rho_o \Phi = \frac{2i\bar{\rho}\kappa\tilde{Q}(1+\Lambda)}{m\Omega(\Lambda^2 - k^2)} \left[e^{-\Lambda\zeta} + \frac{2(1+\Lambda)\kappa - 1 - k^2}{(1+\Lambda)(1 - ik - 2\kappa)} e^{-ik\zeta} \right] e^{-\zeta} \quad (30)$$

At the surface $(\zeta = 0)$, this yields

$$p(0) = |p|e^{-i\tau_m} = \frac{2i\bar{\rho}\kappa\tilde{Q}(1 - ik)}{m\Omega(\Lambda + ik)[1 - ik - 2\kappa]} \quad (31)$$

In the case of the semidiurnal wave $(2, 2; -2)$ $\epsilon_2^2 = 11.2$ and $k \simeq -0.42$. The negative sign on k means that the wave propagates to the west (see (27)). With $\Lambda \simeq 2$ (Forbes, 1982), and $\tilde{Q} = 25$ mW/kg one obtains at the surface $|p| \simeq 113$ Pa, and $\tau_m \simeq 9.35$ h the solar local time of maximum amplitude if the solar heat input has its maximum at local noon. This estimate agrees reasonably well with the observations (see (1)). The value for the heat input is also consistent with the estimate made in (9) if one takes into account the number in (9) as a mean value averaged over the whole atmosphere while $\tilde{Q}$ is the number on the earth's surface. The height integrated energy is then

$$\int_0^\infty \rho_o Q dz = \frac{2H\bar{\rho}\tilde{Q}}{1 + \Lambda} \simeq 170 \text{ W/m}^2 \quad (32)$$

which is only a fraction of the total solar energy available for the atmosphere in (8) including the greenhouse effect. The horizontal wave structure (Hough function) is (Chapman and Lindzen, 1970)

$$\theta_2{}^2(\phi) = 0.425 P_2{}^2(\phi) - 0.041 P_4{}^2(\phi) \quad \cdots \tag{33}$$

(see also Fig. 8; right panel). It is normalized such that $\theta_2{}^2(0) = 1$ at the equator. Evidently, this structure simulates well the observed horizontal structure in (1) so that the wave $(2, 2; -2)$ is primarily responsible for the observed semidiurnal tide at the surface.

This is also true for its vertical structure. The vertical wavelength of this wave is $\lambda_z = 4\pi H/|k| \simeq 240\,\mathrm{km}$ which means that its phase changes only slowly with altitude in agreement with the observations (see fig. 2). Both terms on the right hand side in (30) are responsible for an almost constant amplitude up to about $40\,\mathrm{km}$ height. Above this height, the second term on the right hand side begins to dominate and increases exponentially with height, in agreement with the observations.

The horizontal structure of the diurnal tidal wave $(1, -2; -1)$ optimally matches the corresponding structure of the solar heat input. Its parameters are $\epsilon_{-2}{}^1 = -7.1$ (Fig. 7) and $k \simeq -1.3i$. With the values of $\Lambda = 2$ and $\tilde{Q} = -50\,\mathrm{W/kg}$ (maximum at noon, and twice as large as for the semidiurnal wave) one obtains a pressure amplitude on the ground of $|p| \simeq 53\,\mathrm{Pa}$ and a phase of $\tau_m = 6\,\mathrm{h}$, again in rough agreement with the observations (see (1)). Certainly, other wave modes must be involved, in particular wave $(1, 1; -1)$, as one can easily verify from a comparison between the meridional structure of the first equation (1) and the Hough functions of Fig. 8 (left).

Wave mode $(1, -2; -1)$ is an external (evanescent) wave of class 2, the propagation of which is suppressed, although it is most strongly excited. On the other hand, the vertical wavenumber of the internal wave $(2, 2; -2)$ is located in the nearer environment of the resonance position of the atmosphere which is at $ik = 1 - 2\kappa \simeq 0.43$. At this value, the denominator in the second term on the right hand side of (30) disappears. Class 2 waves of this kind are called resonant modes or normal modes. They are indictated in Fig. 7 by the symbol 'NM'.

In order to explain the dominance of the semidiurnal tide, it was believed before the discovery of the external class 2 waves that wave mode $(2, 2; -2)$ is resonance amplified (e.g., Chapman and Lindzen, 1970). From a comparison of the numerical values in (31) it is evident, however, that the term $(1 - ik)$ in the nominator causes the suppression of the diurnal wave by a factor of about four as compared to the semidiurnal wave, while resonance amplification of the semidiurnal wave is of minor importance (the numerical value of the denominator in (31)).

In the case of the lunar semidiurnal wave we introduce the tidal gravitational potential $M_2 = \bar{M} = -0.793\,\mathrm{m^2/s^2}$ into (24) and find a pressure amplitude on the ground of (with zero heat input $Q = 0$)

$$p(0) = |p| e^{i\phi_m} = -\frac{\bar{\rho}\tilde{M}(1 - ik)}{\Lambda - ik - 2\kappa} \tag{34}$$

Furthermore, assuming $\Lambda = 1$ and $k = -0.46$, one obtains $|p| = 1.8\,\mathrm{Pa}$ and $\phi_m = 108°$. Here, the agreement between theory and the observation (see (5)) is not so good. The reason for this may be the neglected influence of the ocean.

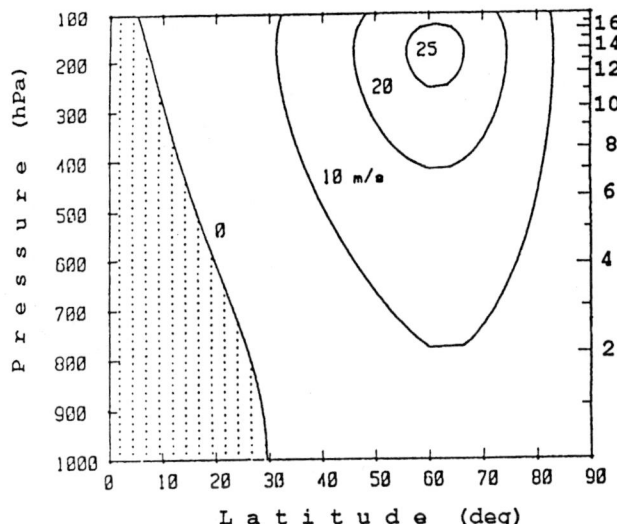

Fig. 11. Height–latitude structure of calculated climatic mean zonal background flow in the northern hemisphere (Volland, 1996). The left height coordinate is in pressure coordinates. The hatched area at low latitudes is the region of eastwinds. In the other regions there are westwinds. Reprint by permission of Kluwer Academic Publishers.

5.2   Seasonal Tides

The zonal background flow considered in section 2.3 exchanges angular momentum with the solid earth. In fact, the observed short–periodic fluctuations of the earth's rotation (changes in the length of the day) with periods between weeks to years are believed to result from changes in the atmospheric axial angular momentum (see also Plag "Chandler Wobble", this volume) mainly due to fluctuations of the zonal wind (Barnes *et al.*, 1983; Eubanks, 1993). The system solid earth–atmosphere possesses an almost constant angular momentum over time intervals of many years. The configuration on the ground of the climatic mean zonal background flow is therefore such that the transfer of angular momentum from the atmosphere to the solid earth at medium latitudes is exactly compensated by the same amount of transfer from the solid earth to the atmosphere at low latitudes.

   Using this condition, a simple model of the climatic mean symmetric zonal wind component ($\nu = 0$) can be constructed. It consists of an external symmetric class 2 mode $(0, -2; \nu_R)$ (see Figs. 9 and 10) and the Rossby–Haurwitz wave $(0, -1; \nu_R)^*$ from (26). The zonal winds of the two waves $(0, -2; \nu_R)$ and $(0, -1; \nu_R)^*$ have the meridional structures

$$U_{-2}{}^0 = A_1\, P_1^1 + A_3\, P_3^1 \cdots$$
$$U_{-1}{}^0 = B_1\, P_1^1 \tag{35}$$

Only the superrotation component of (18) (the terms with the spherical function P_1^1) contributes to an effective transfer of angular momentum between solid earth and atmosphere. The sum $A_1 + B_1$ in (35), as well as its height derivative, must therefore disappear at the surface. This implies that the zonal wind on the ground consists only of the spherical functions of higher degrees of wave $(0, -2; \nu_R)$. In particular, the function P_3^1 is already a good simulation of the observed zonal wind distribution on the ground.

Figure 11 shows the result of a numerical calculation based on these boundary conditions. A height–latitude cross–section is presented with east winds in the tropics (the dotted area) and westwinds at medium and higher latitudes culminating in the westwind jet near 12.5 km altitude and 55° latitude. The actual position of the jet is at somewhat lower latitudes (near 45°), indicating the limitations of such a simplified model. Adding higher wave modes could improve the agreement with the observations.

The antisymmetric zonal wind component evident in Fig. 3a (upper panel) can be simulated reasonably well by the wave mode $(0, -1; \nu_R)$ (see Fig. 10). This mode is responsible for the two oppositely flowing jetstreams at middle atmospheric heights during solstice conditions.

The solar heat input of planetary waves of wavenumber one considered in section 2.4 is a standing wave with a basic period of one year. A standing wave can be composed of the sum of a westward and an eastward migrating wave:

$$Q = \sum Q_n{}^m(\phi)\cos(m\lambda)\ \cos(j\Omega_A t)$$
$$= 0.5 \sum Q_n{}^m(\phi)\left[e^{i(m\lambda+j\Omega_A t)} + e^{i(m\lambda-j\Omega_A t)}\right] \tag{36}$$

Such a heat configuration must be developed into a sum of Hough functions of eastward and westward propagating wave modes. The observed wave structures have maximum amplitudes at medium and higher latitudes (see Fig. 4). These waves migrate within the zonally averaged basic westerly wind (Fig. 3, upper panel). Therefore, for the determination of their eigenvalues in Fig. 7, the Doppler–shifted frequency ν_D (see (19)) must be used. The frequency of the annual wave, which is $\nu = 1/366$, is much less than $\tilde{U}/(a\Omega) \simeq 0.04$ (with $\tilde{U} \simeq 20$ m/s the amplitude of the superrotation component of the zonal background flow (see Fig. 3)). Thus, the Doppler–shifted frequencies of both waves differ only slightly and are negative:

$$(\nu_D)_\pm = -\frac{\tilde{U}}{a\Omega} \pm \nu \simeq -\frac{\tilde{U}}{a\Omega} \tag{37}$$

and the Hough functions of the two waves possess nearly identical meridional structures, so that it is possible to add the vertical structure functions of the eastward and westward migrating waves of zonal wavenumber $m = 1$ to a quasi-standing wave. Its geopotential is then

$$\Phi \simeq \sum S_n{}^1\theta_n{}^1(\phi, \nu_D)\cos(\lambda + \Delta k\zeta)\ \cos(\tilde{k}\zeta + \Omega_A t)\ e^\zeta \tag{38}$$

with $\tilde{k} = (k_{n+} + k_{n-})/2$ and $\Delta k = (k_{n+} - k_{n-})/2$ (with $\tilde{k} \gg \Delta k$) sum and difference of the vertical wavenumbers of both waves.

Westward migrating waves with sufficiently small positive eigenvalues are the internal class 2 Rossby waves (indicated in Fig. 7 by the symbol 'R'): The two most important waves are the symmetric mode $(1, -2; \nu_D)$ and the antisymmetric mode $(1, -3; \nu_D)$. The sum of these is a quasi–standing wave, the phase of which moves slowly downward (few kilometers per month), and the amplitude increases exponentially with altitude according to $\exp(\zeta)$. Both waves interfere constructively in the northern hemisphere and destructively in the southern hemisphere, in agreement with the observations (see Fig. 4). The annual antisymmetric wave $(1, -3; \nu_D)$ is mainly responsible for the seasonal variation of polar motion of the solid earth (Volland, 1996).

The Hough functions of the Rossby waves have almost the same meridional structures as the Rossby–Haurwitz waves with equal wavenumbers. It is, therefore, often sufficient in numerical treatments to replace the Hough functions of the Rossby waves by those of their corresponding Rossby–Haurwitz waves from (26).

References

Barnes, R.T.H., R. Hide, A.A. White, and C.A. Wilson. 1983. Atmospheric angular momentum fluctuations, Length–of–Day changes and polar motion. *Proc. R. Soy. London Ser.* **A 387**: 31.

Bartels, J. und W. Kertz. 1952. Gezeitenartige Schwingungen der Atmosphäre. In: Landoldt–Börnstein *Zahlenwerte und Funktionen aus Physik, Chemie, Astronomie, Geophysik und Technik* **3**: 674.

Chapman, S., and R.S. Lindzen. 1970. Atmospheric Tides. Reidel, Dordrecht.

Charney, J.G., and P.G. Drazin. 1961. Propagation of planetary–scale disturbances from the lower into the upper atmosphere. *J. geophys. Res.*. **66**: 83.

Eubanks, T.M. 1993. Variations in the orientation of the Earth. In: Contributions of Space Geodynamics: Earth Dynamics. *Geodynamics Series.* **24**: 1. Smith, D.E., and D.L. Turcotte (eds.). American Geophysical Union, Washington, D.C..

Forbes, J.M. 1982. Atmospheric tides– I and II. *J. geophys. Res.*. **87**: 5222 - 5241.

Forbes, J.M. 1984. Middle atmospheric tides. *J. Atm. Terr. Phys.*. **46**: 1049.

Fröhlich, C. 1994. Irradiance observations of the Sun. In: The Sun as a Variable Star. *Proc. IAU Symposium.* **143**: 28. J.M. Pap, C. Fröhlich, H.S. Hudson, and S.K. Solanki (eds.). Cambridge University Press, Cambridge.

Haurwitz, B. 1956. The geographic distribution of the solar semidiurnal pressure oscillation. *Meteorol. Papers.* **2**. New York University.

Hines, C.O. 1960. Internal gravity waves at ionospheric heights. *Can. J. Phys.*. **38**: 1441.

Hollingworth, A. 1971. The effect of ocean and earth tides on the semidiurnal lunar air tide. *J. Atm. Terr. Phys.*. **28**: 1021.

Holton, J.R. 1983. An Introduction to Dynamic Meteorology. Academic Press, New York.

Kiehl, J.T. 1994. Clouds and their effects on the climatic system. *Physics Today.* **11**: 36.

Labitzke, K. 1981. Stratospheric–mesospheric midwinter disturbances: a summary of observed characteristics. *J. geophys. Res.*. **86**: 9665.

McAvaney, B.J., W. Bourke, and K. Puiri. 1978. A global spectral model for simulation of the general circulation. *J. Atm. Sci.*. **35**: 1557.

Möller, F. 1973. Einführung in die Meteorologie. Bibliographisches Institut, Mannheim.

Murgatroyd, R.J. 1969. The structure and dynamics of the stratosphere. In: The Global Circulation of the Atmosphere. p. 159. G. A. Corby (ed.). Roy. Met. Soc., London.

Reed, R.J. 1972. Further analysis of semidiurnal tidal motions between 30 and 60 km. *Mon. Wea. Rev.*. **100**: 579.

Siebert. M. 1961. Atmospheric tides. *Adv. Geophysics.* **7**: 105.

Volland, H. 1988. Atmospheric Tidal and Planetary Waves. Kluwer, Dordrecht.

Volland, H. 1996. Atmosphere and Earth's rotation. *Surveys Geophys.*. **17**: 101.

Long-Period Variations of Solar Irradiance

Helmut Wilhelm

Geophysical Institute, Karlsruhe University
Hertzstrasse 16, D-76187 Karlsruhe

Abstract. Long-period variations of solar tides on earth and of solar irradiance received at the top of the atmosphere are produced by changes of the earth's orientation and its orbit in space, i. e. by the precession of the earth's axis of angular momentum and by variations of the obliquity and eccentricity of its orbit. The Milankovic-hypothesis states that ice ages and interglacials are caused by these long period changes of solar irradiance. Prominent climatic variations of much shorter period detected in Greenland ice-cores cast doubts on the validity of this hypothesis.

1 Introduction

The temporal variations of the solar irradiance received at the top of the earth's atmosphere are caused by the celestial motion of the earth around the sun. This is also responsible for the solar tidal variations in gravity, potential hight, inclination, stress and strain, ocean currents, atmospheric winds, or groundwater level: While the rotating earth moves along its orbit through the inhomogeneous solar gravitational field it is exposed to a correspondingly varying radiational field from the sun. The rotation of the earth, its motion around the sun and the variations of its orbit due to nutations, precession and long period changes in eccentricity and obliquity are the sources of the solar tidal variations of the gravitational potential as well as of the solar irradiance. The simplest term in the tidal potential and the irradiance, have both a $1/r^2$ dependence on the distance from their source. The irradiance is a scalar field like the tidal potential, varying only on the sunlit side, whereas it is zero on and behind the dark side of the earth. The spectrum of the variations of solar irradiance is therefore somewhat different from the corresponding spectrum of the solar tidal potential, but both spectra extend from very low frequencies, caused by the long-period variations of ellipticity, obliquity and precession, to diurnal frequencies and higher harmonics. As these temporal variations have the same origin, it would be justified to call these variations of solar irradiance 'tidal variations', but as tides in the strict sense are commonly restrained to comprise the temporal changes in gravity of celestial origin and to the corresponding gravitationally induced variations of related physical properties, preference will be given to the term 'temporal variations' of solar irradiance.

2 Temporal Variations of Solar Irradiance

The earth's orbit around the sun is elliptical; its eccentricity and orientation in space changes slowly with time, whereas its semimajor axis remains invariable to second order pertubations (Roy, 1988). For example, the direction of the earth's axis of rotation with respect to the ecliptic varies with a period of about 26 ka because of the luni-solar precession.

The total amount of solar radiation received by the earth during its annual orbital motion depends on the time for passages between one perihelion to the next (365.25964 days), i. e. on the anomalistic year, because of the invariability of the semi-major axis a, but it also depends on the seasonal variation at a point on the surface of the earth which is determined by the change in the declination of the sun, i. e. by the tropical year (365.24220 days). These periods are slowly changing with time because of corresponding changes of the eccentricity and orientation of the earth's orbit and because of the precession of the equinoxes. Thomson (1995) has shown that in many cases the annual temperature variation is controlled by the perihelion.

The importance of the solar irradiance for the climate of the earth suggests a coherence between variations in climate and solar irradiance. Croll (1864) showed that variations in the annual energy influx only depend on eccentricity and are on the order of e^2, or 0.1 percent. But he suggested that variations of the seasonal influx might trigger a substantial climatic response, namely the cycle of Pleistocene ice ages. Milankovic (1930) performed the first quantitative calculations of the long-period variations of radiation at different latitudes based on astronomical theory and concluded that the intensity of irradiance at high northern latitudes during the summer was critical for the growth and decay of ice sheets, because these variations exceed eight percent of the mean values. Short period variations in the emitted solar radiation are correlated with the sunspot cycle, but a long-period trend of the emittance has not yet been detected (Foukal, 1987, 1994).

The solar irradiance is the main heat source for all natural processes operating near the earth's surface, except for volcanism and tectonic activity. The radiational power of the sun is $3.9 \cdot 10^{23}$ kW, from which the earth gleans $1.74 \cdot 10^{14}$ kW in the mean. However, only 40 - 50 % of this value reach the earth's surface because of scattering, absorption and reflection in the atmosphere, compare Volland (Atmospheric Tides, this volume). The solar constant, i.e. the rate of solar radiation energy S_0 received at a surface of unit area perpendicular to the earth-sun direction, situated at the mean earth-sun distance r_m is presently

$$S_0 = (1366 \pm 5) \text{ W/m}^2 \tag{1}$$

according to satellite measurements (Lean, 1991). At the distance a (semi-major axis of the earth's elliptical orbit, constant to second order pertubations) from the sun the irradiance is

$$S_a = S_0 (r_m/a)^2. \tag{2}$$

The mean distance r_m is varying as a function of eccentricity e(t) according to

$$r_m^2 = a^2 \sqrt{1 - e^2} \tag{3}$$

so that

$$S_0 = S_a \sqrt{1 - e^2} \tag{4}$$

is also varying with time whereas S_a is constant because a remains constant. Therefore it would be more convenient to refer to S_a as „solar constant" (Berger et al., 1993).

The irradiance W(r,ψ) received at distance r from the sun located at zenith angle ψ, is given by

$$W(r, \psi) = \begin{cases} S_a \left(\dfrac{a}{r}\right)^2 \cos \psi & 0 \le \psi < \dfrac{\pi}{2} \\ 0 & \dfrac{\pi}{2} \le \psi < \pi. \end{cases} \tag{5}$$

The decomposition of this expression into Legendre polynomials yields

$$W(r, \psi) = S_a \left(\frac{a}{r}\right)^2 \left\{ \frac{1}{4} + \frac{1}{2} P_1(\cos \psi) + \frac{5}{16} P_2(\cos \psi) - \frac{3}{32} P_4(\cos \psi) - + ... \right\} \tag{6}$$

In order to reveal the correspondig periods this expression has to be written in celestial coordinates, compare for example Berger et al. (1993). The elliptical orbit of the earth around the sun, i.e. the apparent orbit of the sun around the earth, is given by

$$\rho = \frac{r}{a} = \frac{1 - e^2}{1 + e \cos \upsilon} . \tag{7}$$

Here, υ ist the true anomaly of the earth, i.e. the angle between perihelion P and the earth, expressed by the corresponding ecliptical longitudes $\tilde{\omega}$ and λ, measured from the true equinox of date (Roy, 1988)

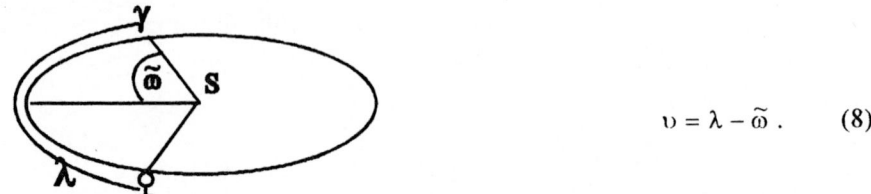

$$\upsilon = \lambda - \tilde{\omega} . \qquad (8)$$

In the equatorial system the zenith angle ψ is expressed by the geographical latitude φ of the observation point and the declination δ of the sun, and by the hour angle H measured westward from the local meridian to the meridian of the sun

$$\cos \psi = \sin \varphi \sin \delta + \cos \varphi \cos \delta \cos H \qquad (9)$$

with $$H = H_0 + \lambda_E, \qquad (10)$$

H_0 Greenwich hour angle, westward positive, λ_E east longitude of observation point, eastward positive. A variation of 2π in H corresponds to a mean solar day, i.e. to 86400 s.

From these expressions the instantaneous insolation is calculated as a function of the position υ of the earth and the eccentricity e of its orbit, and of the positions of the observation point given by φ and λ_E, and of the sun, H_0 and δ, respectively, measured in the equatorial system.

$$W = W(\varphi, \lambda_E, H_0, \delta; \upsilon; e)$$
$$= S_a \left(\frac{1 + e \cos \upsilon}{1 - e^2} \right)^2 \left(\sin \varphi \sin \delta + \cos \varphi \cos \delta \cos(H_0 + \lambda_E) \right). \qquad (11)$$

By introducing the ecliptical mean longitude $h = \lambda_S$ of the sun according to

$$\sin \delta = \sin \varepsilon \sin h, \qquad (12)$$

ε obliquity of the celestial equator with respect to the ecliptic, the slowly varying parameters ε, e and $\tilde{\omega}$ explicitly appear in the expression

$W = W(\varphi, \lambda_E, H_0; h, \tilde{\omega} ; \varepsilon, e)$.

The variation of h is given by

$$h = 279°.696678 + 36{,}000°.768925\, T + 0°.0003025\, T^2, \qquad (13)$$

T expressed in Julian centuries from 1899 December 31, 12 UT, compare Bartels (1958), and of ε, e and $\tilde{\omega}$ by

$$\varepsilon = \varepsilon_0 + \Sigma\, A_i \cos (f_i t + \delta_i)$$
$$c = c_0 + \Sigma\, E_i \cos (\lambda_i t + \phi_i)$$
$$e \sin \tilde{\omega} = \Sigma\, P_i \sin (\alpha_i t + \zeta_i) \qquad (14)$$
$$e \cos \tilde{\omega} = \Sigma\, P_i \cos (\alpha_i t + \zeta_i)$$

with coefficients A_i, E_i, P_i, f_i, λ_i, α_i, δ_i, ϕ_i, ζ_i , $\varepsilon_0 = 23°.320556$ and $c_0 = 0.028707$, listed in Berger (1978) for t = 0 at 1950.0.

In order to get an idea about the amount of the effects of absorption, reflection and scattering in the atmosphere on the spectrum of the radiation reaching the earth's surface Friederich (1984) investigated a three years' data set of the visible

Table 1. Amplitude and phase of extraterrestrial solar radiation and radiation measured at the earth's surface near Karlsruhe (49.0°N, 8.4°E), both determined by HYCON method (Schüller, 1976), * values determined from harmonic analysis.

Doodson group			Tidal Symbol	Amplitude extraterr. radiation W/m²	Amplitude measured radiation W/m²	Phase extraterr. radiation degrees	Phase measured radiation degrees
m	j	k					
0	0	0	S_0	293.84	81.42 *	0.00	0.00 *
0	0	1	S_a	210.90	47.85 *	251.80	85.35 *
0	0	2	S_{sa}	1.68	- *	22.30	- *
0	0	3	S_{ta}	0.17	23.78 *	300.00	86.07 *
0	0	5		0.05	10.73 *	63.72	82.92 *
0	0	7		-	12.73 *	-	82.83 *
0	0	8		-	7.03 *	-	158.38 *
1	1	-8		-	9.94	-	355.24
1	1	-7		-	12.71	-	78.53
1	1	-4		2.14	22.87	92.40	350.37
1	1	-3	π_1	4.08	23.71	348.14	81.25
1	1	-2	P_1	114.77	13.18	237.08	327.49
1	1	-1	S_1	434.50	115.17	188.38	67.12
1	1	0	K_1	122.16	54.63	57.78	179.35
1	1	1	ψ_1	18.55	31.10	313.90	264.15
1	1	2	φ_1	11.38	11.84	144.82	165.44
1	1	4		0.13	11.22	278.30	166.84
1	1	5		0.03	8.32	201.12	259.29
2	2	-8		-	9.52	-	161.40
2	2	-4	$2T_2$	13.29	19.64	238.28	163.12
2	2	-3	T_2	0.48	6.58	13.99	257.82
2	2	-2	S_2	151.30	17.13	16.80	164.67
2	2	-1	R_2	1.39	12.03	105.58	248.90
2	2	0	K_2	26.10	23.29	155.33	2.37
2	2	2		1.48	2.74	292.84	172.73
2	2	4		0.08	9.68	56.91	343.53
3	3	-5		0.01	2.80	79.28	249.54
3	3	-4		21.26	6.48	225.94	156.83
3	3	-2		18.55	3.16	184.47	159.94
4	4	-5		0.41	5.00	73.88	244.64
4	4	-4	S_4	11.16	2.74	213.60	336.90

radiation (1979/12/1 to 1982/11/30), measured at the Research Centre Karlsruhe, Germany. He analyzed the spectrum of the measured influx and made a comparison with the spectrum of the unaffected solar radiation incident at the observation point. The results are given in Table 1 (from Friederich and

Wilhelm, 1986). For explication of the Doodson group numbers m,j,k see Bartels (1958).

The regional meteorological situation of Karlsruhe is rather well understood by simulations using the model KAMM (Adrian and Fiedler, 1991), and global influence is characterized by westerly winds and variable atmospheric conditions, corresponding to the geographic position near 49° latitude in western Europe. Three main differences between both spectra are conspicuous:

- While the energy of the solar radiation incident from space is concentrated around the solar diurnal and semi-diurnal waves S_1 and S_2 , the spectral power of the actual influx on the ground is smeared out across the whole diurnal and semi-diurnal frequency-band; S_2 even cannot be recognized as a peak from the background.

- There is no systematic relation between the phase spectrum of the measured influx and the undisturbed solar irradiance

- The mean (constant) value of the measured radiation is only 36% of the corresponding undisturbed input.

Obviously the insolation measured on the ground differs significantly from the undisturbed irradiance which would reach the earth without atmosphere. Regarding the strong variations in atmospheric conditions it is therefore doubtful, whether the small variations in the undisturbed influx due to the long-period variations of e, ε and $\tilde{\omega}$ are transmitted unchanged to the ground.

3 Long-Period Variations

The low frequency part of the spectrum is determined by the variations of $\tilde{\omega}$, ε and e which are inherent in the expression

$$\rho^{-2} = (1 + e\cos\upsilon)^2 / (1 - e^2)^2 \cong 1 + 2e\cos\upsilon + 0.5e^2 \cos 2\upsilon$$
$$+ 2.5e^2 + 4e^3 \cos\upsilon + ... \quad (15)$$

with $\cos\upsilon = \cos(h - \tilde{\omega}) = \cos h \cos \tilde{\omega} + \sin h \sin \tilde{\omega}. \quad (16)$

The corresponding terms in the development of W are varying approximately with the following periods

term	period [ka]	origin
$e \cos \tilde{\omega}$	19, 23	precession
$e^2\cos \tilde{\omega}$	9.5, 11.5	precession
$\sin \varepsilon$	41	obliquity
e	100, 413	eccentricity.

Presently, e = 0.017 and ε = 23° 26' 22''. The eccentricity varies between 0.0 and 0.06 with periods of about 100 and 400 ka and the obliquity between 22.1° and 24.5° with a period of 41 ka (Hays et al., 1976). The obliquity is currently decreasing at a rate of -46''.85 per century leading to an expansion of the world's temperate zone at the expense of the tropical and arctic zones at a rate of some 1500 km² per year (Chao, 1996). The general precession consists of two components: (1) The axial precession, i. e. the rotation of the axis of angular momentum of the earth around the pole of the ecliptic due to the torque of the sun, moon and planets on the earth's equatorial bulge with a period of 26 ka, and (2) the corresponding precession of the equinoxes resulting in the precession of the perihelion in the earth's orbit with a period of 21 ka. The modulation by the variation of eccentricity yields periods around 19 and 23 ka (Crowley and North, 1991).

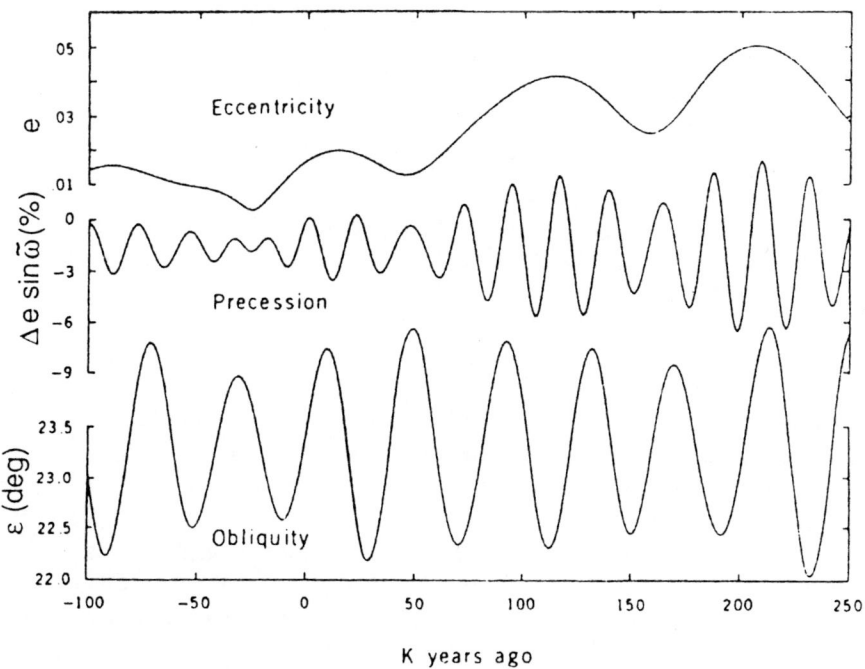

K years ago

Fig. 1. Variation of eccentricity e, precession index Δe sin ω̃ and obliquity ε with time passed before 0 A.D. in ka, *reprinted with permission from Imbrie and Imbrie (1980). Copyright 1980 American Association for the Advancement of Science.*

The corresponding periods are also contained in the solar tidal potential. From the long period changes, however, only the variation of ω̃ is included in the harmonic development as described by H.-G. Wenzel (Tidal Analysis, this volume), because for tidal considerations the longer periods of ε and e are not yet important.

In Fig. 1 the variations of e, Δe sin $\tilde{\omega}$ (deviation of the 1950 A.D. value of the precession index e sin $\tilde{\omega}$) and ε, calculated after Berger (1978), are displayed as functions of time passed before 0 A.D. between - 100 ka (future) to 250 ka (past), clearly revealing the long-period periodicities. Corresponding long-period variations of the intensity of the received insolation on the northern hemisphere have been suggested by Milankovic (1930) to trigger the global variations of climate.

4 Other Causes for Long-Period Climatic Variations

The model of Milankovic (1941) has been accepted as a working-hypothesis in climatic research after Hays et al. (1976) had demonstrated the existence of the corresponding periods in the ice volume records. No other hypothesis can be tested as directly as this one (Suarez and Held, 1976). However, there are also other possible causes for global climatic changes:

1. Variations of the solar emittance or influences by varying concentrations of cosmic dust.
2. Variations of the earth's orbit with respect to the invariable plane of the solar system.
3. Internal variations of the climatic system, caused by
- variations of concentration of dust in the atmosphere due to volcanic eruptions
- sea-level changes
- variations in ice-volume and -area
- changes in albedo
- variations of CO_2- and CH_4-content in the atmosphere and oceans
- changes of current systems and deep water circulation in the oceans
- shifts between metastable states of a non-linear and chaotic climatic system
- iceberg discharges.

With the exception of the Milankovic-hypothesis these suggestions cannot be precisely examined. This conjecture therefore has attracted the greatest interest, especially since isotopic data with sufficient accuracy in age dermination and temporal resolution are available from sedimentary records and from ice-cores.

Geological and geochemical indications of climatic variations are provided by the variation $\delta^{18}O$ and $\delta^{16}O$ with respect to the standard water abundances of ^{18}O and ^{16}O, contained in the shells of planktonic foraminifera of marine sediments and in ice-cores (Hays et al., 1976; Johnsen et al., 1992). The method is based on the fact that the calcareous skeletons of the foraminifera preserve the $^{18}O/^{16}O$ ratio of the scawater, in which the plankton has lived bcfore it was deposited in the sea-sediment. This ratio varies with the volume of water frozen in glaciers and ice sheets: Water molecules containing the heavier isotope ^{18}O are more likely to be precipitated near their source in the oceans than on land, so that ice sheets and glaciers are relatively depleted in ^{18}O. Simultaneously ^{18}O is enriched in seawater, if the volume of ice is increasing.

The isotope ratio is depending on other factors, too, for example on the temperature of the water bearing the plankton, or on the height, where the water vapor condenses. By dating leading horizons in the sedimentary record with radioactive methods and with certain assumptions about sedimentation rates, it was possible to demonstrate that the global ice-volume had several peaks in the last 800,000 years and that it shows a quasiperiodic behaviour.

Fourier analysis of the δ^{18}O-variations yields a spectrum of climatic variations with peaks at 10^5, 43 10^3, 24 10^3, and 19 10^3 years, compare Fig. 2, from Hays et al. (1976). The most astonishing fact of this spectrum is the prominent peak at 10^5 years, corresponding to the main period of eccentricity, which has only a small influence on the variation of insolation. If these peaks would be linearly related to the incident solar radiation, the 10^5-years peak should be distinctly smaller than the other three peaks. On the other hand there is no doubt that the climate system is a non-linear system, as is demonstrated by the ice-volume curve, which has a saw-tooth character, although the radiation input is quasiperiodic. This follows from the fact that ice sheets melt faster than they grow (Broecker and van Donk, 1970).

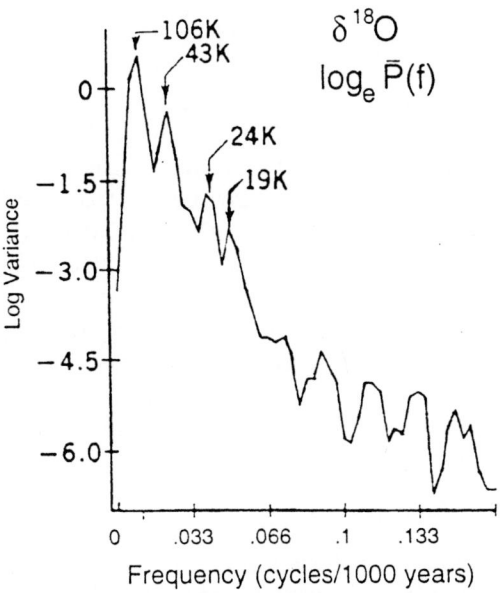

Fig. 2. Spectrum of variance of δ^{18}O, determined from two subantarctic deep-sea cores, *reprinted with permission from Hays et al. (1976). Copyright 1976 American Association for the Advancement of Science.*

Aside from the saw-tooth character of the glacier-building curve there are two additional facts which cannot be explained by a purely linear Milankovic-hypothesis, i. e. the large climatic effect at a 100 ka cycle and the correlation between summer insolation in northern latitudes and the global effect of glaciation, which has already been stated by Milankovic.

Recently, Muller and MacDonald (1995) claimed that the inclination of the earth's orbit with respect to the invariable plane of the solar system varies with a period of 100 ka. They demonstrate that whenever the earth's orbit has a minimal inclination with respect to this invariable plane the δ^{18}O-data show a maximum within the last 600 ka. If cosmic dust would cause sufficient cooling to trigger a glacial cycle the change in eccentricity could have a negligible effect, as expected. This suggestion is intensively tested by a search for extraterrestrial ^{3}He and iridium in sediments and ice samples (Newman, 1995). If cosmic dust is the dominant trigger mechanism for glacial cycles, the 100 ka cycle, which is practically absent for times more than 1 Ma ago, could be a transient phenomenon on geological time scales. For the interval 1.2 to 5.2 Ma ago, i. e. for a time span directly before the Late Pleistoce, Clemens and Tiedemann (1997) find 400 ka and 100 ka climate cycles in a marine δ^{18}O-record which correlate with the spectral power of a truncated July 65° N insolation.

It is generally accepted that the climate system is intrinsically nonlinear involving complex interaction between atmosphere, oceans, land, and biosphere. An important part in this network of interactions is played by the thermohaline circulation of the oceans (Broecker and Danton, 1989) and its coupling to fresh water input by surging ice sheets. The global pattern of ocean circulation can flip between several modes which are triggering or are triggered by climate variations. Small fluctuations in insolation, evaporation and precipitation may turn the regulation mechanism so that the ocean circulation pattern changes with a global climate variation as a result. Recent deep-sea core investigations have shown that iceberg discharges from the Laurentian ice sheet may have turned the conveyor-belt like circulation in the North Atlantic on and off by the input of fresh water to the sea (Broecker, 1995). These so-called Heinrich-events are mainly caused by surges in the Hudson Straight and have a recurrence period of about 7 ka to 10 ka. In addition, high resolution studies of North Atlantic deep sea cores have shown that in iceberg calving even shorter periods of 2 ka to 3 ka can be reveiled, which correlate with warm-cold oscillations, the so-called Dansgaard-Oeschger events, in Greenland ice-cores (Bond and Lotti, 1995). It appears that these short period variations of iceberg discharges have been triggered by a change in climate-related conditions rather than vice-versa. Recent investigations of the role which tropospheric dust may have played in the development of glacial climates indicate that episodic dust loading and associated snow darkening can provide the warming needed to trigger these Dansgaard-Oeschger events (Overpeck et al., 1996). The influence of different forcing perturbations on the thermohaline ocean circulation is presently extensively studied by model calculations (Rahmstorf et al.,1996).

If such short-period oscillations in climate have been present without being resolved in the ice-volume records then, regarding the spectrum, the possibility of aliasing must be taken into account. In this case long period climate variations would appear in the spectrum, which acutally have been short period variations. This misinterpretation is especially menacing if observations are tuned to Milankovic-cycles, because then an artificial regularity is introduced into the records.

From geochemical investigations of ocean sediments it has been concluded that during the last ice age essential parts of the general circulation of the North Atlantic did not exist and that the present system of ocean currents has evolved only 14 ka ago. The main driving force for the oceanic convection is the deep-water formation in the North Atlantic and in the Wedell and Ross Sea. For the North Atlantic the corresponding heat fed into the atmosphere equals 30 % of the annual insolation. Aside from the action of the Gulf Stream this heat transfer into the atmosphere is the main reason for the present mild climate in the North Atlantic region. It is therefore of outmost importance for the interpretation of climate variations to understand the causes for the variability of ocean circulation and ocean-atmosphere interaction.

5 Unsolved Problems

At the present stage the role the Milankovic-hypothesis in the history of climate variations is not quite clear:
  - the $\delta^{18}O/\delta^{16}O$-variations in the foraminifera shales can be explained by ice-volume changes, but also by changes in ocean temperature and both effects cannot be clearly separated,
  - the growth of ice sheets cannot be modelled in correspondence to the Milankovic orbital perturbations for a broad range of conditions (Rind et al., 1989), although for special assumptions successful models of ice sheet generation and conservation have been produced (Weertman, 1976; Otto-Bliesner, 1996),
  - there is no generally accepted explanation for the preponderance of the 100 ka period in the climate data (Wigley, 1976; Liu, 1992; Liu, 1995),
  - the 100 ka period is absent for times 1 Ma B. P., although the quaternary ice ages started already 2 Ma ago; the Milankovic-periods are however quasi-stationary, so that they should be visible during the whole quaternary period for a linear response of the system,
  - orbital tuning of the $\delta^{18}O$-records which is applied as a means to determine the sedimentation rate (Imbrie et al.,1984; Bassinot et al., 1994) is based on the assumption that the Milankovic-periods are reflected in the sedimentation rate; with regard to the proof of the validity of the Milankovic-hypothesis, this procedure implies a vicious circle argumentation (Neeman et al., 1988a),
  - to improve the correlation between the orbital changes and the measured $\delta^{18}O$-variation various expression related to the instantaneous irradiance have

been calculated; an improved correlation is however not necessarily indicative of a causal relationship.

- variations of the concentration of greenhouse gases like CO_2 and CH_4 are likely to cause favourable conditions for the formation or elimination of ice sheets (Neeman et al., 1988b).

These considerations suggest that not only the insolation and its variation by orbital forcing but also the greenhouse gas content is of decisive importance for the evolution of the climate. Presently about 60 times as much CO_2 as in the atmosphere is contained in the oceans. Therefore, the CO_2-exchange between atmosphere and ocean is of great importance to the climate. For a conclusive progress in the discrimination of climate forcing processes it is necessary to perform model calculations which take the effects of greenhouse gases, cloud formation and vegetation within a dynamicly coupled earth-ocean-atmosphere-biosphere system into account.

References

Adrian, G. and F. Fiedler. 1991. Simulations of instationary wind and temperature fields over complex terrain and comparison with observations. *Contrib. Atmosph. Physics* **64**: 27-48.

Bartels, J. 1958. Tidal forces. pp. 734-774. **In**: Encyclopedia of Physics, Vol. 48. Geophysics II, S. Flügge (ed.). Springer Verlag, Berlin, 1046 pp.

Bassinot, F.C., L.D. Labeyrie, E. Vincent, X. Quidelleur, J.F. Shackleton, and Y. Lancelot. 1994. The astronomical theory of climate and the age of the Brunhes-Matuyama magnetic reversal. *Earth Planet. Sci. Let.* **126**: 91-108.

Berger, A.L. 1978: Long-term variations of daily insolation and quaternary climate changes. *J. Atm. Sci.* **35**: 2362-2367.

Berger, A., M.-F. Loutre and C. Tricot. 1993. Insolation and earth's orbital periods. *J. Geophys. Res.* **98**: 10,341-10,362.

Bond, G.C. and R. Lotti. 1995. Iceberg discharges into the North Atlantic on millennial time scale during the last glaciation. *Science* **267**: 1005-1010.

Broecker, W.S. 1995. Massive iceberg discharges as triggers for global climate change. *Nature* **372**: 421-424.

Broecker, W.S. and G.H. Denton. 1989. The role of ocean-atmosphere reorganizations in glacial cycles. *Geochimica et Cosmochimica Acta* **53**: 2465-2501.

Broecker, W.S. and J. van Donk. 1970. Insolation changes, ice volumes, and the [18]O record in deep-sea cores. *Rev. Geophys. Space Phys.* **8**: 169-198

Chao, B.F. 1996. ''Concrete'' testimony of Milankovitch cycle in earth's changing obliquity. *EOS Transactions. Am. Geophys. Union* **44**: 434.

Clemens, S.C. and R. Tiedemann. 1997. Eccentricity forcing of Pliocene-Early Pleistocene climate revealed in a marine oxygen-isotope record. *Nature* **385**: 801-804.

Croll, J. 1864. On the eccentricity of the earth's orbit and its physical relations to the glacial epoch. *Phil. Mag.* **33**: 119-131.

Crowley, T.J. and G.R. North. 1991. Paleoclimatology. Oxford University Press, Oxford, 339 pp.

Foukal., P. 1987. Physical interpretation of variations in total solar irradiance. *J. Geophys. Res.* **92**: 801-807.

Foukal, P.1994: Study of solar irradiance variations holds key to climate questions. EOS *Transactions Am. Geophys. Union* **75**: 377-382.

Friederich, W. 1984. Strahlungseffekte bei Gezeitenmessungen. Diploma Thesis, Karlsruhe University, Germany.

Friederich, W. and H. Wilhelm. 1986. Solar radiational effects on earth tide measurements. pp. 865-879. In: Proc. 10th Int. Symp. Earth Tides. R. Vieira (ed.). Consejo Superior de Investigaciones Cientificas, Madrid.

Hays, J.D., J. Imbrie and N.J. Shackleton. 1976. Variations in the earth's orbit: Pacemaker of the ice ages. *Science* **194**: 1121-1132.

Imbrie, J. and J.Z. Imbrie. 1980. Modeling the climatic response of orbital variations. *Science* **207**: 943-953.

Imbrie, J., J.D. Hays, D.G. Martinson, A. McIntyre, A.C. Mix, J.J. Morley, N.-G. Pirias, W.L. Prell, and N.J. Shackleton. 1984. The orbital theory of Pleistocene climate: support from a revised chronology of a marine $\delta^{18}0$ record. pp. 269-305 In: Milankovitch and Climate, part I. A. Berger, J. Imbrie, J. Hays, G. Kukla and B. Saltzman (eds.). Reidel Publ. Comp., Dordrecht, 895 pp.

Johnsen, S.J., H.B. Clausen, W. Dansgaard, K. Fuhrer, N. Gundestrup, C.U. Hammer, P. Iversen, J. Jouzel, B. Stauffer, and J.P. Steffensen. 1992. Irregular glacial interstadials recorded in a new Greenland ice core. *Nature* **359**: 311-313.

Lean, J. 1991. Variations in the sun's radiative output. *Rev. Geophysics* **29**: 501-535.

Liu, H.-S. 1992. Frequency variation of the earth's obliquity and the 100-kyr cycles. *Nature* **358**: 397-399

Liu, H.-S. 1995. A new view on the driving mechanism of Milankovitch glaciation cycles. *Earth Planet. Sci. Let.* **131**: 17-26

Milankovic, M. 1930. Mathematische Klimalehre und Astronomische Theorie der Klimaschwankungen. Bornträger, Berlin. 176 pp.

Milankovic, M. 1941. Kanon der Erdbestrahlung und seine Anwendung auf das Eiszeitproblem. *Königl. Serb. Akad. Beograd, Spec. Publ.* **132**. 633 pp.

Muller, R.A. and G.J. MacDonald. 1995. Glacial cycles and orbital inclination. *Nature* **377**: 107-108

Neeman, B.U., G. Ohring, and J.H. Joseph. 1988a. The Milankovitch theory and climate sensitivity. 1. Equilibrium climate model solutions for the present surface conditions. *J. Geophys. Res.* **93**: 11,153-11,174.

Neeman, B.U., G. Ohring, and J.H. Joseph. 1988b. The Milankovitch theory and climate sensitivity. 2. Interaction between the northern hemisphere ice sheets and the climate system. *J. Geophys. Res.* **93**: 11,175-11,191.

Newman, A. 1995. New mechanism proposed for glacial cycles. *EOS Transactions Am. Geophys. Union* **76**: 489-490.

Otto-Bliesner, B.L. 1996. Initiation of a continental ice sheet in a global climate model (GENESIS). *J. Geophys. Res.* **101**: 16,909-16,920.

Overpeck, J., D. Rind, A. Lacis, and R. Healy. 1996. Possible role of dust-induced regional warming in abrupt climate change during the last glacial period. *Nature* **384**: 447-449.

Rahmstorf, S., J. Marotzke, and J. Willebrand. 1996. Stability of the thermohaline circulation. pp. 129-158. In: The Warmwatersphere of the North Atlantic Ocean. W. Krauß (ed.). Gebrüder Bornträger, Berlin, 446 pp.

Rind, D., D. Peteet and G. Kukla. 1989. Can Milankovitch orbital variations initiate the growth of ice sheets in a general circulation model? J. *Geophys. Res.* **94**: 12,851-12,871.

Roy, A.E. 1988. Orbital Motion. 3rd edition. Adam Hilger, Bristol, 532 pp.

Schüller, K. 1976. Ein Beitrag zur Auswertung von Erdgezeitenregistrierungen. Deutsche Geodätische Kommission Reihe C., No. **227**, München.

Suarez, M.J. and I.M. Held. 1976. Modelling climatic response to orbital parameter variations. *Nature* **263**: 46-47.

Thomson, D.J. 1995. The seasons, global temperature, and precession. *Science* **268**: 59-68.

Weertman, J. 1976. Milankovitch solar radiation variations and ice age ice sheet sizes. *Nature* **261**: 17-20.

Wigley, T.M.L. 1976. Spectral analysis and the astronomical theory of climatic change. *Nature* **264**: 629-631.

Geomagnetic Tides and Related Phenomena

Nils Olsen

Geophysics, Univ. Copenhagen, Juliane Maries vej 30, DK-Copenhagen, Denmark

Abstract. Daily variations of the geomagnetic field with amplitudes of about 50 nT at middle latitudes have been known for almost three centuries. Electrical currents in the conducting upper atmosphere are responsible for their main part. They are caused by tidal winds, which are excited by solar heating of the atmosphere as well as by gravitational forces of the moon. In addition, smaller contributions to magnetic variations arise from electrical currents in the magnetosphere, and from tidal currents in the oceans.

1 Historical Notes

Geomagnetic daily variations are of regular nature: they recur on each day in a similar manner. The phenomenon was described for the first time by Graham (1724) although presumably known even before. Graham observed in London a compass needle through the microscope and found regular, slow variations on some days, and much faster, irregular oscillations on other days. From this investigation he distinguished magnetically quiet and magnetically disturbed days. Celsius (1740) made similar observations in Uppsala, and because he classified almost the same days as respectively magnetically quiet and disturbed as Graham did in London, they concluded, that geomagnetic variations are rather a global than a local phenomenon.

A correlation of the amplitudes of geomagnetic daily variations with the eleven year solar cycle was soon noticed. This led Stewart (1883) to the suggestion that geomagnetic daily variations are produced by electric currents in the upper atmosphere. Wind systems caused by atmospheric tides – which were already well known – could be the source for these currents. Stewart suggested that the upper parts of the ionosphere are electrically conducting. The tidal winds would move conducting matter across the Earth's magnetic field and geomagnetic daily variations would then be observable at ground level as the magnetic field of the resulting electric currents.

To check this assumption, Schuster (1889) applied the method of spherical harmonic analysis of the magnetic field, developed by Gauß about 50 years earlier for analysis of the main field, to geomagnetic daily variations. This method enables a separation of the magnetic field into one part produced by currents outside the Earth and another part produced by currents inside the Earth. Schuster confirmed Stewart's suggestion of the external origin of geomagnetic daily variations. But in addition to the predominantly external part, he found a small internal part of about 30%. Lamb showed in an appendix to Schuster's work, that this internal part can be explained by means of electromagnetic induction in the conducting Earth. They concluded that the upper layers of the Earth have higher resistivity than the lower layers. In a later paper, Schuster (1908) developed ionospheric dynamo theory quantitatively.

In addition to geomagnetic variations with a period of 24 h, called solar variations (S-variations), there are small lunar variations (L-variations) with a period of 24.48 h. Kreil (1839) observed "an influence of the moon on the magnetic declination" at the magnetic observatory Milano, and Sabine (1853) confirmed this for several other observatories. The first spherical harmonic analysis of lunar variations was performed by Van Bemmelen (1912) who showed that – in agreement with solar variations – about 70% of the observed variation is due to external sources, and the remaining 30% of internal origin can be explained by electromagnetic induction.

Lunar variations are caused by gravitational forces of the moon. But such a gravitational excitation plays only a minor role for solar variations: due to the diurnal variation of solar irradiation which causes a heating of the atmosphere on the daylight side (respectively cooling on the nightside), thermally driven wind systems occur. They are much stronger than those driven by gravitational forces. Solar geomagnetic variations are therefore about 10 times stronger than lunar variations (although lunar gravitational forces are about twice as strong as solar gravitational forces).

2   Classification of geomagnetic daily variations

Classification of geomagnetic daily variations is not consistent in the literature, and often there is no distinction between a *descriptive, statistical* classification and a *causal, physical* classification.

The *statistical* classification originates from the dependence of geomagnetic daily variations on geomagnetic activity. Q-days are defined as the five quietest days of each month, whereas the five most disturbed days are denoted as D-days. Following the original definition (Chapman and Bartels 1940, for example), S_q is the mean daily variation estimated from Q-days only, and S_d is the mean daily variation at D-days. But besides this, the notation S_q is often used for denoting the magnetic daily variation caused by the ionospheric dynamo. This is a *physical* rather than a *statistical* definition. And sometimes the current system of the ionospheric dynamo is called S_q current system, although these currents occur not only on Q-days, but on all days.

In this paper, a physical classification will be used. In classifying atmospheric tides it is common to denote the solar pressure variations as tides although they are mainly produced by thermal and not by gravitational excitation. This led Chapman and Lindzen (1970) to distinguish *thermal tides* and *gravitational tides*. However, common for both is the periodic excitation which produces tidal wind systems and – by means of the ionospheric dynamo – *geomagnetic tides*.

But the observed *geomagnetic daily variation* is caused by other mechanisms, too, and goes beyond the usual definition of *geomagnetic tides*. Depending upon geomagnetic latitude, the following processes contribute to geomagnetic daily variations:

- At middle and low latitudes, the ionospheric dynamo is the main cause for daily variations; only these variations will be denoted as *geomagnetic tides*. Lunar variations are solely produced by the ionospheric dynamo. But in addition to the contribution from the ionospheric dynamo, there is a small contribution to solar variations from magnetospheric current systems. They produce a magnetic field variation at ground level with the same dependence on latitude, longitude and season as that of the ionospheric dynamo, but with only 20% to 30% of its amplitude.

- In the polar oval, at about $\pm 67°$ geomagnetic latitude, the polar electrojets (strong east-west directed currents in the ionosphere) produce magnetic distortions with a duration of about 1-3 h. These occur preferably during night and are called *polar substorms*. The cause for the currents are charged particles which move along open field-lines into the ionosphere. It is necessary to distinguish between S_q and S_d at polar latitudes. S_d, the variation estimated from a superposition of the variation at disturbed days, exists only in a statistical sense, but not as a pure variation at individual days. This is because S_d is determined by averaging over physical processes which occur sporadically opposite the regular nature of the field variation produced by the ionospheric dynamo. But polar substorms do not occur stochastically, and therefore they do not average out when applying the method of superposition and averaging over several days. Although this method is powerful in detecting hidden periodicities and has been successfully applied in non-polar latitudes, its wrong application causes problems: The method requires the presence of a ground state (to be found) which is superposed by *stochastic* distortions (which are to be eliminated). But the assumption of stochastic occurrence of the distortions brakes down at polar latitudes. This is often neglected, and leads to confusions about the "real nature" of S_d.

- An additional variation occurs in the polar caps – poleward of $\pm 75°$ geomagnetic latitude – and is denoted as S_q^p (respectively as *DP2* on disturbed days). This variation is caused by electric fields and currents of magnetospheric origin, which are projected along field-lines into the ionosphere even on magnetic quiet days. In contrast to the *dynamo action* at middle and low latitudes, where mechanical energy (tidal winds) is transformed into electrical energy and into heat, this process is a *motor*: electric energy of magnetospheric origin moves the ionospheric plasma (which is coupled to the neutral gas) and daily variations in the electric currents, magnetic field and ionospheric winds are hereby produced.

3 Observational results

Fig. 1 presents the variation of the D-component (positive eastward) at the observatory Fürstenfeldbruck (Germany) for four selected months. A daily variation is obvious, but the amplitudes during years of high solar activity (1958) are about twice as large and more disturbed compared to years of low solar activity (1964). However, not only the 11 year solar cycle leads to a modulation of the amplitudes: seasonal effects (larger amplitudes during summer than during winter) as well as the 27-day recurrence of solar activity can be found.

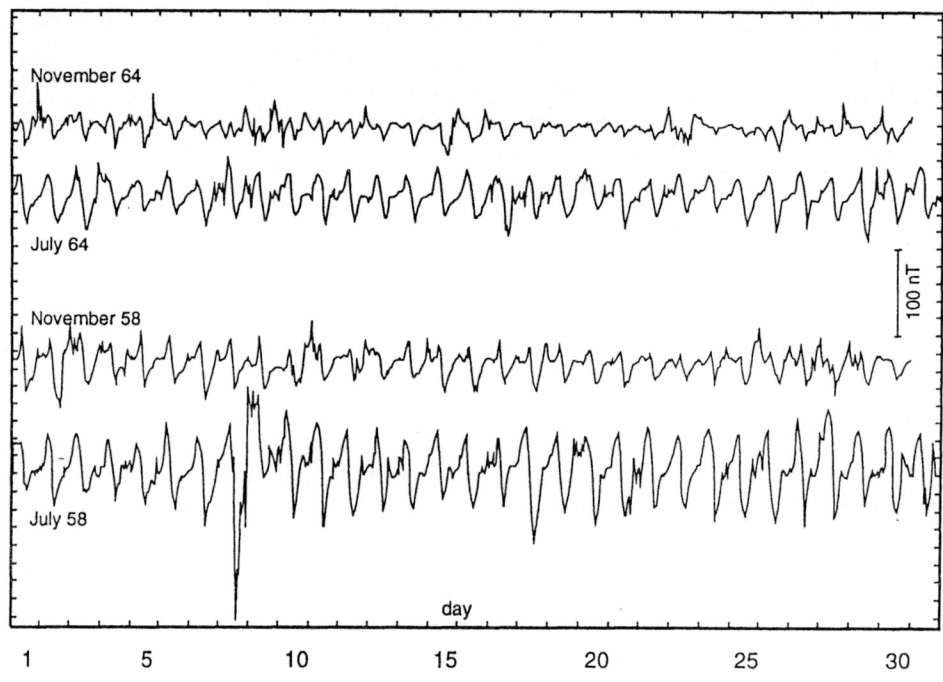

Fig. 1. D-component at Fürstenfeldbruck for July and November during a year of maximum solar activity (1958), respectively minimum solar activity (1964).

Fig. 2 shows this for the diurnal variation of the H-component at the observatory Tucson. The spectrum is estimated by Fourier-analysis of 60 years of hourly mean values. The largest peak is at 24 h and represents the diurnal variation S_1. The second largest peaks are seasonal modulations at frequencies $f = 1 \pm \frac{1}{365} d^{-1}, 1 \pm \frac{2}{365} d^{-1}, \ldots$. Modulations up to the 7th annual harmonic (52,1 days) are found. Even shorter periods are due to the rotation of the sun (one turn in about 27 days) which produces a recurrence of solar activity. Contrary to the pure periodicity of seasonal variations, the 27-day recurrence in solar activity is quasi-periodic, and therefore the spectrum contains no sharp spectral lines in this frequency range, but is "smeared out". Modulation with the 11 year solar cycle and its subharmonics concerns not only S_1, the "main period" at 24 h, but also its seasonal modulations; hence we have a "modulation of the modulation". This is best visible in the enlarged period range between 23,85 h and 24,15 h shown in the upper right part of the figure.

In an about 500 km wide band around the dip equator, the effective ionos-

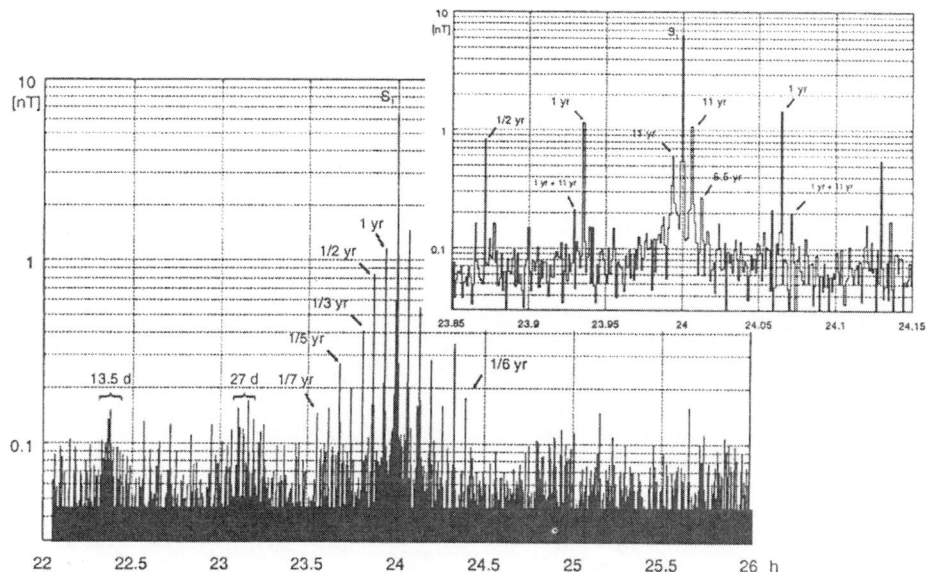

Fig. 2. Raw-spectrum of the H-component at Tucson, determined from hourly mean values of 60 years. The numbers denote the periods by which the diurnal variation S_1 is modulated.

pheric conductivity is enlarged due to the horizontal direction of the main field, which causes an amplification of the current system of the ionospheric dynamo, called the *Equatorial Electrojet*. Due to this amplification it is possible to observe lunar variations in the magnetogram of individual days, as can be seen from Fig. 3, which presents one month of the horizontal component at the observatory Huancayo in Peru. During night-time (hatched), the magnetic field is nearly constant, but there is a variability of the amplitudes during daytime, caused by lunar contributions. The observed variation (denoted by $S + L$) can be split up into a solar part S and a luni-solar part L. Both are suppressed during nighttime (between 18^h and 6^h local time) due to the drastically reduced ionospheric conductivity.

Four time harmonics are usually sufficient for representing the time-dependence of solar daily variations:

$$S(T) = \sum_{p=1}^{4} S_p = \mathrm{Re}\{\sum_{p=1}^{4} s_p e^{ipT}\}$$

T is local time measured in radians and s_p is the complex amplitude of the pth diurnal harmonic ($p = 1$ corresponds to diurnal variations, $p = 2$ corresponds to semi-diurnal variations, and so on).

The luni-solar variation contains mainly a semi-diurnal lunar contribution L_2 (with period 12,42 h), which is modulated by the (solar) daily variation of ionospheric conductivity. M_2 is the largest lunar excitation in the atmosphere;

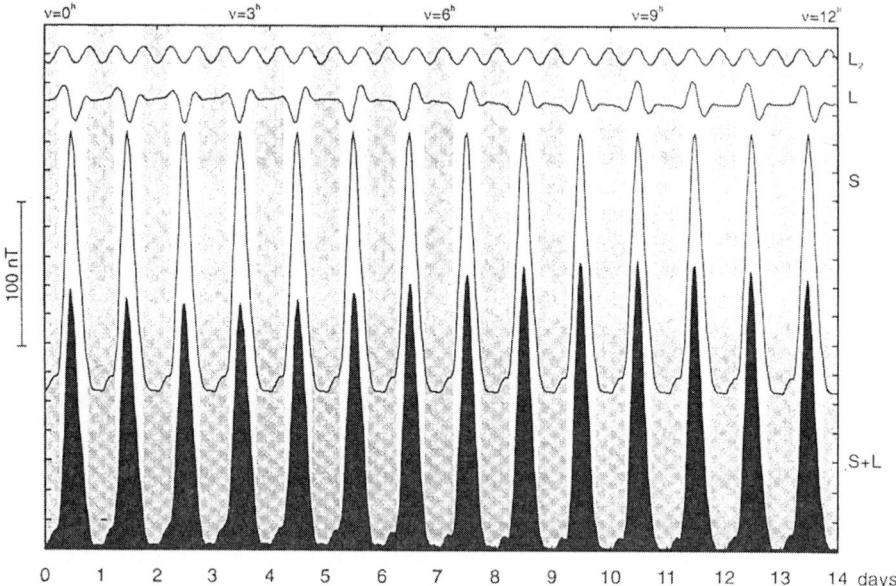

Fig. 3. H-component at Huancayo for 14 days during a year solar maximum (1957). Bottom to top: Sum of solar and lunar variation $(S+L)$; solar part (S), luni-solar part (L), and semidiurnal lunar part (L_2).

diurnal tidal waves are not able to propagate and are therefore negligible. It is a common practice in analyzing lunar geomagnetic variations to assume a pure semidiurnal lunar M_2-excitation which – different from the usual notations of the gravity potential (see Wenzel "Tide-Generating Potential", this volume) – is denoted as L_2. Time variation is given by $L_2(\tau) = \mathrm{Re}\{l_2 e^{i2\tau}\}$, with $\tau = T - \nu$ as lunar time, T as local solar time, and ν as lunar age ($\nu = 0$ during new moon, $\nu = \pi$ full moon, etc.) The multiplication of semidiurnal lunar tidal winds proportional to $\cos 2\tau$ with the ionospheric conductivity, whose time dependence is taken as $\sigma(T) = a_0 + a_1 \cos(T + \alpha_1) + a_2 \cos(2T + \alpha_2) + \ldots$ leads (if restricted to four time harmonics) to *Chapman's phase law*:

$$L(T,\nu) = \sum_{p=1}^{4} L_p = \mathrm{Re}\{\sum_{p=1}^{4} l_p e^{i(pT - 2\nu)}\}$$

Fig. 4 shows the lunar geomagnetic daily variation versus its latitudinal variation. The curves were synthesized from the result of a spherical harmonic analysis assuming that the variations depend only on local time (which means that observation sites of equal latitude, but different longitude observe similar variations in their respective local time). X, Y and Z are the magnetic field components to geographic north, east, and down, respectively. The 5 to 10 fold magnification due to the Equatorial Electrojet has not been taken into account, and hence the variations denoted with $0°$ latitude represent the conditions slightly away from the dip equator, for example at $\pm 5°$ dip latitude.

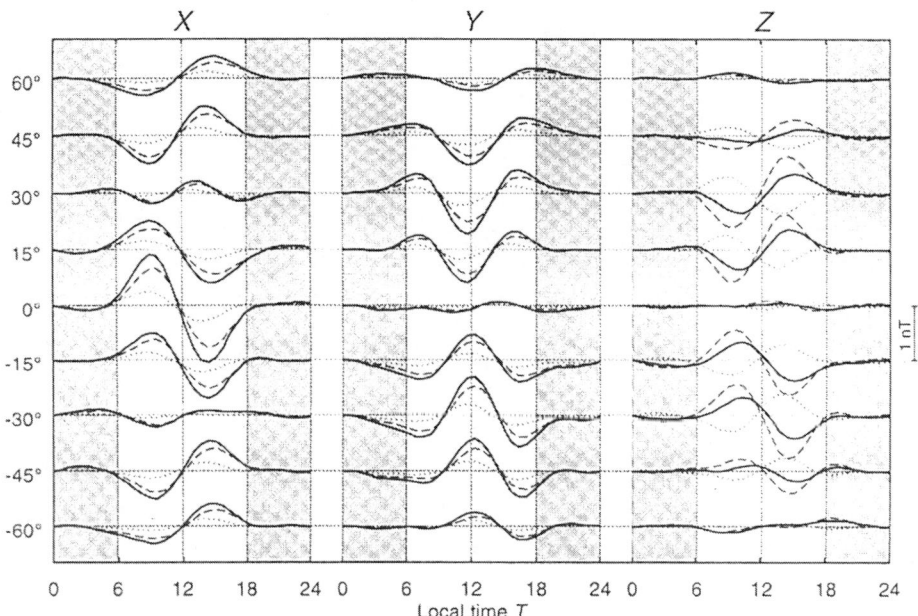

Fig. 4. Latitudinal dependence of L-variations during full moon and solar maximum (sunspot number $R = 100$). Total variation is shown by a solid line, external part is shown by a dashed line, and internal contribution is shown by a dotted line.

The observed variation at ground level contains primary, external contributions (mainly due to the ionospheric dynamo) and secondary, internal contributions (due to electromagnetic induction in the Earth). Both parts are superposed constructively in the horizontal components X and Y, but destructively in the vertical component Z. This difference enables a separation of external and internal contributions, as first developed by Gauß (1838). There is an obvious phase shift between internal and external parts: the internal, induced part lags the external, inducing part due to the electric resistivity of the Earth.

The latitudinal dependence of solar variations is shown in Fig. 5. Typical peak-to peak variations are 20-40 nT, and therefore solar variations are more than 10 times larger than lunar variations. Especially when considering the Y- and Z-component at low latitudes it becomes obvious that lunar variations contain a predominant semidiurnal variation, whereas a diurnal variation is predominant in solar variations. This is due to the period of the corresponding excitation: gravitational excitation (semidiurnal) for lunar, and mainly thermal forcing (diurnal) for solar variations.

Geomagnetic daily variations in the polar cap (poleward of 75° geomagnetic latitude) are totally different from those at middle and low latitudes. The depen-

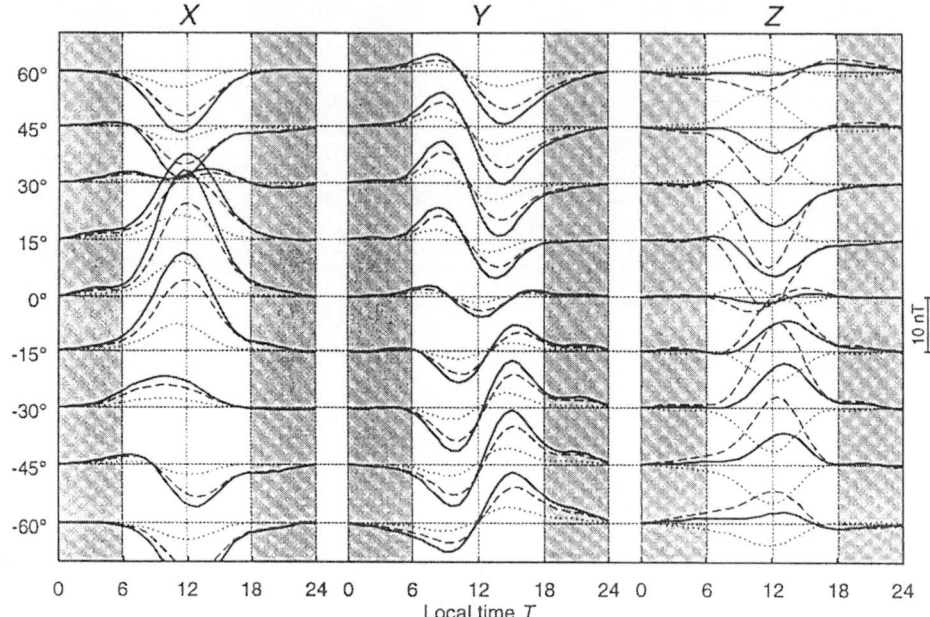

Fig. 5. Latitudinal dependence of S-variations during solar cycle maximum (sunspot number $R = 100$). Total variation is shown by a solid line, external part is shown by a dashed line, and internal contribution is shown by a dotted line.

dence on the direction of the interplanetary magnetic field (IMF) is perhaps the largest effect. Fig. 6 shows the daily variation at Resolute Bay (geomagnetic latitude 83°) estimated separately for days when the IMF is directed away from the sun, respectively directed towards the sun. In both components the dependence upon the IMF polarity is evident particularly between 12 and 24 UT, which is centered around local noon. This effect is so distinct even on individual days that it is possible to infer the direction of the IMF on a day by day basis for times when a direct measurements of the IMF with satellites was not available.

4   Geomagnetic Tides: The Ionospheric Dynamo

The ionospheric dynamo is the only cause for lunar variations and the main cause for the solar variations at middle and low latitudes. Geomagnetic daily variations caused by the ionospheric dynamo will be called *geomagnetic tides*. The physical principles will be sketched in short; the reader is referred to Kelley (1989), Richmond (1989) and Richmond (1995) for a more detailed description of the ionospheric dynamo.

Let us first look at the excitation mechanisms (cf. the scheme in Fig. 7). The rotation of the Earth in the gravitational field of the moon and sun as well as in the radiation field of the sun produces tidal winds in the whole atmosphere.

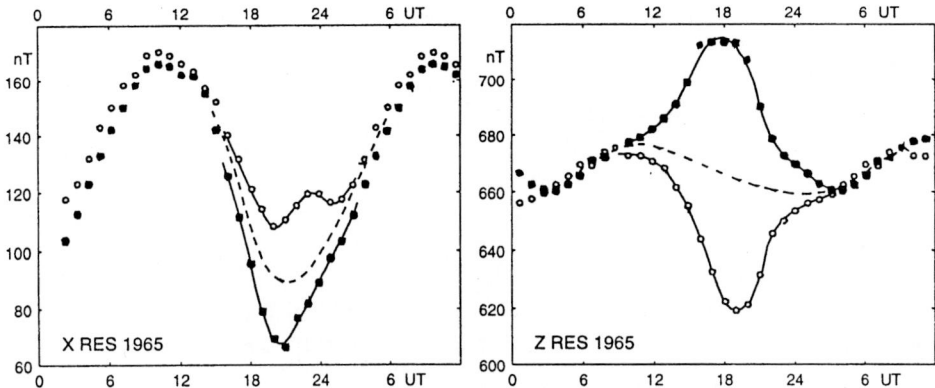

Fig. 6. Daily variation of horizontal component (left) and vertical component (right) at Resolute Bay for 1965. Open circles show the variation on days when the IMF is pointing away from the sun. Solid circles show the variation when the IMF is pointing towards the sun. The dashed curve is the average variation on all days irrespective of the IMF direction. After Svalgaard (1973).

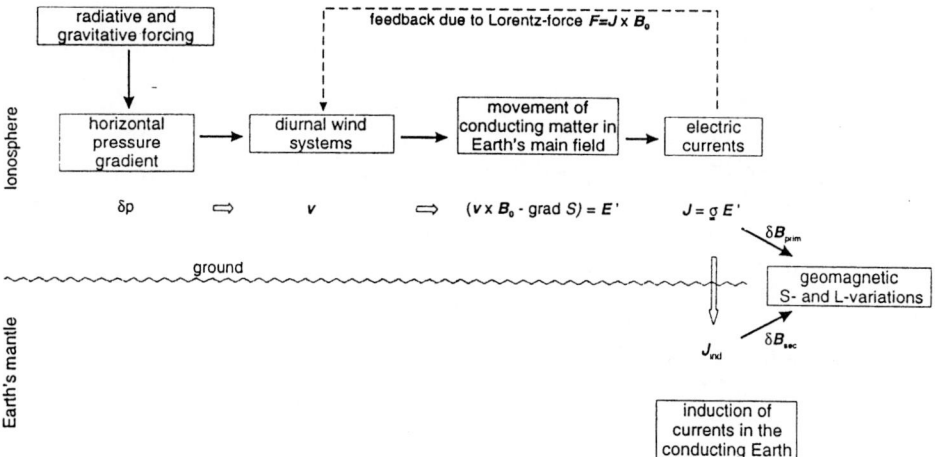

Fig. 7. Scheme of the ionospheric dynamo.

Because of the ionizing effect of solar UV radiation, the atmosphere above 80 km is electrically conducting, and tidal winds $\mathbf{v}$ move conducting matter across the field lines of the main field $\mathbf{B_0}$. This causes an electric field $\mathbf{v} \times \mathbf{B_0}$ as well as electric currents and magnetic field variations at ground.

In contrast to a self-exciting dynamo, as for example in the Earth's core, the ionospheric dynamo corresponds to a bicycle dynamo: following Stewart (1883), the moving atmosphere corresponds to the rotating anchor; the electrically conducting layers of the ionosphere corresponds to the dynamo coils, and the stationary magnet of the bike dynamo corresponds to the Earth's main field.

However, the conductivity in the ionosphere is highly anisotropic: it is much larger in the direction parallel to the main field than it is in the two directions perpendicular to it. This anisotropy is caused by the different behaviour of electrons and (positive) ions in a magnetic field and it is the key point for understanding the principle of the ionospheric dynamo. Although the number of ionized particles is much smaller than the number of neutral particles in the ionosphere (about $1 : 10^7$ in the ionospheric E-layer), the behaviour of the ionospheric plasma is strongly influenced by the charged particles. Electrons and ions cannot move freely in contrast to neutral particles, but are forced to circle in planes which are perpendicular to the main field lines with the frequency of gyration $\Omega_{i,e} = eB_0/m_{i,e}$. B_0 is the strength of the main field, e is the elementary charge, and m_i, m_e are the masses of ions and electrons, respectively.

The left part of Fig. 8 shows typical height profiles of the frequencies of gyration $\Omega_{i,e}$ and of the collision frequencies between neutral particles and ions (ν_i), respectively between neutral particles and electrons (ν_e). Both frequencies of collisions are higher than the respective frequencies of gyrations below 80 km. This means that the influence of the main field is weak and electrons as well as ions participate in the movement of the neutral component – no electric current is produced. On the other hand: both frequencies of gyration higher are than the respective frequencies of collision at heights above 130 km. This means that electrons and ions are strongly influenced by the main field, they are "stuck" to its field lines and cannot follow the movement of the neutral component. Again no current is produced. However, electrons and ions behave quite differently in the region between 80 and 130 km, due to $\Omega_i < \nu_i$, but $\Omega_e > \nu_e$: the ions follow the movement of the neutral component, whereas the electrons are bound to the field lines of the main field. This yields a charge separation and a flow of electric currents, and therefore that height region (hatched in Fig. 8) is called *dynamo layer*.

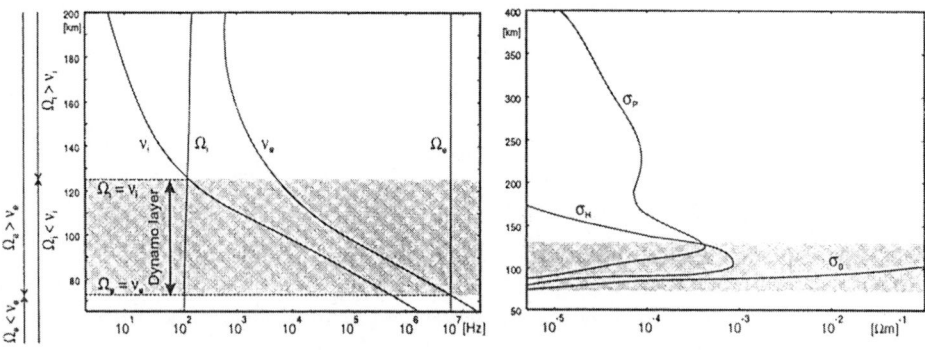

Fig. 8. Left side: Collision frequencies $\nu_{i,e}$ and frequencies of gyration $\Omega_{i,e}$ for ions and electrons, respectively. Right side: Parallel-conductivity σ_0, Pedersen-conductivity σ_1 and Hall-conductivity σ_2. Noon-values are shown, determined from the International Reference Ionosphere for low latitudes during moderate solar activity.

Ohm's law is the basic equation for a mathematical formulation of the ionospheric dynamo problem. It connects the current density $\mathbf{J}$, the conductivity σ and the electric field $\mathbf{E}'$, which is further divided into the Lorentz-field $\mathbf{v} \times \mathbf{B_0}$ and the polarization field $\mathbf{E}$:

$$\mathbf{J} = \underline{\sigma}\mathbf{E}' = \sigma_0 \mathbf{E}_\| + \sigma_P(\mathbf{E}_\perp + \mathbf{v} \times \mathbf{B}_0) + \sigma_H\, \mathbf{e_B} \times (\mathbf{E}_\perp + \mathbf{v} \times \mathbf{B}_0) \ . \tag{1}$$

σ_0 is the parallel conductivity, whereas σ_P and σ_H are Pedersen- and Hall-conductivity. $\mathbf{E}_\|$ and $\mathbf{E}_\perp$ are the components of the polarization field parallel and perpendicular to the main field $\mathbf{B}_0$, and $\mathbf{e_B}$ is a unit vector in the direction of $\mathbf{B}_0$. In deriving this equation, the magnetic field $\delta\mathbf{B}$ of geomagnetic daily variations caused by the current density $\mathbf{J}$ has been neglected compared to the main field $\mathbf{B}_0$ due to $|\delta\mathbf{B}| \approx 100$ nT $\ll |\mathbf{B_0}| \approx 50\,000$ nT.

The right part of Fig. 8 shows the height dependence of the conductivities, calculated from the collision frequencies shown in the left part of the figure. Below about 80 km, all three conductivities are negligibly small. The parallel-conductivity σ_0 is so much higher above 100 km, that usually $\sigma_0 = \infty$ is assumed in model calculations. Field lines of the magnetic main field are therefore equipotential lines, which leads to a coupling of the electrodynamic effects between the different ionospheric layers and between the northern and southern hemisphere. Pedersen-conductivity σ_P has a maximum at about 130 km; the Hall-conductivity σ_H becomes only important in the dynamo layer where it is about twice as large as the Pedersen conductivity.

The polarization field $\mathbf{E}$ is assumed to be curl-free and thus derivable from an electrostatic potential S:

$$\mathbf{E} = -\mathrm{grad}\ S \ . \tag{2}$$

Displacement currents can be neglected for periods larger than 1 s, as is the case for geomagnetic daily variations. This leads to

$$\mathrm{div}\ \mathbf{J} = 0. \tag{3}$$

In the existing literature, mostly a kinematic dynamo is treated, which means that the winds and conductivities are given and the feedback of the electric fields and currents upon the winds (dashed arrow in *Fig. 7*) is neglected. Substitution of Eqs. (2) and (1) into Eq. (3) yields a second order partial differential equation for the electrostatic potential S. Because of the high parallel conductivity, S is assumed to be constant along the field lines of the main field. It is therefore suitable to make the computations in a coordinate system where one coordinate points in the direction of the main field $\mathbf{B}_0$; often a dipole field is assumed for the main field and the equations are solved in dipole coordinates. An integration along the field lines reduces the problem to a two-dimensional differential equation, which is solved numerically, cf. Takeda and Maeda (1980) and Richmond and Roble (1987). Once S is known, the electric polarization field $\mathbf{E}$ is calculated from (2), and the current density $\mathbf{J}$ is determined from (1). In a final step, the magnetic variations $\delta\mathbf{B}$ produced by these current density are calculated.

Although $E_\parallel = 0$ is assumed, field-aligned currents $J_\parallel = \sigma_0 E_\parallel$ might be present because of $\sigma_0 = \infty$. A main field dipole axis tilted with respect to the axis of rotation, as well as a dynamo calculation for solstitial conditions yield field-aligned currents at middle latitudes of 2 to $5 \cdot 10^{-9}$ A/m², which is about two orders of magnitude lower than the horizontal current density. During the summer, current direction is from the northern to the southern hemisphere in the morning and in opposite direction in the evening. Such a current system has been proposed 30 years ago and was detected recently using satellite data.

Although there were substantial progresses during the last years in modelling the ionospheric dynamo, many problems, as for example the discrepancy between observed and calculated electric fields, are yet unsolved. The magnetic field describes always an integrated effect over all currents, and therefore a comparison of measured and calculated electric fields is a much stronger test than a comparison of the magnetic variation at ground. However, ground observations of the magnetic field are available over longer time periods and at more locations than observation of the ionospheric electric field and the magnetic ground variation is therefore much better known.

5   Geoelectric Tides: The Oceanic Dynamo

Finally, a mechanism is presented which is based on dynamo action, too. Although very small, this effect is an interesting extension of the usual interpretation of geomagnetic tides as of primary external origin.

As sketched in Fig. 7, external current systems induce secondary currents in the Earth's mantle; their magnetic field is observed at the Earth's surface as internal contributions. The corresponding geoelectric currents are well correlated with the magnetic variations of ionospheric origin.

In addition, there are geoelectric daily variations which are not correlated with the magnetic variation at the Earth's surface, because they are caused by tidal motion in the oceans. This phenomenon is yet only rudimentary analysed, mainly because of the tiny amplitudes.

Fig. 9 shows the D-component of the magnetic field as well as the North-component of the geoelectric field near Göttingen, Germany. The first five daily harmonics are clearly visible in the magnetic field, but a lunar semi-diurnal L_2 contribution is below the noise level. However, the geoelectric field contains a clear lunar L_2 signal. This difference indicates that the lunar part in the geoelectric field cannot be due to secondary, induced currents caused by the ionospheric dynamo.

If that part of the electric field, which is correlated with the magnetic field, is subtracted from the observed electric field, the power spectrum of the residual electric field (shown in the right part of the Figure) contains a predominant semidiurnal lunar variation and a weaker semidiurnal solar variation. They can be explained by means of tidal motion in the oceans (Junge 1988): In analogy to the ionosphere, tidal motion in the oceans moves conducting matter (seawater) across the field lines of the main field, and electric currents are produced. A

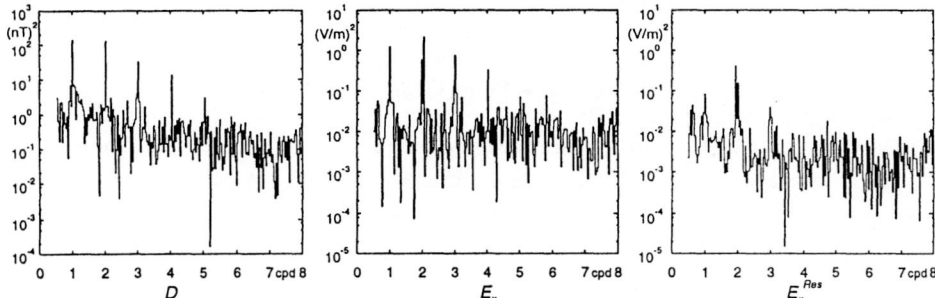

Fig. 9. Power spectrum of the magnetic East component (left) and of the geoelectric North component (middle), respectively, near Göttingen. The right part shows that part of the electric field, which is not correlated with the magnetic field. After Junge (1988)

fraction of these currents is closed on land and can be measured far from the coast (Göttingen is located about 300 km away from the North Sea). In the magnetic field, a lunar oceanic contribution is only observable near coastlines (for example Malin (1970)), because the magnetic field decreases with increasing distance from the coast in a stronger fashion than the electric field. Therefore the influence of the oceanic dynamo at locations like Göttingen can only be found in the geoelectric field, but not in the geomagnetic field.

References

Celsius, A. 1740. Bemerkungen über der Magnetnadel stündliche Veränderungen in ihrer Abweichung. *Svenska Vet. Acad. Handl.* 296 - 299.

Chapman, S. and Bartels, J. 1940. Geomagnetism, vol. I+II. Clarendon Press, Oxford.

Chapman, S. and Lindzen, R. S. 1970. Atmospheric tides, thermal and gravitational. Reidel Publ. Comp., Dordrecht.

Gauß, C. F. 1838. Allgemeine Theorie des Erdmagnetismus, Resultate aus den Beobachtungen des magnetischen Vereins im Jahre 1838. Göttingen & Leipzig.

Graham, G. 1724. An Account of Observations Made of the Variation of the Horizontal Needle at London in the Latter Part of the Year 1722 and Beginning 1723. *Phil. Trans. Roy. Soc. London.* **383**: 96 - 107.

Junge, A. 1988. The telluric field in northern Germany induced by tidal motion in the North Sea. *Geophys. J.* **95**: 523 - 533.

Kelley, M. C. 1989. The Earth's Ionosphere. Plasma Physics and Electrodynamics. *Int. Geophysical Series.* **43**. Academic Press.

Kreil, K. 1839. Resultate dreijähriger magnetischer Beobachtungen zu Mailand und Einfluss des Mondes darauf. *Ann. Physik.* **46**: 443.

Malin, S. R. C. 1970. Separation of lunar daily variations into parts of ionospheric and oceanic origin. Geophys. J. R. astr. Soc.. **21**: 447 - 455.

Richmond, A. D. 1989. Modeling the Ionospheric Wind Dynamo: A Review. In: Quiet Daily Geomagnetic Fields. W. H. Campbell (ed.). Birkhäuser Verlag, Basel.

Richmond, A. D. 1995. Ionospheric Electrodynamics. In: Handbook of Atmospheric Electrodynamics, Vol. 2.. H. Volland (ed.). CRC Press, Boca Raton.

Richmond, A. D. and Roble, R. G. 1987. Electrodynamic Effects of Thermospheric Winds from the NCAR Thermospheric General Circulation Model. *J. geophys. Res.*. **92**: 12365 - 12376.

Sabine, E. 1853. On the Influence of the Moon on the Magnetic Declination at Toronto, St. Helena and Hobarton. *Phil. Trans. Roy. Soc. London.* **A143**: 549 - 560.

Schuster, A. 1889. The Diurnal Variation of Terrestrial Magnetism. *Phil. Trans. Roy. Soc. London.* **A180**: 467 - 518.

Schuster, A. 1908. The Diurnal Variation of Terrestrial Magnetism. *Phil. Trans. Roy. Soc. London.* **A208**: 163 - 204.

Stewart, B. 1883. Hypothetical Views Regarding the Connection between the State of the Sun and Terrestrial Magnetism. *Encyclopedia Britannica.* **16**: 181 - 184.

Svalgaard, L. 1973. Polar cap magnetic variations and their relationship with the interplanetary magnetic sector structure. *J. geophys. Res.* **78**: 2064 - 2078.

Takeda, M. and Maeda, H. 1980. Three-Dimensional Structure of Ionospheric Currents − 1. Currents Caused by Diurnal Tidal Winds. *J. geophys. Res.* **85**: 6895 - 6899.

Van Bemmelen, W. 1912. Die lunare Variation des Erdmagnetismus. *Meteorol. Z.* **29**: 218 - 225.

Tidally Induced Phenomena

Tides in Water Saturated Rock

Hans-Joachim Kümpel

Geological Institute - Section Applied Geophysics, University of Bonn
Nussallee 8, D-53115 Bonn

Abstract. Analysis of water table records from wells or boreholes often reveals the presence of tidal fluctuations. Amplitudes of 'well tides' can attain several centimeters when the well or borehole is open to a confined aquifer. The phenomenon reflects extension and compression cycles of the aquifer rock, i.e. volume strain tides of a water saturated formation. Besides tidal fluctuations, barometric pressure changes and pure loading effects may be observed in well level records. If the forcing functions are known, these signals can be used to constrain petrohydraulic aquifer parameters of the connected formations. Linear poroelasticity is the commonly used rheology to describe the underlying physical process; or elastic deformation of fluid filled fractures in case of low permeable rock with a few fractures that are open to the well.

1 Introduction

Tidal fluctuations of the ground water level are frequently observed phenomena in confined or sufficiently deep aquifers. Sometimes they are simply termed 'well tides', because a well is needed to observe the fluctuations. Likewise, the flow of artesian springs or the production rate of underground reservoir fluids may vary with tidal periods. Since the amplitudes of well tides are usually of centimeter magnitude, monitoring does not require sophisticated instruments.

Figures 1 to 3 show some examples. Figure 1 is a 20 days recording of the water level in the 4 km deep pilot hole of the Continental Deep Drilling Programme, KTB, Bavaria. The borehole is cased except for the lower 150 m, where it is in contact with a metamorphic, amphibolitic formation. Tidal variations have peak to peak amplitudes up to 12 cm, nontidal variations are mainly of barometric origin (Endom and Kümpel, 1994). Figure 2a is a 20 days recording obtained in a 36 m deep well near the village of Dokurçun, North-Anatolia, Turkey. The well is open to water saturated sands from 32 to 34 m depth (Büyükköse *et al.*, 1989). The tidal signal is masked by barometric effects, but reveals peak to peak amplitudes of 0.25 cm after nontidal effects have been reduced through filtering (Fig. 2b), compare Westerhaus (Tidal Tilt Modification, this volume). The presence of tidal constituents O_1 and M_2 in a well level signal is a good indicator for the confinement of an aquifer (i.e. that the aquifer is fully submerged in the saturated groundwater regime). Figure 3 displays the spectrum of level fluctuations in a 250 m deep well drilled in a sequence of Deccan trap basalts near Koyna, Maharashtra, India. The well is uncased for the lower 218 m and reveals also the presence of solar constituents S_1, S_2 with modulations S_3, S_4 caused by thermal and/or barometric effects.

Debating the physical process of well tides means that rocks have to be considered as porous media. Compact rocks of (almost) zero porosity do also undergo tidal deformation, but since wells drilled into such formations will be dry, they are of no interest here. Full water saturation - or saturation with any kind of liquid - is equally important for well tides to develop, because the compressibility of gas as the potentially additional pore filling is orders of magnitudes higher than that of liquids; if gas is present, compression of pores caused by a change in confining presssure is mostly absorbed by volume reduction of the gaseous fraction, hence will result in a heavily attenuated pore fluid flow, if at all.

The most elementary rheology dealing with small deformations in porous, water saturated rock is linear poroelasticity. Since it is not widely applied yet, a short introduction is given here. The characteristics of well level changes due to tidal forces, but also surface loading and barometric pressure variations are presented next. This appears necessary, because tides are generally not an isolated signal in well level data. Schemes to deduce rock parameters from observed well tides, under simplifying assumptions, are compiled in the concluding section.

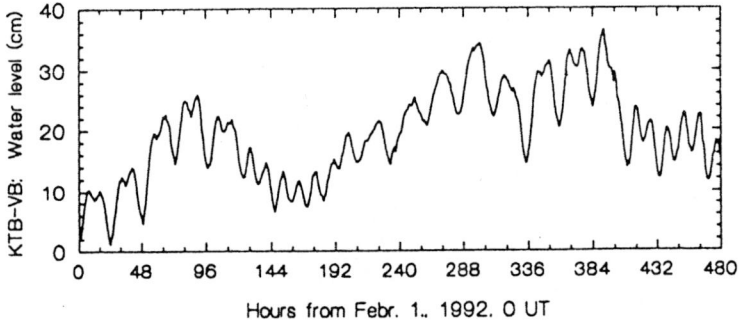

Fig. 1: Raw data of water level fluctuations in 4.0 km deep KTB pilot hole obtained by W. Kessels, J. Kück, and M. Sowa (pers. communication). Tidal constituents O_1, K_1, and M_2 have 2 to 3 cm amplitudes (Endom and Kümpel, 1994). Arbitrary reference level.

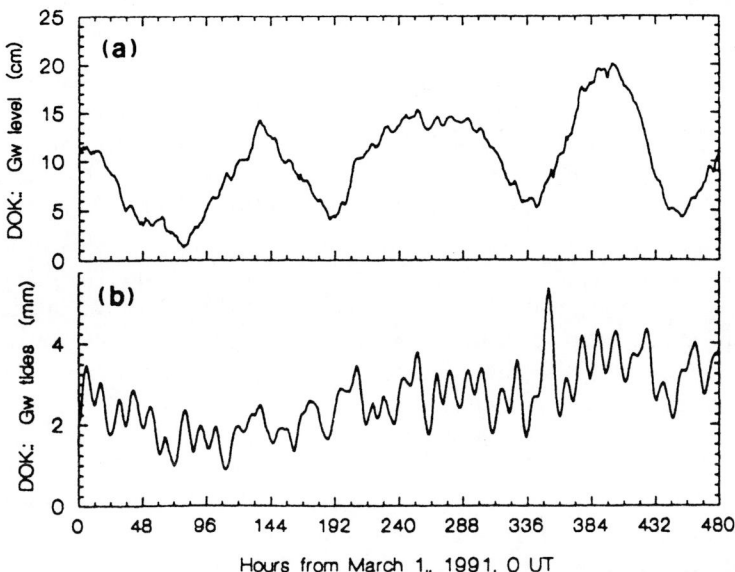

Fig.2: **a** Raw data of well level reflecting barometric effects in the well and pore pressure variations in ground water between 32 and 34 m depth at a test site near Dokurçun, North-Anatolia, Turkey, obtained by M. Westerhaus, W. Welle, and J. Zschau (pers. communication).   **b** Same data after application of tide enhancing band pass filter. Note that tidal amplitudes are two orders of magnitude smaller than in Fig. 1. Arbitrary reference levels.

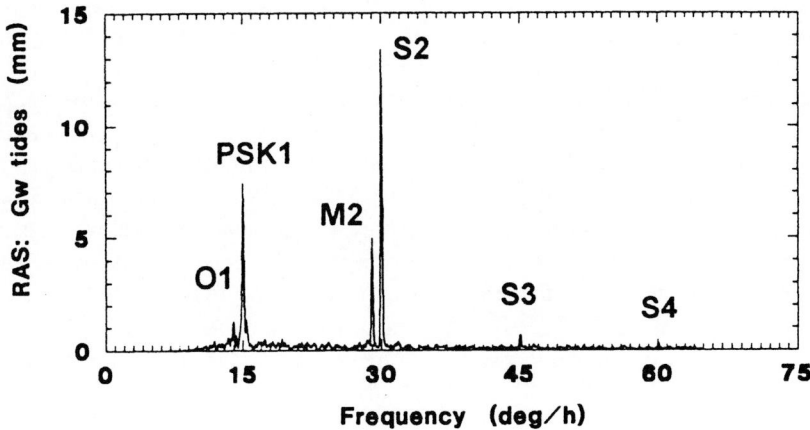

Fig.3: Amplitude spectrum of 5 months of continuous well level recording in a 250 m deep well near Koyna, Maharashtra, India (Well 'Rasati', 24.07.-27.12.95; Gupta *et al.*, 1996). Computation with ETERNA 3.0 of Wenzel (1993; bandpass filtered).

2 Poroelasticity

Fundamental contributions to the theory of linear poroelasticity have been derived by Terzaghi (1923), Biot (1941), and Rice and Cleary (1976). Terzaghi presented solutions for one-dimensional compression, Biot extended the theory to cases of three-dimensional deformation, and Rice and Cleary reformulated Biot's theory in that the mechanical parameters of porous media become more easily accessible through laboratory tests.

Linear poroelasticity is an extension of linear elasticity with the pore pressure as a new variable, in addition to the displacement variables of a rock particle. The role of pore pressure is very similar to that of temperature in thermoelasticity. For many applications, however, coupling between pore pressure and deformation is much stronger than coupling between temperature and deformation: Let us consider a rock sample that is internally heated, e.g. by a point-like thermoelement. The resulting thermoelastic deformation may be significant, comparable perhaps to the poroelastic deformation caused by a point-like injection of a certain quantity of pore fluid. A change in confining pressure applied to the surface of that rock sample, on the other hand, may have no significant effect on the rock's temperature, whereas the pore pressure inside the (water saturated) rock sample will change considerably. Or, stress induced pore pressure change usually dominates over adiabatic temperature change. That is why many of the analytical solutions known in thermoelasticity can not be applied to poroelastic problems. Closed solutions do only exist for very elementary problems, e.g. for a point source in the full space.

In Terzaghi's and Biot's classical theories of poroelasticity, the rock matrix with its pores is treated as a macroscopically uniform medium. Similar to the coefficient of thermal expansion, the thermal conductivity and the heat capacity in thermoeleasticity, the coefficient of *effective stress*, α, the *hydraulic diffusivity*, D, and the *storage compressibility*, S, are characteristic 'effective' rock parameters here. Porosity is an implicit quantity only, reflected in an expression of various compressibility parameters (see below).

It is necessary to distinguish between drained and undrained situations. Accordingly, elastic rock parameters like the Young's modulus, the compressibility, and the Poisson ratio are defined for drained and undrained conditions (or unconfined and confined, respectively); except the shear modulus is identical for both (Biot, 1941). Since drainage is a spatio-temporal diffusion process, appropriate usage of drained or undrained parameters in a particular case depends on (a) the hydraulic diffusivity, (b) the distance Δx at which excess pore pressure can be absorbed (e.g. in a sink for the pore fluid, like the free water table), and (c) the time characteristic of the forcing function. Roeloffs (1996) suggests to use the dimensionless quantity $\omega (\Delta x)^2/D$ as a measure of confinement, with ω being the (dominant) frequency of a strain signal. For $\omega (\Delta x)^2/D < 0.2$, the induced pore pressure disturbance is less than $1/e$ of its value in a perfectly confined aquifer.

The coefficient of effective stress, α, equals $1-c_s/c$, where c_s is the *compressibility of the solid constituents* forming the rock matrix, and $c > c_s$ is the *matrix* or *drained rock compressibility* (Nur and Byerlee, 1971). The notion c_u is used for the *undrained formation compressibility*. The hydraulic diffusivity D is the coefficient in the diffusion equation of the term $P-BP_c$, with P and P_c being the absolute changes in pore and confining pressure, respectively (Rice and Cleary, 1976). $B = (\partial P/\partial P_c)_u$ denotes the *Skempton ratio*, i.e. the change in pore pressure per unit change in confining pressure for undrained media, and, like α, takes values between 0 and 1; for unsaturated rock, B is close to 0. Brown and Korringa (1975) showed that $B = (c-c_u)/(c-c_s)$. D itself depends on various rock parameters, so on B, on the *Poisson ratios*, v and v_u, for drained and undrained conditions, the rock's *shear modulus*, G, and, most important for practical aspects, on the *Darcy conductivity*, κ, which is intrinsic matrix permeability over dynamic pore fluid viscosity (in length^3 × time × mass^{-1}). The corresponding relation, derived by Rice and Cleary (1976), reads $D = [2(1-v)(1+v_u)^2\kappa GB^2]/[9(1-v_u)(v_u-v)]$. Note that always $0 \leq v \leq v_u \leq 0.5$.

When rock deformation is neglected, compensation of a pore pressure imbalance reduces to pure pore pressure diffusion. This approach is usually adopted in hydrology. Then, D equals the ratio of aquifer *transmissivity* (in units length^2 × time^{-1}) over the *dimensionless storage coefficient* after Theis (1935), or the ratio of *hydraulic conductivity* (length × time^{-1}) over the *specific storage* (length $^{-1}$) . In the literature, the storage coefficient is used in various forms. When noted in units of pressure^{-1} it would better be termed *storage compressibility*. For three-dimensional deformation, the storage compressibility, S, can be shown to equal $(c- c_s)/B$. More detailed explanations of poroelastic rock parameters may be found in Kümpel (1991) and Wang (1993).

The governing equations for three-dimensional poroelastic deformation are obtained by combining (a) Darcy's law for laminar fluid flow with (b) Hooke's generalized linear law for elasticity, extended for a pore pressure term, and (c) the equilibrium conditions as used in continuum mechanics. They take the form

$$G \nabla^2 u_i + \frac{G}{1-2v} \frac{\partial \Delta_V}{\partial x_i} = \alpha \frac{\partial P}{\partial x_i} \tag{1}$$

$$(S-\alpha^2 c)\frac{\partial P}{\partial t} + \alpha\frac{\partial \Delta_V}{\partial t} = \kappa \nabla^2 P \tag{2}$$

with x_i = coordinate in an orthogonal, spatially fixed system (i =1,2,3), ∇^2 = Laplacian operator, t = time, u_i = displacement of matrix particle parallel to x_i , and Δ_V = volume dilatation. The same equations hold for plane strain conditions (i = 1,2). In the vicinity of cylindrical sources like wells, it is convenient to choose an axi-symmetric system.

Then, Eq.(1) turns to

$$G\left(\nabla^2 u_r - \frac{u_r}{r^2}\right) + \frac{G}{1-2\nu}\frac{\partial\Delta_V}{\partial r} = \alpha\frac{\partial P}{\partial r} \qquad (3a)$$

$$G\,\nabla^2 u_z + \frac{G}{1-2\nu}\frac{\partial\Delta_V}{\partial z} = \alpha\frac{\partial P}{\partial z} \qquad (3b)$$

with r, z and u_r, u_z being the radial and axi-parallel coordinates and displacements, respectively. Equation (2) remains unchanged. Rice and Cleary (1976) noted the instructive relation $\Delta_V = c(\alpha P - P_c)$ with the confining pressure $P_c = -(\sigma_{11} + \sigma_{22} + \sigma_{33})/3$. Compressive stress, σ_{ii}, is negative in this notation and has opposite sign than pore or confining pressure.

3 Natural Strain in Aquifers

Why do well level fluctuations reflect redistributions of stress and pore pressure in a rock formation connected to the water column in the well? The well screen (open section of the well) allows pore fluid to enter the well, or well water to enter the formation. In general, the well screen is placed at a depth of relatively high permeability, meaning that the well is in contact with an aquifer. If the flow resistance through the well screen can be neglected, the hydrostatic pressure of the water column in the well equals the pore pressure in the connected formation. Thus, a well can be termed a simple 'pore pressure meter' or, because of the coupling between pore pressure and rock matrix deformation, a volume strainmeter (Bodvarsson, 1970). Depending mainly on the formation's permeability, on the persistance of a pore pressure disturbance, and on the time that passed since it occurred, the balance between pore pressure and well level height holds for smaller or larger rock volumes in the vicinity of the well screen.

To understand why the response of an aquifer to stress changes may be attenuated, and why it may be delayed, let us consider the hydraulic conditions in the surroundings of a well. Figure 4 displays five types of characteristic diffusivities (analogous to Rojstaczer, 1988a):
- D_1 is the diffusivity for vertical flow in the refilled annular space, outside the well casing and the permeable screen. A well should have gravel or coarse sand alongside the screen, and clay barriers atop and beneath, so that fluid flow is radial across this zone and D_1 rather low. Without such barriers, vertical flow in the annular space can be substantial, reducing the strain sensitivity of the well.
- D_2 is the diffusivity of the intact formation that is in close contact with the open part of the well. The diffusivities of the well screen and of the gravel alongside are usually much

higher than D_2. Is the length of the well screen less than the aquifer thickness, flow to the well has a vertical component in that formation.

- D_3 is the effective diffusivity for vertical flow in the original formation, from the well screen to the ground water table. If D_3 is similar to D_2, the aquifer is unconfined; it is confined if $D_3 << D_2$ and semiconfined if $D_3 < D_2$. (Since in natural rocks the hydraulic diffusivity, like the permeability, takes values from a wide range of magnitudes, '<' means generally one order of magnitude less and '<<' two orders or more.) For wells in unconfined aquifers, the free ground water table - not the well itself - is the dominant sink for excess pore pressure. Strain signals are therefore strongly attenuated.
- D_4 is the diffusivity in the unsaturated formation atop the ground water table where pores contain mainly gas. As gas does not resist the flow of water, the water table is usually regarded as a zero pore pressure boundary in poroelasticity.
- D_5, finally, is the bulk hydraulic diffusivity for flow to the nearest lateral sink. This could be a spring, artesian well, river, or a drained formation of higher diffusivity than D_2.

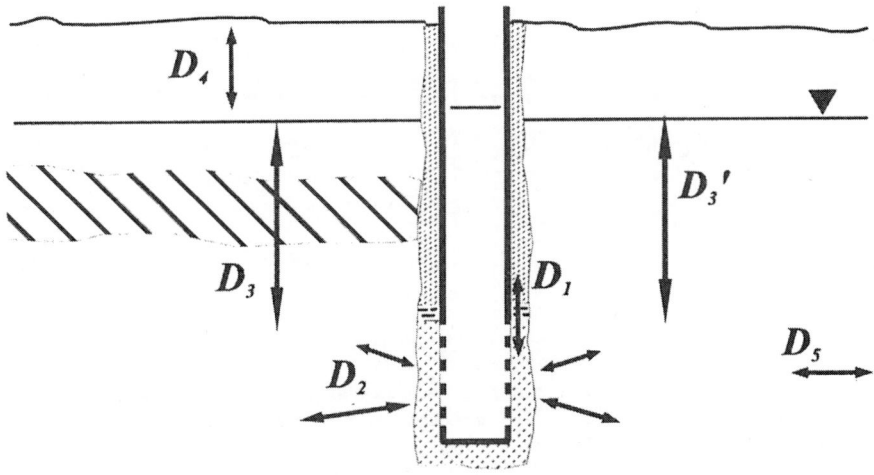

Fig. 4: Types of hydraulic diffusivity D controlling fluid flow conditions in the vicinity of a well. D_1: vertical flow in refilled annular space outside well casing and screen; D_2: flow in aquifer in vicinity of the well screen; D_3, D_3' : vertical flow to free ground water table; left: confined aquifer, sealed by low permeable aquiclude; right: unconfined aquifer; D_4: flow in unsaturated top layers; D_5: flow to lateral sinks.

Obviously, only when D_1, D_3 and D_5 are much smaller than D_2, i.e. when the well is the dominant sink for induced excess pore pressure in the formation, can the well level effectively reflect strain changes in the aquifer. Supposing that is given, three types of well level phenomena can be distinguished. Considering first the static reactions these are, following the elaboration of Rojstaczer and Agnew (1989):

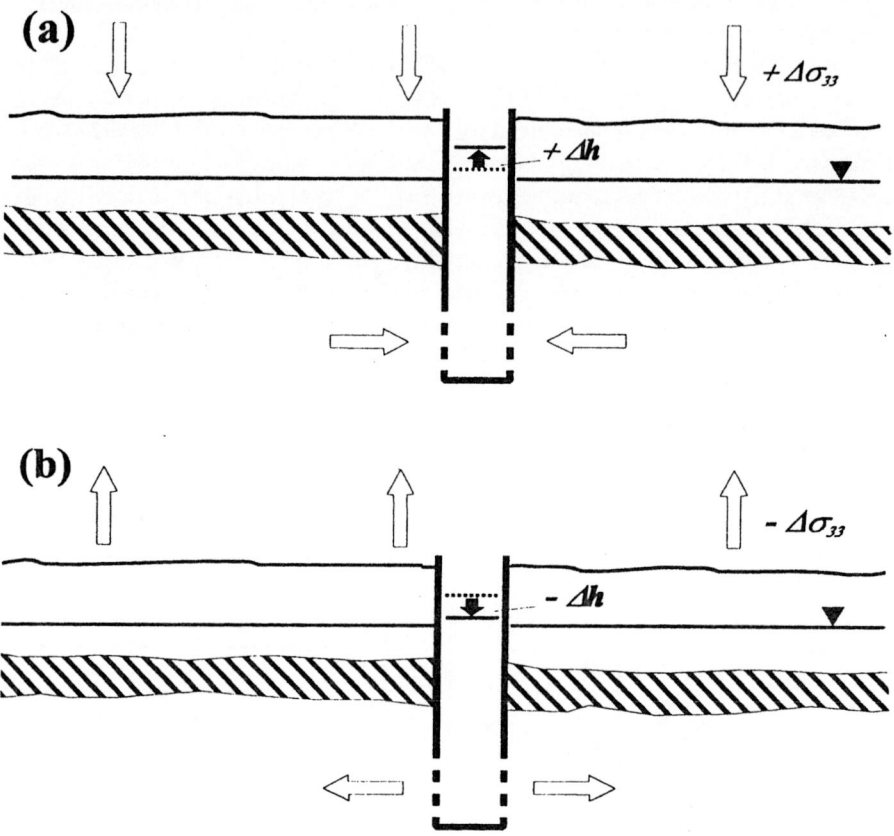

Fig. 5: Loading efficiency of confined aquifers. **a** Pore pressure increase induced by a surface load $\Delta\sigma_{33}$ forces the well level to rise by amount Δh. **b** Unloading with poroelastic rock inflation forces the well water back into the aquifer. The process occurs with atmospheric loading, precipitation, or with marine tidal loading in coastal, offshore areas.

The static loading response. If the surface is loaded by a uniform, extended mass (Fig. 5), the rock matrix transfers the stress change to depth with no delay (in fact, with seismic velocity). Accordingly, the pore pressure in the aquifer rises by an amount ΔP, where for a surface load $\Delta\sigma_{33}$ we have $\Delta P = L_e\Delta\sigma_{33} = g\rho\Delta h$ with the *loading efficiency*

$$L_e = \frac{B(1+H)(1+v_u)}{3[1-(1-H)v_u]} \qquad (4)$$

(g – gravitational acceleration, ρ – density of the pore fluid, Δh – induced well level rise). H, a parameter introduced by Rojstaczer and Agnew (1989), can be understood as the ratio of areal to vertical strain for a finite, though potentially big lateral extension of the load. It takes a value between 0 and 1, where $H=0$ is the traditional assumption in hydrology (vertical strains only) and $H=1$ holds for the half-space solution of finite extended loads.

The static tidal response. With the surface stress free ($\sigma_{33} = 0$), tidal strain can be equated to areal strain $\varepsilon_a = \partial u_1/\partial x_1 + \partial u_2/\partial x_2$. The coefficient T_e in $\Delta P = T_e\, \varepsilon_a / c_u = g\rho\Delta h$ is called *tidal efficiency* and can be shown (Fig. 6) to be

$$T_e = \frac{1-2\nu_u}{1-\nu_u}\, B \qquad\qquad (5)$$

Well sensitivity to tectonic strain changes can be treated in the same way, the difference being that the wavelengths of strain signals are regional rather than global, as for tides. Onshore strain from marine tides in adjacent seas is again a regional phenomenon.

(a)

(b)

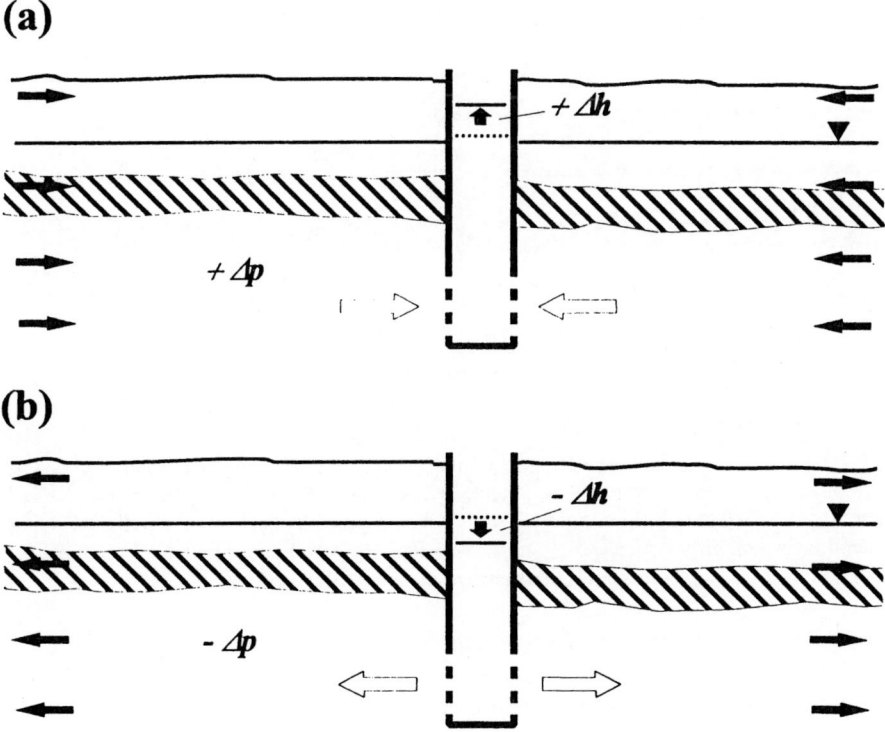

Fig. 6: Tidal efficiency of confined aquifers. **a** Pore pressure rise ΔP during compressional phase forces fluid into the well. **b** Inverse process during dilatational phase. Phase lags may result from moderate diffusivities D_2 (see transient tidal response).

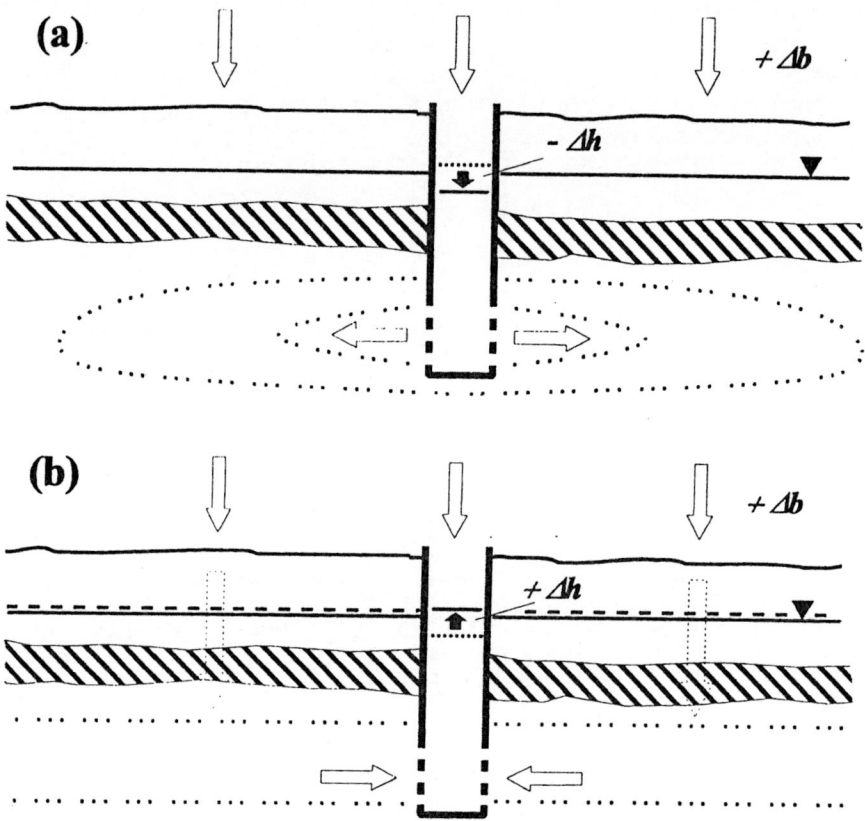

Fig. 7: Barometric efficiency of confined aquifers - without effect from atmospheric loading. **a** Barometric pressure rise Δb in the well forces well water into the aquifer. **b** Pore pressure rise in aquifer due to delayed diffusion from the surface forces pore fluid back into the well. Delay times depend mainly on diffusivities D_3, D_4 and D_5 (see Fig. 4). Dotted lines in aquifer sketch possible isobars of increased pore pressure. The barometric loading effect adds to the situation above.

The static barometric response. This is defined for wells that are open to the atmosphere (usually all non-artesian wells). A change Δb in barometric pressure acts twofold: as a load $\Delta \sigma_{33}$ on the surface (Fig. 5), and as a load on the water column in the well (Fig. 7a), but only the latter is fully and immediately effective at the well screen. Without the surface loading effect, the net well level change is $\Delta h = -\Delta b/g\rho$, also called the *inverse barometric effect*. Together with Eq. (4), the bulk effect or *barometric efficiency* $A_e = -g\rho\Delta h/\Delta b$ is

$$A_e = 1-L_e \qquad\qquad (6)$$

After some time, the incremental barometric pressure has diffused through the confining top layers, rising the pore pressure in the aquifer by Δb. Now, pore fluid is expelled back into the well (Fig. 7b), and only the loading effect remains. Accordingly, the part of the barometric effect that depends on pore pressure diffusion from the surface to the aquifer is transient.

The transient tidal response. Tidal variations are superpositions of harmonics. This facilitates formulation of a closed form of their appearance in well level signals. Following Hsieh *et al.* (1987), a periodic pore pressure change $\Delta P = \Delta P_0 \exp(i\omega t)$ in the connected aquifer causes a well level change $\Delta h = \Delta P/[g\rho(E+iF)]$ with

$$E = 1-\beta \; [\Psi \; ker(\eth) + \Phi \; kei(\eth)] \tag{7a}$$

$$F = \beta \; [\Phi \; ker(\eth) - \Psi \; kei(\eth)] \tag{7b}$$

$$\Phi = \frac{-[ker_1(\eth) + kei_1(\eth)]}{2^{1/2} \; \eth \; [ker_1^2(\eth) + kei_1^2(\eth)]} \tag{8a}$$

$$\Psi = \frac{-[ker_1(\eth) - kei_1(\eth)]}{2^{1/2} \; \eth \; [ker_1^2(\eth) + kei_1^2(\eth)]} \tag{8b}$$

Herein, *ker, kei, ker₁* and *kei₁* are Kelvin functions of orders 0 and 1 (Abramowitz and Stegun, 1964); β, $\eth$ are dimensionless terms of the radian frequency ω, the Darcy conductivity κ, the storage compressibility S, the radii r_c, r_w of the well casing and the screen, and of the area a of the screen, namely,

$$\beta = \frac{\pi \; \omega \; r_c^2 \; r_w}{a \; g \; \rho \; \kappa} \tag{9}$$

$$\eth = (\omega \; S/\kappa)^{1/2} \; r_w \tag{10}$$

Note that $g\rho\kappa$ in Eq.(9) is the frequently used hydraulic conductivity. If ΔP and the well geometry are known, and horizontal flow in the aquifer is prevailing - a condition assumed in deriving Eqs.(7)-(10) - κ and S can be obtained from those equations (see also Roeloffs *et al.*, 1989).

4 Rock Parameters from Well Level Signals

The main practical purpose of analysing well tides is to estimate rock parameters of the connected aquifer. Both, phase lags of the pore fluid flow and attenuation of stress or pressure changes depend critically on the diffusivities in a well's surroundings. Since the flow of pore water takes time, a well's strain sensitivity may be reduced to a certain frequency band (Rojstaczer, 1988a,b). Two parameter sets can be approached:

Mechanical poroelastic rock parameters, e.g. a combination of B and ν_u , when values for H or c_s are adopted. The latter may be guessed from the composition of the rock matrix, or from seismic P- and S- wave velocities and the rock's density. From Eqs. (4)-(6) is obvious that $0 \leq T_e \leq B$ and $1-B \leq A_e \leq 1-2B/3$, $(1-B/3)$ in case of $H=1$, (0).

The lower bounds hold for incompressible undrained media ($\nu_u =0.5$), the upper for infinitely compressible ones ($\nu_u =0$). Typical values of ν_u for rocks range from 0.3 to 0.45 (higher values for less consolidated rocks), so that $B/5 \leq T_e \leq 3B/5$ is a realistic range for the tidal efficiency, and $3B/5 \leq A_e \leq B$ for the barometric. Consequently, a value of the Skempton ratio B can rather well be estimated from such analyses.

Hydraulic flow properties, i.e. the hydraulic diffusivity and, adopting values for the fluid viscosity and for the mechanical rock properties, the intrinsic permeability. The phase delay in tidal signals mainly reveals a value for D_2 , the recovery time for the inverse barometric effect reflects the lower value of D_3 and D_5 (Fig. 7b).

A combined analysis of tidal and barometric efficiencies reduces the inherent ambiguity. Recent examples for application of this technique with more detailed descriptions were given by Rojstaczer (1988b), Roeloffs et al. (1989), and Quilty and Roeloffs (1991). Studies where full sets of poroelastic properties have been deduced are those of Beavan *et al.* (1991) and Evans *et al.* (1991). Generally, the forcing functions like ε_a in the case of tides and Δb for making use of the barometric effect have to be known too. The latter is easily available. The areal strain tide ε_a can be calculated theoretically on a global scale, but for the inhomogeneous crust is determined within a factor of 2 only (Beaumont and Berger, 1975).

In the literature, various attempts are described in which the rock *porosity*, n, is deduced from tidal and barometric efficiencies. This is controversial because, as mentioned before, the porosity does not explicitly enter in poroelasticity. Considering the relation $B = (c-c_s)/S = (c-c_s)/[(c-c_s)+n(c_f-c_s)]$ of Brown and Korringa (1975) with S denoting the storage compressibility and c_f the compressibility of the pore fluid, where a homogeneous rock matrix and fully interconnected pore space is assumed, it is seen that n depends on the rather badly known difference $c-c_s$. Moreover, the sensitivity of natural well level fluctuations with respect to S is low (Hsieh *et al.*, 1987). Some of the ambiguity can be reduced by applying an iterative algorithm (Rojstaczer and Agnew, 1989).

The aforementioned methods assume that poroelasticity is a tenable approach to the rheology of water saturated formations. This appears justifiable for many rocks, consolidated or not, when fluid flow is dominated by matrix permeability. The situation changes, however, when fracture permeability prevails, i.e. when a system of interconnected fissures or fractures is hydraulically more conductive than the porous matrix. Massive fractures cannot be regarded as single pores in a (piecewise) macroscopically uniform rock. The increasing demand for modelling fluid flow in fractured rocks has led Bower (1983) and Hanson (1983) to elaborate schemes to interpret well tides in such media. Both authors consider pore pressure changes in a single, fluid filled fracture in an otherwise impermeable or poorly permeable medium. Bower (1983) has analysed the effect of a plane, penny shaped fracture that symmetrically intersects a well with less than 60° inclination. Hanson (1983) has looked into effects occurring in a bi-wing fracture intersecting a well along its axis. Fundamental to their schemes is the continuity equation for fluid flow in an open, elastically deforming fracture (Duguid and Lee, 1977). Amplitudes and phase lags of well tides in the connected well depend on (a) the orientation of the fracture with respect to the forcing external stress field, (b) the lateral extension of the fracture, and (c) its width (or separation), and can be used for inversion. With increasing number of degrees of freedom, as for several intersecting fractures, the observed signals become difficult to interpret.

In view of the complexity of structures in real rock formations (Kümpel, 1992), well level deduced poroelastic rock or fracture parameters should generally be taken as equivalent or effective rock parameters only. However, since it is difficult to obtain quantitative assessments of effective in-situ poroelastic and fluid flow properties through other geophysical techniques, a well in contact with a saturated confined formation and exposed to known tidal forces and barometric pressure changes offers a simple and efficient way for just this.

Acknowledgements. My research in the field of tides has benefitted from various grants of the Deutsche Forschungsgemeinschaft and of the German Ministry of Research and Technology (now Ministry of Education and Research), from a Killam Trust scholarship (Canada), as well as from support through the Deutsche Akademische Austauschdienst and the Universities of Kiel and Bonn. All this is gratefully acknowledged. Moreover, I am particularly thankful to participants of the Working Group 'Geodesy/Geophysics', who regularly meet once a year and have shaped my understanding for tides in many ways.

References

Abramowitz, M., and Stegun, I.A. (Eds.), 1964. Handbook of Mathematical Functions, *Appl. Math. Ser.* **55**, U.S. National Bureau of Standards, Gaithersburg, Md.

Biot, M.A., 1941. General theory of three-dimensional consolidation. *J. Applied Physics* I.12: 155-164.

Beaumont, C., and Berger, J., 1975. An analysis of tidal strain observations from the United States of America, I. The laterally homogeneous tide. *Bull. Seismol. Soc. Am.* **66**: 1613-1629.

Beavan, J., Evans, K., Mousa, S., and Simpson, D., 1991. Estimating aquifer parameters from analysis of forced fluctuations in well level: An example from the Nubian Formation near Aswuan, Egypt; 2. Poroelastic properties. *J. Geophys. Res.*, **96**: 12,139-12,160.

Bodvarsson, G., 1970. Confined fluids as strain meters. *J. Geophys. Res.* **75**: 2711-2718.

Bower, D.R., 1983. Bedrock fracture parameters from the interpretation of well tides. *J. Geophys. Res.* **88**: 5025-5035.

290   Kümpel

Brown, R.J.S., and Korringa, J., 1975. On the dependence of the elastic properties of a porous rock on the compressibility of the pore fluid. *Geophysics* **40**: 608-616.

Büyükköse, N., Kümpel, H.-J., Westerhaus, M., and Zschau, J., 1989. Well level data at six multi-parameter stations in the Mudurnu-Abant Valley. In: Turkish-German Earthquake Research Project, J. Zschau and O. Ergünay (eds.), Inst. of Geophysics, Univ. of Kiel, 74-81.

Duguid, J.O., and Lee, P.C.Y, 1977. Flow in fractured porous media. *Water Resour. Res.* **13**: 558-566.

Endom, J., and Kümpel, H.-J., 1994. Analysis of natural well level fluctuations in the KTB-Vor-bohrung: Parameters from poroelastic aquifer and single fracture models. *Scientific Drilling* **4**: 147-162.

Evans, K., Beavan, J., Simpson, D., and Mousa, S. 1991: Estimating aquifer parameters from analysis of forced fluctuations in well level: An example from the Nubian Formation near Aswuan, Egypt; 3. Diffusivity estimates for saturated and unsaturated zones. *J. Geophys. Res.* **96**: 12,161-12,191.

Gupta, H.K., Kümpel, H.-J., Radhakrishna, I., Chadha, R.K., and Grecksch, G., 1996. In-situ pore pressure studies in an area of continuously high induced seismicity. IASPEI Regional Assembly in Asia (abstract), Aug.1996, Tangshan, China.

Hanson, J.M., 1983. Evaluation of subsurface fracture geometry using fluid pressure response to solid earth tidal strain. Terra Tek Research Techn. Report 83-26, Salt Lkae City, Utah.

Hsieh, P.A., Bredehoeft, J.D., and Farr, J.M., 1987. Determination of aquifer transmissivity from earth tide analysis. *Water Resour. Res.* **23**: 1824-1832.

Kümpel, H.-J., 1991. Poroelasticity - parameters reviewed. *Geophys. J. Int.* **105**: 783-799.

Kümpel, H.-J., 1992. About the potential of wells to reflect stress variations within inhomogeneous crust. *Tectonophysics* **211**: 317-336.

Nur, A., and Byerlee, J.D., 1971. An exact effective stress law for elastic deformation of rock with fluids. *J. Geophys. Res.* **76**: 6414-6419.

Quilty, E.G., and Roeloffs, E.A., 1991. Removal of barometric pressure response from water level data. *J. Geophys, Res.* **96**: 10,209-10,218.

Rice, J.R., and Cleary, M.P., 1976. Some basic stress diffusion solutions for fluid-saturated elastic porous media with compressible constituents. *Rev. Geophys. and Space Phys.* **14**: 227-241.

Roeloffs, E.A., 1996. Poroelastic techniques in the study of earthquake-related hydrologic pheno-mena. *Advances in Geophysics* **37**: 135-195.

Roeloffs, E.A., Schulz Burford, S., Riley, F.S., and Records, A.W., 1989. Hydrologic effects on water level changes associated with episodic fault creep near Parkfield, California. *J. Geophys. Res.* **94**: 12,387-12,402.

Rojstaczer, S., 1988a. Intermediate period response of water levels in wells to crustal strain: sensitivity and noise level. *J.Geophys. Res.* **93**: 13,619-13,634.

Rojstaczer, S., 1988b. Determination of fluid flow properties from the response of water levels in wells to atmospheric loading. *Water Resour. Res.* **24**: 1927-1938.

Rojstaczer, S., and Agnew, D.C., 1989. The influence of formation material properties on the response of water levels in wells to earth tides and atmospheric loading. *J. Geophys. Res.* **94**: 12,403-12,411.

Terzaghi, K., 1923. Die Berechnung der Durchlässigkeitsziffer des Tones aus dem Verlauf der hydrodynamischen Spannungserscheinungen. Sitzungsber. Akad. Wiss. Wien, Math-Naturwiss. Kl., Abt. IIA, Nr. 132, 125-138.

Theis, C.V., 1935. The relation between the lowering of the piezometric surface and the rate and duration of discharge of a well using groundwater storage. *EOS, Trans. Am. Geophys. Un.* **16**: 519-524.

Wang, H., 1993. Quasi-static poroelastic parameters in rock and their geophysical applications. *Pure Appl. Geophys.* **141**: 269-286.

Wenzel, H.-G., 1993. Earth tide analysis program system ETERNA, Vers. 3.0, Manual. Geod. Inst., Univ. Karlsruhe.

Tidal Triggering of Earthquakes and Volcanic Events

Dieter Emter

Black Forest Observatory, Heubach 206, D-77709 Wolfach, Germany

Abstract. Since the end of the last century many investigations have been carried out concerning the question if stresses due to the tidal forcing of the earth do trigger earthquakes or volcanic activities. Since the publications are so numerous and the results so controversial a complete review is impossible. An attempt is made to describe the different methods of investigation applied in most of the searches for tidal triggering. Summaries of the most important papers and their results are given separately for the different types of geophysical events such as earthquakes, volcanic eruptions and volcanic shocks.

1 Introduction

There was always a tendency to blame extraterrestrial forces to be the reason for catastrophic events inflicting mankind. Especially the moon was often attributed with a mysterious role. Ocean tides were discussed as a cause for earthquakes already in the 17th century. But only with the detection of earth tides at the end of the last century a physical causality became possible. Estimates with realistic earth models show that earth tides are the reason for the largest periodic stress variations within the earth's crust and deeper layers. These cyclic stress variations reach amplitudes between 10 and 100 hPa which are small compared to tectonic stresses. Stress drops during earthquakes f.i. reach values which are two or three orders of magnitude higher than tidal stresses. The rate of tidal stress changes, however, can be around 10 hPa/hour (McNutt and Beavan, 1981) and is comparable to or even higher than tectonic stress rates. Stress accumulations between earthquakes can have rates of 0.2 hPa/hour (e.g. Curchin and Pennington, 1987) and the stress rate during the inflation of the Campi Flegrei caldera was estimated between 0.15 and 1.2 hPa/hour (Rydelek et al., 1992). As the tidal stresses with their large rates are always present and superimposed on the slowly increasing tectonic stress they may finally act not as a cause but as a trigger of a geophysical event. A simple trigger model works only during a special time segment of tectonic stress building (just before failure) and with a well defined critical stress level (Curchin and Pennington, 1987; Rydelek et al., 1992).

The real conditions for the release of an earthquake or the beginning of a volcanic eruption may differ in most cases from such a simple model but the possibility of tidal triggering of at least some events cannot be excluded. This is certainly one of the reasons for the numerous investigations since the end of last century which deal with tidal triggering of earthquakes, geysir activity and volcanic events. The results of these investigations, in short, are very controversial.

In contrast, investigations of the seismicity of the moon demonstrated a clear correlation with lunar tides (e.g. Lammlein et al.,1974). The reason could be that tidal stresses on the moon are about an order of magnitude larger than on the earth and at the same time tectonic stresses are much lower than on the earth. Another case of perfect correlation was found for ice quakes on an Antarctic ice shelf (Kobarg and Lippmann, 1986). But in this case the events were rather caused by tides than triggered since without the uplifting of the shelf by ocean tides these ice quakes would not happen.

As already mentioned, the number of investigations on tidal triggering is very high. It increased to somewhat near 100 within the last 100 years to the best of my knowledge. It is therefore impossible to give a complete account of all the papers and their results without filling a book. The attempt here to select the more important papers cannot be completely without subjectivity. The methods and results of most of the selected papers are represented in a condensed form in tables and only some very typical and remarkable investigations will be described in some detail.

2   Methods of investigation

A variety of methods is used to find possible correlations between earth tides and geophysical events. But fortunately always some of the methods have common features which allows to collect them into groups. Some authours simply try to find tidal and other periodicities in time series of geophysical events. Others relate time series of events to simultaneous time series of different tidal phenomena by simple visual comparison or by more or less sophisticated statistical methods. All methods used in the papers listed in tables 1 to 3 will be mentioned (with abbreviations) in the following subsections. Abbreviations for the different tidal phenomena are also given in the text.

2.1   Search for periodicities

Tidal periodicities in time series of events could be found directly by Fourier transforms (FT). For time series with continuous activity f.i. series of volcanic shocks FT can be performed straightforward. But for time series of discrete events like earthquakes some modifications have to be applied (see Morgan et al., 1961). Furthermore those continuous spectra are supposed to be difficult to be interpreted in the case of a weak correlation (Knopoff, 1964). By these reasons many investigators look only for some special periods in their time series. These can be direct tidal periods like semidiurnal M_2 or fortnightly M_f or periods of phenomena closely related to tides as lunar and solar hour, position (pos.) of moon and sun or altitude (alt.) of moon and sun. One method to find significant correlations at those a priori well known periods is the Schuster test using a statistical method of Rayleigh (1880) and first applied by Schuster (1897) to geophysical time series. Another common method is that of Chapman-Miller (Malin and Chapman, 1970).

2.2   Comparison with tidal phenomena

Instead of searching for single tidal periods, sometimes of minor importance, in the occurrence times of events it makes much more sense physically to relate those to complete time functions of tidal phenomena.

2.2.1 Tidal time functions: Tidal time functions should represent the sum of tidal forces acting on an earthquake-fault or on a volcano or are at least closely related to them. Time series of tidal potential, tidal vertical acceleration (g), tidal horizontal acceleration or tilt (ϕ), tidal strains (ε) as well as of the time rates of these variables ($\dot{g}, \dot{\phi}, \dot{\varepsilon}$) are used. It is very often not realized that time functions of these tidal phenomena (g,ϕ, ε) are in most cases not directly proportional to those of tidal stresses because of geometrical reasons and the tensor relation between strain and stress. Tidal normal stress on a plane perpendicular to the horizontal direction with azimuth ψ depends on all components of the tidal strain tensor:

$$\sigma_\psi = 2\mu\varepsilon_\psi + \lambda(\varepsilon_{NN} + \varepsilon_{EE} + \varepsilon_{ZZ}) \tag{1}$$

with μ and λ the Lamé's constants and ε_ψ the tidal strain or extension in direction ψ. From equ.(1) it can already be seen that σ_ψ is not proportional to ε_ψ as is assumed in many investigations. The linear strain ε_ψ itself depends also on the azimuth and the tensor components:

$$\varepsilon_\psi = \varepsilon_{NN} \cos^2 \psi + \varepsilon_{EE} \sin^2 \psi + \varepsilon_{NE} \sin 2\psi \tag{2}$$

The strain tensor components $\varepsilon_{NN}, \varepsilon_{EE}$ and ε_{ZZ} are in phase with tidal potential (and g), whereas the shear component ε_{NE} is 90° out of phase to the potential. Therefore σ_ψ is only for NS-direction ($\psi = 0^o$) and EW-direction ($\psi = 90^o$) in phase with ε_ψ and with the potential. For all other directions there exist phase differences between stress,strain and potential that can reach up to 60° (Varga, 1983). Since the tensor components of tidal strain vary in a different way with the latitude these phase differences are strongly latitude dependent (Varga and Grafarend, 1996). Because of these complicated relations (see also Fig.1, Zürn "Earth Tide Observations",this volume) the best way to look for a trigger mechanism that should be stress dependent is to calculate the tidal stresses themselves. In the case of volcanoes this can be volumetric stresses ($\sigma(vol)$) and for earthquakes with known focal mechanism stresses normal (σ_N) and/or parallel (τ) to the fault would be adequate. Some authors use a combination of both (σ_c) corresponding to Coulomb's failure criterion (Jaeger, 1969). Since the high rates of tidal stresses have always been the strongest argument for the possibility of tidal triggering the stress rates $\dot{\sigma}_N, \dot{\tau}$ and $\dot{\sigma}(vol)$ are taken as a reference by some authors. It is well known that close to the coasts crustal deformations (and stresses) due to ocean loading can be of the same order of magnitude or even larger (e.g. Jentzsch, 1995) than the earth tides with considerable phase differences. Therefore in the most careful investigations strains and stresses due to ocean loading (ocean) are added to those of the earth tides before they are compared with time series of geophysical events.

2.2.2 Statistical methods: The problem is now to find a correlation or lack of it by comparing a time sequence of events with a time series of one of the tidal phenomena described above. Some authors believe to find a correlation simply by a visual comparison (vis.comp.) f.i. by having more events in the vicinity of a tidal maximum than in other time intervals. A better way is to check by statistical methods whether the events are randomly distributed within tidal cycles or not. One method is to divide the tidal cycles into bins or boxes and test if the events are uniformly distributed within these bins (binom. distr.). A more frequently used method is to determine the tidal phases of the events and test their distributions. The tidal phase of an event is its relative time with respect to two subsequent maxima or minima of the tidal time function. The resulting phase distributions can be compared to distributions generated by random sampling or tested for uniformity by the Schuster test. This test cannot only be used to test for single periodicities as already mentioned but also may be applied for quasi-periodic cyclic variations like the complete tidal time function. Schuster's test is described in detail in many papers (e.g. Heaton, 1975; Young and Zürn, 1979; Rydelek et al., 1992).

Since the Schuster test is used in nearly half of the investigations listed in the tables 1 to 3, it will be described in some detail here. The first step is to determine the tidal phase for each event in a time series or catalogue. The time between two subsequent maxima or minima of the complete tidal time function is scaled to 2π or $360°$. Each event happening between the times of the two maxima is assigned a phase on a linear scale from $0°$ to $360°$ corresponding to its time of occurence with respect to the maxima. For example, an event occuring at a time just in the middle between two tidal maxima will have a phase of $180°$. This will be repeated for all events within the complete time span covered by the catalogue. The tidal phases φ_i of all N events of a catalogue, all between $0°$ and $360°$, can be plotted in rose diagrams for demonstration (see Fig.1). They are collected in bins ($30°$ here) for better visualisation and statistical stability. In the next step this phase distribution will be checked for non-randomness. Each event can be represented by a phasor of unit length with the tidal phase angle as determined above. The phasors of all N events are summed up vectorially, similar to a random walk, resulting in a phasor sum R according to :

$$R = \sqrt{(\sum_{i=1}^{N} \sin \varphi_i)^2 + (\sum_{i=1}^{N} \cos \varphi_i)^2} \tag{3}$$

The probability P_R that for N random phases a resultant equal or larger than R is obtained can be determined according to:

$$P_R = e^{-R^2/N} \tag{4}$$

The null hypothesis of randomly distributed tidal phases can be rejected or not with a significance level of $(1 - P_R)$. For the example of the phase distribution with the resultant R shown in Fig. 1a the null hypothesis can be rejected with a significance of 99.8%. For the phase distribution shown in Fig. 1b the significance is only 32%. In most investigations significance levels of 95% or even lower are accepted as a proof for tidal triggering. This seems to be very optimistic regarding the high sensitivity of the Schuster test. Rydelek et al. (1992) have shown theoretically that only 7 to 10% of tidally triggered earthquakes in a catalogue of 3000 randomly (Poissonian) distributed events will be sufficient to give a 99% significance with the Schuster test.

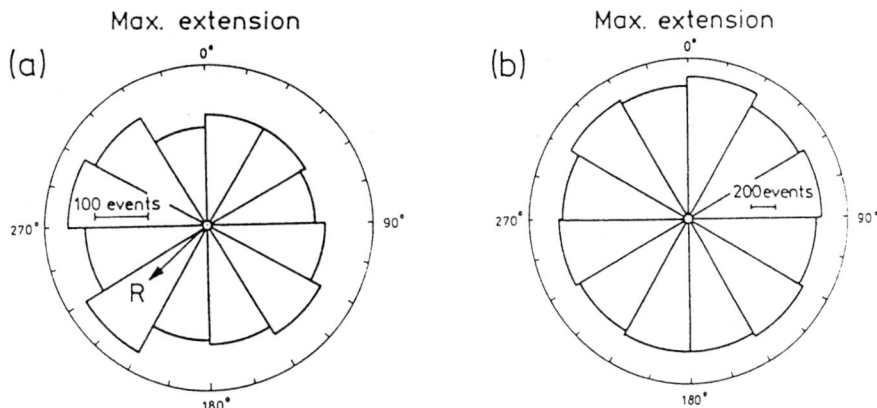

Fig. 1. Examples for tidal phase distributions modified after Emter et al. (1986): (a) for a time series of 2658 volcanic shocks from Mt. Etna with resultant R and (b) for a time series of 13920 shocks from Stromboli (R too small to be shown).

The tests mentioned and described so far use only the time relation between the occurence of an event and a tidal time function f.i. that of tidal stress. Information about the total or relative amplitudes (A) of tidal phenomena at the occurence times of the events is used only by a few authors. Methods like linear regression analysis or cross correlation (cross corr.) automatically take amplitudes into account. In other investigations the stress amplitudes at the occurence times of events are determined and combined to histograms. Those histograms are compared to normal distributions by classical methods like the χ^2-test or the Kolmogorov-Smirnow (K.S.) test (e.g. Taubenheim, 1969). It should be noted here that amplitude related tests include the influence of the fortnightly tidal maxima which are often used as a separate reference in many investigations.

3   Earthquakes and tides

The majority of investigations on tidal triggering deals with earthquakes. A group of authors (e.g. Hunter, 1978) searches only for astronomical (sidereal) periods in earthquake series that are not directly related to tides, these papers will not be considered here. General searches for periodicities in global earthquake catalogues reveal no tidal periods at all (e.g. Lomnitz, 1966) or exclude periods between 2 and 365 days (Shlien and Toksöz, 1970).

Table 1. Earthquakes and tides

First author	Year	Region	N	Correlation with	Test-method	Res.
Allen	1936	Calif.	1200	lunar hour	Schuster	+
Brazee	1957	global	100	g,ϕ	random phase?	−
		local	189	g,ϕ	random phase?	−
Cotton	1922	global	316	pos. sun/moon	binom.distr.?	+
Curchin	1987	Global	7539	$\sigma(vol)$	Schuster,χ^2	−
Gao	1981	China	70	$\tau, \sigma(vol)$	Schuster	±
Gougenheim	1961	global	19	potential	random phase?	−
Hartzell	1989	Calif.	151,834	fortnightly M_f	Schuster	−
		global	189,385	fortnightly M_f	Schuster	−
Heaton	1975	global	107	σ_N, τ	Schuster	±
Heaton	1982	global	330	σ_N, τ	Schuster	−
Jeffreys	1938	Japan	1071	tidal periods	Schuster, χ^2	−
Kilston	1983	Calif.	35	lunar,solar hour	χ^2(Poissonian?)	±
				lunar phase,σ_N		
Klein	1976	7 regions	swarms	$\varepsilon(M_2), ocean$	Schuster, binom.?	±
Klein	1976	8 dams	9-23	$\varepsilon(M_2)$	Schuster	±
Knopoff	1964	Calif.	9000	g,A.	cross-corr.	−
Lopes	1990	global	541	alt. moon/sun	random distr.?	+
Morgan	1961	global	1933	lunar periods	FT	−
Palumbo	1986	global	9157	lunar/solar hour	Chapm.-Miller	−
		Italy	16300	lunar/solar hour	Chapm.-Miller	+
Ryall	1968	Calif.	7800	g,A.	cross-corr.	+
Rykunov	1985	Kamchatka	swarm	g,A.	cross-corr.	+
Sadeh	1978	global	42000	lunar periods	FT	+
Schuster	1897	global	7500	lunar/solar periods	Schuster	−
Shirley	1986	global	13	M_f	Schuster	+
Shirley	1988	Calif.	48	potential, M_f	Schuster	±
		Alaska	24	potential, M_f	Schuster	±
Shlien	1972	global	5000	potential	Schuster	∓
		Japan	1000	potential, A	cross-corr.	−
Shudde	1977	Calif.	2600	$g, \phi, \dot{g}, \dot{\phi}$,A.	χ^2,K.S.	−
Simpson	1967	g, global	22561	g	random distr.?.	−
Souriau	1982	Pyrenees	380	σ_N, τ, σ_c,ocean	binom.distr.?	±
Tsuruoka	1995	global	998	σ_N, τ,ocean	Schuster	∓
Ulbrich	1987	Germany	1530	σ_N	Schuster, χ^2	∓
VanRuymb.	1983	Global	Katalog	potential(M_2)	Schuster	+
Young	1979	SW-Germ.	259	σ_N, τ,A.	Schust.,χ^2,K.S.	∓

Investigations dealing directly with relations between earthquakes and tides are summarized schematically with their methods and their results in Table 1. Because of the great number only those papers are included that contain at least some kind of a statistical test. The number of events used for the investigation is N. Concerning the results, + means that the authors conclude on tidal triggering, - means that they cannot find triggering and ± means that they find triggering only for parts of their time series or only for special tidal periods.

Investigations by Bloxom (1974), Sauck (1975), Mohler (1980), Ding and Wang (1983), Oike and Taniguchi (1988) and Zugravescu et al. (1989) are not included in the table as no statistcal tests are applied. All these authors believe to find a correlation. A sequence of papers by one and the same author, Tamrazyan (e.g.,1968), is excluded for the same reason as he finds only by visual comparison accumulations of earthquakes at special configurations of the moon and the sun relative to the earth. In one of these investigations (Tamrazyan, 1968) that will be mentioned here as a nice example for possible errors, he finds some accumulations of strong global earthquakes around times at which lunar perigee coincides with full or new moon. Since tides are maximized at these times he concludes on tidal triggering. Knopoff (1970), however, could find the same accumulations with an artificially generated series of earthquakes randomly distributed within the same time intervall. Knopoff found as a reason that the time scales chosen by Tamrazyan, the anomalistic month (perigee) and the synodic month (phases of the moon), are astronomically not independent from each other. The observed correlation is therefore a consequence of the correlation of perigee with full or new moon.

Schuster (1897) first applied a statistical method to an earthquake catalogue without finding any lunar or solar periodicities. Cotton (1922) makes the first attempt to relate tidal stresses in the lithosphere to the positions (f.i. zenith distances) of moon and sun. Furthermore he gives a complete outline of the early literatur on tidal triggering until that time.

From the point of view of a physical model for tidal triggering including the total sum of tidal stresses it is difficult to understand why special, less important tidal periods should show up alone in some of the tests. This holds for half the sidereal month corresponding to the fortnightly lunar declinational tide M_f (amplitude about 17% of M_2) which Sadeh and Wood (1978) claimed to have found in a global catalogue. Hartzell and Heaton (1989) tested other catalogues for this period with a negative result. Shirley (1986, 1988) again found this periodicity in much smaller regional earthquake series but he failed to find correlation with the tidal time function for the same series. Kilston and Knopoff (1983) found correlations of a small earthquake series with different lunar and solar periodicities but not for the lunar hour angle which should have the strongest effect and not fore the complete time funtion of tidal stresses.

The relation of earthquakes to the separate positions and motions of sun and moon bears furthermore the danger of wrong interpretations with respect to tidal stresses. For instance tidal extensional stresses have a maximum when moon or sun are overhead and not when they are on the horizon as stated by Cotton

(1922) and Kilston and Knopoff (1983). Palumbo (1986) explains his positive correlation with maximum E-W extensional stresses at 3 hours local lunar time. These are in fact zero at this time whereas E-W horizontal accelerations have their maximum. But also complete tidal time functions can be misinterpreted in terms of tidal stresses. Maximum tidal extension is often interpreted as maximum tidal stress (e.g. Klein, 1976) which is not always correct as already mentioned in context with equation (1).

Physically the best way is to correlate earthquakes directly with the time functions of tidal stresses related to the focal mechanisms (Heaton, 1975; Gao et al.,1981; Heaton, 1982; Ulbrich et al., 1987; Tsuruoka et al., 1995) or even better with the tidal stress amplitudes (Young and Zürn, 1979). Mohler (1980) first introduced a combination of tidal stress normal to the fault and shear stress parallel to the fault (Coulomb stress) in the way that increasing shear with si-multaneous reduction of friction by normal stress could favour tidal triggering. This model was also applied by Souriau et al. (1982) for the careful statistical investigation of an aftershock sequence. In contrast, Ding and Wang (1983) ac-cept tidal triggering for each separate earthquake out of a series as a fact when tidal Coulomb stress is favorable at the occurence time of the earthquake. In this way they found 61% of 70 severe earthquakes in China tidally triggered, somewhat better than by chance.

Berg (1966) brought the ocean load effect into discussion. He realized that the great Alaskan earthquake of 1964 and a majority of its 17 aftershocks with $M \geq 5.5$ happened at low ocean tide or close to it. This is an interesting working hypothesis as ocean tides in the Gulf of Alaska cause crustal strains much higher than those due to earth tides, but more observations and statistical tests are needed. Ocean load effects in addition to earth tidal stresses have been considered later in the statistical investigations of Klein (1976), Souriau et al. (1982) and Tsuruoka et al. (1995).

There remains last not least the problem of statistical significance which is not independent of the number of events used for the tests. Some of the authors take relatively small series of events and others divide their series into small subsets. The dangers involved in such a procedure has shown up convincingly in two subsequent papers of one and the same author. In the first, already very careful investigation Heaton (1975) tested a global catalogue of 107 earthquakes ($M \geq 5.5$) with known focal mechanisms for possible tidal triggering. He grouped the earthquakes according to their focal mechanism and focal depth and cor-related with fault related tidal stresses, f.i. shear stress in direction of the slip vector. Neither the complete set nor most of the subsets gave significant results with one exception. However for a subset of 34 shallow earthquakes with dip-slip or oblique-slip motion triggering by tidal shear stress could not be excluded be-cause of a high significance with the Schuster test. In his second paper Heaton (1982) examined an earthquake catalogue updated from 107 to 328 events and found no significance at all with the same methods as earlier. Especially the enlarged group of now 100 shallow dip-slip or oblique-slip events was not longer significant. Heaton (1982) realized as a mistake in his first investigation that he had divided his data set until a certain pattern was visible and then applied a statistical test . In that case conclusions from a test can be misleading. The very sober discussion of this result by Heaton (1982) should be read by everybody embarking on such investigations. Already Gao et al. (1981) had divided their set of 70 earthquakes into groups as small as 9 events, and the very smallest group was the only one which turned out to be significant. But also after Heaton's negative experience some authors have not learned the lesson. Kilston and Knopoff (1983) deal with subsets as small as 8 out 35 Californian earthquakes and Shirley (1988) divided 48 Californian earthquakes into subsets down to 5 events. Even in one of the most careful investigations within the last years by Tsuruoka et al. (1995) the data set of 998 global earthquakes is divided. Their tests against fault related tidal stresses including the ocean load effect is only significant for one group which contains again the smallest number of events (75). Shudde and Barr (1977) pointed out another fairly frequently commited crime: many different tidal time series are sometimes used within the same test, without readjusting the probabilities, when "throwing the dice more than once".

4   Volcanic events and tides

Volcanic activity involves many different phenomena: eruptions with lava flow, explosive ash eruptions, seismic activity of different kind and geysir activity. A group of investigations (e.g. Rinehart, 1972) on possible tidal influences on geysir activity will not be considered here as only long period, less important tides are found with non-statistical methods. The other papers will be grouped according to eruptive or seismic activity.

4.1   Eruptions

General studies on catalogues of volcanic eruptions are rare. Hayakawa (1995) found a random occurence of the 306 eruptions in Japan within the last 2000 years. Eggers and Decker (1969) realized seasonal variations in the starting dates of 1792 historic volcanic eruptions and Casetti et al. (1981) found monthly cycles in the historic eruptions of Mt. Etna, both periods are certainly not of tidal origin.

Volcanic eruptions are difficult to be tested for possible tidal triggering for two reasons. Firstly only small numbers of eruptions are available for statistical tests, especially for one and the same volcano. Secondly the start times of eruptions are in most cases not precisely enough defined that they allow a comparison with the semidiurnal tides. Therefore in most of the investigations which are summarized in Table 2, eruptions are related to the fortnightly tidal maximum which corresponds to the lunar phases. Nearly all authors find correlations between eruptions and the fortnightly tidal maximum independently of dealing with global or local activity or using statistical methods or not. This is astonishing as the fortnightly maximum is only an amplitude modulation of the semidiurnal tides which cause the strongest tidal pressure variations within a volcano.

Table 2. Volcanic eruptions and tides

First auth.	Year	Volcano/region	Numb.	Correlation with	Test-method	Res.
Dzurizin	1980	Kilauea(Hawaii)	52	g(fortnightly max.)	Schuster	+
		Mauna Loa(Haw.)	37	g(fortnightly max.)	Schuster	−
Golombek	1978	El Salvador	6	g(fortnightly min.)	vis. comp.	+
Johnston	1972	Stromboli(Italy)	33	g(s.d., fortn. max.)	vis.comp.	∓
Hamilton	1973	global	140	moon phases	vis. comp.	+
Martin	1981	Fuego (Guatem.)	48	g (s.d.,fortn. max.)	binom.distr.?	+
Mauk	1973	global	680	g(fortnightly max.)	Schuster	+
McNutt	1987	Pavlov(Alaska)	9	g?	vis.comp.	−
Michael	1975	Ngaurahoe(N.Z.)	12	g(fortnightly max.)	binom.distr.?	+
Ramos	1985	Mayon (Philip.)	11	g(fortnightly max.)	vis. comp.	+

A model explaining triggering by the fortnightly maximum would require an extremely well defined trigger level. The eruptions of only two volcanoes could be correlated with the semidiurnal tide (s.d.). Johnston and Mauk (1972) found triggering of the major eruptions of Stromboli by the fortnightly tidal maximum but not by the semidiurnal tide whereas Martin and Rose (1981) found Fuego eruptions to be triggered by the semidiurnal tide and its fortnightly modulation.

It is interesting to see how the results of some of these investigations are appreciated. Dzurisin (1980) accepted a 89% significance with the Schuster test as a proof for tidal triggering of Kilauea (Hawaii). This is a rather low significance if we keep in mind the results of Rydelek et al. (1992) concerning the sensitivity of the Schuster test. Nevertheless the fortnightly tidal maximum is included in the eruption prediction of this volcano (Klein, 1984).

4.2   Volcanic shocks

Since more and more volcanoes are monitored systematically the number of observed volcanic seismic events has increased dramatically. Besides stronger earthquakes of tectonic origin the overwhelming part of the observed events are microearthquakes or shocks of the B-type after Minakami's (1974) classification. They are attributed to shallow explosions or pressure surges inside the magma conduits and could therefore be sensitive to small pressure variations f.i. by tidal deformations. Volcanic shocks are distinguished by their impulsive character from the so called tremors which are more persistent. Especially around active phases of a volcano they are so frequent that the observed time series contain the high numbers of events needed for a reasonable statistical test. Investigations on tidal triggering of volcanic earthquakes or shocks on different volcanoes are summarized with their methods and results in Table 3.

In most of the papers the onset times of the shocks have been related to volumetric stress which seems to be the most likely trigger source for volcanoes. Correlations with tidal potential, g and volumetric strain ($\varepsilon(vol)$) are also adequate as these time functions are in phase with volumetric stress. In cases where tectonic faults are involved, normal stresses, shear stresses or Coulomb stresses are taken as a reference. In one paper (Patanè et al., 1994) a correlation of volcanic tremors with E-W tidal acceleration is explained by a model with E-W tidal stresses acting on N-S trending faults; this interpretation is not correct since these two tidal phenomena are $90°$ out of phase. Only two authors refer to the lunar hour alone instead of the complete tidal time function with all the disadvantages already discussed. It is interesting that two investigations (Berrino and Corrado, 1991; Rydelek et al., 1992) of one and the same time series of volcanic shocks arrive at opposite results.

The example of Fig.1a shows the histogram of tidal phases of 2658 shocks observed at Mt.Etna within 15 days in Febr.,1981 with reference to tidal volumetric stress (Emter et al., 1986). As already mentioned the Schuster test for this phase distribution resulted in a high probabilty (99.8%) for tidal triggering. As amplitude related tests gave the same high significance, tidal triggering for this special time series could not be excluded from the statistical point of view. Three other time series of shocks at Mt.Etna between Jan. and March, 1981 did not show this high significance. It should be mentioned that the significant series of shocks was observed just after a flanc eruption of Mt.Etna and that McNutt and Beavan (1981) found evidence for tidal triggering of shocks at Pavlov volcano just before and just after eruptions and not at other times. Time series of a large number of explosive shocks at Stromboli (Emter et al., 1986) did not show any correlation with tides (see also Fig. 1b).

Table 3. Volcanic shocks and tides

First. auth.	Year	Volcano/region	Numb.	Correlation with	Test-method	Res.
Ambeh	1991	Mt Cameroon	175	g	Schuster	−
Berrino	1991	C.Flegrei(Italy)	8000	g	Chapm.-Miller	+
Emter	1986	Mt.Etna (Italy)	10,927	$\sigma(vol), \sigma_N, A.$	Sch.,χ^2,K.S.,FT	∓
		Stromboli(Italy)	29,857	$\sigma(vol)$	Schuster,FT	−
Fadeli	1991	Merapi(Java)	189,576	potential	Schuster,FT	−
Filson	1973	Galapagos	200	ε_{zz}	vis. comp.	+
Golombek	1978	El Salvador	300	g	binom.distr.	+
Imbo	1955	Vesuvius(Italy)	high	lunar hour	vis. comp.	+
McNutt	1981	Pavlov(Alaska)	549	$\dot{\sigma}_N, \dot{\sigma}(vol), A.$	linear regr.	±
McNutt	1984	St.Helens	4770	$\varepsilon(vol), \dot{\varepsilon}, A.$, ocean	linear regr.	±
				ε, ocean	Schuster	±
Palumbo	1985	Hawaii	130	lunar hour	Chapm.-Miller	+
		C.Flegrei(Italy)	1064	lunar hour	Chapm.-Miller	+
		Vesuvius(Italy)	high	lunar hour	Chapm.-Miller	+
Patanè	1994	Mt.Etna(Italy)	tremor	$g, \phi, A.$	cross.corr.	±
Rydelek	1988	Kilauea(Hawaii)	1719	$g, \sigma_N, \tau, \sigma(vol), A.$	FT, corr.,regr.	+
Rydelek	1992	C.Flegrei(It.)	8000	σ_N, τ, σ_c	Schuster	−

A careful investigation could be carried out for shocks at Merapi (Fadeli et al., 1991) with a catalogue of nearly 200,000 events counted within about 3 years time. A Schuster test against tidal potential (which is in phase with volumetric stress) resulted in an incredibly high significance for tidal triggering ($P_R \leq 10^{-20}$). This would have been accepted if no other information was available. Because of the continuous shock activity at Merapi during such a long time intervall a high resulution spectrum could be computed. Parts of this spectrum, in the semidiurnal (a) and diurnal (b) frequency band are shown as an illustrative example in Fig.2. It is obvious that spectral lines of lunar tides are missing and that solar terms, especially S_1, have much larger relative amplitudes than they should have in the tidal spectrum. These findings, especially the missing of the strongest tidal component M_2 excludes any tidal influence , especially at a place that close to the equator.

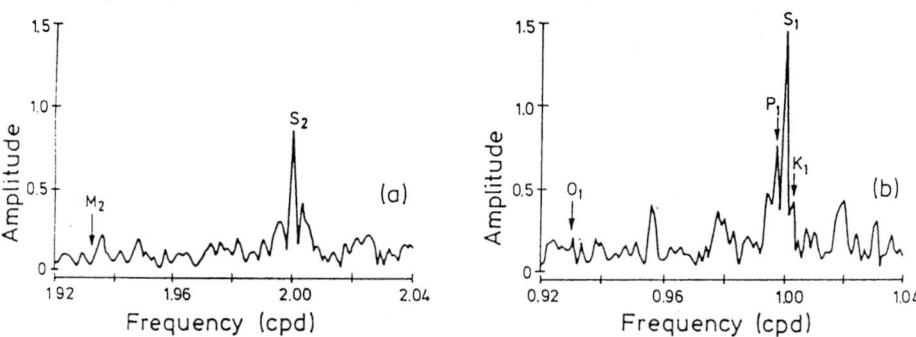

Fig. 2. Semidiurnal (a) and diurnal (b) frequency band from a 3 years time series with nearly 200,000 shocks at Merapi (modified after Fadeli et al., 1991).

But what about the high significance of the Schuster test? Noise-free synthetic earthquake catalogues containing only solar periods were tested with the same method against tidal potential resulting in similar high significances. It could be shown that the solar term S_2 alone as well as the terms P_1 and K_1 can strongly affect the Schuster test. The presence of S_1 contamination in a catalogue producing S_2 as a harmonic and P_1, K_1 by seasonal modulation will affect the Schuster test against tides to an extent of producing a false positive result (see also Rydelek et al., 1992). The reasons for the solar influence on the occurence of shocks at Merapi are not yet clear, meteorological effects are the most likely ones, because no man-made source could be identified.

5   Conclusions

If we take all the investigations on tidal triggering of geophysical events summarized in this review without any weighting of the quality of the data and the methods, positive and negative correlations occur about equally often.

   Concerning the earthquakes there seems to be a tendency for positive results for regional catalogues or earthquake swarms in contrast to global catalogues. But there exists also a tendency for negative results with increasing quality of the data sets and the statistical methods. In the 8 papers out the 32 listed in Table 1 that deal with earthquakes in California, 4 positive and 6 negative correlations are reported. It may be a reality, as Burton (1986) states in a critical review, that the coherency between earth tides and global and local seismicity is usually within the noise level of any analysis. There might be really some events within an earthquake series be triggered by earth tides and show up with a certain significance in one of these sensitive tests. Conclusions for modeling the release of earthquakes or even prediction are not possible. Recent investigations (King et al. 1994; Gomberg and Davis, 1996) about the triggering of seismicity by the Landers, California, earthquake ($M_w = 7.3$) have revealed that strains in the order of 10^{-6} were responsible for triggering, nearly 2 orders of magnitude larger than tidal strains. After all there remains a nice physical model for the triggering which is clearly seen on the moon.

Most investigations about tidal triggering of volcanic eruptions come to positive results but nearly all of them suffer from poor statistics. Large series of volcanic shocks offer the best possibility for careful tests. For some of the series or at least for parts of them tidal triggering cannot be excluded from the statistical point of view. These results cannot be taken directly as a proof for a real physical effect but only as a basis for a model. Volcanism on Io (see Spohn "Tides of Io", this volume), however, is attributed completely to the extreme tidal deformations of this Jupiter moon.

The investigations on volcanic shocks additionally have manifested a more general problem. Contamination of catalogues with periods of the solar day can lead to falsified results concerning tidal triggering. Daily periods in earthquake catalogues are a well known phenomenon (Shimshoni, 1971; Knopoff and Gardner, 1972) which have to be removed before further investigations. Series of small events which are only locally observed as small earthquakes or volcanic shocks should be tested carefully for daily periods which can be introduced by human activities (e.g. quarry blasts) or meteorological effects.

References

Allen, M.W. 1936. The lunar triggering effect on earthquakes in southern California. *Bull. seism. Soc. Am.* **26**: 147-157.

Ambeh, W.B. and J.D. Fairhead. 1991. Regular, deep seismicity beneath Mt.Cameroon; lack of evidence for tidal triggering. *Geophys. J. Int.* **106**: 287-291.

Berg, E.. 1966. Triggering of the Alaskan earthquake of march 28, 1964, and major aftershocks by low ocean tide loads. *Nature.* **210**: 893-896.

Berrino, G. and G. Corrado. 1991. Tidal signal in the recent dynamics of Campi Flegrei caldera (Italy). *J. Volcanol. geotherm. Res.* **48**: 93-101.

Bloxom, D. 1974. San Fernando fault earthquakes and earth tides. *Bull. seim. Soc. Am.* **64**: 2005-2006.

Brazee, R.J. 1957. Earth tides and earthquakes. *Earthquake Notes.* **28**: 1-3.

Burton, P.W. 1986. Is there coherence between earth tides and earthquakes? *Nature.* **321**: p.115.

Casseti, G., G. Frazetta and R. Romano. 1981. A statistical analysis in time of the eruptive events on Mt.Etna (Italy) from 1323 to 1980. *Bull. Volcanol.* **44**: 283-294.

Curchin, J.M. and W.D. Pennington. 1987. Tidal triggering of intermediate and deep focus earthquakes. *J.geophys. Res.* **92**: 13957-13967.

Cotton, L.A. 1922. Earthquake frequency with special reference to tidal stresses in the lithossphere. *Bull. seism. Soc. Am.* **12**: 47- .

Ding, Z. Jia.J. and R. Wang. 1983. Seismic triggering effect of tidal stress. *Tectonophysics.* **93**: 319-335.

Dzurizin, D. 1980. Influence of fortnightly earth tides at Kilauea volcano, Hawaii. *Geophys. Res. Lett.* **7**: 925-928.

Eggers, A.A. and R.W. Decker. 1969. Frequency of historic volcanic eruptions. *EOS Trans. Am. geophys. Un.* **44**: p.343.

Emter, D., W. Zürn, R. Schick and G. Lombardo. 1986. Search for tidal effects on volcanic activities at Mt.Etna and Stromboli. In: *Proc. 10th Int. Symp. on Earth Tides.* pp. 213-219. R.Vieira (ed.). Consejo Superior de Investigaciones Cientificas, Madrid.

Fadeli, A., P.A. Rydelek, D. Emter and W. Zürn. 1991. On volcanic shocks on Merapi and tidal triggering. In: Volcanic Tremor and Magma Flow. pp. 165-181. R. Schick and R. Mugiono (eds.). Scientific Series of the International Bureau, **Vol.4**. Forschungszentrum Jülich.

Filson, J., T. Simkin and L. Leu. 1973. Seismicity of caldera collapse: Galapagos Islands 1968. *J. geophys. Res.* **78**: 8591-8622.

Gao, X., Z. Yin, W. Wang, L. Huang and J. Li. 1981. Triggering of earthquakes by the tidal stress tensor. *Acta Seismologica Sinica.* **3**: 264-275.

Golombek, M.P. and M.J. Carr. 1978. Tidal triggering of seismic and volcanic phenomena during the 1879-1880 eruption of Islas Quemadas volcano in El Salvador, Central America. *J.Vocanol. geotherm. Res.* **3**: 299-307.

Gomberg, J. and S. Davis. 1996. Stress/strain changes and triggered seismicity at The Geysers, California. *J. geophys. Res.* **101**: 733-749.

Gougenheim, A. 1961. Confirmation, par l'observation, du rôle négligeable de la marée terrestre dans la production des séismes. *Compt. Rend.* **252**: 3313-3314.

Hamilton, W.L. 1973. Tidal cycles of volcanic eruptions - fortnightly to 19 yearly periods. *J. geophys. Res.* **78**: 3363-3375.

Hartzell, St. and T. Heaton. 1989. The fortnightly tide and the tidal triggering of earthquakes. *Bull. seism. Soc. Am.* **79**: 1282-1286.

Hayakawa, Y. 1995. A catalog of volcanic eruptions during the last 2000 years in Japan. *EOS Trans. Am. geophys. Un.* **Vol.76, No.46**: p.F670.

Heaton, T.H. 1975. Tidal triggering of earthquakes. *Geophys. J. R. astr. Soc.* **43**: 307-326.

Heaton, T.H. 1982. Tidal triggering of earthquakes. *Bull. seim. Soc. Am.* **72**: 2181-2200.

Hunter, R.N. 1978. An explanation for sideral periods in earthquake aftershock sequences. *J. geophys. Res.* **83**: 1253-1256.

Imbo, G. 1955. Considerations relevées de l'étude séismique du dernier paroxysme vesuvien. *Bull. Volcanol.* **16**: 161-169.

Jaeger, J.C. 1969. Elasticity, Fracture and Flow. Methuen, Chapman and Hall, London. 268 pp.

Jeffreys, H. 1938. Aftershocks and periodicities in earthquakes. *Gerl. Beitr. Geophys.* **53**: 111-139.

Jentzsch, G. 1995. Mayon volcano: Ocean tidal triggering activities ? In: *Proc. 12th Int. Symp. on Earth Tides.* pp. 487-499. H.T. Hsu (ed.). Science Press Beijing, New York.

Johnston, M.J.S. and F.J. Mauk. 1972. Earth tides and the triggering of eruptions from Mt.Stromboli, Italy. *Nature.* **239:** 266-267.

Kilston, S. and L. Knopoff. 1983. Lunar-solar periodicities of large earthquakes in Southern California. *Nature.* **304:** 21-25.

King, G.C.P., R.S. Stein and J. Lin. 1994. Static stress changes and the triggering of earthquakes. *Bull. seism. Soc. Am.* **84:** 935-953.

Klein, F.W. 1976. Earthquake swarms and the semidiurnal solid earth tides. *Geophys. J. R. astr. Soc.* **45:** 245-295.

Klein, F.W. 1976. Tidal triggering of reservoir-associated earthquakes. *Engineering Geology.* **10:** 197-210.

Klein, F.W. 1984. Eruption forecasting at Kilauea volcano, Hawaii. *J.geophys. Res.* **89:** 3059-3073.

Knopoff, L. 1964. Earth tides as a triggering mechanism for earthquakes. *Bull. seim. Soc. Am.* **54:** 1865-1870.

Knopoff, L. 1970. Correlation of earthquakes with lunar orbital motions. *The Moon.* **2:** 140-143.

Knopoff, L. and J.K. Gardner. 1972. Higher seismic activity during local night on the raw world-wide catalogue. *Geophys. J. R. astr. Soc.* **28:** 311-313.

Kobarg, W. and E. Lippmann. 1986. Gezeitenmessungen auf dem Ekström-Schelfeis, Antarktis. *Z. Polarforschung.* **56:** 1-21.

Lammlein, D.R., G.V. Latham, J. Dorman, Y. Nakamura and M. Ewing. 1974. Lunar seismicity, structure and tectonics.*Rev. Geophys.* **12:** 1-21.

Lomnitz, C. 1966. Statistical prediction of earthquakes. *Rev. Geophys.* **4:** 377-393.

Lopes, R.M.C., S.C.R. Malin, A. Mazzarella and A. Palumbo. 1990. Lunar and solar triggering of earthquakes. *Phys. Earth planet. Inter.* **59:** 127-129.

Malin, S.R.C. and S. Chapman. 1970. The determination of lunar daily geophysical variations by the Chapman-Miller method. *Geophys. J. R. astr. Soc.* **19:** 15-35.

Martin, D.P. and W.I. Rose. 1981. Behavioral patterns of Fuego volcano, Guatemala. *J. Volcanol. geotherm. Res.* **10:** 67-81.

Mauk, F.J. and M.J.S. Johnston. 1973. On the triggering of volcanic eruptions by earth tides *J. geophys. Res.* **78:** 3356-3362.

McNutt, S.R. and R.J. Beavan. 1981. Volcanic earthquakes at Pavlof volcano correlated with the solid earth tide. *Nature.* **294:** 615-618.

McNutt, S.R. and R.J. Beavan. 1984, Patterns of earthquakes and the effect of solid earth and ocean load tides at the Mt. St. Helens prior to the May 18, 1980, eruption. *J. geophys. Res.* **89:** 3075-3086.

McNutt, S.R. and R .J. Beavan. 1987. Eruptions of Pavlov volcano and their possible modulation by ocean load and tectonic stresses. *J. geophys. Res.* **92:** 11,509-11,523.

Michael, M.O. and D.A. Christoffel. 1975. Triggering of eruptions of Mt. Ngaurahoe by fortnightly earth tide maxima, January 1972 - June 1974. *N.Z. J. Geol. Geophys.* **18:** 273-277.

Minakami, T. 1974. Seismology of volcanoes in Japan. In: Physical Volcanology. pp.1-27. L. Civetta, P. Gasperini, G. Luongo and A. Rapallo (eds.). Elsevier Sc. Publ. Comp., Amsterdam.

Mohler, A.S. 1980. Earthquake/earth tide correlation and other features of the Susanville, California earthquake sequence of June-July 1976. *Bull seim. Soc. Am.* **70:** 1583-1593.

Morgan, W.J., J.O. Stoner and R.H. Dicke. 1961. Periodicity of earthquakes and the invariance of the gravitational constant. *J. geophys. Res.* **66:** 3831-3843.

Oike, K. and K. Taniguchi. 1988. The relation between seismic activities and earth tides in the case of the Matsuhiro earthquake swarm. *Bull. Disas. Prev. Res. Inst. Kyoto Univ.* **38:** 17-28.

Palumbo, A. 1985. Lunar tidal triggering in volcanic areas. *Lett. Nuovo Cimento.* **44:** 563-568.

Palumbo, A. 1986. Lunar and solar tidal components in the occurence of earthquakes in Italy. *Geophys. J. R. astr. Soc.* **84:** 93-99.

Patanè, G., A. Frasca, A. Agodi and S. Imposa. 1994. Earth tides and Etnean volcanic eruptions: an attempt at correlation of the two phenomena during the 1983, 1985 and 1986 eruptions. *Phys. Earth planet. Inter.* **87:** 123-135.

Ramos, E.G., C.B. deTorres and A.C. Calderon. 1985. Earth tide influence on the recent activities of Mayon volcano. *Philippine J. of Volc.* **2:** 156-171.

Rayleigh, L. 1880. On the result of a large number of vibrations of the same pitch and of arbitrary phases. *Phil. Mag.* **10:** 73-78.

Rinehart, J.S.. 1972. Fluctuations in geysir activity by variations in earth tidal forces, barometric pressure and tectonic stresses. *J. geophys. Res.* **77:** 342-350.

Ryall, A., J.D. VanWormer and A. Jones. 1968. Triggering of microearthquakes by earth tides and other features of the Truckee, California, earthquake sequence of September 1966. *Bull. seim. Soc. Am.* **58:** 215-248.

Rydelek, P.A., P.M. Davis and R.Y. Koyanagi. 1988. Tidal triggering of earthquake swarms at Kilauea volcano, Hawaii. *J. geophys. Res.* **93.** 4401-4411.

Rydelek, P.A., I.S. Sacks and R. Scarpa. 1992. On tidal triggering of earthquakes at Campi Flegrei, Italy. *Geophys. J. Int.* **109:** 125-137.

Rykunov, A.L. and V.B. Smirnov. 1985. Variations in seismicity under the influences of lunar-solar tidal deformations. *Izvestiya, Earth Physics.* **21:** 71-75.

Sadeh, D. and K. Wood. 1978. Periodicity in lunar seismic activity and earthquakes. *J. geophys. Res.* **83:** 1245-1249.

Schuster, A. 1897. On lunar and solar periodicities of earthquakes. *Proc. R. Soc.Lond.* **61:** 455-465.

Sauck, W.A. 1975. The Branley, California earthquake sequence of January 1975 and triggering by earth tides. *Geophys. Res. Lett.* **2:** 506-509.

Shimshoni, M. 1971. Evidence for higher seismic activity during the night. *Geophys. J. R. astr. Soc.* **29:** 97-101.

Shirley, J.H. 1986. Lunar periodicity in great earthquakes, 1950-1965. *Gerlands Beitr. Geophys.* **95:** 509-515.

Shirley, J.H. 1988. Lunar and solar periodicities of large earthquakes: Southern California and the Alaska-Aleutian Islands seismic region. *Geophys. J. Int.* **92:** 403-420.

Shlien, S. 1972. Earthquake-tide correlation. *Geophys. J. R. astr. Soc.* **28:** 27-34.

Shlien, S. and M.N. Toksöz. 1970. A clustering model for earthquake occurences. *Bull. seism. Soc. Am.* **60:** 1765-1787.

Shudde, R. and D. Barr. 1977. An analysis of earthquake frequency data. *Bull. seism. Soc. Am.* **67:** 1379-1386.

Simpson, J. F. 1967. Earth tides as a triggering mechanism for earthquakes. *Earth planet. Sci. Lett.* **2:** 473-478.

Souriau, M., A. Souriau and J. Gagnepain. 1982. Modelling and detecting interactions between earth tides and earthquakes with application to an aftershock sequence in the Pyrenees. *Bull. seism. Soc. Am.* **72:** 165-180.

Tamrazyan, G.P. 1968. Principal regularities in the distribution of major earthquakes relative to solar and lunar tides and other cosmic forces. *Icarus.* **9:** 574-592.

Taubenheim, J. 1969. Statistische Auswertung geophysikalischer und meteorologischer Daten. Akademische Verlagsgesellschaft Leipzig. 368 pp.

Tsuruoka, H., M. Ohtake and H. Sato. 1995. Statistical test of the tidal triggering of earthquakes: Contribution of the ocean tide loading effect. *Geophys. J. Int.* **122:** 183-194.

Ulbrich, U., L. Ahorner and A. Ebel. 1987. Statistical investigations on diurnal and annual periodicity and on tidal triggering of local earthquakes in Central Europe. *J. Geophys.* **61:** 150-157.

VanRuymbeke, M., B. Ducarme and M. DeBecker. 1983. Attempt to model tidal triggering of earthquakes. In: *Proc. 8th Int. Symp. on Earth Tides.* pp. 651-661. J. Kuo (ed.). E.Schweizerbart, Stuttgart.

Varga, P. 1983. Connection between lunisolar and loading effects and the outbreak of earthquakes. In: *Proc. 8th Int. Symp. on Earth Tides.* pp.663-668. J. Kuo (ed). E.Schweizerbart, Stuttgart.

Varga, P. and E. Grafarend. 1996. Distribution of the lunisolar tidal elastic stress tensor components within the earth's mantle. *Phys. Earth planet. Inter.* **93:** 285-297.

Young, D. and W. Zürn. 1979. Tidal triggering of earthquakes in the Swabian Jura? *J. Geophys.* **45:** 171-182.

Zugravescu D., J. Fatulescu, D. Enescu, D. Danchiv and O. Haradja. 1989. Peculiarities of the correlation between gravity tides and earthquakes. *Rev. Roum. Géol., Géophys. et Géogr.* **33:** 3-10.

Tidal Tilt Modification Along an Active Fault

Malte Westerhaus

GeoForschungsZentrum Potsdam, Bereich 2.1
Telegrafenberg A31, D-14473 Potsdam

Abstract. Inhomogeneous deformation in the vicinity of a lateral contrast between different poroelastic structures in the crust causes local tidal tilt disturbances. A number of model approaches have been published since more than 20 years in order to give a quantitative estimation of this effect which is known as 'geologic effect'. Most of them correctly predict the general behaviour of tidal tilt modification near an elastic contrast but underestimate the large disturbances that have been observed experimentally. In principle, the geologic effect can be used to infer changes in the state of deformation with increasing tectonic stress. Several experiments have been conducted in earthquake prone areas in order to detect temporal variations of tidal tilt parameters related to seismotectonic events; the results, however, so far are inconclusive.

Tidal tilt observations along a western strand of the North-Anatolian Fault Zone, integrated within an interdisciplinary earthquake research project, reveal considerable deviations of the tidal tilt parameters from the response of a laterally homogeneous Earth. The most prominent temporal phenomenon during the years 1988 to 1994 are seasonal variations of the order of 8% with respect to the tidal tilt residual amplitudes. They are superimposed on a secular drift of the order of 5%, which starts or intensifies in the beginning of 1991. Both classes of signals are correlated with similar changes in independent observables like microseismic activity, the traveltimes of longitudinal waves, the accumulation of tectonic strain and the temperature of ground water. They are interpreted as being caused by changes of effective stress due to fluctuations of internal pore fluid pressure and/or the ambient stress level.

1   Introduction

The local tidal tilt and strain response of the earth is sensitive to lateral inhomogeneities in the crust, pore fluid diffusion, topography and loading effects. Consequently, it often deviates considerably from the response expected for theoretical earth models. A tidal tilt experiment aimed to study seismotectonic processes in the crust focusses on the 'geologic effect', i.e. the lateral contrast between the poroelastic parameters of an active fault zone and the intact host rock. The elasticity contrast leads to an inhomogeneous tidal strain field since the equilibrium conditions require that stress is continuous across the boundary. The discontinuity in strain induces additional tilts that are superimposed on the tilt tides of the solid earth. The process is conveniently described by factors coupling the anomalous tilt to the well known tidal strains in the farfield of the discontinuity (King et al., 1976).

The modification of the local tidal tilt response functions close to a contrast of two different geological units comprises a lot of information useful in the study of seismotectonic processes. It is known from laboratory experiments that elastic moduli, like many other rock parameters, are stress dependent. Thus, a continuous monitoring of tilt tides is supposed to be indicative for changes in the state of deformation of the fault zone material and/or the host rock with increasing tectonic stress or, equivalently, with time (e.g. Beaumont and Berger, 1974). This approach assumes that any other source of tidal tilt modification either is negligible or is stable in time. In principle, a fit of the observed tidal parameters by numerical or analytical models could yield quantitative estimates of the elastic moduli inside and outside of the fault zone. This task, however, requires that, in contrast to the monitoring of temporal changes, any source of tidal tilt modification in addition to the geologic effect is quantitatively known. In general, the complex interaction of different tilt generating sources will only be incompletely known. This, in combination with an oversimplification of the model approaches, may be one reason for the fact that most attempts to model the geologic effect tend to underestimate the observed anomalies, at least on the basis of a 'plausible' contrast of poroelastic properties (e.g. Meertens, 1989).

Calibration of the local tilt response to an applied far-field strain using the tides allows the recovery of tectonic strain events from the tilt drift (Zschau et al., 1991). Any change in the state of deformation that leads to a temporal modification of tidal tilt parameters, should induce correlated changes of the nontidal tilt. Investigation and interpretation of transient tilt signals, however, is difficult due to the manifold influences of meteorological parameters. One of the advantages in using strongly periodical tidal signals is the fact that meteorological disturbances are concentrated largely on the solar peaks in the spectrum. The lunar tides which are exclusively used in the present study for the monitoring of seismotectonic processes, are much less affected.

2   Strain-Tilt Coupling

2.1 Lateral Inhomogeneities

The existence of a zone with elastic properties differing from those of the surrounding rocks, often called 'inclusion', modifies tilt and strain observations as a result of an applied far-field strain. Beaumont and Berger (1974) quantified the effect of an annular anomalous inclusion with a lateral extension of 40 km and a depth of 20 km using the Finite Element Method. Assuming a 15% reduction of v_p within the anomalous zone they derived a 60% change of tidal tilt and strain amplitudes. The maximum change is found at the edges of the geological inhomogeneity; moving into the anomalous zone the tidal anomalies decrease. The sign of the tilt anomaly depends on the location of the observation point with respect to the inclusion. The amplitudes of the tilt and strain tide anomalies are proportional to the contrast in Poisson's ratio and inverse effective areal bulk modulus. The model calculations of Beaumont and Berger are dedicated to earthquake prediction research. They call their anomalous zone 'dilatant': a reduction of the p-wave velocity along with unchanged shear wave velocity results in a decrease of Poisson's ratio inside the dilatant zone. Thus, the inclusion actually is less compressive than the surrounding medium. Dilatant hardening is an essential feature of deformation of saturated rocks and is assumed to account for a variety of intermediate and short term precursors. Further FEM models, based on different sets of contrasting elastic parameters, have been published by Harrison (1976), Gerstenecker et al. (1986) and Meertens (1989).

Using the minor parameter method Molodensky (1983) derived analytical solutions for the effect of a geological inhomogeneity. He showed that, at a given place, amplitudes and phases of tidal tilts and deformations depend on the orientation of the anomalous crustal zone relative to the principal stress axes of the tidal wave and on the contrast of elastic moduli. He claimed that due to the phase conditions between tidal strains and tilts the main effect of a narrow tectonic fault with long horizontal extension is an amplitude disturbance in the NS-tilt direction and a phase anomaly in the EW-tilts. Molodensky's approximate solutions are in good agreement with the FEM models published by Beaumont and Berger (1974). Comparing his results to tidal tilt observations across the Kondar fault, USSR, he concluded that the analytical model describes the essential features of the observed spatial variations of the tidal parameters. However, the model can explain only about one third of the observed amplitude variations.

Milkereit (1988) included quadratic terms in the stress-strain relation in order to model nonlinear elastic deformation mathematically. This approach leads to stress dependent, anisotropic elastic moduli. The relative volume increase and the concomitant variation of the seismic traveltimes are essentially described by his method. He extended the Eshelby formalism (Eshelby, 1957) on the anisotropic case and calculated the effect of an anisotropic, weak inclusion on tidal tilt measurements at the surface.

The general limitation of model calculations using elastic inclusion models is that they often are unable to fit the the magnitude of the observed tidal tilt anomalies, at least if 'plausible' elastic contrasts are assumed. An exception is the approach of Kirsch and Zschau (1986). They used the Eshelby formalism to calculate the effect of an anomalous crustal zone having an ellipsoidal shape. Their 3-dimensional analytical method allows the calculation of phase differences between the tilt and strain disturbances and tidal potential. Phase anomalies depend on the contrast of shear moduli in the homogeneous half space where the ellipsoid is embedded, and the anomalous inclusion. The authors applied the dilatancy-diffusion model to estimate temporal variations of elastic parameters inside the inclusion in order to predict tidal modifications that would have been observed prior to the Blue Mountain earthquake of 1973. They arrived at maximum values of several hundred per cent for tilt and strain modifications, respectively. Such large tidal tilt disturbances, however, caused by a relatively small contrast in elastic properties, have not been confirmed by other modelling methods.

2.2 Fluid Pressure Variations

The purely elastic effect of a parameter contrast in dry rocks is modified by migration of pore fluid in the case of porous, fluid saturated media. The connection between internal pore pressure disturbances and rock deformation is governed by Hooke's generalized law, extended for poroelastic media (originally derived by Biot (1941)):

$$\varepsilon_{ij} = \frac{c}{3}\left(\frac{1+v}{1-2v}\sigma_{ij} - \frac{v}{1-2v}\sigma_{kk}\,\delta_{ij} - \alpha\,p\delta_{ij} \right) \tag{1}$$

where e_{ij} and σ_{ij} ($i \neq j$) are shear strains and shear stresses, respectively; σ_{kk} is the mean stress, p is the pore pressure, c is the matrix compressibility, n is Poisson's ratio, a is the coefficient of effective stress and δ_{ij} is the Kronecker symbol. A different poroelastic response of two materials leads to a different fluid pressure at both sides of the boundary. In its endeavours to equilibrate the pressure difference, pore fluid migrates from the region of high pressure into the region of low pressure. Friction between the immobile rock matrix and the mobile fluids causes deformations of the rock which are measured by sensitive strain- or tiltmeters.

The influence of moving pore fluids on tilt measurements at subtidal periods is known from observations near pumped wells. An extensive discussion of these phenomena, including model calculations of the relevant deformation components, has been given by Kümpel (1989), compare also Kümpel (Well Tides, this volume). Westerhaus and Zschau (1989) pointed out that an internal pore pressure imbalance may be the result of an inhomogeneous poroelastic response of the fluid saturated formation to tidal volumetric strain. In the vicinity of a lateral inhomogeneity, both the elastic effect as well as pore fluid migration would be able to produce tilts. There is, however, an important difference. The elastic effect can be treated as frequency independent as long as the wavelength of the applied strain is much larger than the inhomogeneity itself. For the scope of the elastic effect St. Venants principle can be applied. This states that the effect will be felt over a distance which is comparable to the size of the inhomogeneity. In contrast to this, the migration of pore fluid is a diffusional process depending on the poroelastic parameters and on the frequency of the applied farfield strain. Thus, the scope of the second effect will vary considerably for tidal and for secular signals.

3   Dilatancy and Rock Friction

In a uniaxial compression test the rock deforms in a nearly linear elastic manner according to its intrinsic elastic properties until the applied stress reaches about half the fracture stress. With further increasing stress rock deformation deviates from the linear behaviour. This is accompanied by a reduction in the axial modulus and an inelastic lateral expansion due to the opening of microcracks (Brace et al., 1966). The inelastic volume increase as a result of the application of a deviatoric stress is called dilatancy. It is assumed that, in principle, the results from the laboratory measurements can be applied to the deformation of rocks under tectonic stresses.

Beaumont (1978) discussed tidal interactions with a nonlinearly elastic crust. He assumed that the tectonic stress-strain curve follows a curve either convex or concave depending on which strain component is considered (Fig. 1). Tidal stresses, being superimposed on the tectonic ones, induce tidal strains proportional to the elastic tangent modulus of the nonlinear stress-strain curve. Due to the changing gradient the strain response will change with increasing tectonic stress. Since tidal stresses are much smaller than the range of tectonic stresses, they will sample an essentially linear elastic behaviour proportional to the elastic tangent moduli of the nonlinear stress-strain curve. Therefore, the tidal tilt response functions remain linearly elastic. However, as Beaumont (1978) points out, they are almost certainly anisotropic because some components of the strain tensor may increase while others decrease, and amplitude as well as phase variations will occur with increasing tectonic stress.

Lateral expansion and, thus, dilatancy of a compressive test specimen is resisted by applying 'confining' pressure to its side. With increasing confining pressure the shape of the stress strain curve remains essentially the same, but the yield point (where transition from linear towards nonlinear behaviour occurs) and the fracture strength (the maximum of the stress-strain curve) are shifted along the ordinate towards higher compressive stresses. Nur (1972), Scholz et al. (1973) and others combined the effects of dilatancy with the diffusion of pore fluids and developed the dilatancy-diffusion model. They claimed that during the initial stages of rapid crack growing the pore pressure continuously drops until the rock is partially undersaturated. The pore pressure decrease is accompanied by a steady increase of effective pressure (ambient pressure minus pore pressure), resulting in a strengthening of the rock ('dilatant hardening'). The enhanced resistivity of the rock to deformation will decelerate the progress of dilatancy

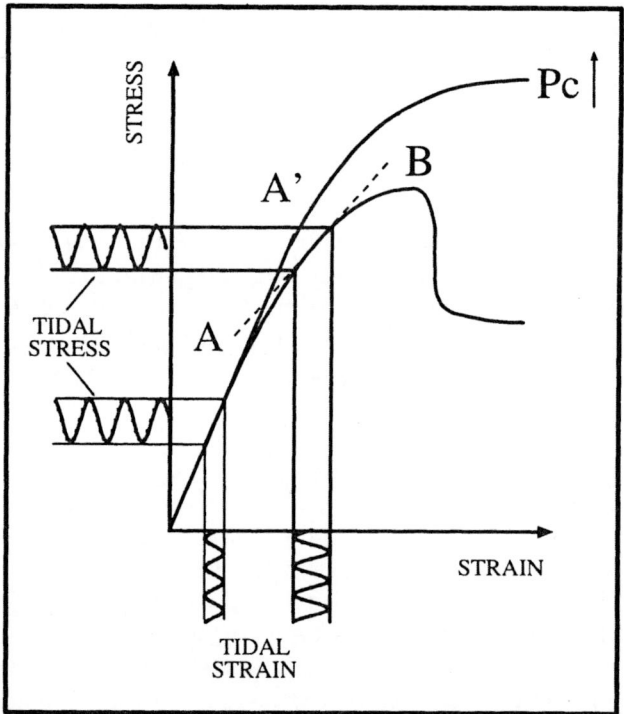

Fig. 1. Typical stress-strain curve showing the influence of confining pressure (P_c) on the deformation of rocks. Points A, A' denote the yield point where deviation from linear elastic behaviour takes place; fracture strength is reached at point B. The dashed line indicates the local elastic tangent modulus (modified after Beaumont, 1978).

until a point is reached at which the influx of pore fluid from adjacent regions exceeds the rate of crack-growing. The pore volume of the rock is refilled again and the pore pressure increases towards the ambient level. During pore pressure increase the effective pressure drops and the rock volume is weakened. A dilatancy cycle is terminated by the rupture of the dilatant volume which is felt by a sequence of (micro)seismic events and a concentration of the epicenters at this place.

The dilatancy-diffusion model was the first closed theory for the generation of earthquakes and was able to explain qualitatively the majority of precursory phenomena known at that time. Undersaturation, for example, should lead to a reduction of p-wave velocity and, thus, to a decrease of the v_p /v_s -ratio, as has been observed prior to several earthquakes in USA (Aggarval et al., 1973) and Garm region, USSR (Neresov et al., 1969). At the meantime, however, reinterpretation of datasets and the lack of precursory phenomena, which should have been observed according to the model, raised some objections against the hypothesis. It seems questionable, for example, that substantial crustal regions (with diameters more than 15km) can undergo dilatancy, as considered by Scholz et al. (1973). However, it is conceivable that prior to earthquakes, dilatancy occurs at limited areas where stress is concentrated (Mogi, 1985). Several authors assume that dilatancy occurs only within the fault zone (e.g. Rice, 1983).

The dilatancy-diffusion model includes two essentially time dependent features: i) the nonlinearity of the stress-strain curve with increasing deviatoric stress and ii) the hardening (or weakening) effect of pore fluid migration, i.e variations of confining pressure. Accordingly, the difference in the elastic properties at the contact of a dilatant and a non-dilatant region is expected to change with the tectonic stress build-up and, thus, with time. Time dependence of the lateral elasticity contrast should lead to temporal variations of the local tidal tilt and strain response functions under the asumption that deformation of the inclusion is fully determined by the elastic tangent moduli of the nonlinear stress-strain curve.

Beaumont (1978) emphasized the importance of hysteresis in addition to nonlinear behaviour, which is known from laboratory experiments. Under cyclic loading the strain paths for increasing and decreasing stress are different. Beaumont argued that in the stage of dilatancy, any small decrease of stress will result in a recovery of elastic strain. The dilatant strain is not recovered until all the elastic strain has been recovered. Consequently, the strain response due to a small periodical stress is controlled by the linear elastic modulus, independent of the deformation state of the probe. Beaumont pointed out that temporal changes of the tidal strain response will occur only if the tectonic stress rate is equal or higher than the mean rate of tidal stress. Normally, the tectonic stress build-up is 1-2 orders of magnitude lower than the mean tidal stress rate; accelerated tectonic stress rates, however, are probable in the final stage of an earthquake cycle (e.g. Tullis, 1988) and may permit (precursory) temporal variations of tidal tilt and strain response functions.

Hysteresis is explaind by the work that has to be done against friction to enable sliding on preexisting cracks, fractures and faults. The effective static frictional strength of an existing fault in a fluid saturated crust may be approximated by a criterion of Coulomb form (e.g. Chen and Nur, 1992):

$$\tau_f = C + \mu_s \left(\sigma_n - ap \right) \tag{2}$$

where C is the cohesive strength of the fault, μ_s is the static friction coefficient (typically about 0.75), p is the fluid pressure, and σ_n is the normal stress on the fault. The cohesive strength of a fault, which has been ruptured by earthquakes in the past, usually is lower than that of the surrounding rocks. Therefore, faults often represent planes of weakness which are reactivated even if they are not optimally orientated with respect to the recent stress system.

To account for hysteresis, Beaumont (1978) assumed that a dilatant inclusion will behave elastic-plastic allowing for frictional slip on preexisting cracks. Gerstenecker et al. (1986) showed that reduced friction on vertical planes considerably increases anomalies of the tidal tilt response function. Introduction of cohesionless fractures with low coefficient of friction has the advantage that deformation is localized, i.e. it is not necessary to assume a strong reduction of elastic properties over a large volume. The authors claimed that this may help to overcome the discrepancies between the rather small tilt anomalies calculated by linear elastic models based on a 'plausible' contrast of elastic parameters and the much larger anomalies in amplitude and phase observed experimentally. In their model they were able to model very large tidal tilt disturbances; the magnitude depending on i) the vertical extension of the fracture, ii) the depth of burial, and iii) the coefficient of friction along the fracture.

Gerstenecker et al. (1986) introduced a low coefficient of friction ($\mu_s = 0.2$ as appropriate for clay gouge) in a depth range of 10 km to 15 km. In a more general approach one could assume a low frictional strength τ_f of the fault zone which could be, for example, caused by low normal stress on the fault or high pore pressure. In models that include frictional slip on existing fractures temporal variations of the tidal tilt response functions can be expected if the frictional strength varies temporally. According to Eq. (2) this will be the case with temporal changes of normal stress, varying pore pressure, changes of the coefficient of friction and changes of cohesion. The dependence of depth dependent friction parameters on increasing tectonic stress or, equivalently, time, has been modelled by Tse and Rice (1986).

It is questionable, however, if the small tidal stresses are capable of initiating inelastic deformation. Presumably, this will be the case for special conditions only, e.g. if the fault zone material deforms plastically at residual strength. In the normal case friction will not be overcome and the fault will deform elastically. According to Goodman and Sundaram (1978) the elastic deformation of a crack is governed by a parameter called 'shear stiffness', k. Model calculations show that a low k results in inhomogeneous deformation around a buried crack which has a considerable influence on the tidal tilt response (Reinders, unpublished results).

No matter whether a model approach assumes (non)linear-elastic or elastic-plastic deformation, numerical calculations should account for the fact that the state of deformation is not only time-, but also depth dependent. Confining pressure, normal pressure on a fault and the coefficient of friction change with depth. In his synoptic model of a shear zone in a quartzo-feldspathic crust Scholz (1990, Fig 3.19) assumes that strength increases with depth following the friction law until peak strength is reached at a depth of about 15 km. This is the depth to which large earthquake that preferably nucleate near the base of the seismogenic layer at about 7 km, can propagate. Below the peak, strength is reduced by increasing temperature. Depth dependence of fracture strength and frictional strength presumably is of considerable importance for the numerical calculation of the effect of a lateral inhomogeneity on local tidal tilt and strain response functions.

4   Observations

The influence of a geological inhomogeneity on the observation of Earth tidal tilt has been proved experimentally. Bonatz et al. (1983) found deviations of the tidal tilt parameters from the global values of up to 100% in amplitude and 90° in phase on a tiltmeter profile crossing the Hunsrück-Fault in Northwest Germany. Meertens (1987) reported tidal tilt anomalies of up to 46% in amplitude and 50° in phase near an anomalous body in Yellowstone Nationalpark, USA. Nishimura (1950), Mikumo et al. (1977), Mao (1989) and Latynina and Rizaeva (1976) reported temporal variations of the tidal tilt and strain parameters in seismoactive regions. Their observations generally support the idea of Beaumont and Berger (1974) who claimed that the Earth's response to tidal forces should be sensitive to temporal changes of the elastic properties inside the hypocentral volume of an impending earthquake. However, the results remain inconclusive and often instrumental effects cannot be excluded. Peters and Beaumont (1985) observed differences between tidal results, obtained from two borehole tiltmeters in the seismically active area around Québec, Canada, as large as 20% in amplitude and 5° in phase. Time variations of the major tidal constituents, however, induced by fluctuations of the loading tide input of a nearby river, were remarkably consistent between the two tilt sensors. The authors claimed that the temporal variations were not in general statistically significant on a 95% level, and that their physical significance would hardly be detected by observations from a single borehole alone.

The tidal tilt experiment included in the interdisciplinary German-Turkish Project on Earthquake Research aims at the systematic search for temporal variations in the tidal parameters as a result of seismotectonic processes in the crust. The investigation area is located about 200 km east of Istanbul in a region where the North-Anatolian Fault Zone ('NAFZ') splits up in several branches

(Fig.2). The deformation along the fault is of right-lateral strike-slip type with some normal faulting in the western part. During the last 100 years, the area has been repeatedly afflicted by earthquakes with magnitudes between 5.5 and 7.2. About 20 institutions from Germany and Turkey are involved in the joint research project covering the fields of geophysics, geology, seismology, geodesy, hydrology, nuclear physics and sociology. The interdisciplinary activities include observations of microseismic activity, water level in wells, ground water temperature, rock deformation, rock magnetization, electric properties of the ground, radon emanation, neotectonics and meteorology.

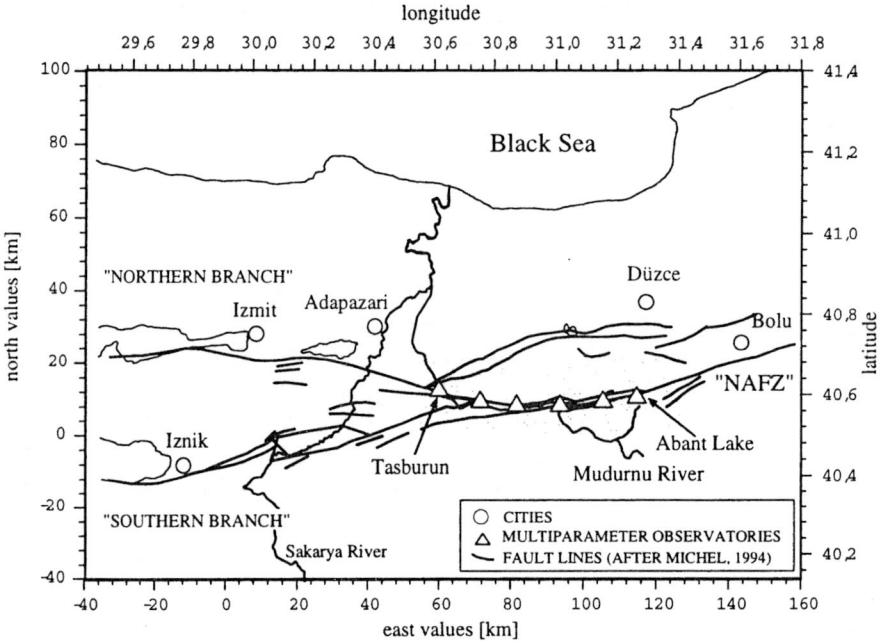

Fig. 2. The investigation area of the German-Turkish Project on Earthquake Research, located about 200 km east of Istanbul. Triangles indicate the six multiparameter observatories; from west to east (abbreviations in brackets): Tasburun (TAB), Samanpazari (SAM), Dokurcun (DOK), Taskesti (TAS), Igneciler (IGN) and Abant (ABA). The shaded area is referred to as 'Mudurnu Area'.

4.1 Data Acquisition and Analysis

Tilt measurements are included in the multidisciplinary experiments concentrated at the six Multiparameter Observatories, located at Abant Lake (ABA) and near the villages of Igneciler (IGN), Taskesti (TAS), Dokurcun (DOK), Samanpazari

(SAM) and Tasburun (TAB). The stations are separated by around 12 km (Fig. 2). Tiltmeters of Askania type vertical pendulums have been chosen which have proven to operate stably at tidal precision and allow an in-situ calibration. They are installed in 30 m deep boreholes. Groundwater wells with depths ranging between 36 m and 58 m are in operation at each station. The water level is continuously monitored by OTT-pressure sensors. Since 1988, several wells are equipped with precise thermometers with a resolution of 10 µKelvin. The tilt and groundwater measurements were started in May 1985. Since autumn 1987 a digital data acquisition system has been employed. Sampling rates are 2 min for tilt and 15 min for groundwater and meteorological data.

Tidal tilt- and well level analysis is done in two steps using the HYCON-MC algorithm (Schüller, 1977). First, a long record (about 4 years) is analyzed to determine the static tidal tilt response function of the Earth at every site. In a second step, the analysis is done by a moving window. A basic interval with a length of 60 days is shifted at constant steps $\Delta T = 48$ hours over the data set. A tidal analysis is carried out for every step. This results in discrete series of tidal parameters γ and κ which are used to investigate the temporal stability of the local tidal tilt response function. For practical purposes, tidal analysis has been restricted on the main diurnal and semi-diurnal lunar waves O_1 and M_2, respectively, which i) have large amplitudes and ii) are not influenced by the effects of solar heating (temperature-, moisture- and barometric pressure changes).

The length of the basic interval is a compromise between the two counterpoles temporal and spectral resolution. A window length of 60 days is sufficient to separate the main lunar waves from other constituents with large spectral amplitudes, and allows for the optional use of a Hanning window with lower spectral resolution than a rectangular window, which is used here. The method implicitly assumes that the tidal tilt response function is constant over the window length. This, in general, will not be the case. Any change of a tidal parameter is convolved with the window function and the results of a moving analysis do not reflect the true values of the tidal parameters γ and κ at a certain point of time. This has to be kept in mind if the moving tidal tilt response functions are compared to non-averaged quantities.

To avoid large interruptions of the sequential tidal estimates data gaps up to 30 days are interpolated. An a-priori-estimate of the tidal parameters within the gap is specified on the basis of a static analysis of an appropriate section of the record at both sides. The results of the moving analysis are checked for an inadequate interpolation. Temporal variations of the tidal tilt response functions with a length equal to the width of a data gap plus the length of the moving rectangular window are reexamined with respect to the interpolating parameters.

Barometric pressure is included as an additional time series into the tidal analysis. On average, barometric disturbances of the lunar tilt tides are small. For a typical pressure sensitive test site like DOK the static influence of the lunar pressure wave on the M_2-tide amounts to approximately 3%. Using time variant spectral analysis methods, however, it can be shown that the lunar air pressure tides are non-stationary, presumably due to aperiodic changes that temporarily influence the tidal band. The apparent amplitude of the local barometric M_2-tide within the test area in Turkey varies between 0.002 and 0.12 hPa, whereas the static value is 0.032 hPa +/- 0.008 hPa. If the tiltmeter site is sensitive to barometric pressure changes, the random barometric signals are superimposed on the tidal tilts and are recovered by a time-variant analysis of the tidal parameters. As an example, Fig. 3 shows the normalized amplitude and the phase of the M_2-tilt tide at DOK before and after the removal of barometric disturbances. Because the disturbing signals are unpredictable, the MC-mode is not applicable with a time variant analysis of gappy data. Therefore, the pressure series is convolved with the regression coefficients obtained by a static MC-analysis over a 4-year-period and subtracted before a time variant analysis on the tilt data is carried out. This procedure does not take into account potential temporal variations of the local pressure response of the site. It is intended to optimize the current tidal analysis routine in future. The need of high-quality, digitally recorded barometric data for this purpose shortens the available data sets to Nov. 1987.

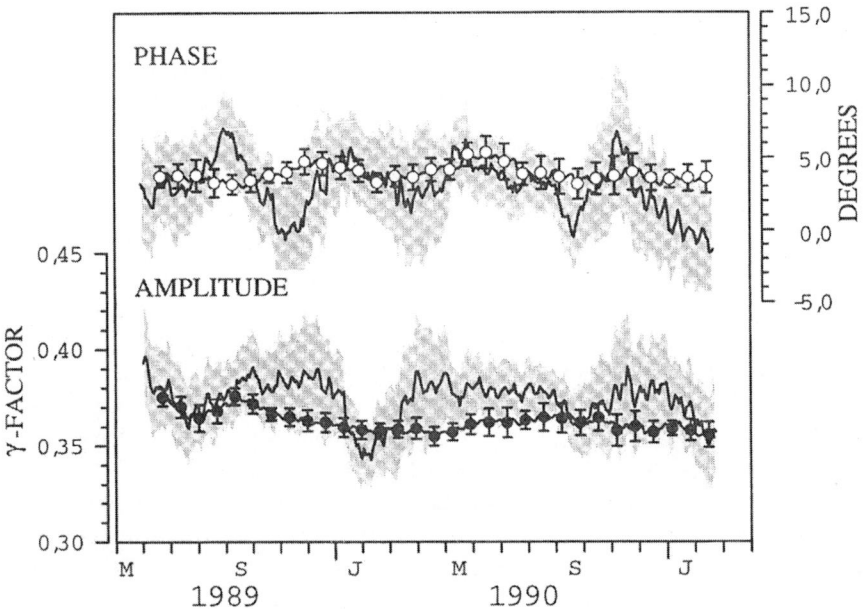

Fig. 3. Normalized amplitude (γ-factor) and phase of the time variant M_2-tidal tilt response function (Dokurcun, NS-component) before and after removal of barometric noise. The corrected functions are marked by circles. Shaded backgrounds and error bars (inserted every 10th value) indicate 95%-confidence limits.

4.2 Static Tidal Tilt Parameters

Amplitudes and phases of the main diurnal (O_1) and semi-diurnal (M_2) lunar constituents composing the average or time-independent local tidal tilt response of the Earth at the six multiparameter sites are summarized in Table 1. The results are obtained from a HYCON-MC-analysis including a parallel air pressure record to remove barometric disturbances. The raw data has been high-pass filtered (Pertzev) to avoid a bias of the tidal estimates by longterm trends.

Table 1. Static tidal tilt parameters observed at the six Multiparameter Observatories in the Mudurnu region

		NS-COMPONENT			
		γ-factor	sig(γ)	phase κ	sig(κ)
				[deg.]	[deg.]
TAB	O1	0.2816	0.2464	-8.26	50.17
	M2	0.7488	0.0143	-0.58	1.10
SAM	O1	1.3697	0.0521	-168.35	2.18
	M2	1.2576	0.0034	-7.88	0.16
DOK	O1	1.4425	0.0202	-2.61	0.80
	M2	0.3562	0.0016	3.47	0.25
TAS	O1	0.5126	0.2002	47.75	22.47
	M2	0.6982	0.0130	-11.19	1.07
IGN	O1	2.5495	0.0957	-17.51	2.15
	M2	0.1803	0.0047	-21.73	1.49
ABA	O1	2.7417	0.0347	-70.68	0.73
	M2	0.9482	0.0026	80.55	0.16

		EW-COMPONENT			
		γ-factor	sig(γ)	phase κ	sig(κ)
				[deg.]	[deg.]
TAB	O1	0.6573	0.0417	-2.86	3.63
	M2	0.7381	0.0083	-4.17	0.64
SAM	O1	0.6486	0.0147	6.57	1.30
	M2	0.7357	0.0020	-0.31	0.16
DOK	O1	0.7393	0.0041	1.88	0.32
	M2	0.6434	0.0007	5.16	0.06
TAS	O1	0.7280	0.0188	3.89	1.48
	M2	0.7388	0.0046	-5.99	0.36
IGN	O1	0.8676	0.0173	22.79	1.14
	M2	0.6812	0.0034	6.99	0.29
ABA	O1	0.5192	0.0067	-59.44	0.74
	M2	0.3258	0.0015	26.56	0.27

Contribution of ocean loading to the tidal tilt parameters in the investigation area, calculated from the model of Schwiderski (1980), is of the order of 3% and 2° in the amplitudes and phases, respectively. The loading tides are not considered further since their influence on the results presented in the following chapters is marginal.

Table 1 reveals a strong scattering of the normalized amplitudes (γ-factor) at all stations along the fault and considerable deviations from the value $\gamma= 0.69$ expected for a laterally homogeneous Earth. With the exception of Abant, where strong anomalies are found in the EW-direction as well, the deviations are concentrated in the NS-direction. A first insight into the mechanism producing the tidal tilt disturbances can be obtained by comparing the amplitude responses for the O_1 - and the M_2-wave at a single station, as for example Dokurcun or Igneciler. At both places, the observed amplitude of the M_2-tidal tilt response in the NS-direction is smaller than on a laterally homogeneous Earth, whereas the γ-factor of the O_1-wave significantly exceeds the global value of 0.69. The difference in the phases of both waves is close to $0\,°$. Thus, it can be concluded that the O_1- and the M_2-response of the Earth at these stations is influenced with opposite sign, respectively. This special behavior of the local tidal tilt response of the Earth is diagnostic for the influence of secondary tilts caused by an inhomogeneous response of the crust to Earth tidal strains. A process like this is known in the literature as strain-tilt coupling.

4.3 Tilt-Site Tensors

Strain-tilt coupling is mathematically described by the tilt-site tensor which linearily relates tilt to strain (King et al. (1976); Zürn et al. (1977)):

$$R_j = A_{jk}\, B_{kl}\, E_l \tag{3}$$

R_j are the observed tilt residuals (i.e. the difference of the observed tide and the tide expected on a laterally homogeneous Earth) in two orthogonal directions. E_l contains the 3 independent components of the surface strain tensor. A_{jk} is the tilt-site tensor depending on the local conditions. It consists of six independent components, called strain-tilt coupling coefficients. As an extension of the procedure of King et al. (1976) the 'strain-site tensor' B_{kl} is inserted in Eq. (3) which relates the components of the surface strain tensor calculated for a chosen earth model to the *in-situ* strains acting on the tiltmeter site. In the phenomenological approach used here the unit matrix is used. B_{kl}, however, will become important when the strain-tilt coupling coefficients are modelled in terms of plausible numerical values of local crustal parameters. Factors expressing harmonic dependence of R_j and E_l on time have been omitted. The potential time dependence of the components of A_{jk} is investigated later on. Eq. (3) is solved by a least squares approach using the main diurnal and semi diurnal lunar constituents.

The approach assumes that the process is frequency independent. This assumption is valid only for the purely elastic effect of crustal structures. If a coupling between pore pressure gradients and tilt exists, the frequency dependence of pore fluid diffusion makes the components of A_{jk} frequency dependent and also leads to a phase shift between the tilt disturbance and the tidal potential. In that case, a determination of the strain-tilt coupling coefficients by use of two tidal species of different period can lead to erroneous results.

Tilt-site tensors are established for each one of the six tilt stations in the Mudurnu area. The calculated strain-tilt coupling coefficients are presented in Table 2. They are of the same order of magnitude as is known from other places, for example from tilt observations at Schiltach mine observatory in SW-Germany (Zürn et al., 1977). The NS-anisotropy of the local tidal tilt response is preserved by the large coefficients coupling NS-tilt to NS-strain at Samanpazari, Dokurcun and Igneciler, which considerably exceed those in EW-direction. The pecularity of the Abant site is accentuated also by the tilt-site tensor. All six strain-tilt coupling coefficients exceed 1 nrad/nstrain, the largest one is 3.72 nrad/nstrain for coupling of NS-tilt to the shear strain. A chi-square test shows that the quality of the solution for Abant is inacceptable, indicating that the simple linear approach of Eq.(3) is not suitable for the special conditions at this site.

Table 2. Coefficients coupling the observed tidal tilt residuals to strain calculated for the Gutenberg-Bullen Earth model with the top 1000 km replaced by the oceanic and continental shield structures of Harkrider (1970). All entries are given in [nrad/nstrain].

		$c(\theta\theta)$	error	$c(\lambda\lambda)$	error	$c(\theta\lambda)$	error
		[nrad/nε]	[nrad/nε]	[nrad/nε]	[nrad/nε]	[nrad/nε]	[nrad/nε]
TAB	NS	-0.116	0.078	-0.219	0.214	-0.035	0.064
	EW	0.267	0.061	-0.070	0.150	-0.311	0.056
SAM	NS	-1.530	0.017	-0.484	0.045	-0.757	0.015
	EW	0.095	0.018	-0.277	0.052	-0.316	0.014
DOK	NS	1.033	0.007	-0.113	0.018	0.080	0.007
	EW	-0.298	0.006	0.115	0.015	0.338	0.005
TAS	NS	0.100	0.067	-0.308	0.175	-0.556	0.052
	EW	0.477	0.032	-0.443	0.068	-0.261	0.030
IGN	NS	1.478	0.028	0.311	0.083	-0.323	0.021
	EW	-0.132	0.026	-0.902	0.062	0.128	0.023
ABA	NS	1.884	0.012	-1.023	0.030	3.721	0.011
	EW	-1.250	0.011	2.089	0.024	2.347	0.010

4.4 A Phenomenological Model

It is assumed that the tilt measurements are carried out in the vicinity of a lateral inhomogeneity with two-dimensional shape, i.e. a long extension in one of the horizontal directions. The structure is loaded by the farfield tidal strains according to a chosen Earth-model. A coordinate system is chosen with the x-axis parallel to the striking of the structure. It is assumed further that the rocks in the vicinity of the tilt stations are isotropic. The existence of the 2-dim. structure induces an effective anisotropy. The strain perpendicular to the striking of the structure, e_{yy}, produces the largest tilt disturbance. If the observing tilt station is placed far away from the edges of the structure, the only effect of deformation in x-direction is a small contribution to the y-disturbance due to cross coupling. The horizontal shear strain, e_{xy}, has no effect since, in isotropic rocks, it does not lead to volume changes. Thus, if the simple model of a 2-dim. structure in an isotropic medium is sufficient to explain the tidal tilt disturbances, an orientation φ must exist for which the coefficient coupling the tilt perpendicular to the structure to shear strain, $C_{y,xy}$, vanishes, the coefficient coupling y-tilt to the strain in x-direction, $C_{y,xx}$, is small and the coefficient coupling y-tilt to y-strain, $C_{y,yy}$, reaches its maximum. In order to verify this, strain tilt coupling coefficients are calculated for any direction. This is done by rotating the observed NS- and EW-residuals together with the strains in steps of 1°. For each step, the strain-tilt coupling coefficients are calculated by use of Eq.(3). The calculations are carried out for the stations Samanpazari, Dokurcun and Igneciler, where the static NS-tidal tilt anomalies are strong. Results from Abant are not used, because the tilt-site tensor indicates very complicate strain-tilt transfer conditions which are beyond the simple model approach used here. At Tasburun and Tashkesti the observed residuals are small indicating that these two sites are less, if at all, sensitive to the geological effect. For each one of the three remaining stations, orientations of the tilt generating structure are individually calculated. It is assumed that the distance between neighbouring stations is so large that the individually fitted structures are long enough to avoid edge effects at the position of the stations.

Azimuth φ of the hypothetical structure for which $C_{y,yy}$ reaches its maximum and $C_{y,xy}$ passes through zero shows a clockwise rotation of 37° and 54°, respectively, from East to West (Fig.4). At Igneciler φ ($|C_{y,yy}|_{max}$) and φ ($C_{y,xy}=0$) agree within 6°, at Dokurcun and Samanpazari, there is a difference of 10° and 11°, respectively. Comparison with the geology of the region shows that the main fault changes its direction in exactly the same manner. Approximate orientations of the main fault are N102° W at Igneciler, N95° W at Dokurcun and N80° W at Samanpazari, i.e. there is a rotation in the striking of the fault by about 22° from East to West. The azimuth of $|C_{y,yy}|_{max}$ fits the striking of the fault segments at Igneciler and Dokurcun; at Samanpazari, there is a difference of 13°. At each site, cross coupling coefficients are about one order of magnitude smaller than $|C_{y,yy}|_{max}$.

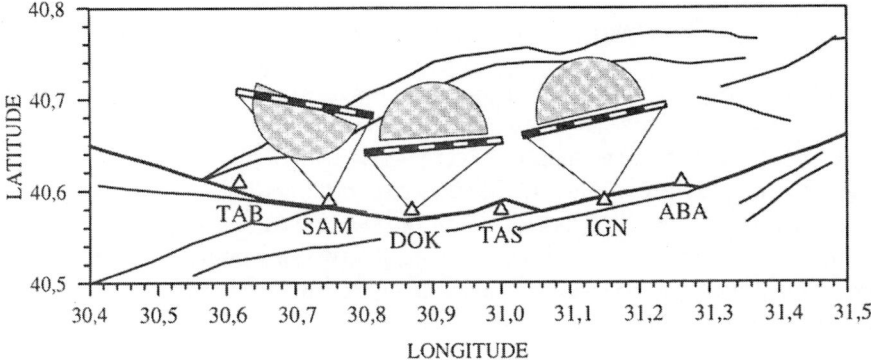

Fig. 4. The rotation of the local direction of strike of the real fault (indicated by the bars) and of a hypothetical fault inferred from the maximum of the coefficients $C_{y,yy}$ coupling the in-line components of tilt and strain (indicated by the flat side of the half-circles). The corpus of the half-circles (upper of lower half) symbolize the different signs of strain-tilt coupling at the sites.

It is concluded that the major part of the tidal tilt anomalies at Igneciler, Dokurcun and Samanpazari can be phenomenologically explained by strain-tilt coupling due to the existence of a 2-dim. crustal structure. The local orientation of the hypothetical structure follows the striking of the main fault, thus, it is reasonable to assume that the physical source of strain-tilt coupling in that area is the lateral elastic contrast between the fault zone material and the host rock. Plausibility considerations show that the contributions of topography and horizontal pore pressure diffusion which are known to be potential sources of tidal tilt modification, supposedly are considerably lower than the geologic effect at the 3 stations (Westerhaus, 1996). However, these effects may be the main cause of the small EW-residuals which cannot be modelled by the simple 2-dimensional approach followed here.

4.5 Temporal Variations

Continuous and undisturbed data as well as the precise knowledge of the calibration factor are required for a proper evaluation of time varying tidal tilt parameters. The longest timeseries, interrupted by minor gaps only, have been recorded at the neighbouring stations Samanpazari and Dokurcun (see Fig. 2). The effective data losses, including recording gaps as well as the effects of filtering and air pressure correction, amount to 16% at Samanpazari (01.01.88 - 31.12.92) and 10% at Dokurcun (01.01.88 - 15.11.92). The mean scale factors at these stations can be determined with a precision better than 1% (Dokurcun) and about

2% (Samanpazari). Slow temporal changes of the calibration factors due to maturation of electronic components of the instrumental equipment have been linearily interpolated.

Quasi-periodical variations with seasonal character are the predominant feature of the time variant tidal tilt response functions shown in Fig. 5. Mean peak-to-peak amplitudes of seasonal time variations in the normalized M_2-tidal tilt response amplitudes (γ-factor) at Samanpazari and Dokurcun are 0.055 (equivalent to amplitude variations of 2.1 nrad) and 0.019 (0.7nrad), respectively. Compared to the static tidal tilt residuals, the relative seasonal variations are about 6% at Dokurcun and 9% at Samanpazari. The annual extrema in the Dokurcun response are delayed by 1-2 months with respect to the variations observed at Samanpazari.

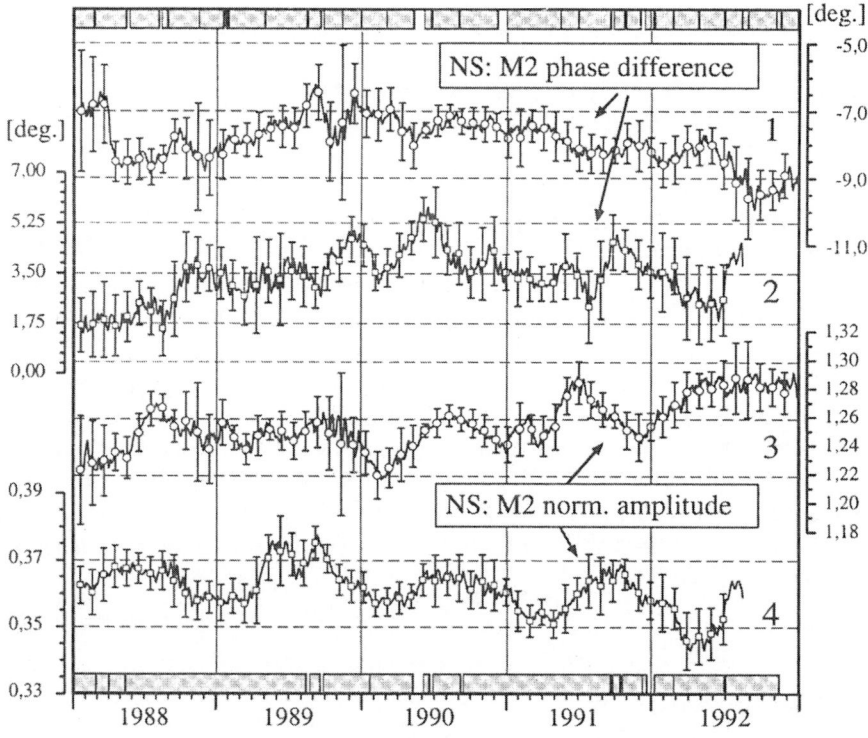

Fig. 5. Normalized amplitudes and phases of the M_2 (NS)-tidal tilt response functions at Samanpazari (curves 3 and 1, respectively) and Dokurcun (curves 4 and 2, respectively). 2σ-error bars are inserted every 15th value. Data gaps are symbolized by the interrupted parts of the grey bars at the lower (Dokurcun) and upper (Samanpazari) edges of the figure.

The seasonal variations seem to be superimposed on slow temporal changes at both stations. Linear trend calculations for the period 19.01.88 - 16.08.92 yield changes of the normalized M_2-amplitude response in NS-direction of -0.015 at Dokurcun and -0.033 at Samanpazari. The relative amplitude changes with respect to the M_2-residual tilts are 4.5% and 6%, respectively. Close inspection of Fig.5 shows that the trend may not be linear but seems to increase in 1990. Correlated, arch-like trend changes are observed in the M_2 phases at both stations. The curves reach the vertex in 1989/1990, at about the same time when the trend in the amplitudes changes. The maximum change of the M_2-phases is about 2° at both sites.

Seasonal and secular changes are persistent in the time-variant strain-tilt coupling coefficients. Fig. 6 shows the coefficient coupling the in-line components of tilt and strain perpendicular to the striking of the fault zone, $C_{y,yy}$, calculated in a direction where the static value is at maximum. The overall shape of the coupling coefficients closely resembles that of the M_2-response amplitudes (Fig. 5). This is because the M_2 strain component in NS-direction, i.e. about

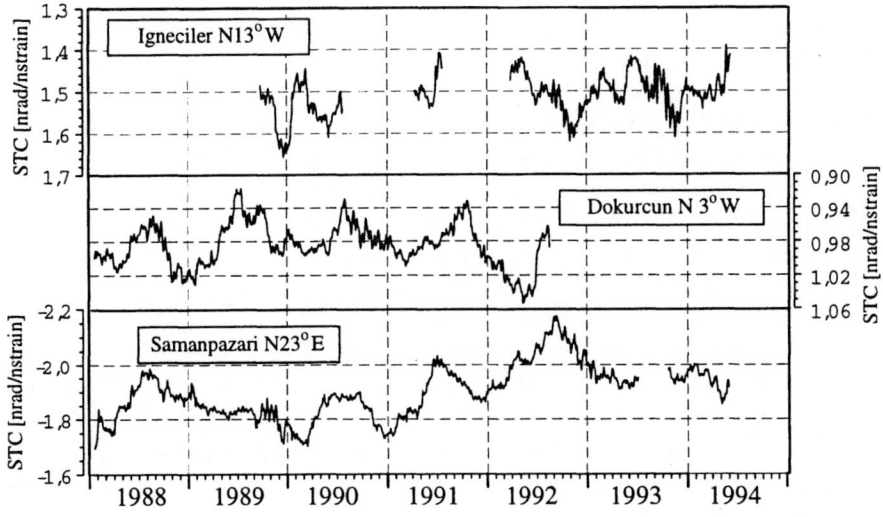

Fig. 6. Coefficients coupling the in-line components of tilt and strain, $C_{y,yy}$, calculated in a direction where the static values are at maximum, i.e. about perpendicular to the local strike of the fault.

perpendicular to the striking of the fault, is largest. The largest O_1-strain-component is directed EW where the effect of the fault is low. Therefore, the M_2-tidal tilt response is supposed to be much more sensitive to temporal changes of elastic properties within the fault zone than the O_1-response. The simultaneous increase of the M_2-residual amplitudes and the absolute values of the strain-tilt coupling factors in a direction perpendicular to the fault at both stations indicates that the influence of the lateral contrast increases with time over a horizontal distance of at least 15km.

5 Discussion

5.1 Seasonal Variations - an Indication for Changes of Effective Stress ?

Fig. 7 reveals a remarkable coincidence between annual variations in the strain-tilt coupling function $C_{y,yy}$ at Dokurcun and Samanpazari, the traveltimes of p-waves and microseismic activity. The discrete information about seismicity has been converted into a smooth curve by applying a 60-day moving average in order to achieve a similar low pass characteristic as in tidal tilt. For the present purpose, the investigation of microseismic activity is restricted on the region extending from 30.50° - 31.60° east longitude and from 40.55° -40.80° north latitude. The region contains the tiltmeter profile, the stations of the active seismic experiment and the geodetic network. The traveltimes of seismic waves are regularly determined on traces crossing the fault zone at different angles. The changes in the p-wave traveltime included in Fig.7 have been observed on a path subparallel to the direction of least principal stress (Lühr, 1991).

Generally, variations of seasonal type are induced into continuously recorded geophysical observables by annual changes of meteorological parameters. Possible sources are: temperature effects on the recording equipment, thermoelastic and -radiation effects with a skin depth up to some hundred meters, annual changes of barometric pressure response of the site which are not included in the multi-channel tidal analysis and pore pressure effects driven by yearly groundwater recharge. Assuming that barometric and thermoelastic effects mainly influence the solar constituents of the tidal spectrum and do not affect the detection of seismic signals, the most probable source for annual variations in tilt and seismic velocities are pore pressure effects. It is known that variations of the ground water table have major influence on non-tidal tilt observations in bore holes which penetrate into saturated layers (e.g. Levine et al., 1989; Kümpel, 1982) and also affects the seismic velocities.

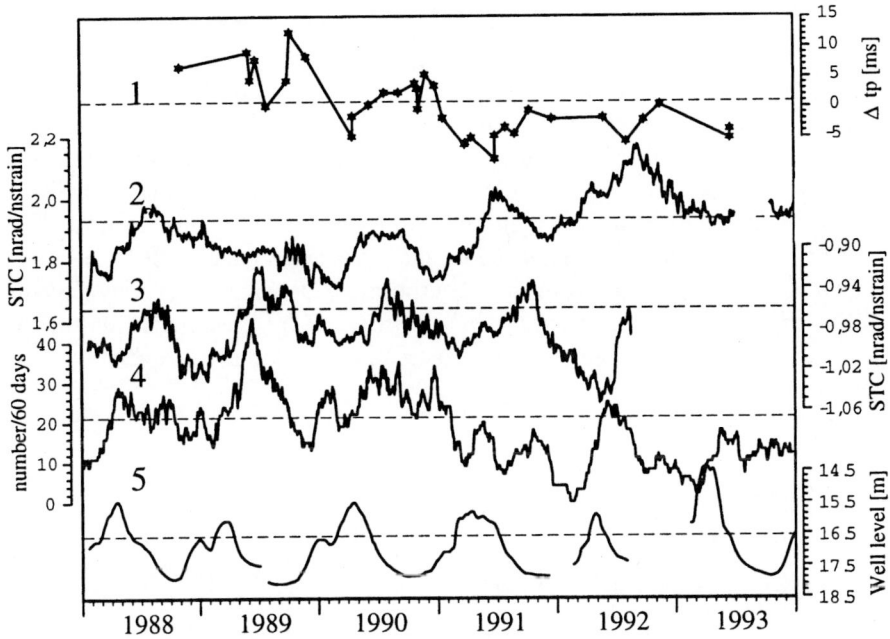

Fig. 7. Strain-tilt coupling functions $C_{y,yy}$ perpendicular to the local strike of the fault at Samanpazari and Dokurcun (curves 2 and 3, respectively), compared to microseismic activity (4), the p-wave velocity observed on a path subparallel to least principal tectonic stress (1) and the level of the unconfined aquifer at Abant (5). A 60-day moving average has been applied to the number of seismic events and well level in order to achieve a similar low-pass characteristic as in the tidal response.

A moving 60-day average of the groundwater level of the unconfined aquifer at Abant, which has proven to reflect recharge events with minor time delay, is included in Fig. 7. Comparison shows considerable time shifts between the maxima in well level and the annual changes in the tidal tilt response and microseismic activity. Quantitative values of the time shifts, calculated from cross correlation, are 90 days for microseismic activity, 118 days and 158 days for the strain-tilt coupling functions at Samanpazari and Dokurcun, respectively (Fig. 8). A time shift of about 3 months exists between annual variations in tidal and non-tidal tilt at the two sites, leading to the conclusion that the tidal tilt response functions are not directly affected by hydrological changes at the surface. No shift can be recognized between the hydrological changes and the variations of p-wave velocity. It can be shown, however, that the magnitude of the velocity variations cannot be explained by the direct influence of the groundwater table variations (Lühr, priv. comm.).

The variations of seasonal type could be explained either by periodically varying tectonic stresses or by varying pore pressure, which seems to be the more realistic assumption. It is proposed that seasonal groundwater recharge causes a pore pressure disturbance that propagates vertically downwards. In a depth region, where elastic strain is accumulated, the pore pressure variation alters the effective stresses. This may affect stress dependent elastic moduli and, thus, may lead to the observed seasonal variations in the tidal tilt response functions and other observables. This implies that the pore space in the fault zone is hydraulically connected to a depth of 10 km and more. Indeed, isotope studies at the San Andreas Fault, California, have indicated that meteoric water must circulate to a depth of 10 km to 15 km (Byerlee, 1993, and references therein). It follows that mean pore pressure is in hydrostatic equilibrium. Deviations from hydrostatic equilibrium (pore pressure disturbances) are reduced by pressure diffusion; the characteristic time for compensation of a pressure disturbance depends on the hydraulic diffusivity, D.

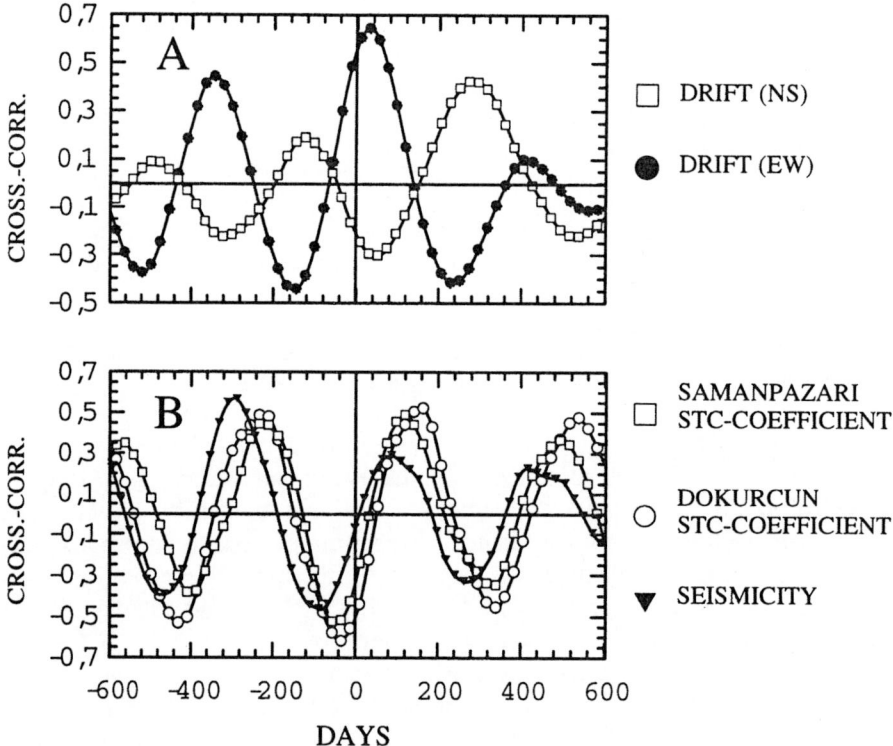

Fig. 8. Cross correlation functions. A: Non-tidal tilt versus Abant well level; B: Strain-tilt coupling functions and microseismicity versus Abant well level.

The pressure, p, at depth, d, caused by a sinusoidally varying pressure disturbance at the surface, p_o, is given by:

$$p = p_0 \cdot \exp\left(-d\sqrt{\omega/2D}\right)\cos\left(\omega t - d\sqrt{\omega/2D}\right) \tag{4}$$

where t is time and ω is frequency. The hydraulic diffusivity, D, is a combination of elastic and hydraulic properties of the rock matrix and the pore fluid. According to Eq. (4), amplitude and phase of the pressure response at depth are controlled by the term $-d\sqrt{\omega/2D}$. The mean hypocentral depth of microseismic events is 8 km. With a phase lag of 90 days, as obtained from cross correlation with the well level at Abant and $\omega = 2\pi/365$days, a mean hydraulic diffusivity of $D = 2.7m^2/s$ is estimated. This lays well in the range of 0.5 m^2/s $< D_S < 50$ m^2/s given by Talwani and Acree (1985) for the seismic hydraulic diffusivity, which is estimated from spreading of reservoir induced seismicity (note that D_S has the same meaning as D), and is near to 1 m^2/s, a value used by Scholz (1990).

For the strain-tilt coupling functions at Samanpazari and Dokurcun cross correlation with the well level at Abant yields phase lags of 118 days and 158 days, respectively. Inserting the diffusivity obtained from the variations in microseismic activity in Eq. (4), it turns out that a depth of 11 km at Samanpazari and 14 km at Dokurcun is assigned to the annual variations of the tidal tilt response functions. The fractional pressure variations at this depth with respect to the disturbance at the surface are 13% and 7%, respectively.

The Abant well is located at the bottom of a valley within the fault zone. Thus, it is used to estimate the fluid pressure disturbance at the surface. Taking a seasonal peak-to-peak amplitude of 350 cm, pressure changes of $8 \cdot 10^3$ Pa in a depth of 8 km and $3 \cdot 10^3$ Pa in a depth of 14 km are calculated from Eq. (4). This is of the same order of magnitude as the yearly increase of tectonic stress. Averaged over the period 1988 to 1993, the accumulation of apparent tectonic strain, calculated from lengthening of geodetic lines in the Tashkesti net, amounts to $2 \cdot 10^{-7}$/year. With a mean shear modulus of $2.5 \cdot 10^{10}$ Pa (granite, limestone) a tectonic stress rate of $5 \cdot 10^3$ Pa/year is obtained.

It is of considerable interest to compare the depths, obtained from the phase lags under assumption of spatially constant diffusivity, with a fault zone model that incorporates depth dependent friction parameters (Scholz,1990). The mean hypocentral depth (8 km) of microevents in the Mudurnu area exactly fits into the Scholz model, under the premise that the geothermal gradients of the San Andreas and the North- Anatolian Fault are comparable. The larger phase lags of the extrema in the M_2-tilts and in the strain-tilt coupling function indicate that the tidal tilt response is affected when the pore pressure front approximately reaches the depth of maximum crustal strength. These observations are temptably explained as follows. Let's assume that the local tidal tilt response is influenced by a narrow, dipping fault that obeys a strength profile as proposed by Scholz (1990). The large tidal tilt anomalies will be mainly caused by the 'low-strength' parts of the fault near to the surface. However, there is also a dependency on the stronger parts of the fault. It seems plausible that the relative effect of a variation in σ_n will be largest in the region of maximum strength where stress is concentrated, because this would affect the movement of the fault as a whole. In contrast, temporal changes of deformation governing parameters within weaker sections of the fault will have less influence on the tidal tilt response. An extreme case illustrating this behavior would be a fault that is allowed to slip freely in its upper and lower parts and deforms elastically only in a limited depth region. Numerical model calculations in order to prove these ideas have started.

5.2 Secular Changes - an Indication for Dilatancy ?

Fig. 7 shows that the secular trend changes in the strain-tilt coupling function at Samanpazari and Dokurcun coincide with a considerable reduction of microseismic activity. The onset of seismic quiescence in the beginning of 1991 is even more obvious in the cumulative number of seismic events with magnitudes above 2.2 (Fig. 9). A significant tendency towards shorter p-wave travel times has been observed in the western part of the investigation area on several profiles crossing the main fault at different angles. A retardation of the velocity increase since 1991 is observed on a path subparallel to the direction of least principal stress (Fig. 7). Another example is a 2°- increase of water temperature in a hot spring at Kuzuluk and a 200 mK increase in the water well at Igneciler which are well correlated with an increase of the strain-tilt coupling function at Samanpazari (Fig. 9). A continuous decrease in the accumulation of apparent tectonic shear strain from $2 \cdot 10^{-6}$ to $2 \cdot 10^{-7}$ is observed over the 6 years period.

The observed secular trend changes indicate temporal variations of effective stress within the investigation area. Possible reasons are temporal changes of the hydrologic situation and the ambient stress level. Following the results of laboratory experiments, the hardening effect due to a pore pressure decline during the opening of new cracks could be an explanation for the decrease of microseismic activity and the accumulation of apparent tectonic strain as inferred from geodetic distance measurements. This would also affect the p-wave velocity provided that pore pressure decreases so fast that the cracks are undersaturated. The v_p variations shown in Fig. 7 have been observed on a path crossing the main fault at an angle close to the direction of least principal stress. It is known from laboratory measurements that p-wave velocity in the direction of least principal stress is most sensitive to dilatancy. Thus, the trend change observed on that path might be taken as indication for the onset of dilatancy in the beginning of 1991. In order to get measurable changes of the velocities of seismic waves, a substantial fraction of the travel path has to be influenced by dilatancy. The width of the fault zone is assumed to be approximately 1000 m, about 1/6 of the horizontal distance between the shotpoint and the receiving station on that path. If dilatancy is assumed to be the main source of the observed change in v_p it must be volume dilatancy, extending at least over major parts of the fault zone in that area.

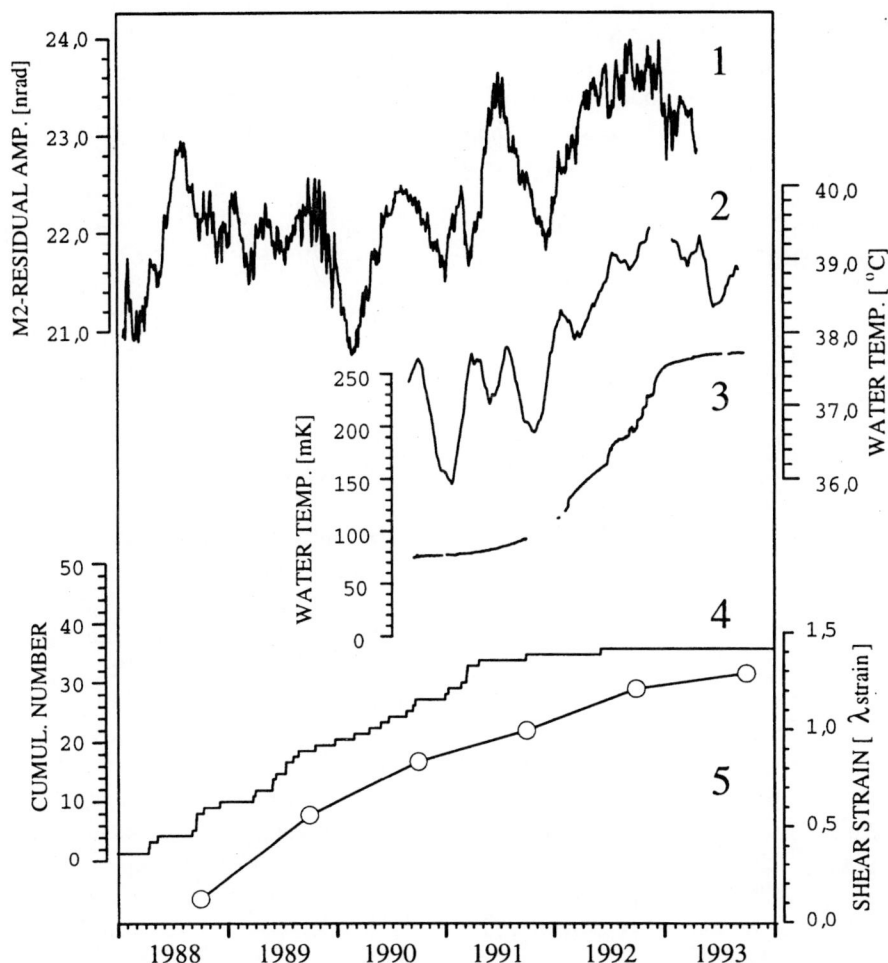

Fig. 9. The strain-tilt coupling function $C_{y,yy}$ at Samanpazari (1), water temperature in a hot spring (2), water temperatur in the well at Igneciler (3), cumulative number of seismic events (4) and accumulation of tectonic shear strain (5). A 60-day moving average has been applied to the temperature time series. Seismic quiescence has been evaluated in a limited area with extensions 30.5⁰-30.9⁰ E / 40.4⁰-40.8⁰ N. Only events with M>2.1 have been used to ensure homogeneity of the catalogue. Apparent tectonic strain has been calculated from repeated geodetic distance measurements

One objection against dilatancy is that the frictional strength of faults would be too low to withstand the stress needed to initiate dilatancy. This is a general result from laboratory experiments. Meissner (1986) pointed out that velocity anisotropy or changes in anisotropy start already at small stresses. Accordingly, the observed temporal variations in the velocities may indicate increasing tectonic stress; deformation, however, may still be in the elastic range. On the other hand, Scholz (1990) argued that, on a geological scale, fracture strength of rocks should be lower than measured in laboratory due to subcritical crack growth, whereas friction should be about the same. Therefore, dilatancy could be expected at much lower stresses relative to the frictional strength.

6   Conclusions

A tidal tilt experiment is included within the interdisciplinary German-Turkisch Project on Earthquake Research. Its goal is to utilize the geologic effect, i.e. the coupling of tilt to the inhomogeneous deformation near a contrast of two different geological units, to investigate changes of elastic properties in the crust related to seismotectonic processes.

Strong deviations of the local tidal tilt response functions from the global values of up to 400% in amplitude and 80° in phase are found in NS-direction, i.e. perpendicular to the main fault; in EW-direction, the response functions take almost normal values. The anisotropy and an apparent frequency dependence of the response functions - diurnal and semi-diurnal tides are influenced with opposite sign- has been interpreted in terms of strain-tilt coupling. An analysis of the azimuthal dependency of the strain-tilt coupling coefficients indicates that the tidal tilt disturbances at Igneciler, Dokurcun and Samanpazari can be reduced mainly on the existence of a two-dimensional lateral contrast which is orientated subparallel to the average striking of the real fault near the 3 sites. Because topography, pore pressure variations and ocean loading can be ruled out as main sources of static tidal tilt modification, it is concluded that the tidal tilt response functions at these sites are influenced by the mechanical contrast between the fault zone and the surrounding rocks. Accordingly, the local tidal tilt response functions potentially are sensitive to stress dependent changes of elastic properties.

Indeed, a moving analysis over 60-day-intervals, shifted in steps of 48 hours, reveals considerable variations of the tidal tilt parameters with time. Seasonal variations are the predominant temporal phenomenon in the tidal tilt reponse functions. They are superimposed on a secular trend that starts or intensifies in the beginning of 1991. Similar variations are observed in the microseismic activity, the traveltimes of longitudinal waves on a path subparallel to least principal stress, the accumulation of apparent tectonic strain and several other observables. The temporal and spatial correlations, observed in independent experiments, are interpreted as being caused by temporal variations of physical rock parameters. Preliminary estimations available so far give rise to the assumption that the tidal tilt response functions, as well as other observables, may be sensitive to even small changes in effective stress. Quantitative model calculations, in combination with repeated examinations of all the data, have to show whether the proposed models can be maintained. Because up to date there are no experiences with medium or large earthquakes in the region, no statement can be made about an impeding earthqake from the change in secular trends. Statistical investigations of seismic activity in the area, however, show that quiescence prior to relatively 'large' events ($M \cong 4$) is common (Wyss et al., 1995).

Acknowledgements. Jochen Zschau provided motivation and encouragement throughout this study. I'm grateful to Wolfgang Welle for providing computer programmes, Hans-Jochen Kümpel, Claus Milkereit and Rongjiang Wang for discussions, and Martin Beisser for proof reading the manuscript. I'm indebted to Ahmet Tarimci, Sinasi Canbay, Cetiner Türküm and Georg Lohr for technical support in the field. Aysel Yatman, Birger Lühr, Peter Franke and Heiko Woith were so kind to provide seismological, seismic, geodetic and hydrologic data. This study was supported by the German Research Foundation.

References

Aggarval, Y. P., L. R. Sykes, Y. Armbruster, and M. L. Sbar. 1973. Premonitory changes in seismic velocities and prediction of earthquakes, *Nature*, **241**: 101-104.

Beaumont, C. 1978. Linear and nonlinear interactions between the earth tide and a tectonically stressed earth. 313-318. In: Applications of geodesy to geodynamics, I. Mueller, ed., Ohio State University Press.

Beaumont, C. and J. Berger. 1974. Earthquake prediction: modification of the Earth tide tilts and strains, *Geophys. J. R. Astron. Soc.*, **39**: 111-121.

Biot, M. A. 1941. General theory of three-dimensional consolidation, *J. Applied Physics*, **1**: 155-164.

Bonatz, M., C. Gerstenecker, C. Kistermann, and J. Zschau. 1983. Tilt measurements across a deep fault zone, Proc. Int. 9th Symp. Earth Tides, New York, 695-704.

Brace, W. F., B. W. Paulding, and C. H. Scholz. 1966. Dilatancy in the fracture of crystalline rocks, *J. Geophys. Res.*, **71**: 3939-3953.

Byerlee, J. 1993. Model for episodic flow of high-pressure water in fault zones before earthquakes, *Geology*, **21**: 303-306.

Eshelby, J. D. 1957. The determination of the elastic field of an ellipsoidal inclusion, and related problems, *Proc. Roy. Soc.*, **241**: 376-396

Gerstenecker, C., J. Zschau, and M. Bonatz. 1985. Finite element modelling of the Hunsrück tilt anomalies - a model comparison, Proc. Int. Symp. Earth Tides, R. Vieira, ed., Madrid, 797-803.

Harkrider, D. G. 1970. Surface waves in multilayered media, 2, Higher mode spectra and spectral ratios from point sources in plane layered earth models, *Bull. Seism. Soc. Amer.*, **60**: 1937.

Harrison, J. C., 1976. Cavity and topographic effects in tilt and strain measurements, Jour. Geophys. Res., 81, 319-328.

King, G. C. P., W. Zürn, R. Evans, and D. Emter. 1976. Site corrections for long period seismometers, tiltmeters and strainmeters, *Geoph. J. Roy. Astr. Soc.*, **44**: 405-411.

Kirsch, R. and J. Zschau. 1986. The influence of a dilatant region in the Earth's crust on the Earth tide tilt and strain, *J. Geophys.*, **59**: 157-163.

Kümpel, H. J. 1982. Neigungsmessungen zwischen Hydrologie und Ozeanographie, PhD thesis, University of Kiel.

Kümpel, H. J. 1989. Verformungen in der Umgebung von Brunnen, Habil. thesis, University of Kiel.

Latynina, L. A. and S. D. Rizaeva. 1976. On Tidal Strain Variations Before Earthquakes, *Tectonophysics*, **3**: 121-127.

Levine, J., C. Meertens, and R. Busby. 1989. Tilt observations using borehole tiltmeters - 1. Analysis of tidal and secular tilt, *J. Geophys. Res.*, **94**: 574-586.

Lühr, B.-G., C. Milkereit, R. Meissner, and N. Büyükköse. 1991. Temporal variations of seismic signals within the active seismic experiment in the Mudurnu Valley, Turkey, Proceedings of the International Conference on Earthquake Prediction: State-of-the-Art, Strasbourg, 336-343.

Mao, W. J., C. Ebblin, and M. Zadro. 1989. Evidence for variations of mechanical properties in the Friuli seismic area, *Tectonophysics*, **170**: 231-242.

Meertens, C. M. 1987. Tilt tides and tectonics at Yellowstone National Park, PhD thesis, University of Colorado.

Meissner, R. 1986. The continental crust - a geophysical approach, International Geophysics Series, 34, Academic Press, London.

Michel, G. W. 1994. 'Neo'-Kinematics along the North-Anatolian Fault, PhD thesis, Tübinger Geowissenschaftliche Arbeiten (TGA), Reihe A, Band 16, University of Tübingen.

Mikumo, T., M. Kato, H. Doi, J. Wada, T. Tanaka, R. Shichi, and A. Yamamoto. 1977. Possibility of temporal variations in Earth tidal strain amplitudes associated with major earthquakes. 123-136. In: Earthquake Precursors: Proceedings of the U.S.-Japan Seminar on Theoretical and Experimental Investigations of Earthquake Precursors, C. Kisslinger and Z. Shuki (eds.), Central Academic Publishers of Japan, Tokyo.

Milkereit, C. 1988. Dilatanz und Anisotropie als spannungsabhängige Phänomene nichtlinearer Elastizität und ihr Einfluß auf seismische Messungen und Erdgezeitenregistrierungen, diploma thesis, University of Kiel.

Mogi, K. 1985. Earthquake Prediction, Academic Press, Tokyo.

Molodensky, S. M. 1983. Local anomalies in amplitude and phase of tidal tilts and deformations, *Izvestiya, Earth Physics*, **17**: 501-505.

Neresov, I. L., A. N. Semyenov, and J. G. Simbireva. 1971. Space-time distribution of the ratio of travel times of compressional and transverse waves in the Garm region. 334-348. **In**: Experimental Seismology, Sadovsky (ed.), Nauka, Moscow, (in Russian).

Nishimura, E. 1950. On earth tides, *Am. Geophys. Union Trans.*, **31**: 357-376.

Nur, A. and J. D. Beyerlee. 1971. An exact effective stress law for elastic deformation of rock with fluids, *J. Geophys. Res.*, **76**: 6414-6419.

Nur, A. 1972. Dilatancy, pore fluids, and premonitory variations of t_s/t_p traveltimes, *Bull. Seism. Soc. Am.*, **62**: 1217-1222.

Peters, J. and C. Beaumont. 1985. Borehole tilt measurements from Charlevoix, Québec, *J. Geophys. Res.*, **90**: 12791-12806.

Rice, J. R. 1983. Constitutive relations for fault slip and earthquake instabilities, *Pure appl. Geophys.*, **121**: 443-475.

Scholz, C. H. 1990. The mechanics of earthquakes and faulting, Cambridge University Press.

Scholz, C. H., L. R. Sykes, and Y. P. Aggarval. 1973. Earthquake prediction: a physical basis, *Science,* **181**: 803-810.

Schüller, K. 1977. Tidal analysis by the hybrid least squares frequency domain convolution method. 103-128. **In:** Proc. Int. 8th Symp. Earth Tides, M. Bonatz and P. Melchior (eds.), Bonn,

Schüller, K. 1985. Computer program HYCON.

Tse S. T. and J. R. Rice. 1986. Crustal earthquake instability in relation to frictional constitutive response, *J. Geophys. Res.,* **91**: 9452-9472.

Tullis, T. E. 1988. Rock friction constitutive behavior from laboratory experiments and its implications for an earthquake prediction field monitoring program, *Pure appl. Geophys.,* **126**: 555-588.

Westerhaus, M. 1996. Tilt and well level tides along an active fault, PhD-Thesis, Scientific-Technical Reports, 5/1996, GeoForschungsZentrum Potsdam.

Westerhaus, M. and J. Zschau. 1989. Tidal tilt modification at the western end of the North-Anatolian Fault Zone: an indication for slow changes of crustal properties. 82-108. **In:** Turkish-German Earthquake Research Project, J. Zschau and O. Ergunay (eds.), Kiel,

Westerhaus, M., W. Welle, N. Büyükköse, and J. Zschau. 1991. Temporal variations of crustal properties in the Mudurnu Valley, Turkey: an indication for regional effects of local asperities?, Proceedings of the International Conference on Earthquake Prediction: State-of-the-Art, Strasbourg, 272-281.

Wyss, M., M. Westerhaus, H. Berckhemer, and R. Ates. 1995. Precursory seismic quiescence in the Mudurnu Valley, North Anatolian Fault Zone, Turkey, *Geophys. J. Int.,* **123**: 117-124

Zschau, J., M. Westerhaus, W. Welle, and N. Büyükköse, N., 1991. Regional strain accumulation from local tidal tilt- and well level data: a new approach, Proceedings of the International Conference on Earthquake Prediction: State-of-the-Art, Strasbourg, 444-453.

Zürn, W., H. Kiesel, H. Otto, and H. Mälzer. 1977. Phenomenological approach to strain-tilt coupling at Schiltach Observatory. 451-465. **In:** Proc. 8th Int. Symp. Earth Tides, M. Bonatz and P. Melchior (eds.), Bonn.

Tides in Outer Space

Satellite Orbit Perturbations Induced by Tidal Forces

Peter Schwintzer

GeoForschungsZentrum Potsdam
Div. 1, Kinematics and Dynamics of the Earth
Telegraphenberg A17, D-14473 Potsdam

Summary. The restitution of the trajectory of near-Earth orbiting satellites with accuracies to some centimeters by evaluating precise tracking data, is essential in satellite geodesy for applications like station positioning, earth orientation parameter determination, altimetry, SAR interferometry, and gravity field recovery from observed satellite orbit perturbations. Besides the stationary gravitational geopotential and non-gravitational forces, tidal induced temporal variations of the potential have to be considered in the numerical integration of the satellite's equation of motion.

Perturbing orbit accelerations have to be modelled which are induced by direct and indirect tidal effects: (a) the tide generating potential through Moon, Sun and planets, (b) changes in the geopotential caused by deformations of the Earth due to Earth tides, and (c) changes in the geopotential caused by mass redistributions and associated loading deformations due to ocean and atmosphere tides.

The induced accelerations for a satellite in about 1000 km altitude amount to 1 (a), 0.2 (b) and 0.03 (c), respectively, in units of 10^{-6} m/s^2. Even the small accelerations caused by the ocean tides in a height of 1000 km lead to periodic perturbations in the satellite's position with an amplitude of several meters and periods of some days to 100 days due to resonant amplifications.

Whereas the tide generating potential and the Earth tides can be modelled with a sufficiently high level of accuracy, the ocean tides are due to the broader spectrum more difficult to describe. Satellite tracking or altimeter data are therefore exploited to recover the principal constituents of the ocean tidal wave spectrum.

The overall spectrum of orbit perturbations due to tides can be analyzed analytically for a specified set of Keplerian orbit parameters with Lagrange's planetary equations in Kaula's formulation when inserting the disturbing tidal potential expressed e.g. by the coefficients of a spherical harmonic expansion. Such analyses are important in order to judge prior to numerical tracking data processing on the parameters which have to be modelled or to be solved for, or can be omitted due to a lack of sensitivity.

References

Cassotto, S. (1989): Nominal Ocean Tide Models for TOPEX Precise Orbit Determination. Ph. D. Dissertation, Univ. of Texas at Austin.

Dow, J.M. (1988): Ocean Tides and Tectonic Plate Motions from LAGEOS. Deutsche Geodätische Kommission, Reihe C, Nr. 344, München.

Kaula, W. (1966): Theory of Satellite Geodesy. Blaisdell Publ., Waltham, Mass.

McCarthy, D.D. (1996): IERS Conventions. IERS Technical Note 21, Observatoire de Paris.

Reigber, Ch. (1989): Gravity Field Recovery from Satellite Tracking Data. In: Sanso, F. and R. Rummel (eds.), Theory of Satellite Geodesy and Gravity Field Determination, 197 234, Springer, Berlin.

Schwintzer, P., Reigber, Ch., Bode, A., Kang, Z., Zhu, S.Y., Massmann, F.-H., Raimondo,J.C., Biancale, R., Balmino, G., Lemoine, J.M., Moynot, B., Marty, J.C., Barlier, F.,Boudon, Y. (1996): Long-Wavelength Global Gravity Field Models: GRIM4S4, GRIM4-C4.Journal of Geodesy, in print.

Tides of Io

Tilman Spohn

Institut für Planetologie
Westfälische Wilhelms-Universität
W. Klemmstrasse 10
D-48149 Münster, Germany

Abstract. Jupiter's satellite Io is the most active earth-like planetary body in the solar system with a surface heat flow of, at least, 2.5 W m^{-2}, a resurfacing rate of 1.3 cm a^{-1}, and, possibly, a self-sustained magnetic field. It is universally accepted that the activity is driven by tidal energy dissipated in Io's mantle. Tides with amplitudes two orders of magnitude larger than the lunar tides on Earth are raised on Io by Jupiter. Since Io rotates synchronously with its orbital revolution, substantial tidal deformation requires an eccentric orbit. The orbital eccentricity is maintained by the Laplace resonance between the inner Jovian satellites against the damping induced by tidal dissipation in Io's interior. Models of tidal dissipation assume a visco-elastic mantle rheology and require a fluid (outer) core to allow sufficiently strong tidal deformation. The mantle most likely is partially molten and there may be an asthenosphere or magmasphere underneath the lithosphere. The energy that is dissipated in Io is drawn from Jupiter's rotational energy and is transferred to Io's orbital energy before part of it is dissipated in the satellite. Tidal dissipation thus is a sink in the orbital energy balance and a source in the energy balance of the interior. The energy balances are coupled through the temperature dependent rheology parameters. Models of the thermal-orbital evolution indicate that a quasi-stationary high dissipation state is possible as well as oscillations of the thermal and orbital parameters. A magnetic field is unlikely in a quasi-stationary state. The time rate of change of orbit parameters such as the mean motion are constrained by astrometrical observation over the past 300 years. These data can be used to constrain the present tidal dissipation rate. These constraints indicate that the present heat flow is an order of magnitude larger than the present dissipation rate. A model of time dependent heat transfer with local hot spots in the mantle where melt is generated by viscous dissipation is proposed. This model may explain the gap between the present heat flow and the tidal dissipation rate.

1 Introduction

Tidal interactions have been ubiquitous in the solar system. According to our present understanding, tidal interactions caused the formation of the asteroid belt, the peculiar spin-orbit coupling of Mercury, the distribution of matter in the rings of the giant planets as well as the resonances between the orbits of some of the giant planets' satellites and their spin-orbit couplings. The best known and studied example of a primary-satellite system is, of course, the Earth-Moon system. The most prominent example of a resonance is the Laplace resonance between the orbits of the Jovian satellites Io, Europa, and Ganymede. Other prominent examples of tidal interactions are the evolutions of the orbits of the Martian satellites Phobos and Deimos and of Neptune's satellite Triton. Finally, a system that most likely underwent significant tidal interactions is formed by Pluto and its satellite Charon.

The motivation of studying tides differs to some extent between geophysics and planetary physics. While on Earth tides are of interest by themselves, not the least because they affect human life, in planetary science the interest focuses on tidal dissipation as an energy source for interiors of planetary bodies and as an important element of the orbital and rotational evolution. Tidal dissipation is an energy sink in the orbital and rotational energy balances and a source for the interiors of planets and satellites. This review focuses on the tidal interactions in the Jovian system between Io and its primary Jupiter and between Io and its neighboring satellites Europa and Ganymede.

2   Io and the Jovian System

The Jovian system consists of the planet Jupiter, its 9 satellites, and of Jupiter's ring. Four of the nine satellites are major size objects with radii between 1570 and 2630 km (compare Tab. 1 and Fig. 1). These are, in increasing distance from Jupiter the Galilean satellites Io, Europa, Ganymede and Callisto named after their discoverer Galileo Galilei who discovered the moons in 1610. The orbits of the inner three of these satellites are in resonance, with their orbital periods being approximately in the ratio of 1:2:4. (Because of dissipation, these resonances can never be exact.) The resonance is named after its discoverer Ferdinand Laplace. The densities range from 3566 kg/m^3 at Io to 1860 kg/m^3 at Callisto and suggest a compositional gradient in the Jovian system similar to the gradient in the solar system. Spectroscopy data of the surfaces show that ice is a major component on Europa, Ganymede, and Callisto and the densities suggest that the latter two contain rock and ice in approximately equal shares. Europa should be mostly composed of rock with a water ice crust, possibly underlain by a liquid water layer. The ice-water layer may have a total thickness of a few 100 km. Water is completely lacking from the surface of Io. Instead, the most volatile components are sulfur and sulfur dioxide which are ubiquitous on Io's surface. Modeling of Io's tidal deformation Segatz et al. (1988) and, most recently of its gravity field (Anderson et al., 1996) together with cosmochemical modeling (e.g., Consolmagno, 1981, Lewis, 1982) suggests that Io has an iron rich core with some sulfur and other light constituents and a mantle whose composition and mineralogy is similar to that of the Earth's upper mantle. The intense volcanic activity of Io which was predicted by Peale et al., (1979) and confirmed later that year by the Voyager mission and reconfirmed most recently by the Galileo mission suggests that there is a basaltic crust.

Fig. 1. The four Galilean Satellites (clockwise from top left) Io, Europa, Callisto, and Ganymede ©NASA/JPL

Nine active volcanic plumes were observed by the Voyagers. These plumes erupted sulfur and sulfur dioxide. Many more plumes have been suggested by Johnson *et al.* (1995) to be present as gas (without condensates) erupting "stealth plumes" which cannot be observed optically. Although the plumes erupt sulfur and sulfur dioxide, it is generally agreed that the activity is driven by silicate volcanism and that the plumes originate when silicate lava comes into contact with sulfuric aquifers and near-surface layers and when the sulfur or SO_2 is vaporized by the heat of the silicates. Estimates of the resurfacing rate by volcanism range from 1 to 10 cm/a, equivalent to a volcanic flow of a few 10^{11} to 10^{12} m^3/a. By comparison, the total rate of volcanic mass flow on Earth has been estimated to be less than 5×10^9 m^3/a (Verhoogen, 1980). Even given that these numbers are uncertain, it must be surmised that the volcanic activity of Io is impressive. The surface heat flow is similarly impressive: heat flow from Io

Table 1. Physical properties of Jupiter and its major satellites (Burns, 1986). Orbit data are given for the satellites only. Since the satellites rotate synchronously with their orbital periods, the rotational periods are the same as the orbital periods.

Parameter	Jupiter	Io	Europa	Ganymede	Callisto
Mass (10^{20}kg)	1.9×10^7	894.	480.	1482.	1077.
Radius (km)	71398.	1816.	1569.	2631.	2400.
Density (10^3kg/m^3)	1.33	3.57	2.97	1.94	1.86
Orbital Period (days)		1.769	3.551	7.155	16.689
Semi-major Axis (10^6m)		421.6	670.9	1070.	1883.
Mean motion (10^{-5}s^{-1})		4.111	2.048	1.016	0.4357
Eccentricity		0.0041	0.0101	0.0006	0.007

can be measured remotely in the infrared (Veeder *et al.*, 1994) since its intrinsic luminosity is larger than the extrinsic luminosity that is due to the reflection and re–radiation of the solar insolation. This is a remarkable phenomenon since on all the other smaller bodies in the solar system the intrinsic luminosity is much smaller than the extrinsic luminosity. The current best estimate gives a total of 10^{14} W or 2.5 W m^{-2} (Veeder *et al.*, 1994). This much heat is not primarily radiated from surface hot spots with temperatures of up to 1000 K but from the large, relatively cool areas with temperatures of about 150 K. (The so called effective temperature, the temperature in equilibrium with the solar radiation is about 100 K.) This suggests that lava flows on Io are widespread. The measured intrinsic luminosity provides a lower bound to the actual heat flow since it does not account for the heat flow from regions at temperatures below 200 K. Io is certainly the most active silicate body in the solar system and it is generally accepted, (e.g., McEwen *et al.*, 1989), that the activity is powered by tidal energy dissipated in Io's interior.

Recent Galileo magnetometer data suggest that Io may generate its own magnetic field and the magnetic moment has been estimated to be 10^{20} A m^2, equivalent to a field strength at the surface of about 1300 nT (Kivelson *et al.*, 1996). A field of that magnitude was predicted by Neubauer (1978) on the basis of Busse's (1976) scaling law. (Actually the value given for the magnetic moment in Neubauer (1978) has to be corrected by applying a factor of 4π.) However, the evidence for a magnetic field is inconclusive since the signal recorded by the Galileo magnetometer may also be explained as caused by ionospheric currents in Io's substantial ionosphere (Frank *et al.*, 1996). Wienbruch and Spohn (1995) have used a thermal-orbital evolution model coupled to a core dynamo model to discuss the possibility of generating a magnetic field in the core and the implications that a magnetic field would have on the present state and the evolution of Io.

3   Interior Structure of Io

It is almost unanimously agreed that Io possesses an iron–rich core, at least the outer layer of which should be molten (e.g., McEwen *et al.*, 1989). A substantial core is required to explain the observed figure of the planet (Segatz *et al.*, 1988) and the value of the gravitational coefficient C_{22} (Anderson *et al.*, 1996), recently determined from the Galileo data. A substantial core is also invoked from cosmo-chemical evidence (e.g., Consolmagno, 1981). A molten outer core is necessary to allow sufficient flexing of the mantle to explain the luminosity through tidal heating. The composition of the core is not very well known. However, it is likely that the core is rich in sulfur. The abundance of elemental sulfur and SO_2 on the surface of Io, according to cosmochemical arguments (Consolmagno, (1981; Lewis, 1982), suggests that the composition of Io is close to that of a volatile depleted C2 or C3 chondrite. The elemental abundance of sulfur would then be about 5 weight–% of the entire planet. Consolmagno (1981) points out that this sulfur can only be present either near the surface where it would form a 50 km thick crust or in the core where it would allow a close to eutectic composition. A 50 km thick sulfur crust is thought to be incompatible with surface elevations of 10 km, judging from the rheology of sulfur. An eutectic core would comprise 20 weight–% of the planet with a radius of roughly 1000 km. Segatz *et al.* (1988) have modeled the core of Io assuming an eutectic composition and allowing for thermal expansion and compression. They find a core radius of 980 km. Moreover, they argue from their modeling of the Love numbers of Io that the planets figure, if it is in approximate hydrostatic equilibrium, requires a core of roughly this size. Anderson *et al.* (1996) estimate a core radius of 950 km.

The widespread volcanism together with the extremely large luminosity and resurfacing rate suggests that the mantle of Io is partially molten. Both tidal dissipation in a melt channel or asthenosphere underneath the lithosphere and in a mantle that is partially molten throughout can explain the luminosity (Segatz *et al.*, 1988). Ross *et al.* (1990) have tried to fit the large scale topography with the figures of Io calculated from an asthenosphere dissipation model and a whole mantle dissipation model of Segatz et al (1988). They find that the topography is best fit if 2/3 of the dissipation occurs in an asthenosphere and 1/3 in the mantle below the asthenosphere. With a luminosity of 10^{14} W, the mantle still contributes 3×10^{13} W which strongly suggests that it should be partially molten. Fischer and Spohn (1990) have argued that Io must be partially molten to allow Io to be locked into a stable state with a sufficiently large dissipation rate. Their argument will be discussed further below.

In Fig. 2 we present a reasonable Iotherm together with the solidus of dry peridotite (a likely candidate for the mantle of Io) and together with a possible range of core liquidi. The melting range of dry peridotite is taken from Takahashi (1986, 1990), the core liquidi have been derived by integrating the data of Us-

selman (1975a; b) and of Boehler (1986; 1987). In the core, the Iotherm follows
the adiabat. The core adiabat has been calculated using the parameter values
for small planets given by Stevenson *et al.* (1983). In the mantle, the Iotherm is
assumed to start at the solidus temperature at the core/mantle boundary from
where it follows the wet adiabat defined by (e.g., Turcotte, 1982),

$$\frac{dT}{dP} = \frac{L}{c_m} \frac{df}{dP} \tag{1}$$

where T is temperature, P is pressure, L is the latent heat of mantle melting,
c_m is the mantle specific heat at constant pressure, and f is the degree of partial
melting, until it hits the 40% degree of partial melting line. This is roughly the
degree of partial melting required in the asthenosphere model of Segatz *et al.*
(1988). It is also roughly the maximum degree of partial melting at which the
solid matrix still dominates the properties of the partial melt. At significantly
larger degrees of partial melting large scale melt segregation through vertical
fluid transport is expected to occur which would render the asthenosphere unstable. Moreover, the observed topography seems to rule out the existence of an
extremely low viscosity (< 100 MPa s) melt layer underneath the lithosphere
(Webb and Stevenson, 1987). df/dP has been estimated by linearly interpolating
between the solidus and the liquidus.

In the asthenosphere, the Iotherm follows the constant degree of partial melting line. In the lithosphere, the temperature gradient is steep enough to approximately balance a heat flow of 10^{14} W from the mantle below by heat conduction.
Account of volcanic heat transfer would result in a less steep gradient and in a
thicker lithosphere. It is possible that the lithosphere is compositionally different from the underlying mantle, e.g., it may be a layer of basalt (Ross *et al.*,
1990), and its thickness may not depend simply on the temperature gradient but
also on the rates of basalt production and, possibly, recycling. An often-quoted
minimum value for the lithosphere thickness is 35 km (e.g., McEwen *et al.*, 1989).

As Fig. 2 shows, the solidus temperature of the mantle at the core/mantle
boundary is significantly larger than the liquidus temperature even of pure iron.
Although the core adiabat is less steep than the liquidus, the temperature in the
center of the core is still superliquidus. This has an important consequence for
the energetics of a possible dynamo: The latent heat and the gravitational energy
liberated by the redistribution of light elements in the core upon freeze out of
an inner core are not available to drive a dynamo. Instead, a dynamo must be
driven by thermal power alone. It should be noted that a purely thermal dynamo
is much less efficient as compared with a dynamo that is driven by chemical
buoyancy (Stevenson *et al.*, 1983). For the latter, the gravitational energy can be
directly made available to drive the dynamo while a thermal dynamo is subject
to a Carnot efficiency factor.

4   The Tides of Io

Jupiter's gravitational field causes Io to deform into a prolate spheroid with the
long axis pointing toward the planet. Because the orbit is eccentric, Io experien-

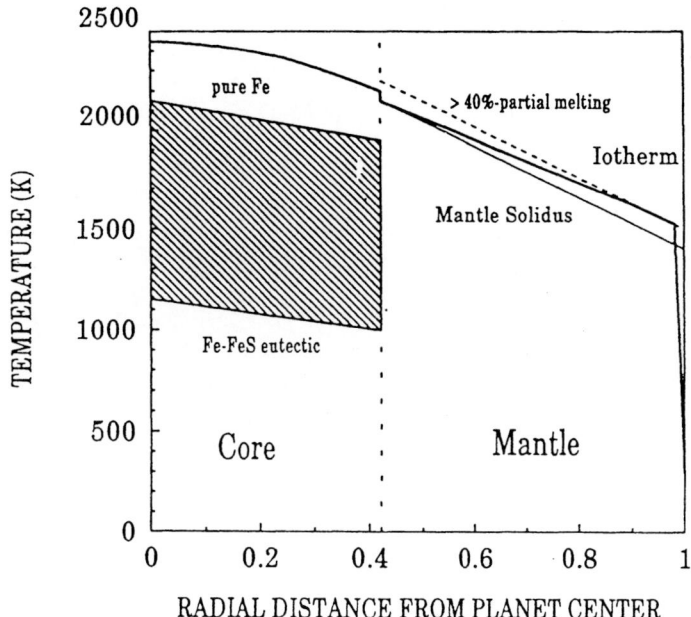

Fig. 2. Thermal structure of the interior of Io. Plotted are a simplified mantle solidus based on the solidus of dry peridotite of Takahashi (1986; 1990), a line of constant 40%-degree of partial melting after Wyllie (1988), a range of core liquidi between the pure iron liquidus and the liquidus of a eutectic Fe-FeS mixture based on data from Usselman (1975a; b) and Boehler (1986; 1992) and an Iotherm. The Iotherm follows the wet adiabat in the lower mantle, the line of constant degree of partial melting in the asthenosphere, and the adiabat in the core. The steep near-surface gradient indicates the lithosphere.

ces a periodic forcing with two components: the first component, the pumping tide is caused by the distance to Jupiter varying along the eccentric orbit. This component has an equilibrium elevation of 28.8 m and a tidal acceleration of 5.7×10^{-5} m s^{-2}. The second component, the librational tide is caused by the variation of the orbital velocity along the orbit which decreases from perijove to apojove and increases again as the satellites moves from apojove to perijove. The orbital velocity is therefore not in perfect synchronization with the rotation of Io. This causes the long axis to oscillate about its mean direction to Jupiter by a total angle of $2e_I$, where e_I is the orbital eccentricity. The latter component has an equilibrium elevation of 38.4 m and a tidal acceleration of 7.6×10^{-5} m s^{-2}. These equilibrium elevations and tidal accelerations are two orders of magnitudes larger than the equivalent values for the tides raised by the Moon on the Earth. In the following mathematical description of the tides and the tidal deformation we follow Segatz et al. (1988) who give a particularly elegant representation because of their choice of the co-ordinate system. This model neglects the effects of the small inclination of Io's rotational axis.

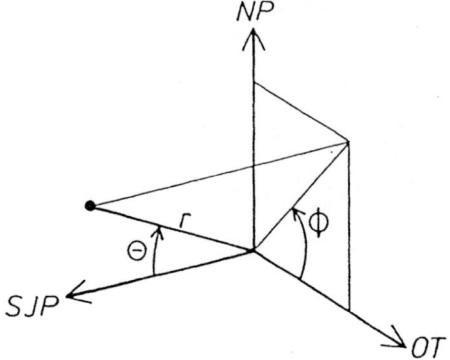

Fig. 3. Spherical coordinate system used to calculate the tidal disturbance potential and the deformation of Io. The polar axis of this coordinate system is the average sub-Jovian anti-Jovian line and the equatorial plane contains the Iographic north and south poles and the orbit tangente points. The direction of zero colatitude θ points towards the sub-Jovian point (SJP) and the longitude φ is measured from the trailing Iographic equator (OT: orbit tangent point; NP: Iographic north pole).

4.1   Mathematical formulation

Consider Io in its synchronous orbit with eccentricity e_I and mean motion n_I and assume that the satellite's interior is in hydrostatic equilibrium and incompressible. The assumption of hydrostatic equilibrium is reasonable because Io's interior is hot and most likely partially molten. Jupiter's gravity gives rise to a tidal disturbance potential Φ_e in the satellite which has been given in general terms by Kaula (1964) and which will be approximated here to first order in eccentricity and lowest order in r/a_I, where r is the radial distance from Io's center of mass and a_I is the length of the semimajor axis of Io's orbit. Upon transformation of this potential into a spherical coordinate system with zero colatitude θ at the average sub-Jovian point and with zero longitude φ at the trailing half of the Iographic equator (Fig. 3), one finds a particularly simple representation

$$\Phi_e(r,\theta,\varphi,t) = \Phi_e^*\, (3\, P_2^0(\cos\theta)\ \cos n_I t + 2\, P_2^1(\cos\theta)\ \cos\varphi\ \sin n_I t) \qquad (2)$$

with

$$\Phi_e^* = (n_I r)^2 e_I \qquad (3)$$

The first of the two bracketed terms in (2) is the radial part of the disturbance potential and is due to the variation of Io's orbital distance to Jupiter. The second term is due to the satellite's geometrical libration and has a phase shift of $\pi/2$ relative to the first term. The two nodal curves of the radial potential are small circles at angular distances of about 55° from the sub-Jovian and anti-Jovian points, respectively. The nodal curves of the librational potential are great circles;

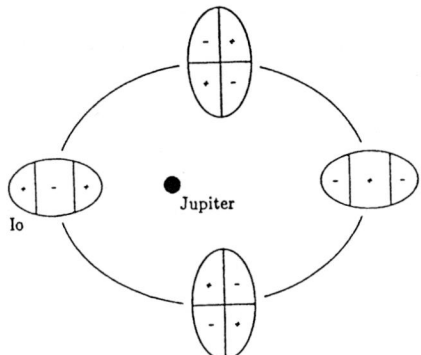

Fig. 4. Tidal deformation of Io along its eccentric orbit (not to scale). Tidal extensions relative to the triaxial hydrostatic figure are denoted by plus signs (+) and compressions by minus signs (-). The resulting tidal bulges circle Io with an angular velocity of $n_I/2$.

one is the time averaged terminator of the Jupiter facing hemisphere and the other separates the leading and trailing hemispheres (Fig. 4). The two maxima of the disturbance potential at the equator circle Io with an angular velocity of $n_{I/2}$ and with amplitudes varying between $1.5\ (n_I r)^2 e_I$ and $3.2\ (n_I r)^2 e_I$. The tidal bulge also circles Io with the same angular velocity as the potential. The total tidal disturbance potential Φ_t is the sum of the external disturbance potential Φ_e and the disturbance potential Φ_i due to the tidal deformation. The potential must satisfy the Poisson equation and the material displacement $\underline{u}$ must satisfy the equation of motion. Like the disturbance potential Φ_e, the displacement and Φ_t are written in terms of second degree spherical harmonics

$$\underline{u} = \sum_{m=0}^{1} U_r(r) P_2^m(\cos\theta)\cos(m\varphi)\hat{\underline{e}}_r$$

$$+ U_\theta(r)\left(\frac{\partial P_2^m(\cos\theta)}{\partial\theta}\cos(m\varphi)\hat{\underline{e}}_\theta - \frac{1}{\sin\theta}P_2^m(\cos\theta)\sin(m\varphi)\hat{\underline{e}}_\varphi\right) \qquad (4)$$

$$\Phi_t = \sum_{m=0}^{1} \Phi_t^*(r)\,P_2^m(\cos\theta)\cos(m\varphi) \qquad (5)$$

In (4) and (5), $\hat{\underline{e}}_r$, $\hat{\underline{e}}_\theta$, and $\hat{\underline{e}}_\varphi$ are unit vectors and $\Phi_t^*(r)$, $U_r(r)$, and $U_\theta(r) = U_\varphi(r)$ are the complex disturbance potential and radial and tangential components of the displacement, respectively. $U_\theta(r)$ is equal to $U_\varphi(r)$ because of the assumed incompressibility.

The amplitude and the phase angle of the tidal deformation depend strongly on the rheology of the interior of the satellite. Dissipation of tidal energy is a viscoelastic phenomenon. If Io were perfectly elastic or a non-viscous fluid, the tide would follow the external disturbance immediately, the phase angle between Φ_e and Φ_i would be zero, and no energy would be dissipated. In planetary physics, the Maxwell rheology is often used because it is comparatively simple while retaining the basic properties of viscoelasticity. The Maxwell time t_{MW}, defined as the ratio between the viscosity ν and the shear modulus μ, is a quantity that characterizes the viscoelastic behavior of the Maxwell-body. For forcing periods much smaller than t_{MW} the Maxwell body responds by deforming elastically. For forcing periods much larger than t_{MW}, the Maxwell body responds by deforming as a viscous fluid. The phase lag between the forcing and the response also varies with the forcing period and is maximal for forcing periods equal to the Maxwell time. In both extreme cases, for forcing periods that go towards zero or infinity, the phase lag goes to zero.

The constitutive relation for the Maxwell rheology is

$$\dot{\tau}_{kl} + \frac{\mu}{\nu}(\tau_{kl} - \frac{1}{3}\tau_{kk}\delta_{kl}) = 2\mu\dot{\varepsilon}_{kl} + \lambda\dot{\varepsilon}_{kk}\delta_{kl} \tag{6}$$

where τ_{kl} and ε_{kl} are the stress and strain tensors, respectively, δ_{kl} is the Kronecker delta, λ is the Lamé constant, and the dot denotes a derivative with respect to time. Both, viscosity and shear modulus are functions of temperature and degree of partial melting. For subsolidus temperatures, a commonly used parameterization of the temperature dependence of ν is

$$\nu(T) = \nu_o \exp\left(\frac{A}{T}\right) \tag{7}$$

where ν_o is a constant and A is the constant activation temperature for flow. The shear modulus μ is assumed to be constant in this temperature range (compare Fig. 8). At temperatures above the solidus Fischer and Spohn (1990) have derived the following parameterization by fitting experimental data on peridotite:

$$\nu(T) = \nu_1 10^{\nu_2/T} \tag{8}$$

$$\mu(T) = \mu_1 10^{\mu_2/T} \tag{9}$$

where ν_1, ν_2, μ_1, and μ_2 are constants. The numerical modeling of the effects of temperature dependent rheology for both tidal deformation and heat transfer for Io is a demanding problem that has not yet been solved. The numerical results of model calculations presented in this review simplify the problem by averaging the temperature in individual layers. This average temperature is a function of time, however, in the evolution calculations presented further below.

Hooke's law is obtained by taking the Laplace transform of (6)

$$\overline{\tau_{kl}} = 2\mu(s)\overline{\varepsilon_{kl}} \tag{10}$$

where $\overline{\tau_{kl}}$ is the Laplace transform of the deviatoric stress tensor, $\overline{\varepsilon_{kl}}$ is the Laplace transform of the strain tensor, $\mu(s) = \mu s/(s + \mu/\nu)$ is the complex shear modulus, and s is the Laplace variable. The imaginary part of $\mu(s)$ gives the phase lag between stress and strain. Application of the correspondence principle formally reduces the viscoelastic problem to an elastic problem with well known methods of solution for layered bodies (e.g., Alterman et al., 1959; Takeuchi et al., 1962).

By using (10) and inserting (4) and (5) into the Poisson equation and the equation of motion one gets an ordinary first order differential equation

$$\frac{dY}{dr} = \mathbf{A}(r)\, Y \tag{11}$$

for the vector $Y = (U_{\rm r}, U_\theta, T_{\rm r}, T_\theta, \Phi_1^*, q)$, where $T_{\rm r}$ and T_θ are the spherical harmonic coefficients of the radial and tangential stresses and q is a quantity related to the radial gradient of the disturbance potential Φ_t. The complex elements of the matrix $\mathbf{A}$ depend on the radius, the acceleration of gravity, and the material parameters. Boundary conditions for the solution of (11) have been discussed in Takeuchi et al. (1962). The radial and tangential stresses are zero at $r = R_{\rm s}$, where $R_{\rm s}$ is the satellite's mean radius, and $q(r = R_{\rm s}) = -5/R_{\rm s}$ if the external potential is appropriately normalized. As r approaches zero, it is required that the displacement and the disturbance potential also approach zero.

The volumetric tidal dissipation rate can most easily be obtained from the integral over one orbital period T of the work function (Love, 1927),

$$\dot{W} = \frac{1}{T} \int_0^T \varrho\, \frac{\partial u}{\partial t} \cdot \nabla \Phi_t\; dt \tag{12}$$

where ϱ is density. The volume integral of $\dot{W}$ can be transformed into a surface integral with the help of Green's first integral transformation (Zschau, 1978; Platzman, 1984):

$$\dot{E} = -\frac{5n_{\rm I}}{8\pi^2 G R_{\rm s}} \int_0^T \int_S \Phi_{\rm i}\, \frac{\partial \Phi_{\rm e}}{\partial t}\; dS\, dt \tag{13}$$

where G is the gravitational constant. Using the definition of the secondary Love number

$$k_{\rm I} = \frac{\Phi_{\rm i}}{\Phi_{\rm e}} \tag{14}$$

and rearranging the integrals, we get

$$\dot{E} = -\frac{5\,\Im(k_{\rm I})}{8\pi^2 G R_{\rm s}} \int_S \int_0^T \left(\frac{\partial \Phi_{\rm e}}{\partial t}\right)^2 dt\, dS \tag{15}$$

where $\Im(k_{\rm I})$ is the imaginary part of $k_{\rm I}$. The integral gives the tidal dissipation rate for an arbitrary satellite. Upon inserting the tidal disturbance potential $\Phi_{\rm e}$ from (2) and integrating, we find the ratio 3/4 between the dissipation rate due to the radial disturbance potential and that due to the librational disturbance potential, in accordance with Yoder and Peale (1981). For the total dissipation rate we get

$$\dot{E} = -\frac{21}{2}\, \Im(k_{\rm I})\, \frac{(n_{\rm I} R_{\rm s})^5}{G}\, e_{\rm I}^2 \tag{16}$$

In order to calculate the tidal dissipation rate for a complicated model of a satellite it is thus sufficient to find the imaginary part of the satellite's secondary Love number which, for a Maxwell rheology, is easily calculated with the help of the correspondence principle. Other authors prefer to use the quality factor Q over $\Im(k)$ as a more hands-on parameter. The quality factor is defined by

$$Q^{-1} \equiv \frac{\Delta E}{E} \tag{17}$$

where ΔE is the energy loss per work cycle and E is the total energy stored in the system. Although a mathematically rigid relation between Q and $\Im(k)$ cannot be given, in many practical cases the following approximate relation is valid

$$\Im(k) \approx \frac{|k|}{Q} \tag{18}$$

where $|k|$ is the absolute value of k.

4.2   Models of tidal dissipation

Figures 5 and 6 show contours of the logarithm of the dissipation rate, for two models. The first model is three layered with a 30 km thick elastic lithosphere, a viscoelastic mantle, and an inviscid core of 980 km radius. The second model has, in addition, a low viscosity asthenosphere of 50 km thickness. These models are similar to the ones presented in Segatz *et al.* (1988) but have been recalculated correcting for an error of 2π in their calculation of the forcing frequency. This error translates into a shift by 2π or roughly one order of magnitude of the lines of constant dissipation rate to smaller values of viscosity. The dashed lines mark the values of the shear modulus and the viscosity for which the Maxwell time equals the orbital period. The largest dissipation rate for the first model of $3 \times 10^{15} W$ is obtained with a viscosity of $4 \times 10^{14} Pa\ s$, close to the value $(2/19)\overline{\varrho} g R_s T$ ($\approx 2 \times 10^{14} Pa\ s$), that maximizes the dissipation rate in a homogeneous satellite, where $\overline{\varrho}$ is the average density.

To balance the observed surface heat flow by tidal dissipation, a mantle viscosity between 4×10^{11} and 6×10^{15} Pa s and a shear modulus in excess of 8×10^7 Pa are required. A preferred mantle-dissipation model of Io could have the parameter values $\nu = 10^{15}$ Pa s and $\mu = 10^{10}$ Pa. The choice of a large viscosity is rationalized by the thermostat effect of convection in a fluid whose viscosity depends on temperature (Tozer, 1965). The chosen shear modulus is about 1/6 of the value derived for the Moon and is reasonable considering the larger interior temperatures postulated for Io. The preferred model of Io plots below the dashed line indicating that Io's tidal response is predominantly elastic. The large surface heat flow requires the Maxwell time of any reasonable model of Io to be close to the orbital period. The magnitude of the Love number k_I for the preferred model is 0.25. Segatz *et al.* (1988) have shown that the dissipation rate depends relatively little on both the core radius and the lithosphere thickness within reasonable variations of these parameters.

While the dissipation rate for the mantle dissipation model depends mainly on the mantle viscosity and shear modulus, dissipation in the asthenosphere depends mainly on the asthenosphere viscosity and thickness. Reasonable values may range from 10^8 Pa s to 10^{12} Pa s and between roughly 10 and 100 km. The shear modulus is required to be larger than about 10^4 Pa. For any given model there are two asthenosphere thicknesses with the same dissipation rate

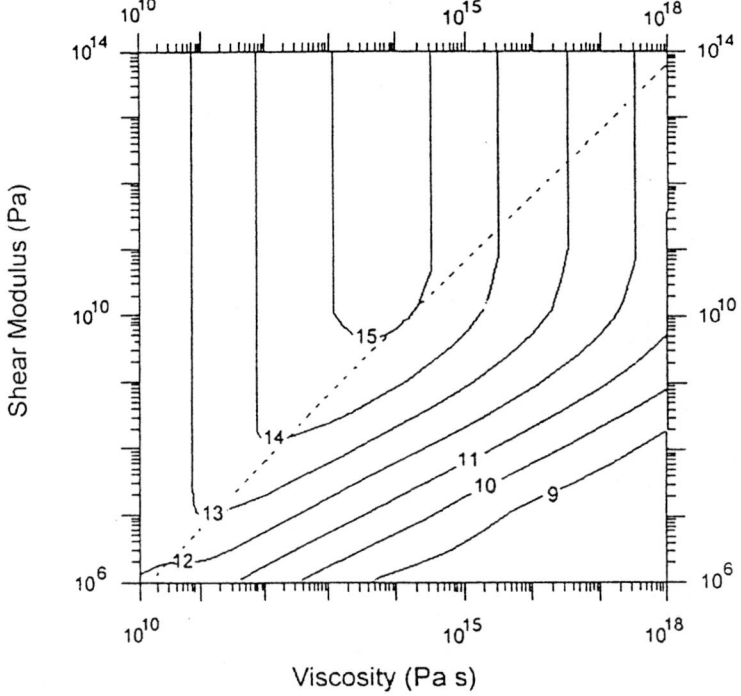

Fig. 5. Contour map of the logarithm of the tidal dissipation rate $\dot{E}$ (W), as function of mantle viscosity and shear modulus for the mantle dissipation model. The dashed line denotes the viscosities and shear moduli for which the Maxwell time equals the orbital period.

(Fig. 7). However, only asthenospheres with a thickness of more than about 10 km thickness are stable, those for which there is a negative slope of the $\dot{E}$ versus thickness curve. With a positive slope, runaway thickening to the larger thickness will occur if the dissipation rate is in excess of the rate of heat removal. (Such a runaway thickening may help in creating an asthenosphere.) The dissipation rate for the asthenosphere model decreases approximately linearly with increasing lithosphere thickness in the range of lithosphere thicknesses from 10 to 100 km. This is because the deformation of the lithosphere decreases with increasing lithosphere thickness which results in decreasing asthenosphere deformation and dissipation rates.

It is interesting to note that the mantle dissipation model shows maxima of the dissipation rate underneath the poles along the core mantle boundary and minima at the sub-Jovian and anti-Jovian points (Segatz *et al.*, 1988). The asthenosphere model shows a belt of maxima and minima circling Io along its equator with maxima at the sub Jovian and anti-Jovian points and with minima underneath the poles. Ross *et al.* (1990) have used the topography of Io (Gaskell and Synott, 1988) to constrain a model of Io that combines dissipation in the asthenosphere and the deep mantle. They assumed that the surface heatflow is

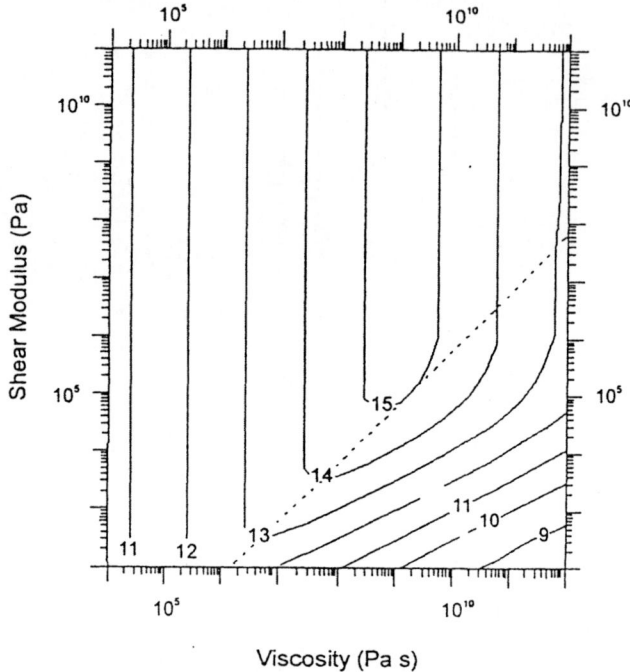

Fig. 6. Contour map of the logarithm of the tidal dissipation rate $\dot{E}$ (W) for the asthenosphere dissipation model. The dashed line denotes the viscosities and shear moduli for which the Maxwell time equals the orbital period.

locally in equilibrium with the dissipation rate integrated over depth and that the heat is conducted through an isostatically compensated lithosphere with a constant lower boundary temperature. By fitting the topography thus calculated from a linear combination of the two dissipation models to the observed topography they concluded that about 2/3 of the dissipation occurs in the asthenosphere and about 1/3 in the lower mantle. However, the observed pattern of basins and swells girdling Io is phase shifted by about 25° with respect to the pattern predicted from the model. Ross *et al.* (1990) speculate that Io's lithosphere may have recently undergone a zonal rotation by about that angle. Their model of the distribution of dissipation between the asthenosphere and the deep mantle is supported by the observed pattern of hotspots on Io that resembles the tidal dissipation pattern of the asthenosphere dissipation model.

5   Thermal and Orbital Evolution

The large rate of dissipation of tidal energy in Io will affect both its orbital and thermal evolution. The latter is easily understood: A heat production rate of about 10^3 times the radiogenic production rate must be a major contribution

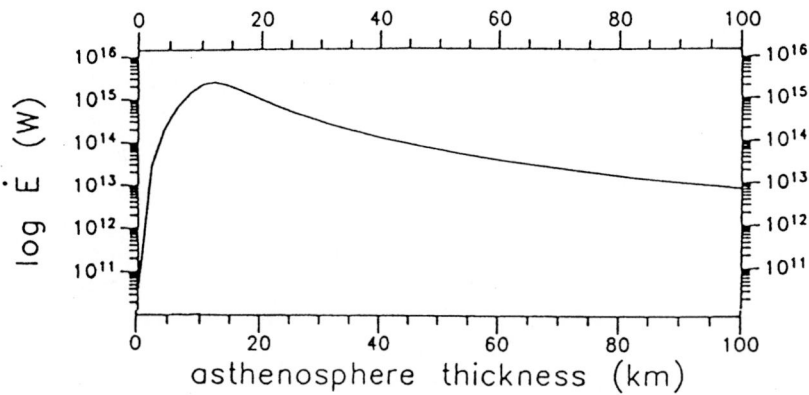

Fig. 7. Dissipation rate $\dot{E}$ versus asthenosphere thickness for the asthenosphere dissipation model.

to the satellite's interior energy balance. The way the tidal dissipation affects the orbital evolution can be qualitatively understood from a consideration of the energy and momentum balances in the Jovian system. The ultimate source of the energy that is dissipated in Io is the rotational energy of Jupiter. The latter is a giant reservoir of about 4×10^{34} J. If Io had removed energy at a constant rate of 10^{14} W throughout the age of the solar system, it would have removed just about one part in 10^4 from the reservoir. Energy is transferred to the orbit of Io via the tides raised on Jupiter by Io. Since Jupiter rotates faster than Io revolves about the planet, the tidal bulge on Jupiter precedes Io just as the tidal bulge on Earth precedes the Moon. The torque exerted by the bulge on Io accelerates the satellite and causes its orbit to expand. Conservation of energy and momentum requires the planet's rotation energy to decrease and the satellite's orbital energy to increase. The rate of energy transfer is proportional to Jupiter's $\Im(k_J)$ where k_J denotes Jupiter's secondary Love number. A significant part of the orbital energy gained is dissipated in the satellite. This is promoted by a torque exerted by Io's tidal bulge on Jupiter. The rate of expansion of the orbit and its eccentricity are decreased as a consequence of the conservation of angular momentum. Another part of Io's orbital energy is passed on to the orbital energies of the other Galilean satellites through tidal interaction which also decreases the rate of orbital expansion. Thus, the rate of expansion of the orbit depends on the dissipation rate in Io which in turn, depends on temperature through its dependence on the viscosity and the shear modulus.

5.1 Orbital evolution equations

The orbital evolution equations in the presence of tidal dissipation have been derived by Yoder and Peale (1981). These equations have been coupled to a model of the thermal evolution of the satellite by Fischer and Spohn (1990). The coupling is promoted by the temperature dependences of the viscosity and the shear modulus. Their set of equations was expanded by Wienbruch and Spohn (1995) to include the thermal evolution of the core and a simple core dynamo model based on an energy balance for the dynamo. In the following we will briefly summarize these equations, simplified by neglecting small quantities to emphasize the roles of tidal dissipation in Jupiter, Io, and Europa. Dissipation in Ganymede is neglected because, at least at present, tidal dissipation in Ganymede and its eccentricity (comp. Table 1) are small and would have a negligible effect on Io. Ganymede and Io interact through Europa as an intermediary. In the past, Ganymede's eccentricity and its dissipation rate may have been much larger (Malhotra, 1991). A modeling of this effect is beyond the scope of the present review, however.

The secular variation of Io's mean motion is, to second order in eccentricity,

$$\dot{n}_I = C_{n_I} \cdot n_I^{16/3} \cdot \left[\Im(k_J) - 47.2 \cdot \Im(k_I) \cdot e_I^2 - 0.69 \cdot \Im(k_E) \cdot e_E^2 \right] \tag{19}$$

$C_{n_I} \approx -1.08 \cdot 10^7 \; s^{10/3}$ is a numerical constant, e_I and e_E are Io's and Europa's orbital eccentricities. Since Europa is in resonance with both Io and Ganymede, its eccentricity is the result of a superposition of the eccentricity forced by Io and that forced by Ganymede. $\Im(k_E)$ is the imaginary part of Europa's Love number. The rate of change of Io's mean motion is mainly determined by the difference between tidal dissipation in Jupiter, the first bracketed term in the above equation, which acts to decrease Io's mean motion and by dissipation in Io, the second term, which acts to increase n_I. Tidal dissipation in Europa is presently of secondary importance. However, it is important for values of e_E significantly larger than the present value. Large values of e_E can occur early and very late in the orbital evolution of the Jovian satellite system and during oscillations of the orbital and thermal variables.

Io's orbital eccentricity is given by

$$e_I = C_1 \cdot \frac{m_E \cdot \gamma}{m_J} \cdot \frac{n_I}{\varepsilon + \omega_I} \tag{20}$$

where $C_1 \approx 1.19$ is a constant of integration, m_E and m_J are the masses of Europa and Jupiter, respectively, $\gamma = 0.63$ is the ratio between the semimajor axis of Io and Europa and of Europa and Ganymede, ω_I is the rate of the secular variation of Io's perijove, and

$$\varepsilon \equiv n_I - 2 \cdot n_E \tag{21}$$

where n_E is Europa's mean motion.

Europa's orbital eccentricity is given by

$$e_E = \left(C_1 \cdot \frac{m_G \cdot \gamma^{5/2}}{m_J} + C_2 \cdot \frac{m_I \cdot \gamma^{3/2}}{m_J} \right) \cdot \frac{n_I}{\varepsilon + \omega_E} \tag{22}$$

where $C_2 \approx 0.430$ is a constant of integration, m_I and m_G are Io's and Ganymede4s masses, respectively, ε is the resonance variable, and ω_G is the rate of the secular variation of Europa's perijove. If the resonance were perfect, ε would be zero. Dissipation causes the variable to be nonzero and its time rate of change is given by

$$\dot{\varepsilon} = C_{\varepsilon_I} \cdot n_I^{16/3} \cdot \left(\Im(k_J) - 64.7 \cdot \Im(k_I) \cdot e_I^2 - 0.95 \cdot \Im(k_E) \cdot e_E^2 \right) \tag{23}$$

where $C_\varepsilon \approx -3.96 \cdot 10^6 \ s^{10/3}$ is a numerical constant. Equation (23) together with Equations (20) and (22) show how dissipation in Jupiter acts to increase the eccentricities while dissipation in both Io and Europa acts to decrease the eccentricities.

5.2   Thermal evolution equations

Since the dissipation rate is temperature dependent as a consequence of the temperature dependence of the rheology parameters, a thermal evolution model is required. To include temperature dependence in Jupiter, Io and Europa would result in an utterly complex model, however. Therefore, only dissipation in Io is modeled as temperature dependent while dissipation in Jupiter and Europa is assumed to occur at a constant rate. This is a reasonable simplification since the effects of tides raised on Jupiter should effect the thermal evolution of Jupiter only negligibly. Of course, there is some secular cooling of Jupiter which should cause $\Im(k_J)$ to decrease with time to some extent. For Europa, the assumption is reasonable even though Europa's present eccentricity is larger than Io's by a factor of roughly 2.5. $\Im(k_E)$ is estimated to be about two orders of magnitude smaller than $\Im(k_I)$ as the numerical factors preceding this term in equations (19) and (23).

Heat transfer in the partially molten mantle of Io must be accomplished by vigorous convection driven by gradients in temperature and in melt concentration. The vigor of the convection is measured by the Rayleigh number Ra here defined as

$$Ra = \frac{g_I \, \alpha \, \varrho_m \dot{E} (R_I - R_c)^5}{\Lambda_m \kappa \nu V_m} \tag{24}$$

where g_I is surface gravity, α is the thermal expansion coefficient, ϱ_m is the mantle density, R_c is the core radius, Λ_m is the mantle thermal conductivity and κ the mantle thermal diffusivity, V_m is the mantle volume, and ν is the kinematic viscosity which depends on temperature through Equations (7) and (8). The current mantle Rayleigh number is approximately

$$Ra \approx 2 \cdot 10^{15} \cdot \frac{\dot{E}}{\nu} \tag{25}$$

$$\approx 10^{14}$$

The heat flow q_s from the mantle is related to Ra via

$$q_s = \Lambda_m \frac{(T - T_s)}{R_I - R_c} a Ra^\beta \tag{26}$$

where T is the temperature of the upper convecting mantle, T_s is the surface temperature, and a and β are constants. The convective heat transfer rate increases with increasing Rayleigh number and, through the dependence of the Rayleigh number on the viscosity, decreases with increasing mantle viscosity. Although the exact form of the convective heat transport parameterization differs among various authors, the approach is well established as a way of investigating the thermal evolution of planets (e.g., Schubert *et al.* (1986)). It has been shown (Spohn and Schubert, 1982; Christensen 1985) that reasonable uncertainties in the parameter values entering the $q_s - Ra$ relation do not significantly affect the results of evolution calculations unless $d\ln q_s/d\ln Ra$ is very much smaller than its accepted standard value for internally heated convection of about 0.25.

In the following, we give the equations for the calculation of the thermal evolution of the core and the mantle. The core energy balance equation is

$$\frac{4}{3}\pi R_c^3 \varrho_c\, c_c\, \eta_c \dot{T}_{\text{cm}} = -4\pi R_c^2\, q_c \tag{27}$$

where ϱ_c is the density of the core, c_c is its specific heat, $\eta_c \approx 1.1$ is a constant that relates the core–mantle boundary temperature T_{cm} to the average core temperature, and q_c is the heat flux from the core to the mantle. Since we assume continuity of heat flow across the core-mantle boundary we can calculate the core-mantle heat flow in the mantle

$$q_c = \Lambda_{\text{m}} \frac{(T_{\text{cm}} - T - \frac{dT}{dz}|_{\text{wa}}(R_{\text{I}} - R_c))}{R_{\text{I}} - R_c} a Ra^\beta \tag{28}$$

where $dT/dz|_{\text{wa}}$ is the slope of the mantle wet adiabat calculated from Equation (1).

The energy balance equation for the mantle is

$$\frac{4}{3}\pi(R_{\text{I}}^3 - R_c^3)\varrho_m c_m \eta_m \left(\frac{L}{c_m}\frac{df}{dT} + 1\right)\dot{T} = -4\pi R_{\text{I}}^2 q_s + 4\pi R_c^2 q_c + \dot{E} + Q \tag{29}$$

where, in addition to previously defined quantities, c_m is the mantle heat capacity, $\eta_m \approx 1.4$ is a constant to relate the upper mantle temperature T to the average mantle temperature, $\frac{df}{dT}$ is the rate of change of the degree of partial melting with temperature, and Q is the (chondritic) radiogenic heat production rate. Equations (27) and (29) may be integrated together with the orbital evolution equations using standard numerical techniques such as fourth order Runge–Kutta schemes or shooting methods.

It is possible to estimate the dipole moment of a possible magnetic field produced by dynamo action in the core as a function of the core-mantle heat flow. Dynamo action in the core requires that there is enough power available for the dynamo to overcome the Ohmic dissipation Φ_{Ohm}. The thermal power P_{d} derived from a convecting, non-crystallizing core is given by

$$P_{\text{d}} = \gamma_{\text{therm}} \left(\dot{E}_{\text{therm}} - F_{\text{cond}}\right) \tag{30}$$

where γ_{therm} is an average Carnot efficiency factor approximately equal to 0.05 (Stevenson *et al.*, 1983), $\dot{E}_{\text{therm}}$ is the time rate of change of core heat content, and

$$F_{\text{cond}} = -4\pi R_c^2 \Lambda_c \frac{dT}{dz}\big|_{\text{ad}} \tag{31}$$

where Λ_c is the thermal conductivity in the core and $dT/dz|_{\text{ad}}$ is the adiabatic temperature gradient in the core. F_{cond} is the conductive heat flow down the core adiabat which is not available to drive the core dynamo.

To estimate the magnetic moment we equate Φ_{Ohm} with P_d and use a simple model of the dynamo introduced by Stacey (1977). The model considers an equatorial toroidal conductor in the core with mean radius $R_c/2$. The dipole moment m_{dipole} associated with the magnetic field that is generated by a current in the toroid is approximately

$$m_{\text{dipole}} = \left(\frac{\pi^2 \Phi_{\text{Ohm}} \left(\frac{R_c}{2}\right)^5}{2\Omega} \right)^{\frac{1}{2}} \tag{32}$$

where Ω is the resistivity in the toroid. Using $3 \times 10^{-6}\,\Omega\text{m}$ for the resistivity and other parameter values as listed in Tab. 2,

$$m_{\text{dipole}} \approx 5 \times (\chi - 1)^{\frac{1}{2}} \; [\times 10^{19}\,\text{A}\,\text{m}^2] \tag{33}$$

where χ is the time rate of change of core heat content relative to F_{cond}. For instance, the dipole moment resulting from a core to mantle heat flow of $2F_{\text{cond}}$ is about $5 \times 10^{19}\,\text{A}\,\text{m}^2$. This dipole moment is of the same magnitude as that suggested by Kivelson et al. (1996). Our estimate of the dipole moment is crude, however, because the model is rather simplistic. The model was developed for Earth and has been applied with some success to this planet (Stacey, 1977). Schubert et al. (1988) have applied the model to Mercury but they found that its dipole moment is overestimated by a factor of 10. This is not necessarily a problem with the model. Rather, it is likely that a thermo–electric dynamo works in the core of Mercury instead of the conventional hydromagnetic dynamo (Schubert et al., 1988).

5.3   Discussion of Io's thermal-orbital evolution

For radiogenically heated convecting planetary mantles the interplay between internal heat production, which is independent of temperature, and the strongly temperature dependent convective heat transfer rate defines a self–regulating system that buffers the mantle temperature. If the mantle temperature is too small, the viscosity will be large and temperature will increase because the heat transfer rate is smaller than the heat production rate. On the contrary, if the temperature is too large, the viscosity will be small, heat transfer will be more effective than heat production and the temperature will be reduced. This buffering of the mantle temperature is also known as Tozer's principle (Tozer, 1965). In the visco–elastic tidal heating model of Io, both the tidal heating and the heat transfer rates are temperature dependent. As Segatz et al. (1988) and Fischer and Spohn (1990) have shown and as we will discuss below, the interior temperature is then buffered such that it is supersolidus if the heat production and transfer rates are in stable equilibrium.

In Fig. 8, the tidal dissipation rate, the surface heat flow, the viscosity, and the shear modulus are plotted as functions of temperature. Both the heat flow and the tidal dissipation rate increase with temperature as the viscosity decreases with temperature up to the solidus. Above the solidus the tidal dissipation rate decreases sharply with increasing temperature because the shear modulus then decreases rapidly with temperature. There are three points of equilibrium between the heat production rate and the heat flow in Fig. 8. For clarity and ease of reference, these points are numbered and marked with circles. The first point of equilibrium occurs at small values of temperature and of tidal dissipation and heat flow rates. Because the heat production rate at this point – essentially the chondritic radiogenic heat production rate – is independent of temperature, this is the point of equilibrium where Tozer's principle works. The second point occurs at a temperature below the solidus temperature where both the tidal dissipation and the heat transfer rates increase with temperature. The equilibrium at this point is unstable to fluctuations of temperature which will result in runaway heating or cooling because $(d\dot{E}/dT)/(dq_s/dT) \approx \beta^{-1}$ is > 1 and both $(d\dot{E}/dT)$ and (dq_s/dT) are > 0. The third point of equilibrium occurs at a temperature above the solidus where the heat transfer rate decreases less rapidly with temperature as compared with the tidal dissipation rate. The equilibrium is stable because $(d\dot{E}/dT)/(dq_s/dT) \approx \beta^{-1}$ is > 1 and both $(d\dot{E}/dT)$ and (dq_s/dT) are < 0. It is evident that a stable equilibrium is possible only at temperatures above the solidus where $\dot{E}(T)$ decreases rapidly with increasing temperature. Any evolutionary path that starts at a mantle temperature above about 1200 K (at significantly smaller temperatures tidal heating is unimportant) will end close to the stable equilibrium point. Io will be locked there as long as the third equilibrium point exists. The third equilibrium point will cease to exist whenever the tidal heating curve drops below the heat flow curve as a consequence of decreases in mean motion and eccentricity.

The equilibrium between surface heat flow and the tidal dissipation rate, however, cannot be exact because of the contributions of the core heat flow and the cooling of the entire satellite to the energy balance. The results of numerical calculations to be presented further below suggest that the difference between the heat flow and the tidal heating rate should be small with cooling and core heat flow providing only a few $10^{-1}\%$ of the total power budget. The inevitable expansion of Io's orbit also implies cooling. Because $\dot{E}$ decreases with decreasing values of n (Equation 16) the $\dot{E}(T)$-curve is lowered as the orbit expands while the q_s-curve stays in place. As a consequence, the point of equilibrium moves towards smaller values of T. We can estimate the cooling rate associated with the expansion of the orbit by considering the changes in tidal heating and in heat flow between two successive points of equilibrium and use this estimate to derive an upper bound on the core heat flow. The incremental change in the tidal heating rate is

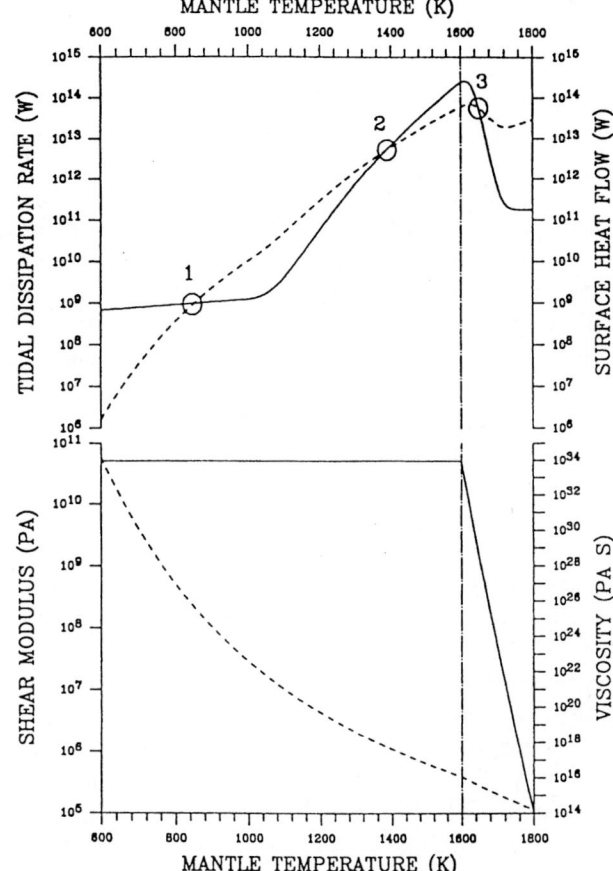

Fig. 8. (upper part) Dissipation rate (solid line) at constant values of mean motion and eccentricity and surface heat flow (dashed line) as functions of temperature. The three points of equilibrium are discussed in the text. The vertical line indicates the characteristic temperature at which the melt begins to dominate the rheology of a partially molten mantle. (lower part) Mantle viscosity (dashed line) and shear modulus (solid line) as functions of temperature.

$$d\dot{E} = \left(\frac{\partial \dot{E}}{\partial T}\right) dT + \left(\frac{\partial \dot{E}}{\partial n_{\mathrm{I}}}\right) dn_{\mathrm{I}} + \left(\frac{\partial \dot{E}}{\partial e_{\mathrm{I}}}\right) de_{\mathrm{I}} \qquad (34)$$

and the change in heat flow is

$$dq_s = \left(\frac{\partial q_s}{\partial T}\right) dT \qquad (35)$$

Dividing by dt and assuming that $d\dot{E} = dq_s$ we get

$$\left(\frac{\partial \dot{E}}{\partial T} - \frac{\partial q_s}{\partial T}\right) \dot{T} = -\left(\frac{\partial \dot{E}}{\partial n_{\mathrm{I}}}\right) \dot{n}_{\mathrm{I}} - \left(\frac{\partial \dot{E}}{\partial e_{\mathrm{I}}}\right) \dot{e}_{\mathrm{I}} \qquad (36)$$

Evaluating the partial derivatives with Equations 34 and 26 at temperatures above the solidus temperature, we get

$$\frac{dT}{dt} = -\frac{5\frac{\dot{n}_I}{n_I} + 2\frac{\dot{e}_I}{e_I}}{2.3\left(\nu_2 - \frac{2}{3}k_I\mu_2\right)(1-\beta)}T^2 \tag{37}$$

The orbital terms $\dot{n}_I/n_I$ and $\dot{e}_I/e_I$ are constrained by observational values (Lieske, 1987) to be about $(-8.56 \pm 28.5) \times 10^{-19}$ s^{-1} and $(-6.93 \pm 31.7) \times 10^{-18}$ s^{-1}, respectively. Reasonable values of the rheological constants μ_2 and ν_2 are given in Tab. 2 and $\beta \approx 0.25$ for internally heated convection. From these values and taking T approximately equal to the solidus temperature of 1600 K we find a cooling rate of $(-8 \pm 14) \times 10^{-16}$ K s^{-1} and, together with the heat capacity of Io's mantle, a heat flow of -8×10^{10} W, about 0.1 % of the mantle heat flow. This value can be taken as an upper limit to the core heat flow and is of the same order of magnitude as the value of the heat flow down the core adiabat F_{cond} calculated to be about -5×10^{10} W using the value of the core thermal conductivity Λ_c given in Tab. 2 and the adiabatic temperature gradient $dT/dz|_{\text{ad}}$ from Fig. 2. The conclusion is that in stable near-equilibrium the core may be cooling conductively in which case hydromagnetic dynamo action would be impossible. In the more detailed numerical model to be discussed below we indeed find the core heat flow to be only -10^{10} W.

Before we present some representative numerical solutions, we will discuss in more general terms how the thermal-orbital variables evolve as Io's orbit expands and as the interior cools. For this purpose, we use Fig. 9 where the heat transfer and tidal dissipation rates are compared at various points in time. In panel A, Io has responded to the chosen initial conditions and runaway warming ensues because the temperature is above the unstable equilibrium point 2 of Fig. 8. In panel B, stable equilibrium is reached and Io evolves steadily being locked near equilibrium point 3. Continuous cooling and orbital expansion results in a slow drop of the tidal dissipation curve until in panel C the dissipation rate and heat transfer curves become tangential. A further drop in temperature as indicated in panel D will initiate runaway cooling. This cooling is accompanied by a rapid decrease in the tidal dissipation rate and a rapid increase in the eccentricity according to Equations (20) and (23). The increase in eccentricity will push the dissipation curve upward since the tidal dissipation rate depends quadratically on eccentricity (Equation (16)). The result is shown in panel E. Runaway warming is initiated and continues until equilibrium is again reached as in panel B. Continued cooling and orbital expansion will then initiate another cycle. The oscillations will stop when the temperature becomes smaller than the equilibrium temperature on the unstable part of the dissipation rate curve as indicated in panel F. From there on Io will cool until it comes into equilibrium with the radiogenic heat production rate.

5.4   Numerical solutions

Numerical solutions of the above set of equations requires the specification of a number of parameters (Tab. 2). Some of these, in particular the rheology and

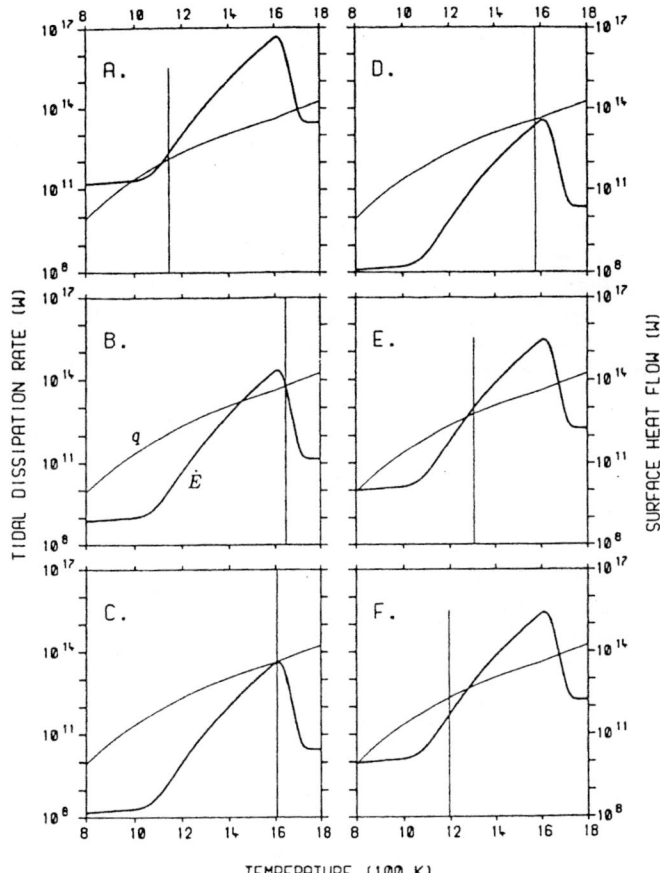

Fig. 9. Tidal dissipation rate (thick line) and surface heat flow (thin line) as functions of temperature at various points in time (compare Fig. 8). The thin vertical line marks the current mantle temperature. In panels 'A' and 'E' the dissipation rate is larger than the rate of heat loss and runaway heating is initiated. In 'B' the model is stable and in 'C' it is marginally stable. In 'D' and at 'F' runaway cooling is initiated because the rate of heat loss is larger than the dissipation rate. For further discussion see the text.

heat transfer parameters, have been varied in previous work (Fischer and Spohn, 1990; Wienbruch and Spohn 1995) to determine their influences on the solutions. We restrict ourselves here to present a representative example.

As for the initial conditions, Fischer and Spohn (1990) have shown that the results of thermal-orbital history calculations are not sensitive to the choice of initial conditions. In their model, Io enters the resonance prior to about 2 Ga b.p. as a completely differentiated (into a core and mantle) body. The age of the Laplace resonance is not very well known. Yoder (1979) has derived a minimum value of the age of the Laplace resonance of about 60 Ma from a consideration

Table 2. Parameter values for the thermal-orbital evolution model of Io

Parameter	Definition	Value
g_I	surface gravity	1.8 $\mathrm{m\,s^{-2}}$
T_s	average surface temperature	100. K
ϱ_m	mantle density	3270. $\mathrm{kg\,m^{-3}}$
c_m	mantle heat capacity	1200. $\mathrm{J\,kg^{-1}\,K^{-1}}$
Λ_m	mantle thermal conductivity	4.0 $\mathrm{W\,m^{-1}\,K^{-1}}$
α	thermal expansion coefficient	$5.\times10^{-5}$ $\mathrm{K^{-1}}$
ν_0	viscosity constant	1.6×10^5 Pas
A	activation temperature	$4.\times10^4$ K
ν_1	viscosity parameter	0.1 Pas
ν_2	viscosity parameter	2.7×10^4 K
μ_1	shear modulus parameter	2.5×10^{-41} Pa
μ_2	shear modulus parameter	8.2×10^4 K
Q	radiogenic heat production rate	10^9 W
L	latent heat of mantle melting	10^5 $\mathrm{W\,kg^{-1}}$
ϱ_c	core density	5161. $\mathrm{kg\,m^{-3}}$
c_c	core heat capacity	800. $\mathrm{J\,kg^{-1}\,K^{-1}}$
Λ_c	core thermal conductivity	32. $\mathrm{W\,m^{-1}\,K^{-1}}$
R_c	core radius	950. km
γ_{therm}	dynamo Carnot efficiency factor	0.005.
$\Im(k_J)$	Jupiter tidal dissipation factor	10^{-5}
$\Im(k_E)$	Europa tidal dissipation factor	10^{-4}
ω_I	rate of secular variation of Io's perijove	3.25×10^{-8} $\mathrm{s^{-1}}$
ω_E	rate of secular variation of Europa's perijove	9.70×10^{-9} $\mathrm{s^{-1}}$

of the rate of decrease of Io's free libration amplitude. But Fischer and Spohn (1990) have argued that the resonance must be more than 500 Ma old and have proposed an age of at least 2 Ga. Their argument is based on the hypothesis that tidal heating must have been sufficiently strong to heat and partially melt a cold Io after it entered the resonance unless Io's mantle temperature was already close to the solidus at the time of formation of the resonance. The age given by Fischer and Spohn (1990) is based on an extrapolation back in time of Io's orbit and brings Io close enough to Jupiter such that even a relatively cold, subsolidus Io can warm to solidus temperatures as a consequence of sufficiently strong tidal heating.

The initial value for the mean mantle temperature is 1250 K. A mean mantle temperature of 1200–1300 K can be expected, as thermal evolution calculations suggest, for an Io-size terrestrial planet heated by radiogenics after 1–2 G of thermal evolution. In addition, an initial potential core temperature of 2300 K is assumed. The potential core temperature is the core temperature extrapolated adiabatically to zero pressure. For simplicity, the potential core temperature is assumed to be superliquidus. It is possible that the core temperature was lower at the time of the formation of the resonance and that an inner core existed at this time. In this case, the core temperature and inner core size would adjust rapidly to the thermal conditions in the resonance. The initial value of the mean motion is the minimum value for which the present surface heat flux can be explained by tidal dissipation if the model is started with an initial temperature as low as 1250 K. The difference between the initial and the present mean motions determines the starting time which for this model is 2.3 Ga before present. The initial value of the eccentricity is chosen to be 2×10^{-4}. This value is between the values of the free eccentricity of $(1 \pm 2) \times 10^{-5}$ (Lieske, 1987) and the present eccentricity.

Results for a representative model are shown in Fig. 10 where the mantle tidal dissipation rate, the surface heat flow, the eccentricity and the mantle temperature are shown as functions of time. The evolution can be divided into two periods. (We do not include the very short interval in time, a few million years, prior to period I during which the model adjusts to the chosen initial conditions.) The boundary between these periods is indicated by a vertical line. In the first period, termed the quasi-stationary period by Fischer and Spohn (1990), the tidal dissipation rate and the surface heat flow are very close to equilibrium. This equilibrium is equivalent to that indicated by the third point in Fig. 8 and panel B in Fig. 9. The eccentricity and the mean motion (not shown in Fig. 10) and, as a consequence, the tidal dissipation rate, surface heat flow, and the potential mantle temperature all decrease monotonously during that period of time.

At the end of the first period, the satellite's thermal and orbital variables become unstable and a cycle is set up as was illustrated in panels C, D, and E of Fig. 9. The onset of the oscillations and the length of period II depends mostly on the rates of change of the viscosity and the shear modulus with temperature (Fischer and Spohn, 1990). We have used parameter values representative of the rheology of the Earth's upper mantle. With these choices of parameter values the present Io is in period I. The time axis in Fig. 10 is gauged such that $t = 0$ is the present time. Fischer and Spohn (1990) have argued for a quasi- stationary state of the present Io noting that this choice would best fit observational constraints on the orbital elements and on the surface heat flow. But as we will discuss below this model still faces some problems with the astrometrical observations of the time rates of change of n_I and ε_I.

Figure 11 shows the results for the core. The large mantle tidal heating rate dominates the energy balances of the mantle and the core through period I and little heat is removed from the core. The core is not convecting because the core to mantle heat flow of about 1 mW m^{-2} is a factor of roughly 5 smaller than the heat flow conducted down the adiabat and dynamo action is impossible.

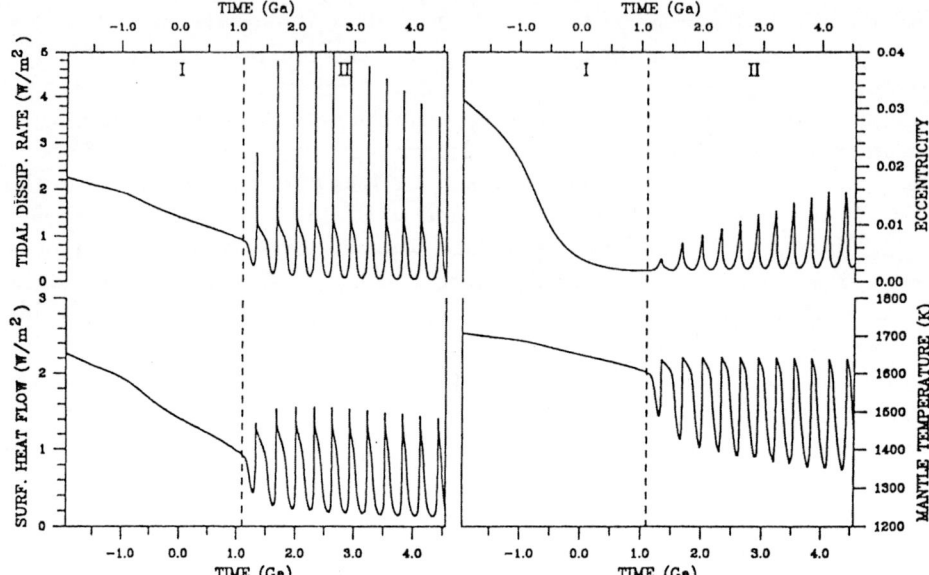

Fig. 10. Thermal–orbital evolution of Io. The figure shows as functions of time (clockwise from top to bottom) the tidal dissipation rate in the mantle, the eccentricity of the orbit, the mantle temperature, and the surface heat flow. Time t=0 is the present time; negative times are times before present. The evolutions are separated into two periods I and II, the characteristics of which are discussed in the text; the boundary between these periods is marked with a vertical line.

This conclusion remains valid even if a reasonable uncertainty by a factor of two for the core thermal conductivity (Stevenson *et al.*, 1983) is taken into account. A dynamo is possible only during period II at times when the mantle tidal dissipation rate is relatively small and when the core to mantle heat flux is relatively large. The magnetic moment shown in Fig. 11 has been calculated assuming that there is no time lag between the increase of the heat flow and dynamo action.

6   Discussion of Tidal Dissipation and Evolution Models

The tidal heating model is capable of providing the dissipation necessary to balance the heat flow and the evolution models provide a self-consistent scenario allowing Io in a stable high dissipation state. Recent modeling (Wieczerkowski and Wolf, 1996) shows that these conclusions are robust against variations in rheology. These authors have expanded the Segatz *et al.* (1988) model by allowing other rheologies such as the Burgers and Caputo bodies and find that the tidal dissipation maps are only moderately modified if the Maxwell rheology is replaced by one of these rheologies.

Several observational constraints suggest, however, that the problem of Io

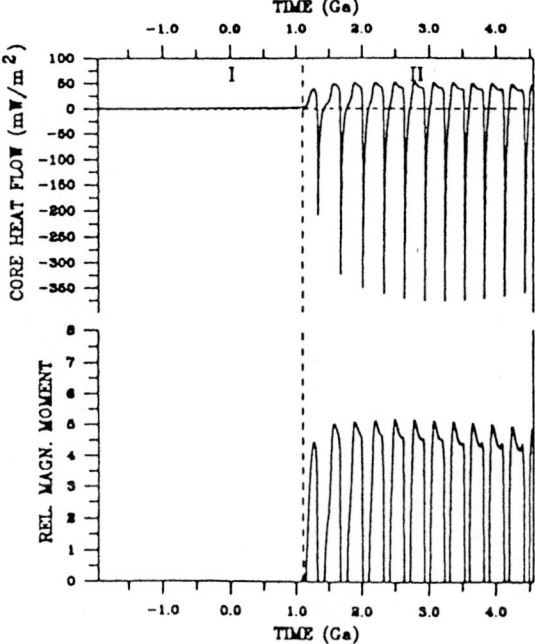

Fig. 11. Thermal evolution of an eutectic Fe-FeS core with a sulfur mass fraction of 24 weight%. The top panel shows the heat flow from the core into the mantle while the bottom panel shows the magnetic moment in units of 5×10^{19} A m^2. Time $t = 0$ is the present time; negative times are times before present.

has not yet been completely solved. First and most importantly perhaps, observations of the of change of the mean motion over the past 300 years do not allow more than 10^{13} W of dissipation (Lieske, 1987). The astrometrical data suggest that

$$\frac{\dot{n}_{\mathrm{I}}}{n_{\mathrm{I}}} = (-0.74 \pm 0.87) \cdot 10^{-11} \, [\mathrm{a}^{-1}] \tag{38}$$

and

$$\frac{\dot{\varepsilon}}{n_{\mathrm{I}}} = (-0.08 \pm 0.42) \cdot 10^{-11} \, [\mathrm{a}^{-1}] \tag{39}$$

The error bounds given are formal and to quote Lieske (1987): "Realistic errors are perhaps a factor of two or three larger". These errors include "possible errors in the astronomical constants, in the theory, or in the elementary model employed for secular accelerations". It is evident and has already been pointed out by Lieske (1987) that his values of $\dot{n}_{\mathrm{I}}/n_{\mathrm{I}}$ and $\dot{\varepsilon}/n_{\mathrm{I}}$ are inconsistent with the hypothesis of the presently observed rate of heat loss being balanced by tidal dissipation. Equations (19) and (23) can be used to calculate maximum equilibrium values of $\Im(k_{\mathrm{I}}) \approx 3 \times 10^{-3}$ and $\Im(k_{\mathrm{J}}) \approx 3 \times 10^{-6}$ that would be consistent with Lieske's data. This value of $\Im(k_{\rm I})$ is by about an order of magnitude smaller than the value required to balance the surface heat flow by tidal dissipation. The value for $\Im(k_{\rm J})$ is more consistent with simple dissipation models for giant planets that account for dissipation in the gaseous envelopes but a value of 10^{-5} does not necessarily create a problem if dissipation in a solid core or through the fall of helium raindrops in the gaseous envelope of Jupiter is accounted for. In any case, the eclipse data suggest that there is disequilibrium between tidal dissipation and heat flow at the present time.

It could be proposed that the present Io is in the oscillatory regime II of Fig. 10. But Fischer and Spohn (1990) have argued that the present observables of Io's thermal and orbital variables can hardly be reconciled in that regime. More random fluctuations of the mantle heat flow may be due to chaotic convection in the mantle. At a Rayleigh number of 10^{14}, mantle convection should be in a regime termed hard thermal turbulence. For a review of the fluid dynamics of hard turbulence see Castaing *et al.* (1982). Thermal hard turbulent convection in the Earth's mantle has been discussed by Malevsky and Yuen (1992) and Yuen *et al.*, (1993). The random fluctuations of heat flow may induce some moderate fluctuations in the eccentricity, in the time rate of change of the mean motion, and in the core/mantle heat flow as numerical calculations performed by U. Wienbruch have shown.

However, little is known about hard thermal turbulence at the extremely large values of the Rayleigh and Prandtl numbers calculated for Io because convection at very high Rayleigh and Prandtl numbers is very difficult to study experimentally and numerically. The Prandtl number $Pr = \nu/\varrho_{\rm m}\kappa$ measures the importance of the inertial relative to the viscous forces. For the mantle of Io the Prandtl number is very large and can be taken as infinite. Experimental work is restricted to low Prandtl number fluids such as Helium. Experiments with low Prandtl number fluids (Castaing *et al.*, 1982) suggest that the fluctuations of the heat transport rate in turbulent convection show a characteristic frequency which scales with Rayleigh number according to $D^2/\kappa\,Ra^{0.491}$ where D is the thickness of the layer. Applying this scaling law to Io a characteristic frequency on the order of 10^{-11} s^{-1} is calculated corresponding to a characteristic period of the order of 10^5 years. The fluctuations are usually attributed to boundary layer instabilities. The amplitudes of these fluctuations decrease according to $Ra^{-1/7}$ at large Rayleigh numbers (Castaing *et al.*, 1982). Applying the scaling laws of Castaing *et al.* (1982) to Io, the heat flow fluctuations would be predicted to be negligible in amplitude.

Fluctuations have also been observed in the results of numerical calculations at infinite Prandtl numbers, (e.g., Hansen and Yuen, 1990; Vincent *et al.*, 1991; Malevsky and Yuen,1992; Yuen *et al.* 1993). Convection at infinite Prandtl number cannot be studied experimentally because the time scales are too large. Numerical calculations profit from the assumptions of infinite Pr because the Navier-Stokes equation can be simplified by neglecting the inertial terms. These calculations are still very demanding in terms of computer resources. The difficulties increase with increasing Rayleigh number because the flow structures become increasingly fine scaled which causes resolution problems that are extremely demanding in terms of computer power. A characteristic frequency seems to also be present in the infinite Prandtl number regime although the physical mechanism that causes the characteristic frequency appears to be different (Vincent et al., 1991). However, while the average amplitude of the fluctuations was reported to decrease with increasing Rayleigh number for very low Prandtl numbers, it increased with Ra between Ra equal to 10^5 and 10^8 in infinite Pr calculations. Extrapolating the results of the calculations referenced above to the relevant Rayleigh number predicts fluctuations by only $\pm30\%$. Although this prediction is based on sparse data, it is likely that the fluctuations are only moderate and may be even smaller. The fluctuations cannot increase indefinitely with increasing Ra. It is conceivable that fluctuations will eventually begin to decrease in amplitude as the convection becomes increasingly chaotic just as in the experimental work with low Prandtl number fluids. A further stabilizing effect may be provided by the pattern of tidal heating which restricts the degrees of freedom of the flow.

An alternative mechanism to explain the up to an order of magnitude difference between the tidal heating rate suggested from eclipse data and the present surface heat flow is proposed in the following. It is motivated by results of Balachandar and colleagues (Balachandar et al. 1993,1995) from numerical models of convection with viscous dissipation in high Rayleigh and Prandtl number fluids. These authors find that convective flows may cause strong local maxima in viscous dissipation. In fact, their calculations show that the local power density may become much larger than the heat production rate per unit volume that drives the flow. Energy is conserved because the hot-spots concentrate a fraction of the total power in a small volume. Simple scaling laws suggest that the power density associated with viscous dissipation scales with $\nu u^2/D^2$ where u is the average convection speed and D is the layer thickness. Since the convection speed increases with $Ra^{2/3}$, the viscous dissipation power density increases with $Ra^{4/3}$. Extrapolating the numerical results obtained by Balachandar and colleagues to the Rayleigh number relevant for Io suggests that the energy density may locally become as large as 10^{-3} W m^{-3}, about 200 times the average energy density associated with tidal dissipation. It should be emphasized that the strong viscous dissipation is a local phenomenon and that the volume of fluid under extreme shear may decrease with increasing Rayleigh number. However, a few hot-spots in the mantle caused by viscous dissipation may well store significant amounts of energy in melt. Energy that may later be released at a high rate once the melt extrudes on the surface or is intruded into near surface layers of the crust.

A simple calculation shows that the model is reasonable. Assume a tidal dissipation rate of 10^{13} W as is suggested by the eclipse data. Further assume for simplicity that the lithosphere is 20 - 30 km thick in which case the conductive heat flow through the lithosphere would approximately balance the tidal dissipation rate. The assumed basal temperature of the lithosphere is 1500 K and the thermal conductivity is 4 W m^{-1} K^{-1} as in Tab. 2. Veeder et al. (1994) have shown that 10^{14} W are radiated from about 8% of the surface of Io with an average temperature of 155 K (see their Fig. 12b) and that the volcanic hot-spots contribute comparatively little to the luminosity. Assume that the lithosphere underneath the 8% of the surface is warmed by volcanic intrusions. A simple calculation by integration of the (linear) temperature profile in the lithosphere to calculate its internal heat content shows that the intrusives must provide about 5×10^{24} J to warm the lithosphere such that the surface temperature is 150 K instead of 100 K which is the average surface temperature. If the intrusives have an average temperature of 1300 K (to account for some cooling as the magma rises) and if the latent heat is 3×10^5 J kg^{-1} then a volume of 10^{15} m^3 of lava would be sufficient. Here we assume that the lava infiltrates the lithosphere such that we can do a simple mixing calculation. To melt this much lava in mantle hot-spots generating 10^{-3} W m^{-3} only a small fraction of the mantle volume is required (less than 0.05 %) for the hot-spots and the power used by the hot-spots would be about 10% of the total tidal dissipation rate. It has been shown in recent calculations to be published elsewhere that the ratio between the viscous dissipation rate and the internal heat production rate in internally heated convection tends towards the value of the dissipation number Di as Ra becomes large. The dissipation number is defined as

$$Di \equiv \frac{g\alpha D}{c_{\mathrm{m}}} \tag{40}$$

and measures the fraction of the potential energy driving the convection that can be dissipated as heat. The value of Di for Io is 0.1. The time required to generate the lava is about 10^4 years accounting only for the latent heat. The hot-spots in the mantle would thus use about 10 % of the tidal dissipation power and store the energy in magma for about 10^4 years. This energy can then be released at an average rate of 10^{14} W from the surface in about a century after the lava has erupted. The long term evolution of this model remains to be studied. Since the time scales of the above mechanism are much smaller than the periods of thermal-orbital oscillations in regime II of Fig. 10 it is likely that oscillations will not be excited by the proposed variations in heat loss. The long term surface heat flow should be close to the dissipation rate but would be augmented for relatively short periods of time by outbursts of intense volcanism. One would also expect that the core to mantle heat flow remains small such that dynamo action would be unlikely.

Acknowledgements. Numerical results presented in this paper have been calculated by M. Segatz and U. Wienbruch. I thank M. Segatz, H. J. Fischer, U. Wienbruch, K. Wieczerkowski, D. Breuer, V. Steinbach, and D. Yuen for stimulating discussions over the past years of this research and U. Wienbruch, F. Sohl and W. Konrad for critically reading the manuscript. The research was funded by the Deutsche Forschungsgemeinschaft.

References

Alterman, Z., H. Jarosch, C. L. Pekeris. 1959. Oscillations of the Earth. *Proc. R. Soc. London* **A 252**: 80 - 95.

Anderson, J. D., W. L. Sjogren, and G. Schubert. 1996. Galileo gravity results and the internal structure of Io. *Science* **272**: 709 - 712.

Balachandar, S., D. A. Yuen, and D. Reuteler. 1993. Viscous and adiabatic heating effects in three-dimensional compressible convection at infinite Prandtl number. *Phys. Fluids A* **5**: 2938 - 2945.

Balachandar, S., D. A. Yuen, D. Reuteler, and G. Lauer. 1995. Viscous dissipation in three dimensional convection with temperature- dependent viscosity. *Science* **267**: 1150 - 1153.

Boehler, R. 1986. The phase diagram of iron to 430 kbar. *Geophys. Res. Lett.* **13**: 1153 - 1156.

Boehler, R. 1992. Melting of the Fe-FeO and the Fe-FeS systems at high pressure: Constraints on core temperature. *Earth Planet. Sci. Lett.* **111**: 217 - 227.

Burns, J. A. 1986. Some background about satellites. In: Satellites. pp. 1 - 38. J. A. Burns and M. S. Matthews (eds.). Univ. Arizona Press, Tucson.

Busse, F. H. 1976. Generation of planetary magnetism by convection. *Phys. Earth Planet. Int.* **12**: 350 - 358.

Castaing, B., G. Gunaratne, F. Heslot, L. Kadanoff, S. Libchaber, S. Thomae, X.-Z. Wu, S. Zaleski, and G. Zanetti. 1982. Scaling of hard thermal turbulence in Rayleigh-Bénard convection. *J. Fluid. Mech.* **204**: 1 - 30.

Christensen, U. R. 1985. Thermal evolution models for the earth. *J. geophys. Res.* **90**: 2995 - 3007.

Consolmagno G. J. 1981. Io: Thermal models and chemical evolution. *Icarus* **47**: 36 - 45.

Fischer H. J., and T. Spohn. 1990. Thermal-orbital histories of viscoelastic models of Io (J1). *Icarus* **83**: 39 - 65.

Frank, L. A., W. R. Paterson, K. L. Ackerson, V. M. Vasyliunas, F. V. Coroniti, and S. J. Bolton. 1996. Plasma observations of Io with the Galileo spacecraft. *Science* **274**: 394 - 395.

Gaskell, R. W. and S. T. Synott. 1988. Large-scale topography of Io: Implications for internal structure and heat transfer. *Geophys. Res. Lett.* **15**: 581 - 584.

Hansen, U. and D. A. Yuen. 1990. Heat transport in strongly chaotic thermal convection. In: *HTD-Vol 149, Heat Transfer in Earth Science Studies*. pp. 43 - 46. C. Carrigan and T. Y. Chu (eds.) American Society of Mechanical Engineers, Book No. G00543.

Johnson, T. V., D. L. Matson, D. L. Blaney, G. J. Veeder, and A. Davies.1995. Stealth plumes on Io. *Geophys. Res. Lett.* **22**: 3293 - 3296.

Kaula, W. M. 1964. Tidal dissipation by solid friction and the resulting orbital evolution. *Rev. Geophys.* **2**: 661 - 685.

Kivelson, M. G., K. K. Khurana, R. J. Walker, C. T. Russell, J. A. Linker, D. J. Southwood, and C. Polanskey. 1996. A magnetic signature on Io: Initial report from the Galileo magnetometer. *Science* **273**: 337 - 340.

Lewis J. S. 1982. Io: Geochemistry of sulfur. *Icarus* **50**: 103 - 114.

Lieske J. H. 1987. Galilean satellite evolution: Observational evidence for secular changes in mean motions. *Astron. Astrophys.* **176**: 146 - 158.

Love, A. E. H. 1927. A treatise on the mathematical theory of elasticity. 4th Ed., Dover, New York. 643 pp.

Malevsky, A. V. and D. A. Yuen. 1992. Strongly chaotic non-Newtonian mantle convection. *Geophys. Astrophys. Fluid Dyn.* **65**: 149 - 171.

Malhotra, R. 1991. Tidal Origin of the Laplace resonance and the resurfacing of Ganymede. *Icarus* **94**: 399 - 412.

McEwen, A. S., J. I. Lunine, and H. C. Carr. 1989. Dynamic geophysics on Io. 11 - 46. In: Time variable phenomena in the Jovian system. M. J. S. Belton, R. A. West, and J. Rahe (eds.) NASA SP - 494.

Neubauer, F. M. 1978. Possible strengths of dynamo magnetic fields of the Galilean satellites and of Titan. *Geophys. Res. Lett.* **5**: 905 - 908.

Peale S. J., P. Cassen, and R. T. Reynolds. 1979. Melting of Io by tidal dissipation. *Science* **203**: 892 - 894.

Platzman, G. W. 1984. Planetary energy balance for tidal dissipation. *Rev. Geophys. Space Phys.* **22**: 73 - 84.

Ross M. N., G. Schubert, T. Spohn, and R. W. Gaskell. 1990. Internal Structure of Io and the global distribution of its topography. *Icarus* **85**: 309 - 325.

Schubert, G., T. Spohn, and R. T. Reynolds. 1986. Thermal histories, compositions and internal structures of the moons of the solar system. 224 - 292. In: Satellites. J. A. Burns and M. S. Matthews (eds.). Univ. Arizona Press, Tucson.

Schubert, G., M. N. Ross, D. J. Stevenson, and T. Spohn. 1988. Mercury's thermal history and the generation of its magnetic field. 429–460. In: Mercury. F. Vilas, C. R. Chapman, and M. S. Matthews (eds.). Univ. Arizona Press, Tucson.

Segatz M., T. Spohn, M. N. Ross, and G. Schubert. 1988. Tidal Dissipation, surface heat flow, and figure of viscoelastic models of Io. *Icarus* **75**: 187 - 206.

Spohn T., Schubert G. 1982. Modes of mantle convection and the removal of heat from the Earth's interior. *J. geophys. Res.* **87**: 4682 - 4686.

Stacey, F. D. 1977. Physics of the Earth. 2ed. Wiley, New York. 414 pp.

Stevenson D. J., T. Spohn, and G. Schubert. 1983. Magnetism and thermal evolution of the terrestrial planets. *Icarus* **54**: 466 - 489.

Takahashi, E. 1986. Melting of a dry peridotite KLB-1 up to 14 GPa: Implications on the origin of peridotitic upper mantle. *J. geophys. Res.* **91**: 9367 - 9382.

Takahashi, E. 1990. Speculations on the Archean mantle: Missing link between komatiite and depleted garnet peridotite. *J. geophys. Res.* **95**: 15941 - 15954.

Takeuchi, H., M. Saito, and N. Kobayashi. 1962. Statical deformations and free oscillations of a model Earth. *J. geophys. Res.* **67**: 1141 - 1154.

Tozer, D. 1965. Heat transfer and convection currents. *Phil. Trans. R. Soc. London* **A 258**: 252 - 271.

Turcotte, D. L. 1982. Magma migration. *Ann. Rev. Earth Planet. Sci.* **10**: 397 - 408.

Usselman, T. M. 1975a. Experimental approach to the state of the core: Part 1. The liquidus relations of the Fe-Ni-S system from 30 to 100 kb. *Am. J. Sci.* **275**: 278 - 290.

Usselman, T. M. 1975b. Experimental approach to the state of the core: Part 2. Composition and thermal regime. *Am. J. Sci.* **275**: 291 - 303.

Veeder G. J., D. L. Matson, T. V. Johnson, D. L. Blaney, and J. D. Goguen.1994. Io's heat flow from infrared radiometry: 1983–1993. *J. geophys. Res.* **99**: 17095 - 17162.

Verhoogen, J. 1980. Energetics of the Earth. National Academy Press, Washington. 139 pp.

Vincent, A. P., U. Hansen, D. A. Yuen, A. V. Malevsky, and S. E. Kroening. 1991. The origin of a characteristic frequency in hard thermal turbulence. *Phys. Fluids.* **A 3**: 2003 - 2006.

Webb, E. K. and D. J. Stevenson. 1987. Subsidence of topography on Io. *Icarus* **70**: 348 - 353.

Wieczerkowski, K. and D. Wolf. 1996. Viscoelastic tidal perturbations: Effects due to density contrasts. *Annal. Geophys.* **14**, **Suppl. I**: 102.

Wienbruch, U. and T. Spohn. 1995. A self sustained magnetic field on Io? *Planet Space. Sci.* **43**: 1045 - 1057.

Wyllie, P. J. 1988. Magma genesis, plate tectonics, and chemical differentiation of the Earth. *Rev. Geophys.* **26**: 370 - 404.

Yoder C. F. 1979. How tidal heating in Io drives the Galilean orbital resonance locks. *Nature* **279**: 767 - 770.

Yoder, C. F. and S. J. Peale. 1981. The tides of Io. *Icarus* **47**: 1 - 5.

Yuen, D. A., U. Hansen, W. Zhao, A. P. Vincent, and A. V. Malevsky. 1993. Hard turbulent thermal convection and thermal evolution of the mantle. *J. geophys. Res.* **98**: 5355 - 5373.

Zschau, J. 1978. Tidal friction in the solid Earth: Loading tides versus body tides. In: Tidal Friction and the Earth's Rotation. 62 - 94. P. Brosche and J. Sündermann (eds.). Springer Verlag, Berlin.

Tidal Effects in Binary Star Systems

Gerhard Schäfer

Max–Planck–Gesellschaft, Arbeitsgruppe Gravitationstheorie,
Friedrich–Schiller–Universität Jena, D–07743 Jena, Germany

Summary

Accelerations in gravitational fields are usually treated as resulting from the gravitational force. The "true" gravitational force, however, is a tidal force which cannot be transformed away by going over to an accelerated reference frame, nor does it need nongravitational forces to act (e.g., the gravitational force acting on a body at rest in the gravitational field of a star is acting in reaction to the nongravitational force which keeps the body at rest). The tidal force is a relative force, acting between neighbouring points. According to the Einstein theory of gravitation, the tidal force is intimately connected with the curvature of spacetime, and gravitation is nothing but this curvature. In spacetimes without curvature only inertial forces (e.g., centrifugal and Coriolis forces) exist; sometimes they are called fictitious.

In binary star systems the tidal force plays an important role. Processes which result from this force comprise capture, stripping, disruption, and dissipation effects as well as resonances between orbital motion and star oscillations. The orbital period damping connected with the gravitational radiation emission is a tidal effect too.

The references cited below allow of a comprehensive insight into the problem of tidal effects in binary star systems.

References

Alexander, M.E. 1973. The weak friction approximation and tidal evolution in close binary systems. *Astrophys. Space Sci.* 23: 459 - 510.

Alexander, M.E. 1987. Tidal resonances in binary star systems. *Mon. Not. R. astron. Soc.* 227: 843 - 861.

Bhattacharya, D. and E.P.J. van den Heuvel 1991. Formation and evolution of binary and millisecond radio pulsars. *Physics Reports.* 203: 1 - 124.

Bildsten, L. and C. Cutler 1992. Tidal interactions of inspiraling compact binaries. *Astrophys. J.* 400: 175 - 180.

Chandrasekhar S. 1987. Ellipsoidal Figures of Equilibrium. Dover Publications. New York. 255 pp.

Clark, J.P.A. and D.M. Eardley 1977. Evolution of close neutron star binaries. *Astrophys. J.*. 215: 311 - 322.

Frank, J., A. King, and D. Raine 1992. Accretion Power in Astrophysics. Second Edition. *Cambridge Astrophysics Series.* **21**. Cambridge University Press. 294 pp.

Hartle, J.B., K.S. Thorne, and R.H. Price 1986. Gravitational interaction of a black hole with distant bodies. In: Black Holes: The Membrane Paradigm. pp. 146 - 180. K.S. Thorne, R.H. Price, and D.A. Macdonald (eds.). Yale University Press, New Haven and London.

Kochanek, C.S. 1992. Coalescing binary neutron stars. *Astrophys. J..* **398**: 234 - 247.

Kokkotas, K.D. and G. Schäfer 1995. Tidal and tidal - resonant effects in coalescing binaries. *Mon. Not. R. astron. Soc..* **275**: 301 - 308.

Kopal, Z. 1978. Dynamics of Close Binary Systems. Astrophysics and Space Science Library. Vol. 68. D. Reidel Publishing Company, Dordrecht, Holland. 510 pp.

Kopeikin, S.M. 1985. General - relativistic equations of binary motion for extended bodies, with conservative corrections and radiation damping. *Sov. Astron..* **29**: 516 - 524.

Kosovichev, A.G. and I.D. Novikov 1992. Non-linear effects at tidal capture of stars by a massive black hole - I. Incompressible affine model. *Mon. Not. R. astron. Soc..* **258**: 715 - 724.

Lai, D., F.A. Rasio, and S.L. Shapiro 1994. Hydrodynamic instability and coalescence of binary neutron stars. *Astrophys. J..* **420**: 811 - 829.

Lincoln, C.W. and C.M. Will 1990. Coalescing binary systems of compact objects to $(post)^{5/2}$ - Newtonian order: Late-time evolution and gravitational-radiation emission. *Phys. Rev. D.* **42**: 1123 - 1143.

Luminet, J.P. 1987. Tidal disruption. In: Gravitation in Astrophysics. pp. 215 - 228. B. Carter and J.B. Hartle (eds.). NATO ASI Series B: Physics Vol. 156. Plenum Press. New York and London.

Papaloizou J. and J.E. Pringle 1980. On the motion of the apsidal line in interacting binary systems. *Mon. Not. R. astron. Soc..* **193**: 603 - 615.

Press, W.H. and S.A. Teukolsky 1977. On formation of binaries by two-body tidal capture. *Astrophys. J..* **213**: 183 - 192.

Reisenegger, A. and P. Goldreich 1994. Excitation of neutron star normal modes during inspiral. *Astrophys. J..* **426**: 688 - 691.

Ruffert, M., H.- Th. Janka, and G. Schäfer 1996. Coalescing neutron stars - a step towards physical models I: Hydrodynamical evolution and gravitational-wave emission. *Astron. Astrophys..* **311**: 532 - 566.

Tassoul, J.- L. 1978. Theory of Rotating Stars. *Princeton Series in Astrophysics.* **1**. Princeton University Press. 506 pp.

Thorne, K.S. 1987. Gravitational radiation. In: 300 Years of Gravitation. pp. 330 - 458. S.W. Hawking and W. Israel (eds.). Cambridge University Press. Cambridge.

Tidal Interactions Between Galaxies

Claus Möllenhoff

Landessternwarte Königstuhl, D-69117 Heidelberg, Germany

Abstract. The basic properties of galaxies and simple models of their formation are discussed and different types of interaction between galaxies are described. The dynamics of galactic interactions, statistical approaches and model calculations are reviewed and the significance of interactions in the evolution of galaxies is presented.

1   Basic properties of galaxies

Our Milky Way is a typical spiral galaxy. It consists of $\approx 10^{11}$ Stars, orbiting in the potential of their common mass. The Milky Way rotates in $\approx 10^8$ years, this corresponds to an orbital velocity of the solar system of about 250 km/sec. The diameter of the Milky Way is about 30000 parsecs (1 pc $= 3.1 \times 10^{18}$ cm). The visible mass of galaxies is dominated by stars, in addition one finds several percents in mass of cold gas, warm gas, and dust (depending on the type of the individual galaxy). The most important types of galaxies are elliptical and spiral galaxies. In elliptical galaxies the stars are moving on all types of irregular orbits in the common potential and thus form a spheroidic or ellpsoidic cloud of stars with no or with only minor global rotation. In contrast to that the spiral galaxies have a much higher specific angular momentum. The stars are in an ordered rotational motion and form a flat rotating disk in these objects, showing the characteristic spiral pattern. The local rotational velocity depends on the radius. The central region of a spiral galaxy is not flat but forms the spheroidic bulge, the kinematics of the stars here is similar as in elliptic galaxies.

The measurement of the rotational velocity of the stars in a spiral galaxy in dependence on the radius showed the astonishing result that the velocity does not decline outside the optical visible galaxy. Obviously there exists a further component of gravity force. From the slow decline of the rotation curves we conclude that this component comprises more than 10 times the mass of the stars! This 'dark matter' halo can only be detected by its gravitational force, it is not detectable by optical radiation or by any other wavelength of the electromagnetic spectrum. The question of the nature of the dark matter is one of the greatest challenges for astrophysics of our times.

The different types of galaxies are determined by different histories of formation and evolution. A simplified image is the following: Galaxies form by the gravitational collapse of a giant gas cloud. In general such a cloud has a small amount of angular momentum. If dissipative processes (gas friction) are dominant during the collapse a rotating gas disk will be formed. Local instabilities lead to the consecutive formation of stars, the stellar disk of a spiral galaxy is formed. If however the cloud collapse occurs in such a way that stars are formed firstly by local instabilities, no more gas dissipation will take place. This will lead to an elliptical galaxy. A totally different scenario of galaxy formation is merging: If two spiral galaxies collide, they will merge and practically always an elliptical galaxy is formed because of the phase space mixing.

2   Phenomenology of galaxy interactions

Galaxies are 'social', i.e. they are found mostly in groups or in clusters, which can have up to several thousand members (e.g. Virgo-Cluster, Coma-Cluster). The typical distances between galaxies are only 10 to 100 times larger than their diameters. Since the typical relative velocities of galaxies are about 100 to 1000 km/sec, they move in about 3×10^7 years by their own diameters (30000 pc $= 10^{22}$ cm). Thus in cosmic times of 10^9 years galaxies may come very close to each other! Example: Our closest normal-sized neighbour, the Andromeda galaxy, has a distance of 700 kpc (Kiloparsec $= 10^3$ pc) and approaches the Milky Way with about 100 km/sec. We will have a collision in 7×10^9 years! The peculiar stochastic motion of the galaxies is superimposed on the general cosmological expansion. However, since the latter is about 75 km/sec for galaxies with a distance of 1 Mpc (Megaparsec $= 10^6$ pc), it plays no role for the dynamics of groups or clusters of today.

Galaxies with disturbed morphology were realized since long times. Only during the last 20 to 30 years it was generally accepted, that these pecularities are the signatures of tidal interactions between the galaxies. The most important signatures of interaction are

- Thin bridges of matter connecting adjacent galaxies.
- Symmetric arms or tails ranging far out from the tidally disturbed galaxy.
- Bursts of star formation due to perturbation and shocks in the gas.

If one of the interacting partners is much less massive, it can be accreted by the larger galaxy. Example: The Magellanic clouds are very nearby dwarf galaxies ($d \approx 50$ kpc), they move around the Milky Way and approach in a spiral orbit. They will finally be swallowed by the Milky Way ('galactic cannibalism').

If the interacting partners have similar mass, they may collide and will finally merge. This dramatic scenario shows in its early phase all the signatures of tidal interactions. The details are very dependent on type, angular momentum and orbit parameters of the partners. Both parent galaxies will totally loose their former identity and form a new object, normally a still fairly irregular elliptical galaxy. The potentials of the stars form a new common potential via a violent relaxation process, part of the stars may be lost. The gaseous matter collides with heavy shocks, huge amounts of newborn bright blue stars are formed, large amounts of gas may also be erupted into the environment. The 'fireworks' of shock emission, young bright stars, line emission of excited gas nebulae, etc, are only a secondary effect. The main thing is the dynamics of the stars.

3   Stellar dynamics

A tidal interaction changes the orbital velocities of the stars in a galaxy. This velocity variation compared to the undisturbed velocity is a measure for the strength of the interaction. The effect is obviously strongest if the galaxies come close to each other, have similar masses, and if the relative velocity between the interacting galaxies is similar to the orbital velocities of the stars in the perturbed galaxy (resonance). The typical orbital velocities in a galaxy are 250 km/sec. Thus the tidal interactions in a rich galaxy cluster with a rms-velocity of 2000 km/sec (Coma cluster) are much weaker than in a poor cluster with 1000 km/sec (Virgo cluster). The most effective interactions occur in small groups of galaxies, where the rms-velocity is only 500 km/sec.

If the perturber moves much faster than the stars, the effective potential of the galaxy will not be changed. The stars suffer only a short momentum increment $m \times dv$ in direction to the perturber (impulse approximation). The galaxy will be deformed towards the perturber, comparable to the tidal elevation of the ocean water due to the moon. The stars get a small increment in kinetic energy. Since these small energy increments are randomly distributed over all stars, the perturbed galaxy gains inner energy. Correspondingly, the perturber looses orbital energy, the encounter is inelastic. The partners come closer and closer and will merge after only a few orbits.

The dynamics of $\approx 10^{11}$ stars in their common potential can be described in a statistical way, similar as the theory of ideal gases. However, in contrast to a normal gas stellar collisions (this means close encounters) are extremely rare in a galaxy (about every 10^{14} years). Therefore the stars do not behave according to an isotropic Maxwellian distribution function. The system can be described by giving the distribution function $f(r,v,t)$ in the phase space, the Poisson equation for the gravitational potential and the equation of continuity in the phase space (called collisionless Boltzmann equation in this special form). In general it is very difficult to find a solution for these basic equations of stellar dynamics. In several aspects the continuum of the stars behaves like a fluid: if at some place the density is enhanced, the stars move in the mean away from that place. The stochastic variation of the velocities of the stars around their mean velocity is called the velocity dispersion. It corresponds to the kinetic temperature of a gas. The main difference between stellar dynamics and fluid dymnamics is that the 'pressure', due to the velocity dispersion is not isotropic. An anisotropic velocity dispersion can e.g. produce a flattened shape of a galaxy without any rotation. This is the case for many elliptical galaxies.

4   Model calculations

The statistical approach of above is mainly suited for rather general statements about the ensemble of stars in a galaxy. If one wants to calculate the stellar dynamics of a single galaxy or of interacting objects in detail, one has to follow the motion of 10^8 to 10^{12} stars under the influence of their common potential. The most useful tool for this task is N-body calculation. One considers N stars as points of given masses, with their space coordinates, and their velocities. Each particle feels the summed-up potential of all other particles. The equations for the potential of each particle and the individual equations of motion are solved simultaneously in a computer. In this way direct numerical solutions for up to $\approx$ 40000 stars are possible. Codes which use some mean potential for larger ensembles of stars can handle up to 10^6 to 10^7 stars.

In the astrophysical literature many impressive examples for the results of N-body calculations can be found:

- Two encountering spherical galaxies, gaining inner energy and loosing orbital energy (Barnes and Hut, Fig 7-1 in Binney and Tremaine 1987), a nice example for the inelastic encounter of galaxies.
- A simulation for the binary galaxy-system M51 (Hernquist, in Wielen 1990).
- The evolution of a galaxy containing gas besides the stars. Due to instabilities in the disk a bar is formed. The bar transports mass and angular momentum and changes thus the overall picture of the galaxy substantially (von Linden et al. 1996).
- The impressive results of J. Barnes and L. Hernquist (1996) about the merging of disk galaxies including gas dynamics. Due to dissipative force the gas collects in the center and makes a strong burst of star formation.

There was great progress in the quality of the N-body calculations during the last years: Stronger computers and better codes allow the calculations with more particles, separately for disk, bulge and an halo of dark matter An adequate treatment of dissipative, hydrodynamical effects (SPH = smoothed particle hydrodynamics) allows the simultaneous calculation of collisionless stars and dissipative gas clouds in a galaxy. These programs need long computing times, even on the latest very fast machines. A very interesting new development is the use of special-purpose machines which are extremely fast since the corresponding equations are hard-wired in them (Sugimoto et al. 1990; Spurzem et al. 1996). Today the simultaneous calculation of a whole group of interacting galaxies is possible, including the hydrodynamics of their gas.

5   Importance of interactions and merging for the evolution of galaxies

Interaction and merging play an important role for the evolution of galaxies. Many galaxies show traces of interactions (bridges, tails, shells, subcomponent with different kinematics, etc), thus tidal interaction is a common process in the lifetime of a galaxy. In the early universe the galaxies were much closer to each other, therefore the interactions were much more frequent. Interactions may change the type of a galaxy drastically. Merging galaxies loose their identity completely, the merger remnant is always an elliptical galaxy. Most massive elliptical galaxies are of merger origin.

In gasrich galaxies interactions cause a burst of star formation, which influences strongly the chemical evolution in these galaxies. Interactions may also trigger the phenomenon of an active galactic nucleus (AGN, e.g. the QUASAR phenomenon) by transporting matter near to the massive black hole. The accretion of this matter to the black hole is the source of the gigantic power which is emitted by AGNs.

Further the interaction and merging play a crucial role for the evolution of galaxy clusters. In the 'hierarchical clustering' scenario groups of galaxies merge to small clusters, and these merge at least to rich clusters of galaxies. This fits the empirical density-type relation: in regions of high galaxy density one finds mainly elliptical galaxies, i.e. objects closely connected to merging processes. In regions of low galaxy density spiral galaxies are dominant, i.e. objects which have not yet experienced a strong merging event.

References
Popular articles:

Barnes, J.E., Hernquist, L., Schweizer, F. 1991. Colliding Galaxies. *Sci. Am.* **265**: 26 - 33.

Bien, R. 1993. Experimentelle Stellardynamik. *Sterne Weltraum.* **32**: 189 - 195.

Keel, W.C. 1989. Crashing Galaxies, Cosmic Fireworks.*Sky Telescope* **77**: 18 - 25.

Schweizer, F. 1986. Colliding and Merging Galaxies *Science.* **32**: 225 - 235.

Schulz, H. 1984. Kosmische Massenkarambolage. *Sterne Weltraum.* **23**: 311 - 316.

Spurzem, R., Einsel, C., Theis, C. 1996. GRAPE und HARP – Spezialrechner für Simulationen von Sternhaufen und Galaxien. *Sterne Weltraum.* **35**: 190 - 195.

Textbooks:

Barnes, J.E. 1995. Interactions and Mergers in Galaxy Formation. In: *Formation and Evolution of Galaxies.* C. Munoz-Tunon, F. Sanchez (eds.). Cambridge Univ. Press.

Binney, J., Tremaine, S. 1987. Galactic Dynamics. Princeton Univ. Press, Princeton.

Contopoulos, G., Spyrou, N.K., Vlahos, L. 1994. Galactic Dynamics and N-Body Simulations. Springer, Heidelberg.

Taylor, R.J. 1993 Galaxies, Structure and Evolution. Cambridge Univ. Press, Cambridge, U.K..

Wielen, R. 1990. Dynamics and Interactions of Galaxies. Springer, Heidelberg.

Selected recent articles in scientific journals:

Barnes, J.E., Hernquist, L. 1992. Dynamics of interacting Galaxies. *Ann. Rev. Astron. Astrophys.* **30**: 705 - 742.

Barnes, J.E., Hernquist, L.: 1996. Transformations of Galaxies II: Gasdynamics in merging disk galaxies. *Astrophys. J.* **471**: 115 - 142.

Von Linden, S., Reuter, H.P., Heidt, J., Wielebinski, R., Pohl, M. 1996. The dynamics of the inner part of NGC 7331. *Astron. Astrophys.* **315**: 52 - 62.

Madejsky, R., Bender, R., Möllenhoff, C. 1991. Morphology and kinematics of the pair of interacting galaxies NGC 4782 and NGC 4783. *Astron. Astrophys.* **242**: 58 - 68.

Moore, B., Katz, N., Lake, G., Dressler, A., Oemler, A. 1996. Galaxy harassment and the evolution of clusters of galaxies. *Nature.* **379**: 613 - 616.

Reduzzi, L., Rampazzo,R. 1996. Surface photometry of binary galaxies I: A multi-colour study of morphologies due to the interaction. *Astron. Astrophys. Suppl.* **116**: 515 - 571.

Smith, E.P., Hintzen, P. 1991. The galaxy activity-interaction connection I: Optical observations. *Astron. J.* **101**: 410 - 433.

Sugimoto, D., Chikada, Y., Makino, J., Ito, T., Ebisuzaki, T., Umemura, M. 1990. A special-purpose computer for gravitational many-body problems. *Nature.* **345**: 33 - 35.

Walker, I.R., Mihos, J.C., Hernquist, L. 1996. Quantifying the fragility of galactic disks in minor mergers. *Astrophys. J.* **460**: 121 - 135.

Weil, M., Hernquist, L. 1996. Global properties of multiple merger remnants. *Astrophys. J.* **460**: 101 - 120.

Subject Index

and

List of Contributors

Subject Index

A

Air pressure 61, 69, 102, 183, 197 - 199,
202, 205, 206, 222, 224, 242, 277,
278, 286, 287, 289, 321, 322, 330
Albedo 230, 242, 254
Amphidrome 123, 124, 134, 136, 138,
146
Amplitude factor: see gravimetric,
dimininishing, tilt factor
Anelasticity 28, 29, 79, 101
Angular momentum 179, 186, 188
— atmosphere 185, 188, 189, 204, 243,
— earth 96, 98, 99, 174, 185, 243, 247,
253
— galaxies 381, 382
— oceans 173, 185
— orbital 178
— tidal 124, 247
Annual wobble 184, 189, 191 - 194, 202,
209, 210
Anti-node 123
Aquifer 85, 277, 280 - 288, 331
Astronomical arguments 18, 19, 175
Atmosphere
— latent heat 222, 228, 231
— superrotation 233, 243, 244
Atmospheric
— excitation 185, 198 - 203, 205
— pressure: see air pressure
— tide 79, 221, 231, 233, 261, 343
—— lunar 222, 225, 226, 233, 243
—— resonance amplification 243
—— solar 79, 221 - 246

B

Barometric
— effects 277, 278, 279, 289
— efficiency 286

— pressure: see air pressure
Body tide 79, 81, 86, 145, 157, 161, 188
Boreholes 47, 78, 86, 87, 88, 162, 277,
321, 330
Boundary-value problem 27, 42, 43,
147

C

Callisto 346 - 348
Cavity effect 29, 47, 79, 86, 87, 162
Chandler wobble 98, 101, 175, 183 -
215
Circulation 254, 256, 257, 259
Climate 200, 201, 205, 248, 254, 255,
256, 257, 258, 259
Compound tides 113, 135, 136
Confined aquifer 277, 280, 283, 284,
285
Core-mantle boundary (CMB) 96 - 98,
100, 105, 106, 350, 357
Core-mantle coupling 96, 97, 176
Core resonance, see NDFW 95 - 106
Crustal structure 85, 145, 155, 158,
161, 162, 327
Cryogenic gravimeter 72, 78, 82, 102,
104, 105

D

Dansgaard-Oeschger event 256
Darcy conductivity 281, 287
Darwin symbol 79
Deformation-induced (tidal) potential
31, 41 - 44
Dilatancy 86, 313 - 317, 334 - 336
Diminishing factor, see tilt factor
Doodson constant 233

List of Contributors

Emter, Dieter
Black Forest Observatory Schiltach (BFO), Universitäten Karlsruhe/Stuttgart,
Heubach 206, D-77709 Wolfach, Germany

Jentzsch, Gerhard
Institut für Geowissenschaften, Friedrich-Schiller-Universität Jena,
Burgweg 11, D-07749 Jena, Germany

Kümpel, Hans-Joachim
Abt. Angewandte Geophysik, Geologisches Institut, Universität Bonn,
Nussallee 8, D-53115 Bonn, Germany

Möllenhoff, Claus
Landessternwarte Königstuhl, D-69117 Heidelberg, Germany

Olsen, Nils
Department of Geophysics, University of Copenhagen, Juliane Maries vej 30,
DK-2100 Copenhagen Oe, Denmark

Plag, Hans-Peter
Institut für Geophysik, Christian-Albrechts-Universität Kiel,
Olshausenstr. 40, D-24118 Kiel, Germany

Schäfer, Gerhard
Max-Planck-Gesellschaft, Arbeitsgruppe Gravitationstheorie,
Friedrich-Schiller-Universität Jena, D-07743 Jena, Germany

Schwintzer, Peter
GeoForschungsZentrum Potsdam (GFZ), Telegrafenberg A 17,
D-14473 Potsdam, Germany

Spohn, Tilman
Institut für Planetologie, Westfälische Wilhelms-Universität,
W. Klemmstr. 10, D-48149 Münster, Germany

Volland, Hans
Radioastronomisches Institut, Universität Bonn, Auf dem Hügel 71,
D-53121 Bonn, Germany

Wang, Rongjiang
GeoForschungsZentrum Potsdam (GFZ), Telegrafenberg A 17,
D-14473 Potsdam, Germany

Wenzel, Hans-Georg
Geodätisches Institut, Universität Karlsruhe, Englerstr. 7,
D-76128 Karlsruhe, Germany

Westerhaus, Malte
GeoForschungsZentrum Potsdam (GFZ), Telegrafenberg A 17,
D-14473 Potsdam, Germany

Wilhelm, Helmut
Geophysikalisches Institut, Universität Karlsruhe, Hertzstr. 16,
D-76187 Karlsruhe, Germany

Wünsch, Johannes
Sternwarte Sonneberg, Sternwartestr. 32, D-96515 Sonneberg, Germany

Zahel, Wilfried
Institut für Meereskunde, Universität Hamburg, Troplowitzstr. 7,
D-22529 Hamburg, Germany

Zürn, Walter
Black Forest Observatory Schiltach (BFO), Universitäten Karlsruhe/Stuttgart,
Heubach 206, D-77709 Wolfach, Germany

Lecture Notes in Earth Sciences

Vol. 37: A. Armanini, G. Di Silvio (Eds.), Fluvial Hydraulics of Mountain Regions. X, 468 pages. 1991.

Vol. 38: W. Smykatz-Kloss, S. St. J. Warne, Thermal Analysis in the Geosciences. XII, 379 pages. 1991.

Vol. 39: S.-E. Hjelt, Pragmatic Inversion of Geophysical Data. IX, 262 pages. 1992.

Vol. 40: S. W. Petters, Regional Geology of Africa. XXIII, 722 pages. 1991.

Vol. 41: R. Pflug, J. W. Harbaugh (Eds.), Computer Graphics in Geology. XVII, 298 pages. 1992.

Vol. 42: A. Cendrero, G. Lüttig, F. Chr. Wolff (Eds.), Planning the Use of the Earth's Surface. IX, 556 pages. 1992.

Vol. 43: N. Clauer, S. Chaudhuri (Eds.), Isotopic Signatures and Sedimentary Records. VIII, 529 pages. 1992.

Vol. 44: D. A. Edwards, Turbidity Currents: Dynamics, Deposits and Reversals. XIII, 175 pages. 1993.

Vol. 45: A. G. Herrmann, B. Knipping, Waste Disposal and Evaporites. XII, 193 pages. 1993.

Vol. 46: G. Galli, Temporal and Spatial Patterns in Carbonate Platforms. IX, 325 pages. 1993.

Vol. 47: R. L. Littke, Deposition, Diagenesis and Weathering of Organic Matter-Rich Sediments. IX, 216 pages. 1993.

Vol. 48: B. R. Roberts, Water Management in Desert Environments. XVII, 337 pages. 1993.

Vol. 49: J. F. W. Negendank, B. Zolitschka (Eds.), Paleolimnology of European Maar Lakes. IX, 513 pages. 1993.

Vol. 50: R. Rummel, F. Sansò (Eds.), Satellite Altimetry in Geodesy and Oceanography. XII, 479 pages. 1993.

Vol. 51: W. Ricken, Sedimentation as a Three-Component System. XII, 211 pages. 1993.

Vol. 52: P. Ergenzinger, K.-H. Schmidt (Eds.), Dynamics and Geomorphology of Mountain Rivers. VIII, 326 pages. 1994.

Vol. 53: F. Scherbaum, Basic Concepts in Digital Signal Processing for Seismologists. X, 158 pages. 1994.

Vol. 54: J. J. P. Zijlstra, The Sedimentology of Chalk. IX, 194 pages. 1995.

Vol. 55: J. A. Scales, Theory of Seismic Imaging. XV, 291 pages. 1995.

Vol. 56: D. Müller, D. I. Groves, Potassic Igneous Rocks and Associated Gold-Copper Mineralization. 2nd updated and enlarged Edition. XIII, 238 pages. 1997.

Vol. 57: E. Lallier-Vergès, N.-P. Tribovillard, P. Bertrand (Eds.), Organic Matter Accumulation. VIII, 187 pages. 1995.

Vol. 58: G. Sarwar, G. M. Friedman, Post-Devonian Sediment Cover over New York State. VIII, 113 pages. 1995.

Vol. 59: A. C. Kibblewhite, C. Y. Wu, Wave Interactions As a Seismo-acoustic Source. XIX, 313 pages. 1996.

Vol. 60: A. Kleusberg, P. J. G. Teunissen (Eds.), GPS for Geodesy. VII, 407 pages. 1996.

Vol. 61: M. Breunig, Integration of Spatial Information for Geo-Information Systems. XI, 171 pages. 1996.

Vol. 62: H. V. Lyatsky, Continental-Crust Structures on the Continental Margin of Western North America. XIX, 352 pages. 1996.

Vol. 63: B. H. Jacobsen, K. Mosegaard, P. Sibani (Eds.), Inverse Methods. XVI, 341 pages, 1996.

Vol. 64: A. Armanini, M. Michiue (Eds.), Recent Developments on Debris Flows. X, 226 pages. 1997.

Vol. 65: F. Sansò, R. Rummel (Eds.), Geodetic Boundary Value Problems in View of the One Centimeter Geoid. XIX, 592 pages. 1997.

Vol. 66: H. Wilhelm, W. Zürn, H.-G. Wenzel (Eds.), Tidal Phenomena. VII, 398 pages. 1997.